Handbook of Research on Advancements in Manufacturing, Materials, and Mechanical Engineering

Leonid Burstein
Independent Researcher, Israel

IGI Global
PUBLISHER of TIMELY KNOWLEDGE

A volume in the Advances in Mechatronics and Mechanical Engineering (AMME) Book Series

Published in the United States of America by
IGI Global
Engineering Science Reference (an imprint of IGI Global)
701 E. Chocolate Avenue
Hershey PA, USA 17033
Tel: 717-533-8845
Fax: 717-533-8661
E-mail: cust@igi-global.com
Web site: http://www.igi-global.com

Library of Congress Cataloging-in-Publication Data

Names: Burstein, Leonid, editor.
Title: Handbook of research on advancements in manufacturing, materials, and mechanical engineering / Leonid Burstein, editor.
Description: Hershey, PA : Engineering Science Reference, 2021. | Includes bibliographical references and index. | Summary: "This book examines emerging obstacles in these fields of engineering and the methods and tools used to find solutions"-- Provided by publisher.
Identifiers: LCCN 2020005692 (print) | LCCN 2020005693 (ebook) | ISBN 9781799849391 (h/c) | ISBN 9781799854418 (s/c) | ISBN 9781799849407 (eISBN)
Subjects: LCSH: Manufacturing processes--Technological innovations. | Materials--Technological innovations. | Mechanical engineering--Technological innovations.
Classification: LCC TS183 .R43 2021 (print) | LCC TS183 (ebook) | DDC 670--dc23
LC record available at https://lccn.loc.gov/2020005692
LC ebook record available at https://lccn.loc.gov/2020005693

This book is published in the IGI Global book series Advances in Mechatronics and Mechanical Engineering (AMME) (ISSN: 2328-8205; eISSN: 2328-823X)

British Cataloguing in Publication Data
A Cataloguing in Publication record for this book is available from the British Library.

For electronic access to this publication, please contact: eresources@igi-global.com.

Advances in Mechatronics and Mechanical Engineering (AMME) Book Series

J. Paulo Davim
University of Aveiro, Portugal

ISSN:2328-8205
EISSN:2328-823X

Mission

With its aid in the creation of smartphones, cars, medical imaging devices, and manufacturing tools, the mechatronics engineering field is in high demand. Mechatronics aims to combine the principles of mechanical, computer, and electrical engineering together to bridge the gap of communication between the different disciplines.

The **Advances in Mechatronics and Mechanical Engineering (AMME) Book Series** provides innovative research and practical developments in the field of mechatronics and mechanical engineering. This series covers a wide variety of application areas in electrical engineering, mechanical engineering, computer and software engineering; essential for academics, practitioners, researchers, and industry leaders.

Coverage

- Intelligent Navigation
- Intelligent Sensing
- Sustainable and green manufacturing
- Medical robotics
- Bioengineering Materials
- Computational Mechanics
- Vibration and acoustics
- Control Methodologies
- Nanomaterials and nanomanufacturing
- Control Systems Modelling and Analysis

IGI Global is currently accepting manuscripts for publication within this series. To submit a proposal for a volume in this series, please contact our Acquisition Editors at Acquisitions@igi-global.com or visit: http://www.igi-global.com/publish/.

The Advances in Mechatronics and Mechanical Engineering (AMME) Book Series (ISSN 2328-8205) is published by IGI Global, 701 E. Chocolate Avenue, Hershey, PA 17033-1240, USA, www.igi-global.com. This series is composed of titles available for purchase individually; each title is edited to be contextually exclusive from any other title within the series. For pricing and ordering information please visit http://www.igi-global.com/book-series/advances-mechatronics-mechanical-engineering/73808. Postmaster: Send all address changes to above address.

Titles in this Series

For a list of additional titles in this series, please visit: http://www.igi-global.com/book-series/advances-mechatronics-mechanical-engineering/73808

Evaluation and Applications for Ultrasonic Methods of Non-Destructive Testing
Ashish Khaira (Samrat Ashok Technological Institute, India)
Engineering Science Reference • ©2021 • 215pp • H/C (ISBN: 9781799845645) • US $195.00

Genetic Algorithms and Remote Sensing Technology for Tracking Flight Debris
Maged Marghany (Syiah Kuala University, Indonesia)
Engineering Science Reference • ©2020 • 334pp • H/C (ISBN: 9781799819202) • US $195.00

Humanizing Cities Through Car-Free City Development and Transformation
Rahma M. Doheim (Assiut University, Egypt & University of Business and Technology, Saudi Arabia) Alshimaa Aboelmakarem Farag (Zagazig University, Egypt) and Ehab Kamel (University of Central Lancashire, UK)
Engineering Science Reference • ©2020 • 305pp • H/C (ISBN: 9781799835073) • US $215.00

Modeling and Optimization of Solar Thermal Systems Emerging Research and Opportunities
Jagadish (National Institute of Technology Raipur, India) and Agnimitra Biswas (National Institute of Technology Silchar, India)
Engineering Science Reference • ©2020 • 170pp • H/C (ISBN: 9781799835233) • US $185.00

Airline Green Operations Strategies Emerging Research and Opportunities
Yazan Khalid Abed-Allah Migdadi (Qatar University, Qatar)
Engineering Science Reference • ©2020 • 231pp • H/C (ISBN: 9781799842552) • US $185.00

Recent Technologies for Enhancing Performance and Reducing Emissions in Diesel Engines
J. Sadhik Basha (International Maritime College Oman, Sohar, Oman) and R.B. Anand (National Institute of Technology, Tiruchirappalli, India)
Engineering Science Reference • ©2020 • 298pp • H/C (ISBN: 9781799825395) • US $215.00

Practical Approach to Substrate Integrated Waveguide (SIW) Diplexer Emerging Research and Opportunities
Augustine Onyenwe Nwajana (University of East London, UK) and Kenneth Siok Kiam Yeo (Universiti Teknologi Brunei, Brunei)
Engineering Science Reference • ©2020 • 171pp • H/C (ISBN: 9781799820840) • US $195.00

IGI Global
PUBLISHER of TIMELY KNOWLEDGE

701 East Chocolate Avenue, Hershey, PA 17033, USA
Tel: 717-533-8845 x100 • Fax: 717-533-8661
E-Mail: cust@igi-global.com • www.igi-global.com

To all my friends…

List of Contributors

Table of Contents

Detailed Table of Contents

Additive manufacturing (AM) is also known as 3D printing and classifies various advanced manufacturing processes that are used to manufacture three dimensional parts or components with a digital file in a sequential layer-by-layer. This chapter gives a clear insight into the various AM processes that are popular and under development. AM processes are broadly classified into seven categories based on the type of the technology used such as source of heat (ultraviolet light, laser) and type materials (resigns, polymers, metal and metal alloys) used to fabricate the parts. These AM processes have their own merits and demerits depending upon the end part application. Some of these AM processes require extensive post-processing in order to get the finished part. For this process, a separate machine is required to overcome this hurdle in AM; hybrid manufacturing comes into the picture with building and post-processing the part in the same machine. This chapter also discusses the fourth industrial revolution (I 4.0) from the perspective of additive manufacturing.

A plethora of experiments were conducted on IDI engine with various biodiesels (e.g., methyl esters of mahua, jatropha, rice bran, pongamia, palm, beef tallow, and waste cooking oils). Review of the results of these endeavors with various additives and blends with or without super charging of the engine are presented in this chapter. All these attempts have been concentrated to arrive at the best yield from a single cylinder engine. The recorded pressure changes during combustion, the derived heat release rates, and exhaust emissions are presented in the form of plots at various loads and at a constant speed. Engine cylinder vibrations (reflect combustion excitation) in the form of FFT and time waves were recorded at radial points and vertical on the cylinder body to assess the combustion propensity in all cases of studies. The results with relative benefits are enumerated.

This chapter is an updated, enhanced, and expanded paper previously published in IJMMME. An attenuation-based mathematical model and computational tool for communication cable design is presented in the chapter. The relationship derived for the model creation connects attenuation with the following parameters of cable design and testing: copper wire, aluminum and insulation thicknesses, cable impedance, test frequency, and lay length. The expression found from numerous measurement data was used to simulate the Monte Carlo prediction of attenuation in the cable. To realize the simulations, a life script and regular MATLAB script programs were developed. The programs produce a special graphical user interface suitable for non-programmers, cable engineers, and technicians. Results of calculations obtained with the developed tool were verified by experiments and indicate that the tool can be used to design cables with optimal parameters.

The present research investigation aims to fabricate a new class of fiber reinforced composite material by using biodegradable materials with epoxy as the strengthening agent. In order to explore the possibilities of using the new class of composite material in required application areas, the research work is carried out mainly in the field of the acoustic absorption properties of these bio fibers as an alternate building material. Also, the utilization of these materials as sound insulation will also provide a good solution to the waste management.

This chapter presents an idea of including the energy cost in the total cost of manufacturing. The main objective of this work is to index and evaluate the process plan based on the energy consumption of different machining processes while considering machining cost to index the process plans based on energy consumption and to select the best process plan based on machining cost. The second portion is the energy evaluation part, in which the process planner, after selecting machining parameters and considering several alternative process plans, analyzes and compares the energy cost and the total machining cost for the part. MPSEL program was used to develop alternative process plans. VISUALMACH program was used to calculate process plan attributes. Different process plans were compared based on the costs of energy, machining, material, and total cost. The obtained results from the proposed methodology showed

and costs for light weighting vehicle bodies. The chapter also exams the joining technologies on their principles, applications for the lightweight materials, and influencing factors for choosing a joining process. Furthermore, this chapter discusses the development trends of material selection and joining technology applications.

E-jet is a high-resolution microfabrication technology. Its operation depends on several process control parameters like applied voltage, flow rate of the material, stand-off height, type of ink material. High-resolution deposition along with high ejection frequency is the performance parameter of E-jet. In the present study, an attempt is made to design the process parameters in such a way that it could simultaneously satisfy both the performance characteristics of the E-jet process. A Taguchi robust design-based utility concept is applied to optimize the multi-response E-jet process through a case study. The optimization result is compared with another multi-objective optimization method called Grey relational grade analysis. The comparative analysis showed that both the methodologies identified the same process parameter combination for improved E-jet performance. Analysis of variance found applied voltage is the most influential control factor in the investigated region. The anticipated improvement in performance shows the successful implementation of the aforesaid methodology.

This chapter proposes a hybrid approach for modelling and optimizing the oblique turning processes. Analytical modelling and statistical regressions are combined for predicting the values of the most important parameters involved in the oblique cutting process. The predictions of the model were validated by using experimental data, showing coincidence for a 95% confidence level. Then, an a posteriori multi-objective optimization is carried out by using a genetic algorithm. Three conflicting objectives, which represent the three pillars of the sustainability as defined in the triple bottom line, are simultaneously considered: the carbon dioxide emissions, the cost, and the cutting time. The outcome of the optimization process is a set of non-nominate solutions, which are optimal in the wide sense that no other solution in the search space can improve one objective without worsening the other one. Finally, the decision-maker chooses the most convenient solution depending on the actual workshop conditions.

on time, pulse frequency, and sample size are presented. Response surface modeling and finite element analysis of the pulsed laser micro-forming process are described and explained. Experimental procedures for conducting pulsed laser micro-forming tests are described and explained. Recent investigations on pulsed laser micro-forming published elsewhere are presented, and possible scopes for future investigations are highlighted.

Additive manufacturing (AM) emerged from rapid prototyping to relinquish sustainable industrial production. The role of AM in the industrial field is to diminish manufacturing pace, functioning cost, and assembling the AM lightweight particles together, which enhances the malleable fabrication of personalized user defined components without symbolic concussion. The automotive manufacturing industry plays a chief role in the aggressive trade field where time to market declines. The engraving design structures with weight reduction materials are the demands faced by the automotive industry that can be ominously resolved by additive manufacturing technology. This research work provides a better understanding of AM technology and its role in automotive sector to enhance modern vehicle designs and enduring features and augments the knowledge of both researchers and industrialists to overcome the efficacy in manufacturing process by fabricating relatively high strength geometries with reduced weight.

Syntactic foam is the porous composite produced by mixing prefabricated hollow spherical particle into the matrix. Syntactic foams are used as energy absorption sandwich core for several applications like marine, automotive, and aerospace. In this work, low density hollow glass microspheres are hybridized with fly ash cenosphere in Bisphenol-A epoxy-based vinyl ester matrix. Hybrid syntactic foams is created with 60% total filler content. Within these hybrid systems internal composition of two fillers were varied in a step of 25 vol% with respect to each other. Hybrid syntactic foams are prepared by the hand lay-up (molding) method. The physical characterization parameter contains density and matrix porosity whereas tensile, quasi-static compression, flexural (3-point bending), Izod impact, and micro Vickers hardness are grouped as mechanical characterization parameters. Scanning electron microscopy was performed on fractured surfaces to examine deformation and fracture mechanisms related with each loading condition.

To further previous works on grain refinement in ferritic stainless steel (FSS) welds via elemental powder addition, microstructure evolution and mechanical properties in non-metallic carbide treated FSS weld is discussed. Silicon carbide (SiC) powder preplaced on medium chromium FSS bevelled joint was fusion melted using tungsten inert gas (TIG) torch at different combination of process parameters. The incorporation of the powder in the weld pool constricted the weld geometry in addition to grain modification from columnar to equiaxed grain structure. Whilst significant improvement was obtained in hardness characteristics in powder treated welds, those of ductility and tensile strength were marginal. The study also established that welds mechanical properties are controlled by a combination of phenomena such as phase fraction and grain morphology.

Preface

Manufacturing, materials, and mechanical engineering (MMME) are the three "pillars", the three central subjects, of modern industry and advanced science. In these areas, research is constantly being conducted with the purpose of improving, developing and creating new industrial areas, technologies and materials. Academic investigation of these directions has a significant impact on production and technology. The results of the research are published in numerous journals and provide a link between scientists and manufacturers worldwide. Journals with wide range of specializations have a substantial advantage because the reader can reference a single source instead of consulting several narrowly-specialized journals. Among the broad-profile scientific journals, the *International Journal of Manufacturing, Materials, and Mechanical Engineering* (IJMMME) stands out and is of interest to professionals in different fields – from aerodynamics, robotics, material science and tribology to bio- and nano-mechanics. Recent studies published in the IJMMME along with some of the latest, unpublished research are selected and presented in this book. The content of all the earlier publications has been updated and enhanced and now covers the latest investigations in the relevant fields. It should also be noted that the investigations presented in this book were carried out in different countries and even on different continents, thus provides a perspective of the directions of interests and presenting promising trends both in the industry of the particular country and in world industry as a whole. Thereby, the book reflects the actual problems and recent advances in the field of MMME and demonstrate the methods and tools applied for real solutions. This opens up opportunities for practical engineers and scientists to resolve various complicated problems that arise in their scientific and industrial fields. The book can also serve as a useful reference for academics, tribology and materials researchers, mechanical, physics and materials engineers and professionals in related industries. Additionally, the materials contained in this book can be used as a tutorial for final undergraduate engineering courses (e.g., mechanics and materials physics) or as a subject on MMME in postgraduate courses.

The chapters of the book represent studies on additive manufacturing (Chapters 1 and 17), testing of fuels for indirect injection engines IDI (Chapter 2), automotive and vehicle component optimization and modeling (Chapters 11, 12, and 15). The book also contains investigations of the acoustic, mechanical and structural properties of materials and their selection (Chapters 4, 8, 13, 18, and 19), digital cable design tool (Chapter 3), fractal surface analysis (Chapter 7) and many other promising areas of the MMME (Chapters 5, 6, 9, 10, 14, and 16).

The book consists of 19 chapters. To provide a more detailed view of the materials of the book and help the reader choose a chapter of interest, the following is a brief description of each chapter of the book provided by the chapter's authors:

Chapter 1. Additive Manufacturing (AM), also known as 3D Printing, classifies various advanced manufacturing processes that are used to manufacture three dimensional parts or components with a digital file in a sequential layer-by-layer. Here, this chapter gives a clear insight into the various popular AM processes under development. AM processes are broadly classified into seven categories based on the type of the technology, such of source of heat (ultraviolet light and laser) and material type (resigns, polymers, metal and metal alloys) used to fabricate the parts. These AM processes have their own merits and disadvantages depending upon the end part application. Some of these AM processes require extensive post-processing in order to achieve the finished part. For this process, a separate machine is required. To overcome this hurdle in AM, hybrid manufacturing has entered into the picture involving building and post-processing the part in the same machine. This article also discusses the fourth industrial revolution (I 4.0) in the perspective of Additive Manufacturing.

Chapter 2. A plethora of experiments have been conducted on IDI engines with various biodiesels including methyl esters of mahua, jatropha, rice bran, pongamia, palm, beef tallow and waste cooking oils. The chapters present a review of the results of these endeavors with various additives and blends with or without super charging of the engine. All these attempts have been concentrated to arrive at the best yield from a single cylinder engine. The recorded pressure changes during combustion, derived heat release rates and exhaust emissions are presented in the form of plots at various loads and constant speed. Engine cylinder vibrations (reflecting combustion excitation) in the form of FFT and time waves were recorded at radial points and vertical on the cylinder body to assess the combustion propensity in all cases of studies. The results with relative benefits are enumerated.

Chapter 3. This chapter is an updated, enhanced and expanded paper recently published in IJMMME. The chapter presents an attenuation-based mathematical model and computational tool for communication cable design. The relationship derived for the model creation connects attenuation with the following parameters of cable design and testing: copper wire, aluminum and insulation thicknesses, cable impedance, test frequency and lay length. The expression found from numerous measurement data was used to simulate the Monte Carlo prediction of attenuation in the cable. To realize the simulations, a life script and regular MATLAB script programs were developed. The programs produce a special graphical user interface suitable for non-programmers, cable engineers and technicians. Results of calculations obtained with the developed tool were verified by experiments and indicate that the tool can be used to design cables with optimal parameters.

Chapter 4. The present research investigation aims to fabricate a new class of fiber reinforced composite material by using biodegradable materials with epoxy as the strengthening agent. In order to explore the possibilities of using the new class of composite material in required application areas, the research work is carried out mainly in the field of the acoustic absorption properties of these bio fibers as an alternate building material. Also the utilization of these materials as sound insulation will also provide a good solution to the waste management.

Chapter 5. This chapter presents an idea of including the energy cost in the total cost of manufacturing. The main objective of this work is to index and evaluate the process plan based on the energy consumption of different machining processes while considering machining cost to index the process plans based on energy consumption and to select the best process plan based on machining cost. The second portion is the energy evaluation part, in which the process planner after selecting machining parameters and considering several alternative process plans, analyzes and compares the energy cost and the total machining cost for the part. MPSEL program was used to develop alternative process plans. VISUALMACH program was used to calculate process plan attributes. Different process plans were compared based on the costs of energy, machining, material, and total cost. The obtained results from the proposed methodology showed a significant contribution toward the effectiveness of energy management function within enterprises to meet targets and goals on energy and emissions reduction.

Chapter 6. Forced vibration analysis is performed on a tapered axially functionally graded beam resting on elastic foundation under externally applied harmonic excitations to present the effect of boundary conditions and taper patterns on the frequency response. The elastic foundation is modelled in the present analysis as Winkler foundation. A displacement based semi-analytical method is adopted for mathematical formulation and the derivation of governing equations is carried out following Hamilton's principle. Von Karman nonlinear strain-displacement relation employed to incorporate geometric nonlinearity. Broyden method is adopted to solve the nonlinear set of equations. Frequency response curves are plotted in non-dimensional frequency-amplitude plane to represent nonlinear forced vibration characteristic of the system. New benchmark results are also provided for different combination of system parameters, i.e., excitation amplitudes, foundation stiffness values, material models, taper patterns and flexural boundary conditions. Operational deflection shapes (ODS) are also presented.

Chapter 7. The present chapter consists of static and dynamic analysis of a fractal rough surface in contact with a rigid flat. The fractal surface is constructed using modified Weierstrass-Mandelbrot function. A rigid flat surface touches the topmost point of the rough surface which moves towards the rough surface and deforms it. Different contact parameters e.g. contact force, contact area, contact stress etc. for varying fractal and material properties are obtained through finite element based static analysis. A parameter denoting the degree of nonlinearity of the contact system is extracted from the force-displacement plot of the surface. This parameter is utilized to explain the dynamic behaviour of the fractal surface which vibrates under the influence of the externally excited rigid flat surface. The dynamic analysis of the contact system is carried out by modelling the contact interface as a single degree of freedom (SDOF) spring-mass-damper system. The dynamic behavior of the system is investigated in terms of frequency response curves, time-displacement plots and phase plots.

Chapter 8. Lightweight is a trend of new vehicle development, driven by government regulations, environmental concerns, and customer's need. A major effort in the automotive industry is on light weighting vehicle bodies. This chapter reviews the various materials, their characteristics, weight reduction potentials, and costs for light weighting vehicle bodies. The chapter also exams the joining technologies on their principles, applications for the lightweight materials, and influencing factors for choosing a joining process. Furthermore, this chapter discusses the development trends of material selection and joining technology applications.

Chapter 9. E-jet is a high-resolution microfabrication technology. Its operation depends on several process control parameters like applied voltage, flow rate of the material, stand-off height, type of ink material. High-resolution deposition along with ejection frequency is the performance parameter of E-jet. In the present study, an attempt is made to design the process parameters in such way, that it could simultaneously satisfy both the performance characteristics of the E-jet process. A Taguchi robust design based utility concept is applied to optimize the multi-response E-jet process through a case study. The optimization result is compared with another multi-objective optimization method called Grey relational grade analysis. The comparative analysis showed that both the methodology identified same process parameter combination for improved E-jet performance. Analysis of variance found applied voltage is the most influential control factor in the investigated region. The anticipated improvement in performance shows the successful implementation of the aforesaid methodology.

Chapter 10. This chapter proposes a hybrid approach for modelling and optimizing the oblique turning processes. Analytical modelling and statistical regressions are combined for predicting the values of the most important parameters involved in the oblique cutting process. The predictions of the model were validated by using experimental data, showing coincidence for a 95%-confidence level. Then, an a posteriori multi-objective optimization is carried out by using a genetic algorithm. Three conflicting objectives, which represent the three pillars of the sustainability as defined in the Triple Bottom Line, are simultaneously considered: the carbon dioxide emissions, the cost and the cutting time. The outcome of the optimization process is a set of non-nominate solutions, which are optimal in the wide sense that no other solution in the search space can improve one objective without worsen the other one. Finally, the decision-maker chooses the most convenient solution depending on the actual workshop conditions.

Chapter 11. High-strength and lightweight methods for vehicle parts include methods such as optimization and application of lightweight materials by reflecting load or material characteristics. Safety regulations have been established in accordance with the loads affecting the vehicle to secure the safety of the vehicle. In order to reduce the weight, high strength materials such as high strength steel (HSS) or high tensile strength steel (AHSS) have been studied. In addition, research on additional lightweight optimization is actively performed by removing parts that do not require high strength or replacing them with plastics. The process of designing a vehicle or part with different properties and considering various loads is costly and time consuming. In order to secure safety and light weight, we propose an approximate model for the optimal design of the seat frame that has a direct impact on occupants among the parts of the vehicle, and reduce the development cost, time, and intuitive design through the procedure.

Chapter 12. The chapter shows the study and simulation of a hydropneumatic suspension to be adopted for a telescopic handler vehicle. The hydropneumatic suspension system with independent wheels and with quadrilateral architecture has been studied to improve comfort and productivity of the existing vehicle, which has a suspended rigid axle on the front and a rigid axle on the rear, limiting the comfort and the grip. After the choice of the architecture and the kind of suspension, the article shows the design of the suspension kinematics. The optimization of the characteristic angles of the suspension has been performed using Adams/Car and Adams/Insight. The kinematic model optimized is subsequently reproduced in Adams/View to simulate the dynamics of the complete vehicle. Simulation results are used to evaluate vehicle performance in terms of comfort and stability according to the methods proposed by the standards.

Chapter 13. Numerical simulation of metal flow behavior was studied using DeformTM3D software. The simulation process was done on X20 steel taken from the software database at 1073-1273K temperature, 10mm/s die speed and 67% height reduction. From the simulation results, forging load, damage and stress/strain distributions were obtained. The results show that the forging load increased with a decrease in temperature or decreased with an increase in temperature. The maximum damage values increased as the temperature increased. The obtained maximum damage values were: 0.42 (1073K), 0.43 (1173K) and 0.45 (1273K). The damage distribution was inhomogeneous in the deformed cylinder. The stress/strain distributions were inhomogeneous in the deformed cylinder. The location of the maximum strain was at the center of the deformed cylinder while the maximum stress occurred at the die-cylinder contact surfaces. The study showed that flow stress behavior can be predicted using finite element method. This shows the feasibility of applying the finite element analysis to analyze the forging process.

Chapter 14. This study considers a fuzzy logic based reasoning approach for control and optimizing performance of overhead gantry crane. The objective of this study is to minimize load swing and to stabilize the crane in the least possible time. The fuzzy controllers were designed using nine Gaussian and triangular shape membership functions. The results clearly confirmed the effect of shape of memberships on performance of fuzzy controllers. Performance of overhead crane was measured in terms of settling time and overshoot ranges. The study also demonstrates the influence of varying mass of the load, mass of crane and length of crane bar on stability of the crane. A mathematical model of the crane system has been derived to develop a Simulink model of proposed system and performing simulations.

Chapter 15. The main objective of this chapter is to investigate the performance of automobile transmission gearbox under the influence of load, rotational speed and lubrication on multi speed gearbox gear surface. Gear oil SAE 80W-90 was used as gearbox lubricant, for cooling of transmission gearbox for high performance. An assumption has been made at the air-gear oil mist within transmission is under steady state condition, in isothermal equilibrium with the transmission gear oil bath of lubricant. The lubrication in multi speed transmission is subjected to thermo-elastohydrodynamic lubrication. The present chapter deals with the thermo-mechanical performance study of multi speed transmission (4 speed, excluding reverse gear) system which combines transient structure analysis of the gear train assembly. The engaged gear teeth pairs transmit torque subjected to thermo-elastohydrodynamic arrangements of lubrication. The study here analyzed transmission in second gear pair.

Chapter 16. This chapter presents study on the deformation behavior of thin sheets in pulsed laser micro-forming. Deformation mechanics of thin sheets under different operating conditions under pulsed laser irradiation are described and explained. Modeling of pulsed laser micro-forming for analyzing and optimizing the process parameters using different techniques mentioned in the published literature are discussed. The estimation of temperature and deformation fields in pulsed laser micro-forming of thin sheets of different shapes considering the effects of different process parameters such as laser power, spot diameter, pulse on time, pulse frequency and sample size are presented. Response surface modeling and finite element analysis of the pulsed laser micro-forming process are described and explained. Experimental procedures for conducting pulsed laser micro-forming tests are described and explained. Recent investigations on pulsed laser micro-forming published elsewhere are presented and possible scopes for future investigations are highlighted.

Chapter 17. Additive Manufacturing (AM) emerged from the Rapid Prototyping to relinquish sustainable industrial production. The role of AM in the industrial field is to diminish manufacturing pace, functioning cost, and assembling the AM lightweight particles together which enhances the malleable fabrication of personalized user defined components without symbolic concussion. The automotive manufacturing industry plays a chief role in the aggressive trade field where time to market declines. The engraving design structures with weight reduction materials are the demands faced by the automotive industry that can be ominously resolved by additive manufacturing technology. This research work provides a better understanding of AM technology and its role in automotive sector to enhance modern vehicle designs and enduring features, and augments the knowledge of both researchers and industrialists to overcome the efficacy in manufacturing process by fabricating relatively high strength geometries with reduced weight.

Chapter 18. Syntactic foam is the porous composite produced by mixing prefabricated hollow spherical particle into the matrix. Syntactic foams are used as energy absorption sandwich core for several applications like marine, automotive and aerospace. In this work, low density hollow glass microspheres are hybridized with fly ash cenosphere in Bisphenol-A epoxy based vinyl ester matrix. Hybrid syntactic foams is created with 60% total filler content. Whereas, within these hybrid systems internal composition of two fillers were varied in a step of 25 vol% with respect to each other. Hybrid syntactic foams are prepared by the hand lay-up (Molding) method.The physical characterization parameter contains density and matrix porosity whereas, Tensile, Quasi-Static Compression, Flexural (3-Point Bending), Izod Impact and Micro Vickers Hardness grouped as mechanical characterization parameters. Scanning Electron Microscopy was performed on fractured surfaces to examine deformation and fracture mechanisms related with each loading condition.

Chapter 19. Further to previous works on grain refinement in ferritic stainless steel (FSS) welds via elemental powder addition, microstructure evolution and mechanical properties in non-metallic carbide treated FSS weld is discussed. Silicon carbide (SiC) powder preplaced on medium chromium FSS bevelled joint was fusion melted using tungsten inert gas (TIG) torch at different combination of process parameters. The incorporation of the powder in the weld pool constricted the weld geometry in addition to grain modification from columnar to equiaxed grain structure. Whilst significant improvement was obtained in hardness characteristics in powder treated welds; those of ductility and tensile strength were marginal. The study also established that welds mechanical properties are controlled by a combination of phenomena such as phase fraction and grain morphology.

In conclusion, the book discusses a number of interesting and relevant areas, and I hope that its readers will find it useful to familiarize themselves with various recent studies in the MMME field.

Leonid Burstein
Independent Researcher, Israel
March 2020

Acknowledgment

The Editor would like to acknowledge IGI-Global Publishing for this fine opportunity and their available and professional support. Of course, I would like to express my sincere thanks to each of the authors of the chapters for their enthusiasm and constructive contributions to this book.

I wish to note the important contributions reviewers to ensuring the quality, coherence, and content presentation of chapters. In addition, some authors also served as reviewers; I highly appreciate their double contribution.

Leonid Burstein
Independent Researcher, Israel
March 2020

Chapter 1
Advancements in Manufacturing Technology With Additive Manufacturing and Its Context With Industry 4.0

Ganzi Suresh
https://orcid.org/0000-0003-1227-0983
Istituto Italiano di Tecnologia, Italy

ABSTRACT

Additive manufacturing (AM) is also known as 3D printing and classifies various advanced manufacturing processes that are used to manufacture three dimensional parts or components with a digital file in a sequential layer-by-layer. This chapter gives a clear insight into the various AM processes that are popular and under development. AM processes are broadly classified into seven categories based on the type of the technology used such as source of heat (ultraviolet light, laser) and type materials (resigns, polymers, metal and metal alloys) used to fabricate the parts. These AM processes have their own merits and demerits depending upon the end part application. Some of these AM processes require extensive post-processing in order to get the finished part. For this process, a separate machine is required to overcome this hurdle in AM; hybrid manufacturing comes into the picture with building and post-processing the part in the same machine. This chapter also discusses the fourth industrial revolution (I 4.0) from the perspective of additive manufacturing.

INTRODUCTION

The innovative thinking of designers develops the product where the manufacturers bring the same into reality by different manufacturing methods using suitable materials, to fill the needs of society. Design and manufacturing are the two important aspects of product development. Problem recognition, conceptualization, synthesis, analysis, design description, optimization and implementation are the steps involved in the design process. Further, proper selection of the material and the process used to manufacture the

DOI: 10.4018/978-1-7998-4939-1.ch001

product play important roles in the life of the component. The manufacturing technology is as old as the invention of the wheel. Several techniques like casting, joining, forming, machining etc. help the manufacturers to bring out the product to meet the requirements. In the year 1920, the powder metallurgy is added to the manufacturing processes by the invention of the sintering process. Later, Computer Numerical Control (CNC) machines made the manufacturing processes easy. Just three decades ago, Additive Manufacturing has been invented, which is a revolution in the manufacturing sector.

The most common manufacturing methods used to fabricate components include casting, machining, forming, joining, sheet metal work and powder metallurgy. The selection of the process is done depending on the applications of the end product. The casting process uses molten material which is poured into a die, prepared to suit the size and shape of the component required. Any complex shape can be made with a casting process if the die is prepared. The dimensional accuracy and finishing of the component depend on the material and accuracy of the die. But the preparation of the die requires much skill, consumes a lot of money and time, sometimes years together. The material properties cannot be retained in the casting process since the material is melted to a liquid state and solidified in the die. The removal of the excess material from the base material is called machining process which is classified as turning, milling, drilling, grinding etc. The material removal in machining is in the form of chips, which shear from the base material. Generally, a hard-cutting tool is used to machine a soft workpiece. More difference in the hardness between the workpiece and tool results in a component with high accuracy and surface finish. With a lesser difference in hardness, it is difficult to machine and the cutting tool may show its traces on the finished components and the total process requires a lot of time. Forming is the process of shaping the components without adding or removing the material but by the application of force, sometimes maybe at elevated temperatures. Rolling, forging, extruding, wiredrawing, punching, blanking etc. are the different types of forming processes. Since the processes are carried out at heavy loads and sometimes at elevated temperatures, the residual stresses induced which affects the properties of the components a lot. The powder metallurgy is the science where solid components are fabricated with the metal powders pressed in a die with heavy loads with or without elevating the temperatures, sometimes in the presence of a binder. Generally, the components made from the powder metallurgy are used as bearings since they possess high porosity, which is essential for self-lubricating components. In sheet metal work, a thin sheet is cut and bent to form the desired components.

A Computer Numerical Control (CNC) machining processes is an autonomous and requires less human intervention, a program is written in the computer to perform one or more of the aforementioned tasks, generally machining, high precision components can be made with high accuracy but it suits well for the mass production. Thus, almost all conventional and computer-controlled processes have limitations to produce customized products. Additive Manufacturing (AM) is the technique which fabricates the components, layer by layer taking the inputs from the CAD data. Any complex shape can be made with AM technology. AM supports many plastics and metals and it is possible to fabricate near net shape components also.

Additive Manufacturing (AM)/ Three-Dimensional Printing (3DP) industry is comprised of a progression of innovations containing computerized methods that can rapidly manufacture any given three-dimensional component/ part with the end goal of testing the structure, fit, and capacity of a configuration (Billiet, et al. 2012). As a layer-based added substance producing system, AM gives an architect the ability to manufacture any possible geometry.

Additive Manufacturing (AM)/ 3D Printing technology refers to processes used to generate a 3D object in which layers of material are successively formed under a computer-controlled program to create a physical object. The 3D file source (CAD data) is usually sliced into several layers, each layer generating a set of computer-controlled instructions. The American Society for Testing and Materials defines Additive Manufacturing as "the process of joining materials to make objects from 3D model data, usually layer upon layer, as opposed to subtractive manufacturing methodologies, such as traditional machining" (ASTM Standard F2792 - 2012a). AM reflect that the technologies share the theme of sequential-layer material addition or joining throughout a 3D work. 3D printing technologies can be split up into two groups: direct and indirect 3D printing. The main difference lies in the fact that the design is directly made from 3D printing (direct) or 3D printing was used in the process of creating a model (indirect). The very first patent of 3D printing was coined in 1984 by Chuck Hull of 3D Systems Corporation. Hull defined the 3D printing process as a system for generating three-dimensional objects by creating a cross-sectional pattern of the object to be formed. His invention consists of a stereolithography fabrication system, in which layers are added by curing photopolymers with ultraviolet light lasers. Stereolithography is still one very popular 3D printing manufacturing technique, also know SLA.

AM frameworks can deliver models from 3D CAD data, Computer Tomography (CT) scan images, X-Ray images and (MRI) scans, and 3D digitizing frameworks (Trevor Boehm). They are typically produced using digital model data from a 3D model or another electronic data source such as a Standard Triangulation Language (.STL) file, one of the most common file types that 3D printers can read. Utilizing an added substance approach, AM join fluid/ liquid, powder or solid/ sheet materials to shape physical models on a layer by layer premise. AM machines process plastic, paper, artistic, metal, metal alloys and composite materials from flimsy, level cross areas of computer-aided models (Gibson, et al. 2015). The objects manufactured through AM/3D printing processes can be of almost any shape or geometry. The workflow in additive manufacturing is shown in the fig.1 and discussed in detail (G. Suresh et al.).

Figure 1.

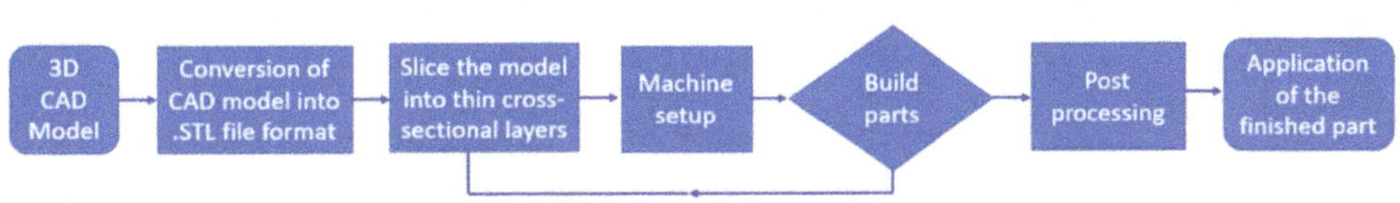

Developing a 3D CAD Model

The first step of the AM system is the modelling of the object by using CAD software. This step is exactly the same for all the AM building programs. The process generally uses solid modellers, like Pro-E, Computer-Aided 3D interactive application (CATIA) and SolidWorks, as these formats epitomize 3D objects precisely in comparison to wire-frame modellers (like AutoCAD), consequently producing better results. Pre-existing CAD files or newly created CAD files are also being considered for prototyping purposes.

Converting CAD Model Into .STL File Format

The second step involves the conversion of a CAD file to a standardized format called .STL file format. This standardization is performed to establish consistency among the model files, which had been saved in different formats. The .STL format comprises a 3D surface, which is a configuration of planar triangles. An increase in the number of triangles improves the result and approximation but also causes an increase in the file size. Since huge and complex files consume extra time for construction, the designer should consider both accuracy and manageability, while creating the .STL file. The universal acceptability of the .STL format makes it common for the entire AM building techniques.

Verification of the .STL File Format

The third phase involves the verification of the .STL file format by using a pre-processing program. Numerous programs are available for this purpose. However, care should be taken in adjusting the size, location and orientation while verifying the .STL file format. Generally, the pre-processing software includes a visualization instrument that permits the user to see and control the parts. Vital tasks, such as the repositioning of parts and a change in orientation (for building in a definite position inside the device), can also be performed by using this software.

Slicing the Object Imported From .STL File Format Into Thin Cross-Sectional Layers

The fabrication of the layers along the *X-Y* plane often makes the model fragile and unreliable along the *Z*-direction. This rendered the building of orientation as an important process. Numerous techniques use part orientation to build the model in such a manner that the minimum dimension lies along the *Z*-direction. This, in turn, helps to reduce the time owing to a decrease in the number of layers. The part orientation uses the pre-processor software and slices the .STL model into multiple layers with thicknesses varying from 0.01-0.7mm, based upon the AM technique and the end applications of the prototype, usually, keeping a nominal layer thickness of 0.1mm.

Machine Setup

AM machine uses certain setup features, which are precise to a particular machine or technique. These selected machines are intended specifically to work with1or2 varieties of materials, however, with a zero difference in a layer of the thickness or some other parameters. These machines are often subjected to a few changes in setup from one build to another. On the other hand, some machines designed to run with diversified materials also exist and contain certain characteristics that need optimization for fitting with the type of part that is being created. Numerous setup options for such machines are provided

Layer by Layer Construction

The layer by layer construction involves the actual construction of the part. The 3D object is made from an AM machine layer by layer. After slicing the part, the machine takes the layers of the object as an input and accordingly constructs the final 3D object. AM machines use paper, polymer or powdered metal to build one layer at a time. The majority of the AM machines work autonomously and require little human intervention.

Cleaning and Finishing

This step involves the cleaning of prototypes taken out from the machine, wherein its supports or substrates are detached and subjected to surface treatment wherever required.

Application of the Part

This final step focuses on providing the additional treatment required to make the prototype a full-fledged and ready-to-use product. For example, priming and painting may be required to get an acceptable finish and surface texture. Depending on the demand for finishing requirements, the treatments may be laborious and lengthy. There can be a need to assemble mechanical/ electronic items to form the final model or product.

Additive Manufacturing/ 3D Printing Technologies

According to ISO/ASTM52900-15 defines seven categories of Additive Manufacturing (AM) processes as shown in fig. 2 [U. M. Dilberoglu et al.].

Figure 2.

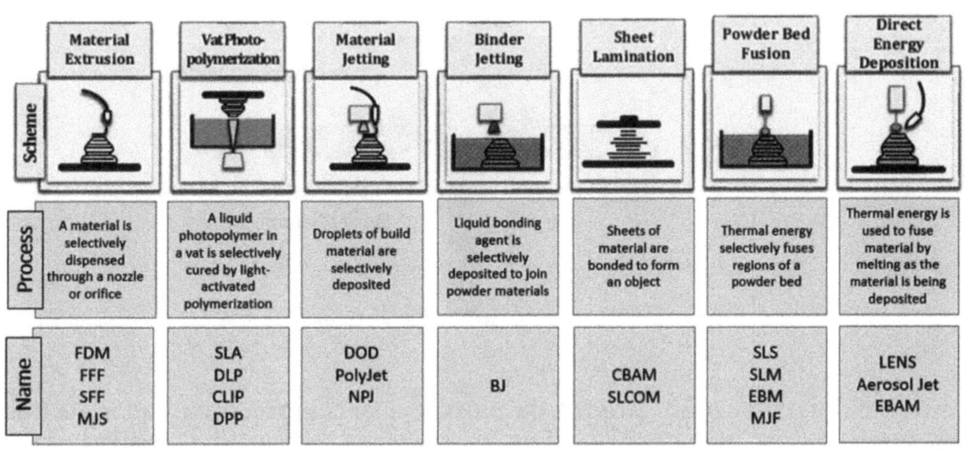

Material Extrusion

Material Extrusion 3D printing technology uses a continuous filament of thermoplastic material as a base material. The filament is fed from a coil, through a moving heated printer extruder head, often abbreviated as an extruder. The molten material is forced out of the extruder's nozzle and is deposited first onto a 3D printing platform, which can be heated for extra adhesion [A. R. Torrado et al.]. Once the first layer is completed, the extruder and the platform are parted away in one step, and the second layer can then be directly deposited onto the growing workpiece. The extruder head is moved under computer control. At least three axes are required for the extruder to move in Cartesian architectures, but polar and delta systems are also becoming increasingly popular. One layer is deposited on top of a previous layer until the object's fabrication is complete.

Material extrusion is known as Fused Deposit Modelling (FDM) /Fused Filament Fabrication (FFF) as shown in fig.3 and is one of the most popular processes for hobbyist-grade 3D printing. The proprietary term Fused Deposition Modelling (FDM) was coined by S. Scott Crump in the late 1980s and was commercialized in 1990 by the Stratasys company. With the expiration of this technology's patent, there is now a large open-source development community called RepRap, as well as commercial and DIY variants, which utilize this type of 3D printing technology. This has led to a measurable price decrease. However, the material extrusion technique has dimensional accuracy limitations and is very anisotropic.

Figure 3.

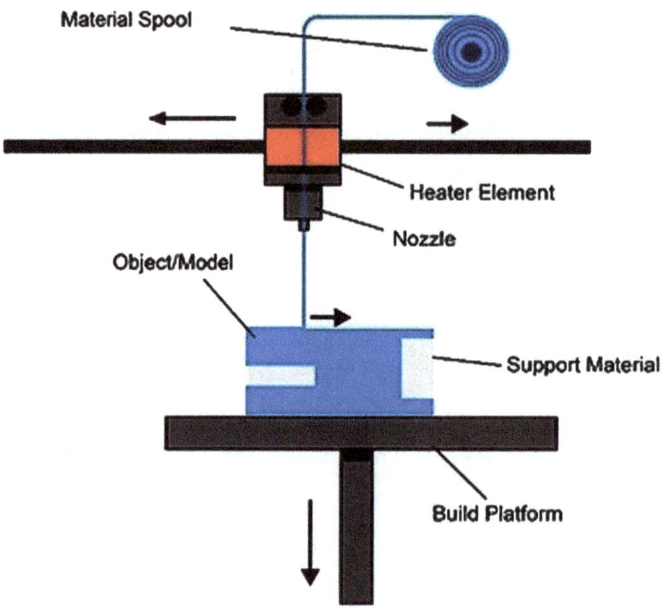

A wide variety of materials can be extruded, the most popular being thermoplastics, such as Acrylonitrile Butadiene Styrene (ABS), PolyLactic Acid (PLA), High-Impact Polystyrene (HIPS), Thermoplastic PolyUrethane (TPU), aliphatic PolyAmides (PA), also known as Nylon, and more recently high-performance plastics such as PolyEther Ether Ketone (PEEK) or PolyEtherimide (PEI). Additionally, paste-like materials such as ceramic, concrete and chocolate can be extruded using this 3D printing technology.

As it became possible to equip a 3D printer with multiple extruders, for speeding up the fabrication process or opening multi-material capabilities, the availability of composite fabrication became possible. Composite Filament Fabrication (CFF) is one of them. This term was coined by the company Markforged and uses two print nozzles. One nozzle operates following the typical material extrusion process; it lays down a plastic filament that forms the outer shell and the internal matrix of the part. The second nozzle deposits a continuous strand of composite fibre (made with carbon, fibreglass, or Kevlar) on every layer. These continuous strands of composite fibres inside 3D printed parts add strength to the built object that is comparable to parts made of metal. In addition to using composite materials for strong parts, the strategy used to lay down layers can affect part strength.

Even composites can be 3D printed with the material extrusion technique on machines equipped with only one extruder. The sole condition is that the base material (a thermoplastic) is present in sufficient quantities to guarantee a fusion between layers. Therefore, a mix of two materials within a single filament made wood 3D printing (wood particles embedded in PLA), metal 3D printing (metal particles embedded in thermoplastic) and even carbon 3D printing (carbon fibres embedded in thermoplastic) possible.

Vat Photopolymerization

Vat polymerisation uses a vat of liquid photopolymer resin, out of which the model is constructed layer by layer as in fig. 4. An ultraviolet (UV) light is used to cure or harden the resin where required, whilst a platform moves the object being made downwards after each new layer is cured (Chartrain, et al. 2018).

Figure 4.

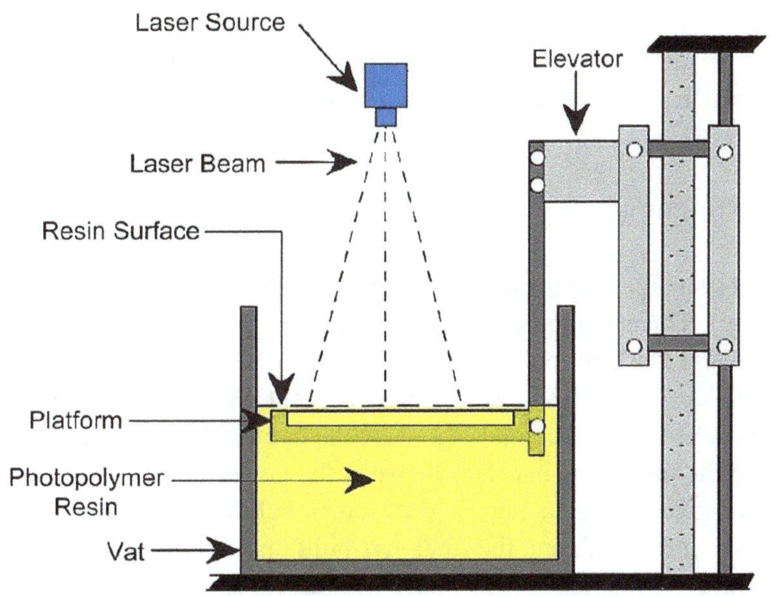

As the process uses liquid to form objects, there is no structural support from the material during the build phase., unlike powder-based methods, where support is given from the unbound material. In this case, support structures will often need to be added. Resins are cured using a process of photopolymerisation or UV light, where the light is directed across the surface of the resin with the use of motor-controlled mirrors. Where the resin comes in contact with the light, it cures or hardens. The most popular vat photopolymerization 3D printing technologies include the following;

Stereolithography (SLA) is an optical fabrication, photo-solidification, or resin printing. During the SLA manufacturing process, a concentrated beam of ultraviolet light or a laser is focused onto the surface of a vat filled with a liquid photopolymer. The beam or laser is focused, creating each layer of the desired 3D object by means of cross-linking or degrading a polymer (Wang, et al. 2016).

Digital Light Processing (DLP)

A digital projector screen is used to flash a single image of each layer across the entire platform at once. Because the projector is a digital screen, the image of each layer is composed of square pixels, resulting in a layer formed from small rectangular bricks called voxels. DLP can achieve faster print times for some parts, as each entire layer is exposed all at once, rather than drawn out with a laser.

Continuous Liquid Interface Production (CLIP)

The CLIP vat photopolymerization technique uses a tank of resin as a base material. Part of the vat bottom is transparent to ultraviolet light and therefore called the window. An ultraviolet light beam shines through the window, illuminating the precise cross-section of the object. The light causes the resin to solidify (photopolymerize). The object rises slowly enough to allow the resin to flow under and maintain contact with the bottom of the object. An oxygen-permeable membrane lies below the resin, which creates a dead zone. This persistent liquid interface prevents the resin from attaching to the window, which means the photopolymerization is inhibited between the window and the polymerizer. Unlike standard SLA, the 3D printing process is continuous and claims to be up to 100 times faster than commercial 3D printing methods.

Daylight Polymer Printing (DPP)

Instead of using a laser or a projector to cure the polymer, the DPP manufacturing process uses an LCD (Liquid Crystal Display). This technique, also called LCD 3D printing, uses unmodified LCD screens and a specially formulated Daylight polymer. The company Photocentric has been able to make this work by developing one of the world's most sensitive daylight resins.

Two different approaches are also competing inside the photopolymerization 3D printing technique, and both of them build the part at the interface between the last layer and the surface of the resin tank. The top-down approach consists of placing the heat source below the vat. The 3D printing platform is then progressively raised into the air and the final part-built upside down. The second technique called bottom-up consists of placing the heat source above the vat.

Photopolymerization is successfully applied to medical modelling, which enables the creation of accurate 3D models of various anatomical regions of a patient, based on data from computer scans. The high resolution of this technique also makes it ideal for all types of prototyping, as well as mass production. Vat polymerization processes are excellent at producing parts with fine details and a smooth surface finish. This makes them ideal for jewellery, investment casting and many dental and medical applications. Material developments have also allowed the printing of low-run injection moulds. The main limitations for vat photopolymerization are the build size and part strength.

Photopolymer resins are available in different colours and exhibit different physical properties, each corresponding to a specific use. The range of resins includes tough resin, low residue resin (for investment casting), transparent resin and flexible polyurethane resin.

Material Jetting

The Material Jetting 3D printing manufacturing technique is often compared to the standard 2D ink-jetting process. Utilizing photopolymers, metals, or wax that solidify when exposed to light or heat (in a similar fashion to stereolithography) ensures that physical objects are built up one layer at a time. The material jetting manufacturing process allows for different materials to be printed within the same part as shown in fig. 5.

Figure 5.

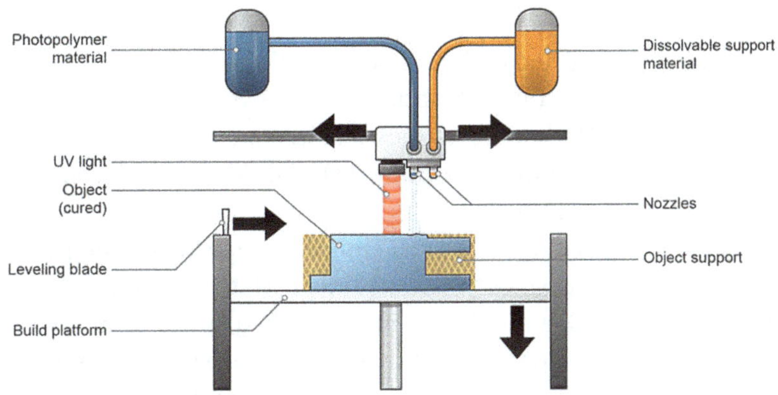

Material jetting dispenses a photopolymer from hundreds of tiny nozzles in a printhead to build a part layer-by-layer. This allows material jetting operations to deposit build-material in a rapid, line-wise fashion, which can be compared to other point-wise deposition technologies that follow a path to complete the cross-sectional area of a layer, also called a slice. As the droplets are deposited to the build platform they are directly cured and solidified using UV light (Yap et al., 2017). Material jetting processes require support, which is often 3D printed simultaneously during the build from a dissolvable material. The support material is then removed during the post-processing step. Several techniques make up the material jetting term, the most popular being:

Drop on Demand (DOD) Printing

DOD material jetting printers have two print jets: one to deposit the build material and another for dissolvable support material. Like all additive manufacturing machines, DOD 3D printers follow a pre-determined path and deposit material in a point-wise fashion to build the cross-sectional area of a component (Park, et al. 2014). These machines also employ a fly-cutter that skims the build area after each layer to ensure a perfectly flat surface before printing the next layer. DOD technology is typically used to produce wax-like patterns for lost-wax casting/investment casting and mould making applications, making it an indirect 3D printing technique.

PolyJet Printing

PolyJet 3D printing technology was first patented by the Objet company, now a Stratasys brand. The photopolymer materials are jetted in ultra-thin layers onto a build tray in a similar fashion compared to inkjet document printing. Each photopolymer layer is cured by UV light immediately after being jetted (Meisel, et al. 2015). The repetition of jetting and curing steps, layer after layer produces fully cured models that can be handled and used immediately. The gel-like support material, which is specially designed to support complex geometries, can easily be removed by hand or by using water jetting.

NanoParticle Jetting (NPJ)

This material jetting technology, patented by XJet, uses a liquid, which contains building nanoparticles or support nanoparticles, that is loaded into the printer as a cartridge and jetted onto the build tray in extremely thin layers of droplets. High temperatures inside the building envelope cause the liquid to evaporate leaving behind parts made from the building material (Kunchala & Kappagantula, 2018). This technique is suitable for metals and ceramics.

Material jetting 3D printing technology is a great choice for realistic prototypes, providing an excellent level of details, high accuracy and smooth surface finish. Material jetting allows a designer to print a design in multiple colours and with a number of materials in a single print. To designate a different material or colour to particular areas of the part, the model must be exported as separate STL files. When blending colours or material properties to create a digital material, the design must be exported as an OBJ or VRML file, because these formats allow the designation of special properties (such as texture or full colour) on a per-face or per-vertex basis.

The main drawbacks to printing with material jetting technologies are the high cost and the fact that UV activated photopolymers lose mechanical properties over time and can become brittle.

Binder Jetting

The binder jetting 3D printing technique consists of the deposition of a binding adhesive agent onto thin layers of powdered material. The powder material or ceramic (glass or gypsum) or metal (stainless steel).

During the binder jetting 3D printing process as shown in fig. 6, the 3D print head moves over the build platform depositing binder droplets, printing each layer in a similar fashion to 2D printers that print ink on paper. When a layer is complete, the powder bed moves downwards and a new layer of powder is spread onto the build area. The process is repeated layer by layer until all parts are complete (Gokuldoss, et al. 2017). After printing, the parts are in a green, or unfinished, state and require additional post-processing before they are ready to use. Often an infiltrant substance is added to improve the mechanical properties of the parts. The infiltrant substance is usually a cyanoacrylate adhesive (in case of ceramics) or bronze (in the case of metals). Another strategy is to put the workpiece, in its green state, inside an oven to achieve sintering of the grains of matter. Interestingly, the term 3D printing originally referred to a process that deposited a binder material onto a powder bed with an inkjet printer-like heads layer by layer.

Figure 6.

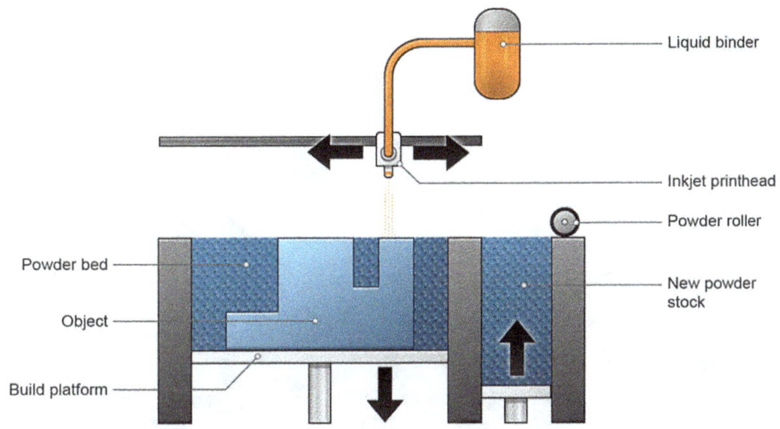

The binder jetting 3D printing process encompasses several techniques honed by different manufacturers. ColorJet Printing (CPJ) by 3D Systems. ColorJet Printing is a trademark of the company Z-Corp, now a 3D Systems company. ColorJet Printing 3D prints are fully coloured. The final parts resemble sandstone and exhibit a somewhat porous surface. The sandstone material is inkjet coloured and glued together during the 3D printing process. At the end of the 3D printing and infiltration is required to solidify and bond the part. Hundreds of thousands of colours are available, almost the full CMYK spectrum. The final prints are not intended for functional applications as they remain porous and must be kept away from humidity to avoid discolouration.

The binder jetting 3D printing process can work with a variety of materials including metals, sands, and ceramics. Some materials, like sand, require no additional processing.

Metal-based binder jetting parts have relatively good mechanical properties thanks to the infiltration process. They can be used as functional components and have relatively good mechanical properties thanks to the infiltration process. They are also more cost-effective than SLM (Selective Laser Melting) or DMLS (Direct Metal Laser Sintering) metal parts but have poorer mechanical properties because the grains of materials do not entirely fuse together.

There are various types of 3D printing binder materials, each suited for a specific application. They can be listed in categories including furan binder (for sand casting applications), phenolic binder (for sand moulds and cores), silicate binder (environmentally-friendly, for sand moulds and cores) and aqueous-based binder (for metals).

Powder Bed Fusion (PBF)

Powder Bed Fusion (PBF) 3D printing technology produces parts/ products with precision. Figure 7 shows the schematic of PBF process. This 3D printing technique enables the manufacturing of a vast array of geometrically complex products using a heat source, mainly laser or electron beams, to fuse powder particles layer-by-layer, therefore forming a solid part (Sutton, et al. 2017). Manufacturers can benefit from substantial freedom of design considering that PBF presents several viable technologies and materials. The different Powder Bed Fusion methods are discussed below.

Figure 7.

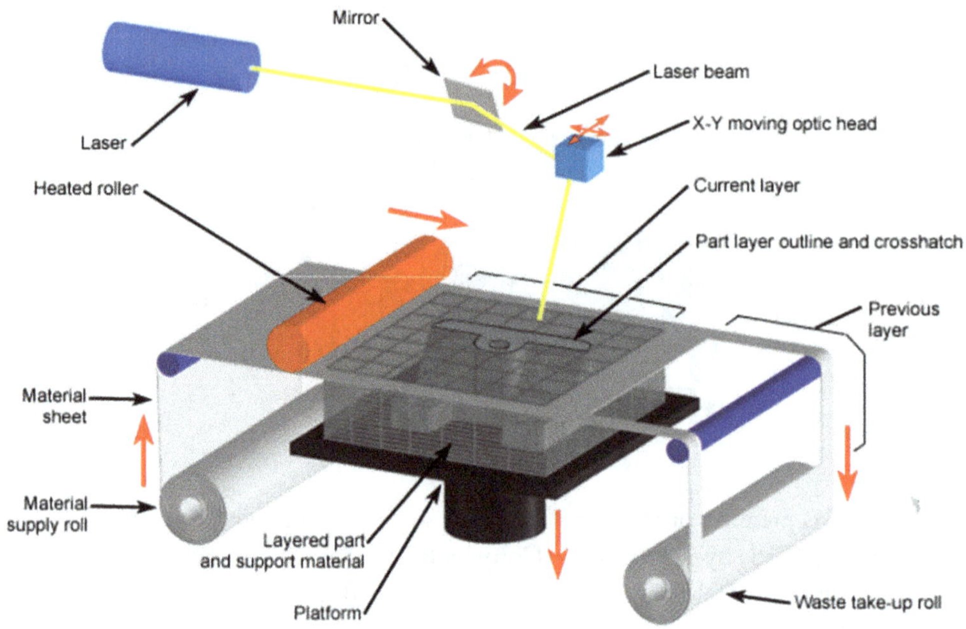

Selective Laser Sintering (SLS)

SLS 3D printing technology originated in the late 1980s at the University of Texas at Austin. Over the years, this technology has experienced remarkable advances. Basically, the process uses lasers to sinter, or coalesce, powdered material layer-by-layer to create a solid structure (Liu, et al. 2013). The final product rendered enveloped in loose powder is then cleaned with brushes and pressurized air. The main materials used in the SLS 3D printing process include polyamide (Nylons), Alumide (a blend of grey aluminium powder and polyamide), and rubber-like materials. Nylons are strong and durable but do feature some flexibility, making them excellent for snap fits, brackets, clips, and spring features. Designers should take the susceptibility for shrinkage and warping of thin parts into consideration during the conceptual phase.

Selective Laser Melting (SLM), also known as Direct Metal Laser Sintering (DMLS) the same technical principle is used to produce Selective Laser Melting (SLM) and Direct Metal Laser Sintering (DMLS) parts but is exclusively used to produce metal parts. SLM achieves a full melt of the powder so that single-component metals, such as aluminium, can be used to create a light, strong spare parts and prototypes. DMLS sinters the powders and is restricted to alloys, including titanium-based alloys. These methods require added support to compensate for the high residual stress and to limit the occurrence of distortion (Atzeni et al. 2016). Applications include jewellery and dental industries, spare parts, and prototypes.

Electron Beam Melting (EBM)

The EBM 3D printing technology as in fig. 8, attains fusion with the use of a high-energy electron beam and produces less residual stress resulting in less distortion (L. E. Murr et al.) . It uses less energy and

Figure 8.

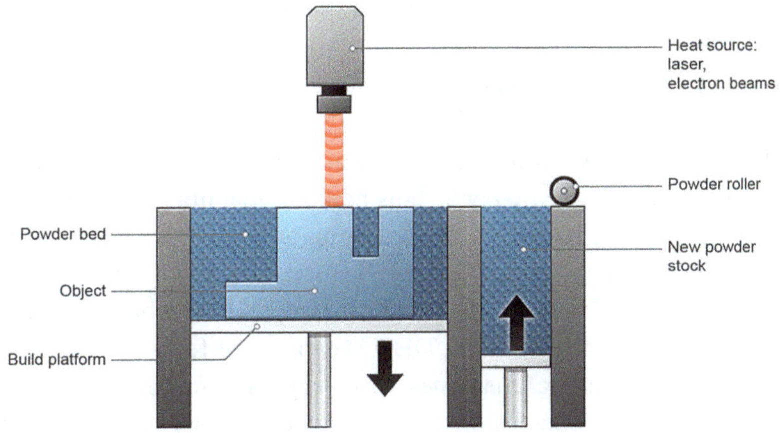

Figure 9.

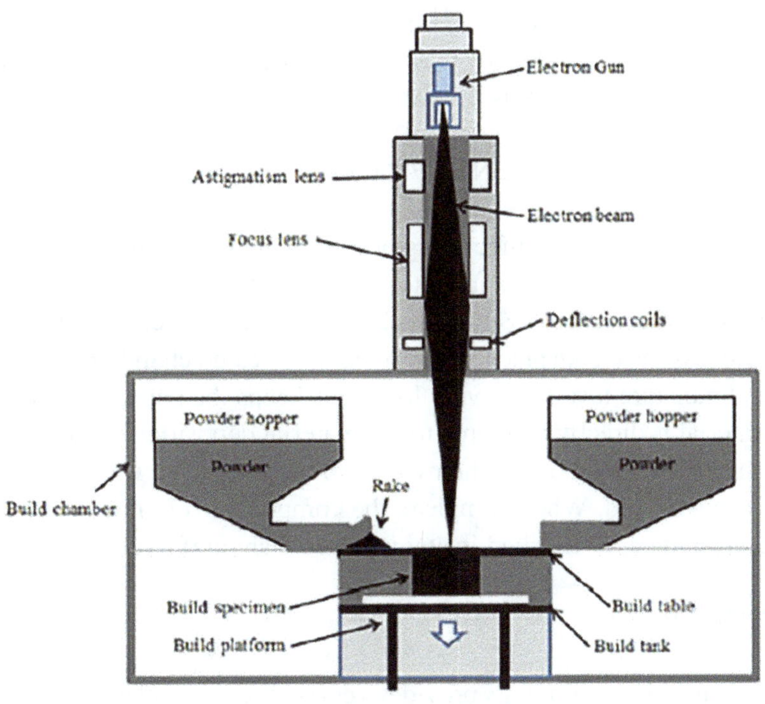

can produce layers faster than SLS. This method is most useful in high-value industries such as aerospace and defence, motorsports, and medical prosthetics. Figure 9 shows the schematic of EBM Process.

Multi Jet Fusion (MJF)

The MJF 3D printing technique differs from the aforementioned methods in that an inkjet array is used to apply to fuse and detailing agents, which are then fused by heating the elements into a solid layer

(O'Connor, et al. 2018). No laser source is used to deposit parts. Detailing agents are jetted around contours to improve part resolution thus opening the possibility for lifelike object production.

Directed Energy Deposition (DED)

The Direct Energy Deposition 3D Printing technology creates parts by directly melting materials and deposing them on the workpiece, layer by layer. This additive manufacturing technique is mostly used with metal powders or wire source materials (Baykasoglu, et al. 2018). Other popular terms for DED include laser engineered net shaping, directed light fabrication, direct metal deposition, Laser Deposition Welding (LDW) and 3D laser cladding. In addition to the capability to build parts from scratch (often with the hybridisation of a mill/turn CNC tool). DED is capable of fixing complex damaged parts.

Most DED 3D printers are industrial machines with very large footprints that require a closed and controlled environment to operate. Therefore, typical Directed Energy Deposition consists of a nozzle mounted on a multi-axis arm inside a closed frame, which deposits melted material onto the workpiece surface, where it solidifies. The process is similar in principle to the material extrusion 3D printing technique, but with DED, a nozzle can move in multiple directions, with up to five different axes compared to the most of the additive machines. The Direct Energy Deposition term can encompass several different technologies. They distinguish themselves by the way the material is fused, each suited for different and specific purposes. The most popular are as follows:

Laser Engineered Net Shaping (LENS) Process

LENS is a promising metal additive manufacturing process used to build near net shape and functionally graded materials. This process uses laser source to build objects layer by layer directly from powdered metals, alloys, ceramics or composites (G Suresh et al. 2019). Figure 10 represents the schematic of LENS process. The LENS process must take place in a hermetically-sealed chamber which is filled with argon so that the oxygen and moisture levels stay very low. This keeps the part clean and prevents oxidation. The metal powder material is directly delivered to the material deposition head. Once a single layer has been deposited, the material deposition head moves on to the next layer. By building up successive layers, the whole part is constructed. When complete, the component is removed and can be heat-treated, hot isostatic pressed, machined, or finished in any required fashion (G Suresh, et al. 2018).

Aerosol Jet Technology

Optomec Aerosol Jet 3D printing technology provides a cost-effective, scalable process to print functional antennas and sensors directly onto consumer and industrial components, making them smart Internet of Things (IoT) devices (T. Seifert). The possible printed antennas include LTE, NFC, GPS, Wifi, WLAN, and BT. This technique is closer to simple deposition techniques but suited for complex compound curved surfaces. Figure 11 shows the aerosol jet technology.

Aerosol Jet systems are ideally suited to develop, fabricate, enhance and repair high-performance electronic and biologic devices for consumer electronics, semiconductor packaging, displays, aerospace/defence, automotive and life sciences end-use products. Aerosol Jet technology can function with a large range of materials including conductive nano-particle metal inks, dielectric pastes, semiconductor, and other functional materials.

Figure 10.

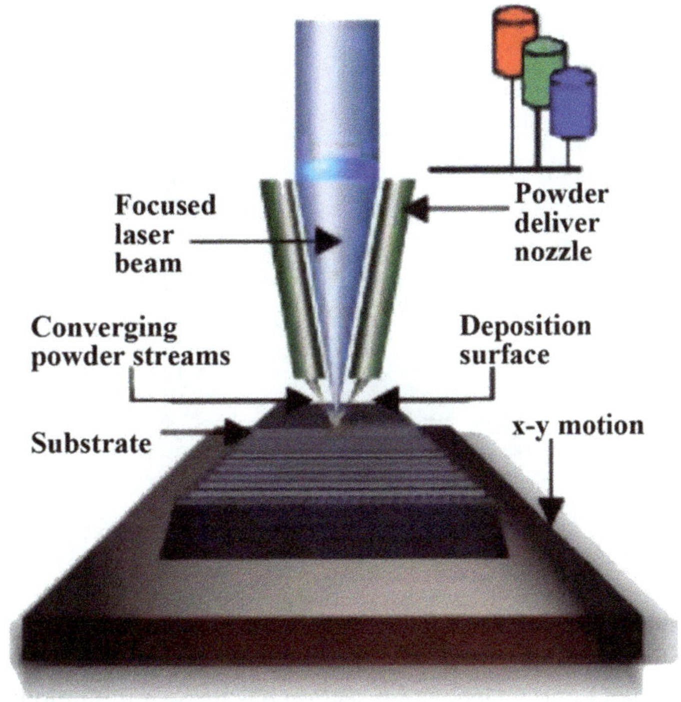

Figure 11.

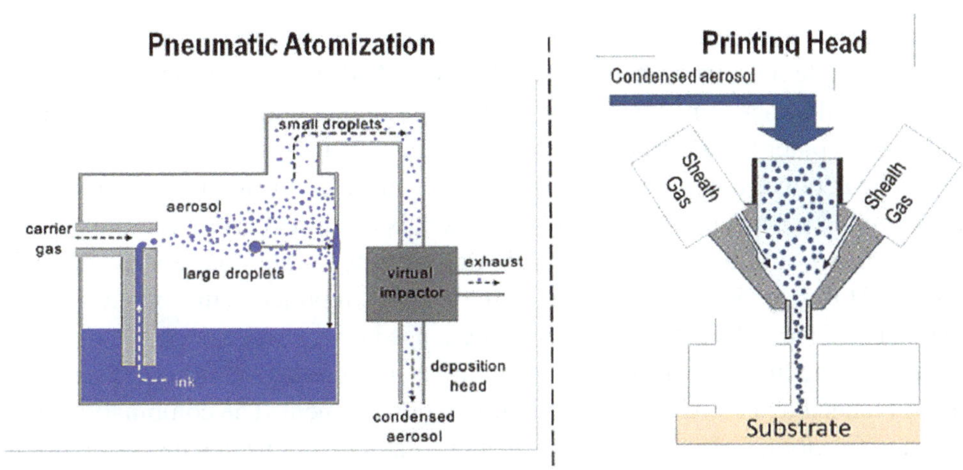

Electron Beam Additive Manufacturing (EBAM)

Electron Beam Additive Manufacturing (EBAM) is an additive manufacturing technology that produces large scale metal structures. Electron Beam (EB) gun deposits metal via a wire feedstock, layer by layer, until the part reaches near-net shape and is ready for finish machining (N. Raghavan et al) . Material deposition rates range from 3 to 9 kg of metal per hour. Compatible metals include titanium, tantalum, and nickel. This DED technique can also be used to repair damaged parts. Figure 12 represents the EBAM process.

Figure 12.

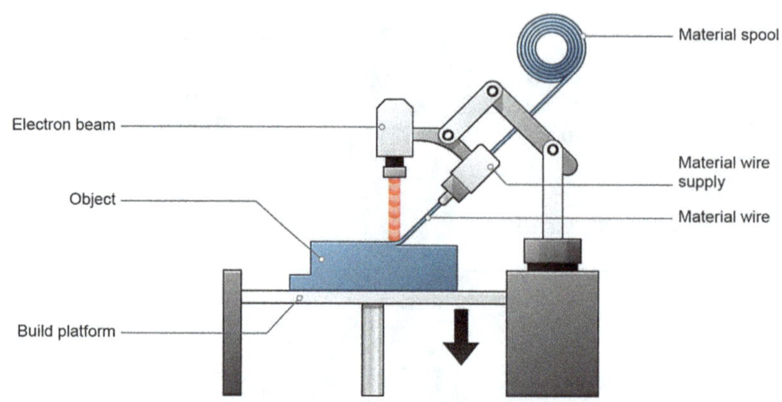

Metals can be 3D printed through the DED additive manufacturing technique and notably include aluminium, copper, titanium, stainless steel, tool steel, copper-nickel alloys, and several steel alloys. Each sub-technique of the Directed Energy Deposition section has its own limitations and compatibilities.

Hybrid Manufacturing Process

Hybrid manufacturing is a term that describes combining additive manufacturing and subtractive manufacturing in a single machine system (Li, et al. 2018). Additive manufacturing, also known as 3D printing, is used to build up a component, part or structure within another part; traditional subtractive manufacturing, or computer numerical control (CNC) milling, is used to fabricate, spot mill, polish and so on. These can be in either order. For example, additive manufacturing in the form of laser cladding could be used to build up a metal form that, when cooled, is machined using traditional methods.

A reverse example would be subtractive machining used to create a base structure, with additive manufacturing used to create thin structures on top of the base. Subtractive manufacturing is used where traditional methods suffice and additive manufacturing is used where a high degree of delicacy is needed.

Since hybrid manufacturing incorporates both additive and subtractive technologies, fabrication and precision can be attained in a more efficient and unified environment. The combination is meant to enable greater design freedom, flexibility and intricacy in creating complicated parts or those with radical geometries that conventional manufacturing was not able to produce.

Hybrid manufacturing seeks to close the gap between the unlimited potential of CAD (computer-aided design) visualizations and the myriad constraints of what can be produced in real life. Other benefits include shorter processing times, the need for fewer machines and easier inspection.

Laser Deposition Welding (LDW) and Hybrid Manufacturing

The Laser Deposition Welding (LDW) additive manufacturing process uses the metal deposition by powder nozzle, which can be up to 10-times faster than PBF technology (Arntz, et al. 2015). Furthermore, the DMG MORI company has integrated their LDW additive manufacturing technology into a 5-axis milling machine. This innovative hybrid solution combines the flexibility of the laser metal deposition process with the precision of the cutting process and thereby allows additive manufacturing in milling quality. This combination renders the manufacturing of high-precision metal parts for various sizes possible.

APPLICATIONS OF ADDITIVE MANUFACTURING TECHNOLOGY

AM technology has been a great advancement in the development of innovative and advanced AM techniques in recent times. This has led to the enhanced application of AM technology in the industry. In comparison to constrictive manufacturing, additive manufacturing has the advantages of creating a low volume of products, particularly components with complicated geometries. AM technologies also offer big potential for customized fabrications, which has been helpful in many areas, one such area being the use in producing implants which are patient-specific, especially in case of replacement of the knee and hip joints.

AM technology extends its applications to various industries such as aerospace industry, automotive industry, building construction, bio-medical field, orthopaedic and dental applications, tissue scaffolds, bio-fabrication, bio-printing of meat and food printing that have been discussed in detail.

Aerospace Industry

Aerospace parts often consist of complicated geometries, usually created using advanced materials such as alloys or superalloys of titanium and nickel, special kinds of steels or Ultra-High-Temperature Ceramics (UHTCs). These materials are small in size, hard to find, expensive and require long manufacturing time. Moreover, the number of aerospace components produced is limited, thereby making AM technology extremely appropriate for this industry.

Automotive Industry

Similar to the aerospace industry, the AM processes have been tailor-made in creating structurally and functionally complicated components of motorcycles, both passenger and commercial vehicles. The major components include gearbox parks, exhausts for the engine, braking systems and driveshaft. In racing cars, extremely low weight alloys are useful.

Building Construction

AM technology has propelled its growth within different industries. Many new-found capabilities relevant to architecture, additive manufacturing is taking another step forward in building construction.

Chinese construction company WINSUN, has constructed 10 single-room houses from an AM module in less than 24 hours in Shanghai, although these were built from components that had been fabricated offsite. WINSUN's material feedstock is a combination of cement, steel, glass fibre and recycled construction debris. The use of waste materials is a clever strategy to diminish the carbon footprint of concrete and the fabrication process ensures that construction workers are less likely to come in contact with potentially hazardous ingredients.

Biomedical Field

Recent innovations in AM technology along with developments in biomaterials, biological sciences and bio-medicine, have extended to bio-medical product sustainability, where this technology is being used to create a new horizon in case of artificial organs, tissue scaffolds, micro-vascular networks, implants of orthopaedic nature, medical devices and biochips.

Orthopedic and Dental Applications

The AM technology has been a breakthrough for patients getting dental and Orthopedic implants as it enables the formation of personalized rough surfaces and complex geometrical structures required for efficient bone integration, which can be tailor-made according to the individual requirements of each patient.

In the case of dentistry, many companies such as Concept Laser and MTT technologies are actively involved in manufacturing dental coping for bridges and crown of the tooth using Selective Laser Melting (SLM) process, whereas Electron Beam Melting (EBM) process was more successful in orthopaedic cases, which was used to produce a wide variety of certifiable implants ranging from acetabular cups to implants of the shoulders, knees, hips and spine. For example, certified Fixa Ti-Porous acetabular cups were launched by The Adler Ortho Group in the European market in 2007 using Arcam EBM technology. The number of successful implants exceeded in 2000.

Tissue Scaffolds

3D-tissue scaffolds are integral to tissue engineering as they act as a base where cells adhere to the extracellular matrix, increase in size to become new tissues for the organs. Conventionally, the synthesis of tissue scaffolds involves various methods such as gas foaming, solvent casting, freeze-drying or melt moulding. These techniques have certain disadvantages which include reproducibility, the requirement of human interaction and difficulty in controlling the complex internal architectures.

There are many types of materials (mostly bio-degradable polymer, bioactive ceramics or bioglass) for creation of scaffolds. The AM methods can be employed directly for effective fabrication of tissue scaffolds in polymers, ceramics and even glass using various methods such as SLA, FDM or SLS and also indirectly as a tool for producing moulds that are needed to cast scaffolds. Some of the examples of tissue scaffolds that have been manufactured utilizing bio-printing include, PCL and PEEK biopolymers, Hydroxy-Apatite (HA) andβ-Tri-Calcium-Phosphate (β-TCP) bioceramics and bioglasses. Bio glasses use 3DP and SLS techniques. The SLS method helps in the creation of bone scaffolds by two processes, i) using a binder, which involves joining the glass-ceramic particles and ii) without the binder, where it is directly melted. The FDM process is conducted for adding mechanical strength to the HA reinforced PCL scaffolds. Moreover, these HA scaffolds have been fabricated through rob casting, bioactive the scaffolds are formed by the deposition of concentrated colloidal suspension in the form of layers through a syringe. The next step involves drying, followed by sintering. Usually, the FDM technique requires a precursor element. However, (Shor, L. et al. 2009) used FDM where, the scaffold was directly extruded in the form of granules, thereby avoiding the use of a filament. This is known as the Precision Extruding Deposition (PED). Other bioglass biomaterials, include Poly-Ethylene Oxide and Poly-Ethylene Glycol (PEG) that use the SLA process for scaffolding live cells that have been encapsulated. Thus, AM has been useful for fabricating scaffolds for tissue engineering and other biomedical purposes.

Bio-Fabrication

The process of bio-fabrication involves the use of biomolecules, the building blocks for the formation of 'biologically functional commodities' through AM technology (Mironov, V. et al. 2009). This area converges both the fields of life sciences with engineering and forms a new application of AM technology. The applications of this process extend from engineering tissues to understanding the pathogenesis of a disease. It also involves drug-related studies, use of biosensors and biochips, targeted drug delivery followed by printing and patterning of cells and organs. Cells can be manipulated using AM methods, which display changes that have been induced optically, electrically, mechanically or magnetically. Some examples include micro-contact printing of microbes, deposition of cells using a syringe for the purpose of constructing tissues, ink-jet based printing on the cells and creating a pattern of cells formed using soft lithography and photo/electro etching. Drexel University, US has bio-printed 3D-liver cell embedded constructs, wherein different cell types and bioactive factors can be deposited in a regulated manner at the precise location using a multi-nozzle technique (Chang, R. et, al. 2010).

Bio-printing of Meat

The process of bioprinting involves the creation of tissues by accurate layer on layer deposition of living cells and biomolecules, without the use of any bio-material for the formation of a scaffold. This is done mostly using either laser-assisted or inkjet printer or micro extrusion.

In 2014, Prof. G. Forgacs and his team from the University of Missouri, US, developed a strip of porcine tissue that was ready for consumption using AM methods. Self-adhering cell types are ideal for this kind of work. In this case, the deposition occurred using an inkjet nozzle on agarose rods to form a final construct that was maintained and matured in a specialized bioreactor, which provided biomechanical characteristics to the maturing graft.

Thus, bio-printing meat has a future as a food alternative even for the vegetarians who ethically avoid meat consumption. Once commercial and affordable to all economic classes, this will open avenues for the production of safe, clean and healthy meat for the masses.

Food Printing

AM techniques have been applied for printing food products with multiple benefits. Not only the number of products can be enhanced by just one step, but it will also be helpful in creating designer food products with increased nutritional potential and new textures. This enhanced nutrition is the result of the synergistic amalgamation of the essential food nutrients such as carbohydrates, fats and protein in the food diet. This was first described by Malone and Lipson in 2007, where FAB @ Home was developed as a personal desktop fabricator which could produce food in liquid form. Grood and Grood further improvised it by making the material dropwise on demand. This technique was commercialized and called as 'Foodjet'. In this case, pneumatic membrane nozzle-jets dispensed the food materials, such as jams, jellies, cheese, icing sugar, liquid dough, chocolate, meat paste, in the form of tiny droplets onto layers, thereby helping in graphical decoration or deposition/filling of surface cavities. As this technique was limited to low viscous materials, it was not useful for the creation of complex food items.

ADDITIVE MANUFACTURING IN THE CONTEXT OF INDUSTRY 4.0

The term "Industry 4.0" is often referred to as the fourth industrial revolution that brings in a new wave of connected manufacturers and smart factories. It a set of technological advances that are having a high impact on the current industrial landscape. It is a current trend in automation and data exchange in manufacturing technologies that include a combination of cyber-physical systems, the Internet of things, cloud computing and cognitive computing. It is the fourth Industrial Revolution, following in the footsteps of computers and the internet (Industry 3.0), mass production and electricity (Industry 2.0) and mechanization and water/steam power (Industry 1.0). The fourth industrial revolution (Industry 4.0) defines as a new level of organization and control over the entire value chain of the life cycle of products, it is geared towards increasingly individualized customer requirements (Dilberoglu, et, al. 2017).

Industry 4.0 combines advanced manufacturing techniques with the Internet of Things to create manufacturing systems that are not only interconnected, but communicate, analyse, and use the information to drive further intelligent action back in the physical world. One of the key components of Industry 4.0 is the Internet of Things and Services (IoTS), which enables services providers to offer their services via the internet. The cloud platform connects the idea of an internet-based service platform with additive manufacturing. Additive manufacturing plays a significant role here as it allows for customization, reduces waste by printing only what is needed, and allows manufacturers to print where (or closer to it) and when items are needed, typically reducing shipping costs and time to market. The emerging new customer demands pose a variety of challenges for manufacturers and designers, but few that can't be overcome with time and technology. Rather than mass production – for which many factories are configured today – the shift toward more flexible manufacturing facilities eventually will become the norm. Additive manufacturing is a natural fit: with 3D printing solutions, a designer or manufacturer needs only change a design file digitally to customize or improve products on the fly.

Industry 4.0 aims at the construction of an open, smart manufacturing platform for an industrial-networked information application. Real-time data monitoring, tracking the status and positions of a product as well as to hold the instructions to control production processes are the main needs of Industry 4.0. The Fourth Industrial Revolution witnesses' innovations like 3D printing, the Internet of Things (IoT), artificial intelligence, and robotics. It's disrupting almost every industry, business model, and country. The key features of Industry 4.0 are shown in figure 13.

Additive manufacturing (AM)/ 3D Printing is considered to be an essential ingredient in this new movement. It is emerging as a valuable digital manufacturing technology in the era of smart manufacturing. Once solely a rapid prototyping technology, today AM offers a huge scope of possibilities for manufacturing from tooling to mass customisation across virtually all industries. The combination of additive manufacturing with Industry 4.0 can help enable the creation of products that are first-to-market, fully customized, and dynamic. These possibilities have the ability to change manufacturing in profound ways. Additive manufacturing works by using digital 3D models to create parts with a 3D printer layer

Figure 13.

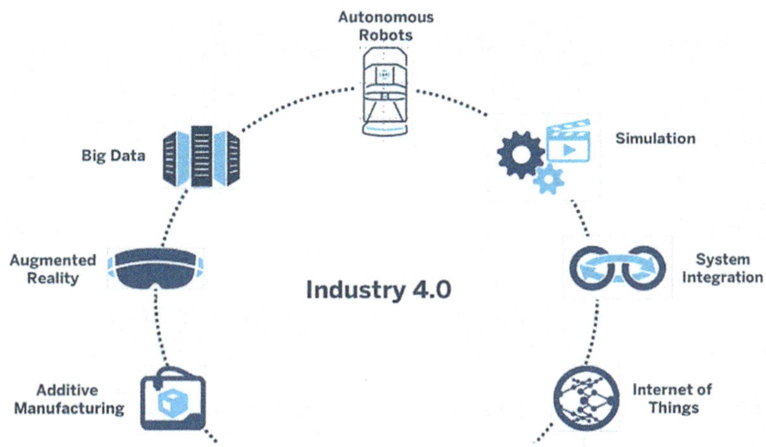

by layer. Additive manufacturing is a perfect fit for countering the emerging new consumers' demands for smart products and their shift towards Industry 4.0. With 3D printing solutions, a designer or manufacturer needs only to change a design file digitally to customize or improve products.

Within the context of Industry 4.0, it enables parts to be stored as design files in virtual inventories, so that they can be produced on-demand and closer to the point of need. This new approach to manufacturing can reduce transportation distances, and hence costs, as well as simplify inventory management by storing digital files instead of physical parts. Both additive manufacturing and Industry 4.0 are evolving fast with greater growth in the present time. Therefore, it is expected increased digitalization of additive manufacturing in the future.

The central objective of Industry 4.0 is fulfilling individual customer needs, which affect areas like order management, research and development, manufacturing commissioning, delivery up to the utilization and recycling of products through digital production. The need for industry 4.0 is to convert the regular machines to self-aware and self-learning machines to improve their overall performance and maintenance management with the surrounding interaction. Additive manufacturing plays a significant role here as it allows for customization, reduces waste by printing only what is needed, and allows manufacturers to print where and when items are needed, typically reducing shipping costs and time to market.

CONCLUSION

This article concludes the various advancements in manufacturing processes to be used in industry to increase the production and efficiency of the shop floor. By adopting the additive manufacturing there would increase the life of the product development cycle, cutting the material and process cost thereby increasing the profits to the manufacturing industry. Here the most popular additive manufacturing processes are discussed in detail along with the Hybrid Manufacturing. The integration of AM with Industry 4.0 (I 4.0) is discussed, in the near future the entire manufacturing sectors will witness tremendous changes where parts or components are made at our doorsteps. It would also give free hand to their customers to take part in product development. The future manufacturing would be part customization, not a mass production

REFERENCES

Arntz, K., Wegener, M., & Liu, Y. (2015). Computer aided manufacturing supported process planning of additive manufacturing by laser deposition welding. *Journal of Laser Applications*, *27*(S1), S14002. doi:10.2351/1.4823748

ASTM International. (2013). ASTM F2792-12a. *Rapid Manufacturing Association*, *1–3*. doi:10.1520/F2792-12A.2

Atzeni, E., Barletta, M., Calignano, F., Iuliano, L., Rubino, G., & Tagliaferri, V. (2016). Abrasive Fluidized Bed (AFB) finishing of AlSi10Mg substrates manufactured by Direct Metal Laser Sintering (DMLS). *Additive Manufacturing*, *10*, 15–23. doi:10.1016/j.addma.2016.01.005

Baykasoglu, C., Akyildiz, O., Candemir, D., Yang, Q., & To, A. C. (2018). Predicting Microstructure Evolution During Directed Energy Deposition Additive Manufacturing of Ti-6Al-4V. *Journal of Manufacturing Science and Engineering*, *140*(5), 051003. doi:10.1115/1.4038894

Billiet, T., Vandenhaute, M., Schelfhout, J., Van Vlierberghe, S., & Dubruel, P. (2012). A review of trends and limitations in hydrogel-rapid prototyping for tissue engineering. *Biomaterials*, *33*(26), 6020–6041. doi:10.1016/j.biomaterials.2012.04.050 PMID:22681979

Chang, R., Emami, K., Wu, H., & Sun, W. (2010). Bio-fabrication of a Three-Dimensional Liver Micro-organ as an In-Vitro Drug Metabolism Model. *Bio-Fabrication, 2*(4).

Chartrain, N. A., Williams, C. B., & Whittington, A. R. (2018). A review on fabricating tissue scaffolds using vat photopolymerization. *Acta Biomaterialia*, *74*, 90–111. doi:10.1016/j.actbio.2018.05.010 PMID:29753139

Dilberoglu, U. M., Gharehpapagh, B., Yaman, U., & Dolen, M. (2017). The Role of Additive Manufacturing in the Era of Industry 4.0. *Procedia Manufacturing*, *11*(June), 545–554. doi:10.1016/j.promfg.2017.07.148

Gibson, I., Rosen, D., & Stucker, B. (2015). *Additive Manufacturing Technologies*. New York, NY: Springer New York; doi:10.1007/978-1-4939-2113-3

Gokuldoss, P. K., Kolla, S., & Eckert, J. (2017). Additive Manufacturing Processes: Selective Laser Melting, Electron Beam Melting and Binder Jetting—Selection Guidelines. *Materials (Basel)*, *10*(6), 672. doi:10.3390/ma10060672

ISO/ASTM. (2015). International standard ISO / ASTM 52900 Additive manufacturing — General principles — Terminology. *International Organization for Standardization*, *5*, 1–26. doi:10.1520/ISOASTM52900-15

Kunchala, P., & Kappagantula, K. (2018). 3D printing high density ceramics using binder jetting with nanoparticle densifiers. *Materials & Design*, *155*, 443–450. doi:10.1016/j.matdes.2018.06.009

Li, L., Haghighi, A., & Yang, Y. (2018). A novel 6-axis hybrid additive-subtractive manufacturing process: Design and case studies. *Journal of Manufacturing Processes*, *33*, 150–160. doi:10.1016/j.jmapro.2018.05.008

Liu, F.-H., Lee, R.-T., Lin, W.-H., & Liao, Y.-S. (2013). Selective Laser Sintering of Bio-Metal Scaffold. *Procedia CIRP*, *5*, 83–87. doi:10.1016/j.procir.2013.01.017

Meisel, N. A., Elliott, A. M., & Williams, C. B. (2015). A procedure for creating actuated joints via embedding shape memory alloys in PolyJet 3D printing. *Journal of Intelligent Material Systems and Structures*, *26*(12), 1498–1512. doi:10.1177/1045389X14544144

Mironov, V., Visconti, R. P., Kasyanov, V., Forgacs, G., Drake, C. J., & Markwald, R. R. (2009). Organ Printing: Tissue Spheroids as Building Blocks. *Biomaterials*, *30*(12), 2164–2174. doi:10.1016/j.biomaterials.2008.12.084 PMID:19176247

Murr, L. E. (2015). Metallurgy of additive manufacturing: Examples from electron beam melting. *Additive Manufacturing*, *5*, 40–53. doi:10.1016/j.addma.2014.12.002

O'Connor, H. J., Dickson, A. N., & Dowling, D. P. (2018). Evaluation of the mechanical performance of polymer parts fabricated using a production scale multi jet fusion printing process. *Additive Manufacturing*, *22*, 381–387. doi:10.1016/j.addma.2018.05.035

Park, J., Kim, B., Kim, S.-Y., & Hwang, J. (2014). Prediction of drop-on-demand (DOD) pattern size in pulse voltage-applied electrohydrodynamic (EHD) jet printing of Ag colloid ink. *Applied Physics. A, Materials Science & Processing*, *117*(4), 2225–2234. doi:10.100700339-014-8650-6

Raghavan, N., Dehoff, R., Pannala, S., Simunovic, S., Kirka, M., Turner, J., ... Babu, S. S. (2016). Numerical modeling of heat-transfer and the influence of process parameters on tailoring the grain morphology of IN718 in electron beam additive manufacturing. *Acta Materialia*, *112*, 303–314. doi:10.1016/j.actamat.2016.03.063

Seifert, T., Sowade, E., Roscher, F., Wiemer, M., Gessner, T., & Baumann, R. R. (2015). Additive Manufacturing Technologies Compared: Morphology of Deposits of Silver Ink Using Inkjet and Aerosol Jet Printing. *Industrial & Engineering Chemistry Research*, *54*(2), 769–779. doi:10.1021/ie503636c

Shor, L., Güçeri, S., Chang, R., Gordon, J., Kang, Q., Hartsock, L., & Sun, W. (2009). Precision Extruding Deposition (PED) Fabrication of Poly-Capro-Lactone (PCL) Scaffolds for Bone Tissue Engineering. *Bio Fabrication, 1*(1).

Suresh, Narayana, & Mallik. (2019). Hardness, Wear and Corrosion Properties of Co-Cr-W Alloy Deposited with Laser Engineered Net Shaping in Medical Applications 1952. *International Journal of Innovative Technology and Exploring Engineering*, *8*.

Suresh, G., Narayana, K. L., Kedar Mallik, M., Srinivas, V., & Jagan Reddy, G. (2018). Processing & Characterization of LENS TM Deposited Co-Cr-W Alloy for Bio-Medical Applications. *276| International. Journal of Pharmacy Research*, *10*(1), 276–285.

Suresh, G., & Narayana, K. L. (2016). *A review on fabricating procedures in rapid prototyping. 3D Printing: Breakthroughs in Research and Practice*. doi:10.4018/IJMMME.2016040102

Sutton, A. T., Kriewall, C. S., Leu, M. C., & Newkirk, J. W. (2017). Powder characterisation techniques and effects of powder characteristics on part properties in powder-bed fusion processes. *Virtual and Physical Prototyping*, *12*(1), 3–29. doi:10.1080/17452759.2016.1250605

Torrado, A. R., Shemelya, C. M., English, J. D., Lin, Y., Wicker, R. B., & Roberson, D. A. (2015). Characterizing the effect of additives to ABS on the mechanical property anisotropy of specimens fabricated by material extrusion 3D printing. *Additive Manufacturing*, *6*, 16–29. doi:10.1016/j.addma.2015.02.001

Wang, J., Goyanes, A., Gaisford, S., & Basit, A. W. (2016). Stereolithographic (SLA) 3D printing of oral modified-release dosage forms. *International Journal of Pharmaceutics*, *503*(1–2), 207–212. doi:10.1016/j.ijpharm.2016.03.016 PMID:26976500

Yap, Y. L., Wang, C., Sing, S. L., Dikshit, V., Yeong, W. Y., & Wei, J. (2017). Material jetting additive manufacturing: An experimental study using designed metrological benchmarks. *Precision Engineering*, *50*, 275–285. doi:10.1016/j.precisioneng.2017.05.015

Chapter 2
IDI Engine With Alternate Fuels

Venkata Appa Rao Basava
Andhra University, India

Aditya Kolakoti
https://orcid.org/0000-0002-7515-8318
Raghu Engineering College, India

Prasada Rao Kancherla
NRI Institute of Technology, India

ABSTRACT

A plethora of experiments were conducted on IDI engine with various biodiesels (e.g., methyl esters of mahua, jatropha, rice bran, pongamia, palm, beef tallow, and waste cooking oils). Review of the results of these endeavors with various additives and blends with or without super charging of the engine are presented in this chapter. All these attempts have been concentrated to arrive at the best yield from a single cylinder engine. The recorded pressure changes during combustion, the derived heat release rates, and exhaust emissions are presented in the form of plots at various loads and at a constant speed. Engine cylinder vibrations (reflect combustion excitation) in the form of FFT and time waves were recorded at radial points and vertical on the cylinder body to assess the combustion propensity in all cases of studies. The results with relative benefits are enumerated.

INTRODUCTION

IDI engines are silent engines with divided combustion chamber to facilitate almost complete combustion with the maximum utilization of the oxygen available. These engines run with higher compression ratios when compared to the open combustion chamber engines. In this chapter, the authors' intention is to try several neat biodiesel fuels with additives and blends to testify the viability of these fuels in the engine without major basic engine component design modification. Normally, the regular experiments on the engine give performance, emissions and the heat release rate calculations based on the combustion pressure plots using the first law of thermodynamics. But as advancement to the conventional data

DOI: 10.4018/978-1-7998-4939-1.ch002

assessment, the engine cylinder vibrations in the direction of piston movement, in the radial piston slap direction, and in a direction perpendicular to crank shaft in the radial direction on the cylinder body are measured in the form of FFT plots and time wave plots. This is to arrive at better understanding of the combustion happening in the engine cylinder with the usage of totally new fuel. Several authors tried tests on their propositions of additives and blends to establish the fact of suitability of the biofuels in the engine without causing side effects which jeopardize the reliability of the engine. Since the combustion excitation is in consonance with engine cylinder vibration, the smoothness of the combustion can be testified by the vibration measurement in different forms like, FFT and time waves. The time verses the combustion accelerated frequency amplitudes can be correlated with the pressure rise and smoothness of the combustion there by interpreting the zones of premixed and diffused combustion for better comparison with the diesel fuel.

BACKGROUND

Experimental studies were conducted to evaluate the performance, emissions, combustion and heat release rate of an Indirect Diesel Injection (IDI) engine with comprehensive engine cylinder vibration proof to substantiate better combustion. The summery of all endeavors yield good evidence that various biodiesel applications either blending with diesel or mixing any additive proved better than the implementation of diesel fuel in a neat form. In the present study, the experimental investigation by K Prasada Rao and B V Appa Rao (2017) used neat Mahua methyl ester (MME) along with Methanol additive blends. Since the auto ignition temperature of methanol is 470^{0}C, it may not be initiator to combustion in the pre combustion chamber but, a part of the biodiesel in the blend may be set to combustion. The authors K Prasada Rao and B V Appa Rao (2017) in the experimental proposal chose to take 1% to 5% methanol blends and finally concluded that 3% methanol blend yielded better results. The author's conception is that the higher latent heat methanol reduces the combustion temperature and there by reduces NO component in the exhaust gas. Normally methanol increases the CO formation. But it is an incidental observation that CO also reduced by 20% at the maximum load, indicates that the presence of methanol increased the air fuel entrainment the evaporation mode at low lower evaporation temperature in both the chambers leading to near totality of the combustion. In another endeavor, by Y Ashok Kumar Reddy (2013) it is not crude oil but the heated Jatropha methyl ester (JME) was tested to verify the results and it yielded better results over the unheated biodiesel or the petroleum-based diesel. The author has observed that the viscosity of biodiesel can be decreased by heating the oil to a level close on parity with the diesel oil's viscosity at 60^{0}C and above this temperature the oil became critical indicating no more further reduction in the viscosity. Thermal efficiency of the engine is under greater control with 60^{0}C heated JME and it has shown betterment at higher operating loads of the engine. At other higher input oil temperatures, the heat losses may be reason for lesser thermal efficiency values comparatively. Predictably, the exergy is better and the irreversibility is contained with preheating of the oil to 60^{0}C.

There is a plethora of refined benefits with reduced cylinder vibration observed by T Vector Babu et al. (2016) which finally led to the conclusion that 2% Isopropanol in rice bran oil biodiesel yielded good results with a competitive parity in the power development and ensuring smoother combustion increasing the reliability of the engine. This time, the additive is injected with a secondary injection system. Isopropanol is introduced at the suction end by secondary injection at a pressure of 3 bar by various quantities and the combustion propensity was assessed investigating the time waves of vibration and

FFT's in log–log graphical form. Combustion heterogeneity is observed to have decreased at all additive percentages. 2% Isopropanol excelled over other percentages. Moreover, the aspects of knock and detonation were clearly abysmal in the range of alcohol percentages implemented here in this experimentation.

The effect of Pongamia Methyl Ester (POME) biodiesel with Isobutanol additive was studied by N Prasada Raju (2015) as an alternative fuel replacing the conventional diesel fuel for the diesel engine. Experimentation was conducted to test the new fuel sample to determine their effect on the engine performance and exhaust gas emissions and cylinder vibrations. 6% Isobutanol blend with POME proved giving good results.

Studies conducted by S.C.V.Ramana Murty Naidu (2015) revealed interesting results in an IDI engine.1,4-Dioxane possesses higher latent heat (413kJ/kg) than that of diesel (280kJ/kg), its blending creates low temperature combustion and reduces the formation of NO_x. 1,4-Dioxane evaporates and ignites first because of its lower boiling point (101.1^0 C) and auto ignition temperature (180^0C). It initiates combustion of rest of the vapor in pre-combustion chamber. 1,4-Dioxane evaporates and ignites first because of its lower boiling point (101.1^0C) and auto ignition temperature (180^0C). It initiates combustion of rest of the vapor in pre-combustion chamber. Four additive blends of 1,4-Dioxane in palm methyl ester (PME) viz. 1%, 2%, 3% and 4% by volume were tested. 3percent 1,4-Dioxane in PME is considered as best choice and engine performance and emission characteristics were evaluated in comparison with neat PME and petro diesel fuels. Brake Specific Fuel Consumption is reduced by 2.64 percent compared with neat PME, an increase of Brake Thermal Efficiency by 9.89 percent, Exhaust Gas Temperature and smoke readings are decreased by 4^0C and 20 percent respectively for 3 percent 1,4-Dioxane in biodiesel with respect to conventional diesel fuel at maximum load. Engine emissions are decreased by 2ppm for HC, 40 percent for CO, 10.89 percent for CO_2 and 56ppm for NOx with 3 percent additive in biodiesel with respect to diesel fuel. The vibration studies reveal the combustion propensity with different fuel samples. Time signal plots are recorded on the engine cylinder head in vertical direction of the piston movement to analyze knocking aspect with the fuels tested. 9.55J/degree and zero peak differences in cumulative heat release rate can be observed with diesel and biodiesel respectively relative to 3percent 1,4-Dioxane in bio-diesel. Since the boiling point and auto ignition temperatures are lower than that of both diesel and biodiesel, combustion pressure reaching maximum is advanced by 5^0 of crank revolution.

Aditya Kolakoti and B V Appa Rao (2019) tried experiments in a supercharged IDI engine with Palm Methyl Ester (PME) biodiesel with Coconut methyl ester (COME) additive as an alternative fuel to replace the conventional diesel fuel for the diesel engine. For this purpose, five different COME additive percentages in PME biodiesel samples containing 2%, 3%, 4%, and 5% were prepared on volume basis and tested in an Indirect Diesel Injection (IDI) engine. This engine is supercharged with 350watt air compressor. Coconut biodiesel increases the saturated fatty acid content in the methyl ester solution and helps in increasing the thermal efficiency and reducing the NO_x emission in the tail pipe gases. This is already a proven fact ascertained from the previous work by Sukumar Puhan et al. (2010) on unsaturated biodiesel application in diesel engines. The percentages from 2 to 5 percent in a dose increase of 1% of COME is tried as an additive with the main biodiesel PME and after analyzing the combustion pressures, heat release rates, engine performance, exhaust emissions, smoke, exhaust temperatures, and the engine cylinder vibration, it is concluded that 3% COME in PME is suitable alternate fuel to replace diesel fuel in the IDI engine.

Ganapathi Vedulla (2017) in their work used Beef tallow oil methyl ester (BTME) to use in a single cylinder, 4-stroke, vertical, variable speed IDI engine with supercharging application. Since it is mooted to study the combustion aspect of the engine only, the clutch and gear box arrangement was forfeited and the load and speed of the engine was fixed arbitrarily at which the engine operated without visible smoke appearance. To encourage low temperature combustion an additive like **hydrated ethyl alcohol** is injected through a separate nozzle at the suction end with 3bar pressure. The investigation includes engine cylinder vibration measurement for combustion performance corroboration and finally it is concluded that 2% additive with BTME yielded best results to totally replace the diesel fuel.

Pentakota Santosh Babji (2018) studied to find an immediate alternative and sustainable energy solution, by using Waste cooking oil (WCO) for biodiesel production. After making the biodiesel, the characterization has been taken up in finding out the property parameters. Gas chromatography revealed the fatty acid content pattern. After this assessment, Waste Cooking Oil Biodiesel (WCOBD) is tested in supercharged single cylinder, 4-stroke, IDI diesel engine. In this case of experimentation, very interesting outcome has been observed by the measurement of engine cylinder vibration, which indirectly inferred that the heat to torque conversion aspect is improved in the case of the new fuel i.e. WCOBD with 6% hydrated methanol as additive *with secondary injection at 3 bar.*

MAIN FOCUS OF THE CHAPTER

IDI engines with several operating factors are being discussed in the background section above. All these topics confine to single laboratory. Notwithstanding the higher heat transfer from the larger area of the two combustion chambers, there is an increase in the NO emission. To tackle this situation, additives (either by blending or by secondary injection) with supercharging are resorted to. The creation of concept has started with methanol additive with higher latent heat to maintain low temperature combustion to contain NO formation. But this heat in the latent heat form should be released at higher pressures reached in the main combustion chamber to utilize it for torque conversion. The combustion in the main chamber and raise in the pressure levels release this latent heat. This technique also contains the unnecessary heat transfer from the cylinder surface area since the absorption of heat takes place at a constant temperature as for the definition of the latent heat. More said and done, the heat release should take place at the right time to get converted to torque. This will be taken care of by the quantity of methanol being injected. This quantity is being optimized by the authors K Prasada Rao and B V Appa Rao (2017) by experimentation. The extent of the contribution of methanol along with biodiesel is also being verified by the cylinder vibration analysis also. Since the combustion excitation will be truly reflected in the cylinder vibration in one direction or other, the vibration measurements are taken in two accessible directions i.e. vertical on the piston head and in the piston slap direction along with the vibration transmitted to the engine foundation. The data collected led to interesting conclusions supporting the author's visualization in the application of the new neat biodiesel fuel with the additive chosen.

Issues, Controversies, Problems

Controlling the Biodiesel Leakage Into the Crank Case

Complete combustion is ensured in two chambers of the IDI engine leading to complete combustion of even low-quality fuel. This phenomenon of complete combustion may be helping in stopping the oil leakage past piston. High induced swirl velocities may increase the heat transfer from the pre chamber leading to reduction in thermal efficiency but the air entrainment rate will be compromised if the swirl decreases. Burn rate of the fuel quantity in the pre-chamber should be controlled and consideration should be given for the type of the fuel used as a replacement in the engine. Oxygenated fuels like biodiesel which contain molecular oxygen may give a part of relief in the aspect of air entrainment. But oxygen in biodiesel will obviously reduce the calorific value of the fuel. That is the reason some researchers tried to increase the fuel quantity by increasing the injection pressure to increase the energy generated in the pre-combustion chamber but increase in the quantity of fuel and availability of the oxygen are at stake now because of unchanged compression ratio. The push that is needed to transfer combustion gas from the pre-chamber to the main chamber is another question since there are some IDI engines that run at higher speeds in which the time of combustion may be inadequate to totally happen in the engine leading to generation of obnoxious pollutants in the exhaust. That is the reason the sharing of combustion and share of clearance volume in between the two chambers play important role in tackling various different fuels that can replace the diesel fuel. Any attempt in introducing the additives also should be taken care of as is the case explained in the back ground aspect.

Control of Heat Wastage From the Chambers Surface Area

Absorbing the heat of combustion by the additive in the pre-combustion chamber to release it when the pressure of the partly burned gases rises to required level in the main chamber is a pertinent aspect to reduce heat transfer. But the properties of the additives will vary from one to another leading to side effects. Lower temperature of evaporation will lead to early burning of the additive which triggers combustion of the main fuel jeopardizing the share of combustion in both the chambers. The burning in the pre-combustion chamber should be limited to such an extent that it produces correct premixed fuel to burn smoothly in the main chamber leading better heat–torque conversion. Additives will have different properties like higher latent heat, lower Cetane number, higher or lower calorific values, different acoustic properties and evaporation rates. Some of the additives are used to thin the main fuel in vapor condition which increase the velocities of semi burned fuel or produce better atomization.

Study of Additive Properties to Meet Conservative Combustion

Most of the additives possess higher auto combustion temperatures. This may lead to increase in the heterogeneity of combustion. There are exceptions like 1,4-Dioxane which possesses different atomization and evaporation qualities and for those fuels the quantity of proportion in the main biodiesel will define the combustion propensity and to fix this the cylinder vibration analysis will be good tool. Small quantities of coconut biodiesel in the main biodiesel will be changing the fatty acid properties of the main biodiesel in the way it increases the saturated fatty acid content leading to betterment in both NO emission in the exhaust and thermal efficiency. There is a recorded cylinder vibration reduction with the blending of coconut biodiesel in the palm methyl ester in a supercharged IDI engine. Supercharging of the IDI engine also saves fuel with adequate oxygen supply to burn completely and contains combustion heat from transferring through the cylinder walls as investigated by Aditya Kolakoti and B V Appa Rao (2017).

Reduction of combustion heat wastage from the surface area of the combustion chambers, reasonable air entrainment with the fuel, controlled combustion in the pre-combustion chamber, adjustment of velocity of flow of the semi-burned gases from the pre-chamber to the main chamber adapting to the speed of the engine, adequate pressure build up in the main chamber to release the latent heat from the burned gases in case of any usage of any additive are the parameters in question to be solved. Measurement of vibration on the cylinder body gives important information about the happening of the combustion in both the chambers; especially the time waves recorded on the cylinder head in the vertical direction render good information about the timely combustion excitation in both the chambers. The FFT measurement portrays the frequencies emanated in order wise with respect to the running frequency of the engine. This information is necessary in deciding the combustion smoothness and the effect of the additive used along with the main biodiesel.

Combustion smoothness varies with the properties of the additive used. The properties viz. latent heat, evaporation temperature, partial pressure of evaporation, auto ignition temperature, Cetane number, and the kinematic viscosity of the additive play important role in deciding the combustion. Most of the times, some of the properties are complementary in supporting the combustion. Instead of nozzle in between the chambers, there is a *grating* in the new designs which splits the ongoing flow creating good turbulence to mix the air and fuel in the main chamber. The grating also acts as miniature nozzles repertoire to convert the pressure flow into the velocity flow. The split combustion chamber converts high frequency orders to low frequency vibration making use of the Helmholtz resonator system principle. These higher amplitude lower frequencies can be easily attenuated with the interposition of good damping structure there by increasing the reliability of the engine with respect to time.

Experimental Setup and Types of Additives Used

Figure. 1 Shows the Experimental test rig comprising loading device and combustion data logger is established in the author's Laboratory. The engine that was used in this study was a four-stroke single cylinder Indirect Diesel Injection (IDI) variable speed engine with specifications outlined in table 1. Reason for this choice is that, the engine constitutes two combustion chambers and it is known for lower exhaust emissions. Notwithstanding the greater heat transfer because of greater surface area of

Figure 1. Experimental test rig

Table 1. Technical specifications of experimental engine

Engine Manufacturer	Bajaj RE Diesel Engine
Engine type	Four Stroke, Forced air and Oil Cooled
No of Cylinders	One
Bore	86.00 mm
Stroke	77.00 mm
Engine displacement	447.3 cm^3
Compression ratio	24 ± 1:1
Maximum net power	5.04 kW at 3000 rpm
Maximum net torque	18.7 Nm at 2200 rpm
Idling rpm	1250 ± 150 rpm
Injection Timings	8.5^0 to 9.5^0 BTDC
Injector	Pintel type
Injector Pressure	142 to 148 kg/cm^2
Fuel	High Speed Diesel
Starting	Electric Start
Arbitrarily fixed power and speed	2.92kW max,1500 rpm

the combustion chambers, the engine is chosen for its clean combustion. Low temperature combustion using an appropriate additive also discourages higher heat loss due to smaller comparative temperature difference and high speed of the engine.

Five additive blends of Methanol in Mahua Methyl Ester (MME) were tested, the blending ratios of 1/99, 2/98, 3/97, 4/96 and 5/95 (by vol.) additive Methanol is added to MME, and five discrete part load conditions viz. No Load, 0.77 kW, 1.54 kW, 2.31 kW, and 2.70 kW loads without gear box and clutch assembly ensuring stable engine operation. Experimentation is carried out at various engine loads with eddy current dynamometer as the loading device. Combustion pressures at each degree of crank revolutions are recorded with C7112 software and heat release rates are computed with pressure data as the input data. Fuel consumption is recorded for characterized fuel to study the performance along with the engine pollution parameters engine and cylinder vibration. Engine cylinder vibration in Fast Fourier Transformation (FFT) is recorded at each load on the cylinder head and the foundation and for each fuel combination to compare the cylinder excitation frequencies with the base line frequencies using diesel oil. Time wave forms on the cylinder head are the potential data to analyze the time of peak pressure estimation, piston slap, leakage past piston etc. The vibration study of the engine cylinder through the measured FFT and time waveforms are the representatives of combustion propensity. The smoke values in Hartridge Smoke Units (HSU), the exhaust gas temperatures and exhaust gas analysis of different components of exhaust are measured and compared.

Figure 2 shows a schematic diagram represents the instrumentation set up for the experiment. The Piezo electric transducer is fixed (flush in type) to the cylinder body (with water cooling adaptor) to record the pressure variations in the combustion chamber. Crank angle encoder sends signals of crank angle with reference to the top dead center position and will be transmitting to the control panel (Data logger). The data logger synthesizes the two signals and final data is presented in the form of a graph on the computer using **C7112** software.

Figure 2. Schematic view of experimental test rig

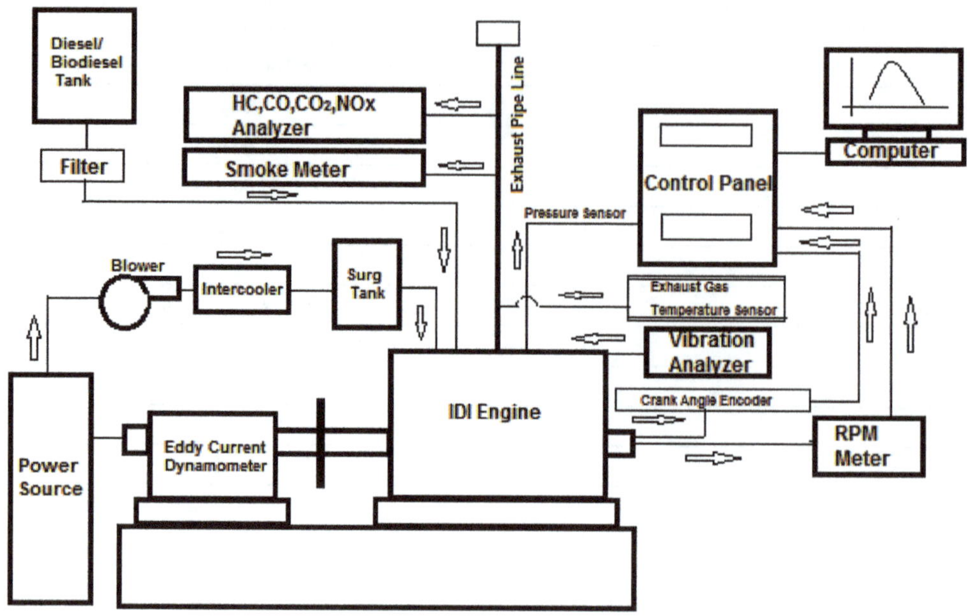

Cylinder vibration is recorded using an accelerometer mounted on the cylinder head in the two desired directions (one is in the piston slap direction and the other being vertical). ***DC-11*** data logger is used to ascertain the spectral data in the form of ***FFT*** and overall vibration levels. The time waveforms and FFT plots are obtained on the cylinder head by **DC-11** in OFF-ROUT mode and are presented in graphic form by ***Vast-an*** doss based software, designed by VAST, Russia.

Aditya Kolakoti and B V Appa (2019) tried on four additive blends of COME in Palm biodiesel and were tested with percentages viz. 2%, 3%, 4% and 5% by volume of COME. Experimentation is carried out following same procedure as mentioned above. Engine emission parameters and engine cylinder vibration are also measured; the average values have been presented.

S.C.V.Ramana Murty Naidu (2015) conducted experiments on four additive blends of 1,4- Dioxane in Palm Methyl Ester (PME) and were tested with percentages viz. 1%, 2%, 3% and 4% by volume. Similar experimentation is carried out on the same engine as mentioned above.

N Prasada Raju (2015) tried with five additive blends of Isobutanol in Pongamia Methyl Ester (PME) and tests were conducted with the blending ratios of 2/98, 4/96, 6/94, 8/92 and 10/90 (by vol.).

Ganapathi Vedulla (2017) in their work used transesterified Beef Tallow oil (BTME) to use the same engine with supercharging application. To encourage low temperature combustion an additive like **hydrated ethyl alcohol** is injected through a separate nozzle at the suction end with 3bar pressure. The investigation includes engine cylinder vibration measurement for combustion performance corroboration and finally it is concluded that 2% additive with BTME yielded best results to totally replace the diesel fuel.

Pentakota Santosh Babji (2018) studied to find an immediate alternative and sustainable energy solution, by using Waste cooking oil (WCO) for biodiesel production. After making the biodiesel, the characterization has been taken up in finding out the property parameters. Gas chromatography revealed the fatty acid content pattern. After this assessment, Waste Cooking Oil Biodiesel (WCOBD) is tested in supercharged single cylinder, 4-stroke, IDI diesel engine. In this case of experimentation, very interesting outcome has been observed by the measurement of engine cylinder vibration, which indirectly inferred that the heat to torque conversion aspect is improved in the case of the new fuel i.e. WCOBD with 6% hydrated methanol as additive *with secondary injection at 3 bar.*

Results Discussion

Engine Performance and Emissions

The performance test results of the engine were compared to diesel, Mahua Methyl Ester (MME) and Methanol additive blends for various percentages by K Prasada Rao and B V Appa Rao (2017)

- Brake Specific Fuel Consumption (BSFC): Shows the Brake Specific Fuel Consumption (BSFC) verses Load in kW. A smoother trend of the curve is observed for the additive blend of 3%. The equivalence ratio for the 3% additive blend is observed on the higher side, i.e. 0.6883, nevertheless, the performance can be rated better when compared to other samples in view of examination of other parameters. This may be because of the reason that the injection plume development and specific heat associated with the 3% blend are congenial for better atomization and evaporation leading to better entrainment with air.
- Brake Thermal Efficiency: gives the details of Brake Thermal Efficiency versus the Equivalence Ratio of neat diesel, biodiesel and additive blends. The figure.5 shows that the thermal efficiency is increasing with Methanol additive percentage even though calorific value of additive is low compared to the main biodiesel. This is due to an improved quality of combustion with the additive. At 3% Methanol additive blend yielded maximum thermal efficiency of 29.33% at equivalence ratio 0.5152.
- Exhaust Gas Temperature: The recorded Exhaust Gas Temperature (EGT) versus Load using neat diesel, biodiesel and additive blends explain the following result. Exhaust temperature decrease of 14^{0}C is observed in the case of 3% additive with biodiesel. This observation of lower exhaust temperature is representative of lower combustion temperatures in both the combustion chambers.
- Engine Smoke Levels: The decrease in the smoke level in exhaust with respect to the neat diesel fuel operation is appreciable (19 HSU reduction at maximum load operation). This is an indication of better combustion with 3% additive in biodiesel.
- Hydrocarbon (HC) and Carbon Monoxide (CO) Emissions: The research on IDI engine indicates lower emissions in the exhaust. Additive mixing further reduced the HC emission as shown in figure.3 and CO emission to greater extent.
- Nitrogen Oxides (NO) Emission: NO_x emission has also reduced at higher loads, especially with the additive as shown in figure.4. 57% HC, 20% CO,14% $NO_,$ 27% smoke reductions were observed at 3% methanol additive at maximum opted load (2.70 kW and 1500 rpm) operation of the engine. Carbon Dioxide (CO_2) Emission: CO_2 emission has increased because of combustion improvement.

Figure 3. HC emissions at varying engine loads

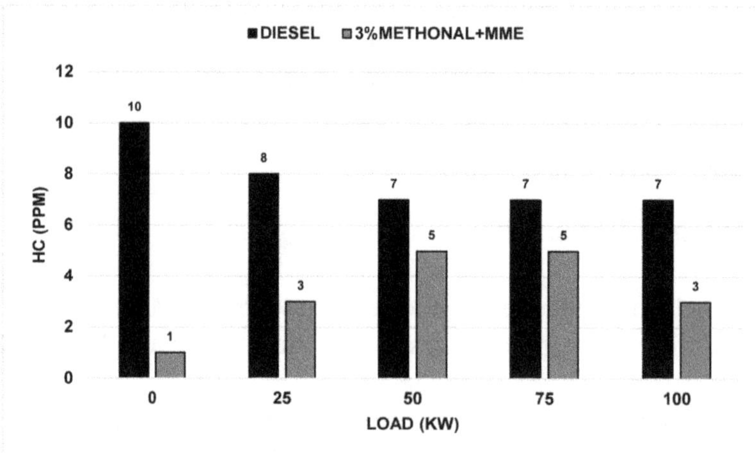

Figure 4. NO emissions at varying loads

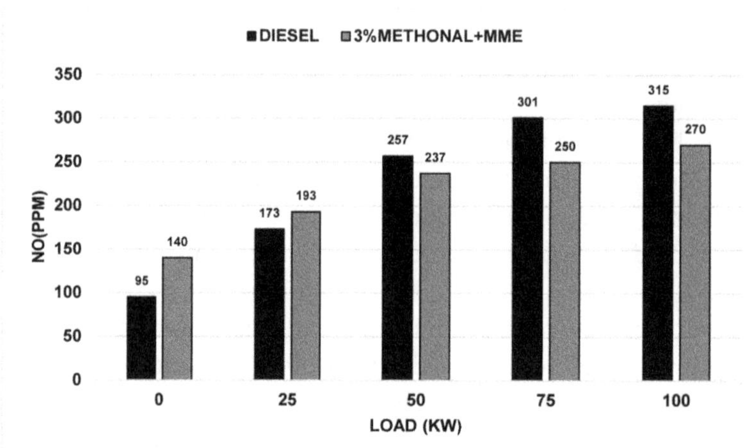

Combustion Characteristics

Combustion pressure values have been logged in the combustion cycle range of 720^0 comprising two crank revolutions. The values have been logged at every single degree of the crank angle interval. In an attempt to improve the performance of the engine with neat biodiesel application and methanol is used as an additive to control the combustion temperatures both in the primary and main combustion chambers. Blends of Mahua methyl ester and methyl alcohol have been used to investigate the engine performance. The chosen blends are 1/99, 2/98, 3/97, 4/96 and 5/95 (by volume) of methyl alcohol with methyl ester. It is observed homogenous blends with all these percentages of alcohol even after longer lapse of time. The creation of low temperature combustion reduces the formation of NO_x and slows down the combustion severity because of the additive's higher auto ignition temperature, lower Cetane value, and higher latent heat. The combustion propensity can also be interpreted from the engine cylinder vibration analysis.

The frequency curves and time waves decipher the vibration trend of the engine with possible knocking and detonation when the fuel sample charged with higher dosages of alcohol. Neat Mahua methyl ester decreased the delay period because of its higher Cetane number and as the addition of alcohol obviously decreases the overall Cetane number increasing the delay period. The Cetane number of methyl alcohol is '4' where as that of biodiesel is nearly '61.4'. This dilution of Cetane number with the addition of alcohol dampens the virulence of the combustion in any heterogeneous state and finally 3% of alcohol has created real time stability of combustion in both the chambers. The figures 5,6&7 showcase the performance, efficient burning capability of the neat Mahua methyl ester with 3% methanol additive enabling smooth combustion pressure variation (figure 6) and sustainable net heat release rate (figure 7) at the maximum load on the engine during premixed and diffused combustion. The time waves are replete with harmonics for the selected fuel (figure 8 (c))

Figure 5. Brake thermal efficiency at varying equivalence ratios

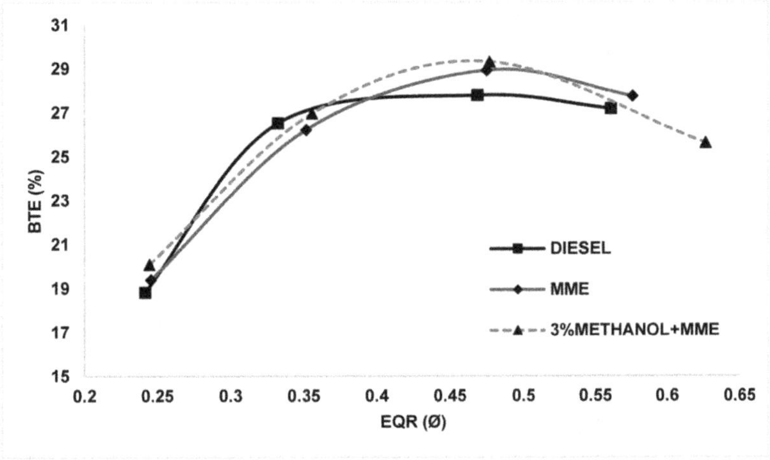

Figure 6. Cylinder Combustion pressure at maximum engine load

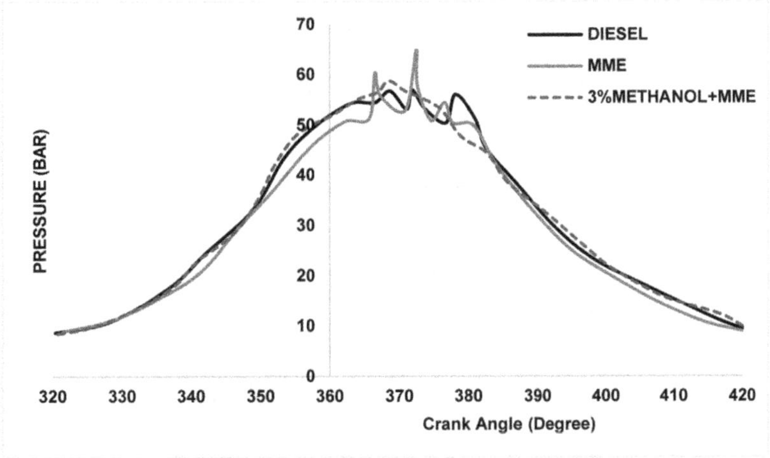

Figure 7. Net heat release rate at maximum engine load

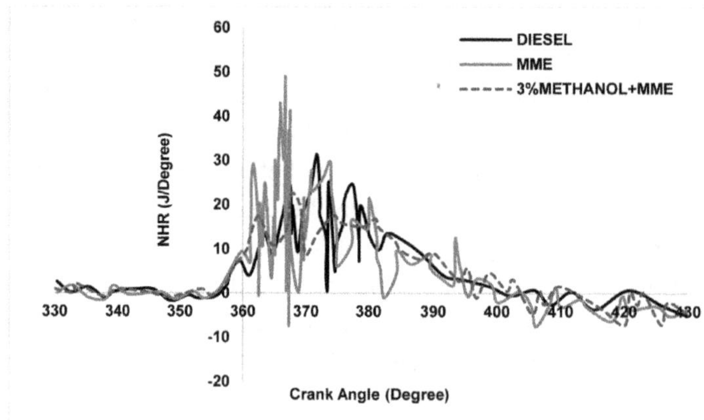

Figure 8. (a) Time wave spectrum for diesel fuel; (b) Time wave spectrum for Biodiesel fuel; (c) Time wave spectrum for 3%methanol in biodiesel

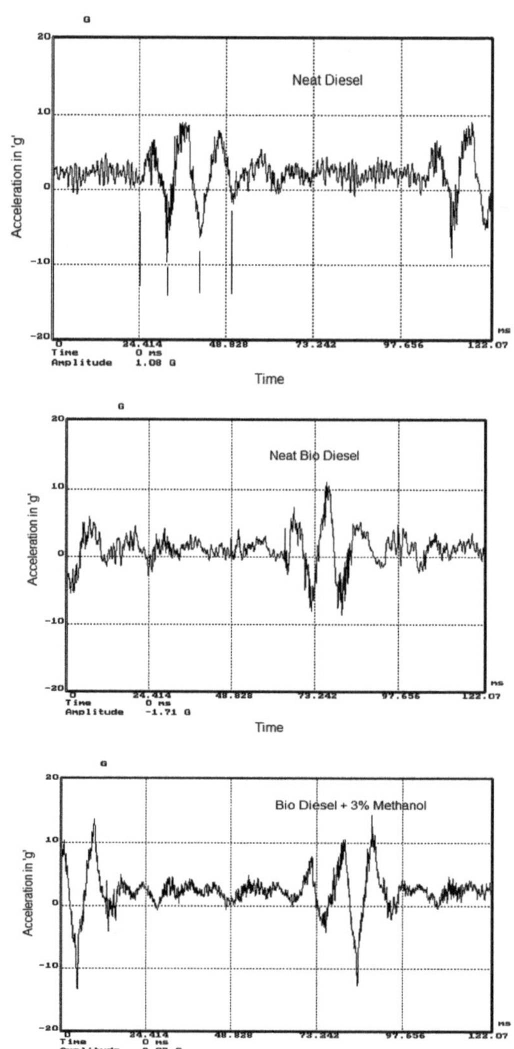

Experimentation was conducted on the same Bajaj IDI auto engine (driver engine for three-wheeler auto) without gear box and clutch has been stationary mounted in the engine laboratory of Department of Marine Engineering, Andhra University as mentioned above. The rated quantities have been redefined to facilitate the study of combustion aspect of the engine with new fuel application totally replacing the diesel fuel. This engine is supercharged and the rated power is fixed at 38 kg (2.92kW) dynamometer spring load @1500 rpm. There is an extra power generation of about 9% when compared to the engine without supercharger at the same speed. This redefined rated specification is taken into consideration viewing the aspects of engine smoke, fuel consumption, and equivalence ratio and supercharging. The single cylinder vertical engine having a bore of 86 mm and stroke 77mm, 4-stroke IDI engine, with a pintel nozzle injection pressure of 142-148 bar, has been upgraded to deliver additional power of 0.5kW by supercharging it with 350Watt air blower. A designed surge tank as shown in the figure.1 is installed in between engine and blower to store air in the strokes other than suction. But for the rated power change, every parameter of the engine is kept as per the company's specification.

Engine Combustion study was taken up by measuring the combustion pressure variation in the main chamber with respect to the crank angle. Grated orifice is the connection between swirl chamber and the main chamber. This latest IDI engine design eliminated the usage of nozzle in between chambers so that the pressure variation due to combustion in both the chambers can be measured by the piezoelectric transducer inserted in the main chamber. Precision was maintained by measuring the combustion pressure at every degree of crank angle. The propensity of the combustion pressure has also been interpreted by measuring the engine cylinder vibration vertical on the cylinder head and in the piston slap direction choosing high frequency range narrow band FFT spectrum and time wave form for more than one cycle. The smoothness of the combustion can also be interpreted with the amplitudes of frequencies generated considering the presence of harmonics in the spectrum and the smoothness of the time waves presenting the type of wavelets generated.

Heat release rate quantities have been estimated and exhaust emissions are recorded along with smoke measurement. The plots of thermal efficiency and specific fuel consumption were compared to analyze the performance aspect of the engine with the new biodiesel fuel, *palm biodiesel with coconut biodiesel as an additive to supplement the missing fatty acids*. In brevity, diesel fuel performance was compared with neat biodiesel, and biodiesel with an additive for a supercharged single cylinder IDI engine and the investigation was taken up in the combustion aspects defined above to establish relative merit in replacing the diesel fuel.

Engine Combustion Pressure Data Presentation and Pressure Rise Trend Interpretation

1. Peak pressure attainment is delayed by few degrees in the case of palm and Coconut biodiesel combined fuel. This is when compared to neat biodiesel operation.
2. Biodiesel with 3% Coconut biodiesel exhibited smooth rise of pressure almost akin to the neat diesel fuel operation and this is consistent with any load on the engine.
3. Pressure crank angle spectrums at maximum load were recorded. It is observed that 3% Coconut biodiesel in palm biodiesel proved efficient.
4. Addition of Coconut biodiesel especially 3% quantity increased the peak pressures in the spectrums in in a progressive manner.

5. Saturated fatty acid content in the Coconut biodiesel is more when compared to the palm biodiesel. When palm methyl ester is mixed with 3% Coconut biodiesel, the saturated fatty acid content in the palm biodiesel increases. This will increase the shelf life of biodiesel and at the same time improves the performance.
6. Linolenic fatty acid reduces by five times lesser when 3percent coconut biodiesel is mixed with the palm methyl ester. This will surely encourage the combustion of the palm methyl ester in the association of Coconut methyl ester additive. Literature says (Sukumar Puhan et al. (2010)) that there is an increase in NO_x and reduction in thermal efficiency with the increase in Linolenic acid.

Performance Comparison

2%COME in the main biodiesel exhibited upper hand over the fuel combination 3%COME in the main biodiesel. Since it is not showing consistency in the heat release plots relatively, it is discounted in favor of 3%COME in the main biodiesel. The specific fuel consumption (kg/kW-h) differences are 0.0184 and 0.0334 at 30 kg load and 38 kg load on the engine in the case of neat diesel and neat biodiesel, biodiesel consuming more fuel because of its lesser calorific value. In the case of 3% COME in Palm biodiesel, the specific fuel consumptions are 0.0186 and 0.013 at 30kg and maximum load. Even though the consumption is more in the case of 3%COME in main biodiesel, the quantities are lesser than the quantities consumed for neat biodiesel at maximum load condition.

When we refer to 2% of COME in palm biodiesel and 3% of COME in palm biodiesel, 3% additive fuel consumption is more by 0.014 and 0.001 at 30 kg and maximum load and the thermal efficiency (%) is reduced by 1.44 and zero at the loads mentioned above respectively. At lower loads the difference is increasing progressively towards no load condition. However, because of the setbacks suffered by the 2% additive in the Palm biodiesel in the aspect of net heat release rates, it is not recommended to accept it as an alternate fuel for the replacement of diesel fuel.

The net heat release rate trend is almost all same with the load of the engine. In the case of diesel fuel the net heat is maintaining lower profile than other combinations with COME additive. 2% and 5%COME in the Palm biodiesel released better heat per degree than the other combinations. Since the net heat generated wave is not totally reversible it indicates the torque conversion is inferior. But in the case of 3% COME in the main biodiesel, the heat generated curves are reversible indicating the better torque conversion. The cumulative heat curves also elevate the performance in the wise of heat accumulation per degree but most of the heat accumulated has gone in the way of exhaust heat as can be observed from the exhaust temperature plots. However, the selected 3% COME in the main biodiesel proved parity at higher loads with better torque conversion. One can observe the neat Palm biodiesel is not delivering goods on par with the combination which is blended with 3% COME as an additive.

Exhaust Emission Analysis

Exhaust gas analysis was done and exhaust temperatures were measured at different loads on the engine within the scope of the load redefined. For different fuel combinations CO, HC, CO_2, NO_x, smoke density in HSU, and exhaust temperature were measured using AVL DIGAS 444N analyzer and AVL 437 C smoke meter and digital temperature sensor.

Carbon monoxide emission for the diesel fuel operation is 0.04% and for palm biodiesel and 3%COME in Palm biodiesel this emission is zero. At other percentages of COME additive in biodiesel, there is marginal release of CO at all loads.

CO_2 emission in the case of palm biodiesel is the lowest when compared to other combinations. When compared to diesel and palm biodiesel there is a decrease of about 2.2%, 3.62% at 3/4th maximum and maximum load in the case of neat diesel. With respect to 3%COME in Palm biodiesel there is reduction of 1.26% and 1.54% at 3/4th maximum and maximum loads respectively.

IDI engines are famous for complete combustion with any fuel as per the literature. In this research findings with the supercharging effect reveal, the HC emissions are low for diesel fuel, Palm biodiesel and Palm with Coconut additive of 3%. Biodiesel with additive at other percentages is producing more HC. NO_x reduction is prominent in the case of palm biodiesel and 3%COME in palm biodiesel sample. At 3/4th maximum load and maximum load, there is reduction of 86ppm and 119ppm in the case of palm biodiesel, and there is reduction of 40 and 44 ppm reduction at the same loads mentioned above when 3% COME in biodiesel is used. There is an obvious increase in O_2 level and λ level in the case of biodiesel and additive usage because of the molecular oxygen in the case of biodiesel. PME increases O_2 levels by 2.79% and 4.47% with respect to the diesel fuel at 3/4th maximum load and maximum load respectively. With respect to the 3%COME in PME biodiesel decreases the oxygen level by 0.18% and 0.25% respectively at the above mentioned loads. COME is utilizing more oxygen than diesel fuel means better combustion.

Smoke levels are more in the case of diesel fuel at maximum load and the load next to the maximum as described earlier. 9.3 HSU & 23.6 HSU difference can be observed in the case of Diesel and neat PME at the two loads 30kg and 38kg spring loads respectively.11.1 HSU & 37.6 HSU in the case of 3% COME in PME at respective loads as mentioned above. Hence 3%COME in PME makes an excellent option as an alternate to the diesel fuel in the aspect of smoke emission.

Diesel exhaust temperatures are more by 21^0 and 74^{0}C with respect to the usage of PME at the two loads as defined and 33 and 78^{0}C in the case of 3% COME in palm biodiesel fuel. Hence COME additive is saving heat energy loss through the exhaust gas. Figure 9 represents the performance aspect with

Figure 9. Comparison plot for brake thermal efficiency and brake specific fuel consumption for selected fuels

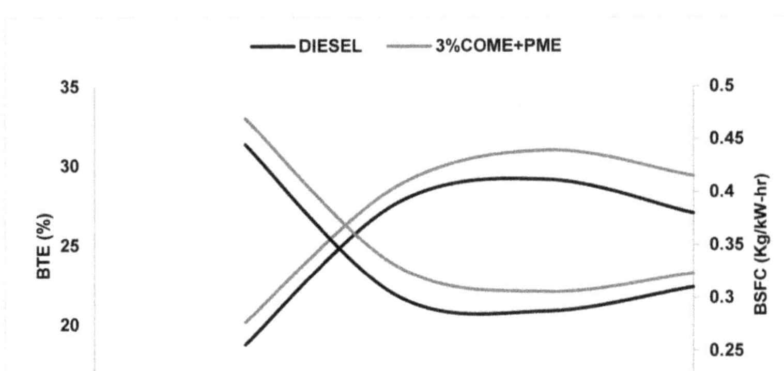

the combo fuel. Thermal efficiency at higher loads is improved at the compromise of brake specific fuel consumption. Figure 10 (b) portrays smooth time wave measured on the cylinder head in the piston moving direction replete with good number of harmonics. Figure 11 represents vibration velocity measured vertical on the cylinder head with lesser comparative spectral density for the combo fuel replete with harmonics. Figure 12 indicates the fatty acid composition of the neat and combo oils and the relevant transformation of the fatty acid composition from saturated fatty acid to monounsaturated and poly unsaturated fatty acids upon mixing the palm methyl ester and coconut methyl ester. This adds advantage to the combo oil to further better the performance to higher levels.

Figure 10. (a) Time wave spectrum for diesel fuel; (b) Time wave spectrum for 3% coconut methyl ester blend with palm biodiesel

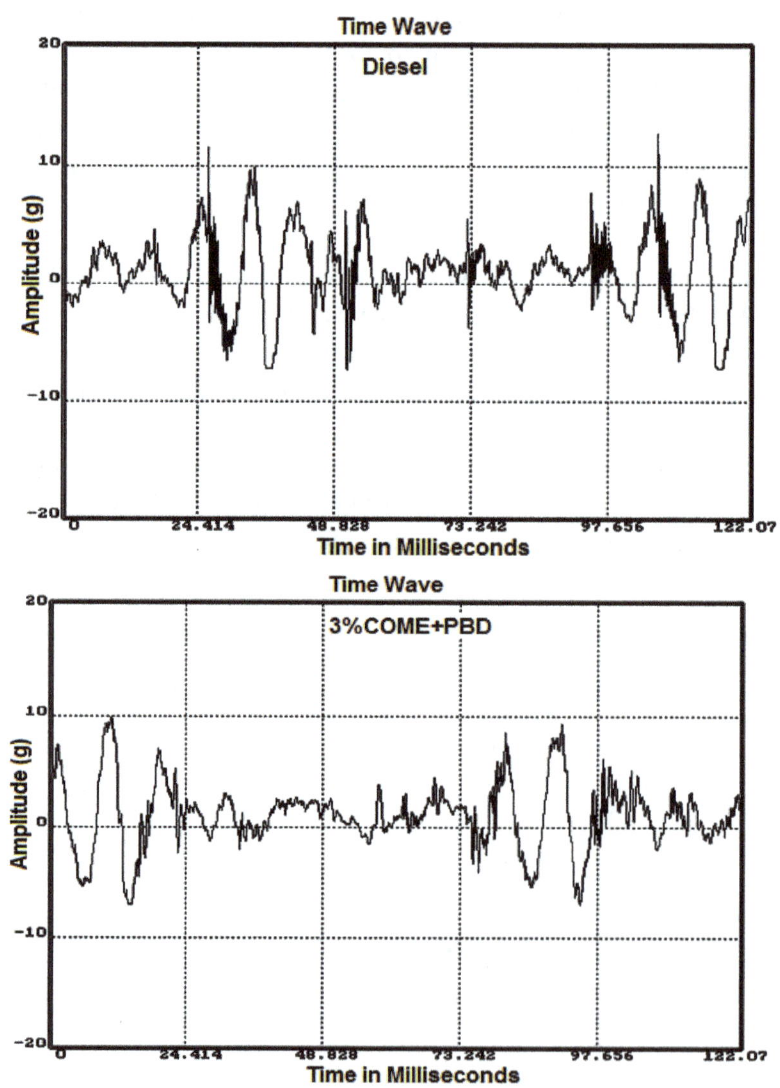

Figure 11. (a) Log-log velocity spectrum for diesel fuel; (b) Log-log velocity spectrum for 3% coconut methyl ester blend with palm biodiesel

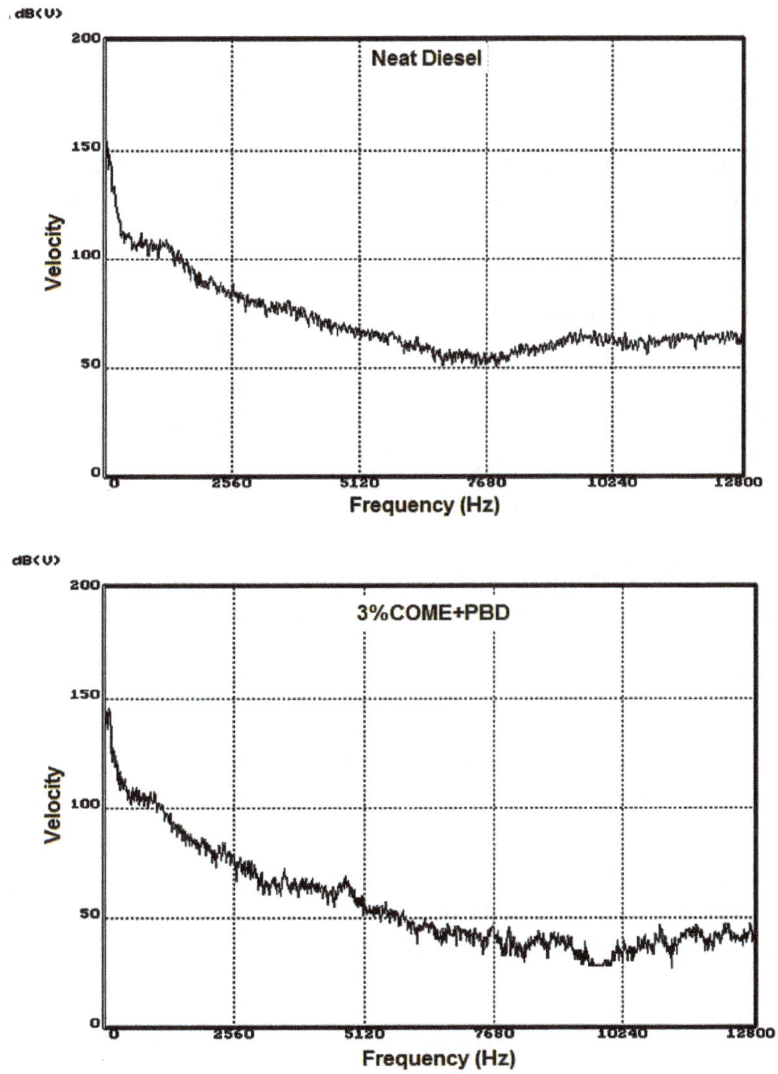

IDI engine is chosen to eliminate the problem of crank case oil dilution with the usage of neat biodiesel. In an attempt to improve the performance of the engine with a different additive with neat biodiesel application, 1,4-Dioxane ($C_4H_8O_2$) *with lower auto ignition temperature (180^{0}C)* is used as an additive to control the combustion of the biodiesel with higher auto ignition temperature(>300^{0}C) and higher Cetane number (65) both in the primary and main combustion chambers. *Blends of Palm methyl ester and 1, 4- Dioxane ($C_4H_8O_2$) in solution* are used to verify the engine performance. The chosen blends are 1%, 2%, 3%, and 4% of *1, 4- Dioxane* with Palm methyl ester. It is observed homogenous blends with all these percentages of *1,4-Dioxane*, since the biodiesel is solvent at room temperature for all these percentages of additive without reaching saturation condition and then leading to separation. Since *1, 4- Dioxane* possesses higher latent heat (413 kJ/kg) than that of diesel (280 kJ/kg), its addition creates low temperature combustion and reduces the formation of NO_x and slows down the combustion severity. The combustion propensity in each combustion chamber can also be understood from the engine cylinder vibration analysis. The frequency curves (FFT) and time waves decipher the vibration trend of the engine diagnosing the knocking and detonation when the fuel sample is charged with higher doses of *1,4- Dioxane*.

Solution making is expected to be more fixed in making proportions at every load than dual fuel injection as was done in the case of Rice bran biodiesel and Isopropanol injection, T Vector Babu et al. (2016). In the case of dual fuel injection, the reference is with respect to 1 liter /hr fuel consumption and the engine fuel consumption will be varying with respect to the load on the engine there by changing the percentage of the additive with respect to the main fuel.

Brake Specific Fuel Consumption (BSFC)

The BSFC verses Equivalence Ratio and brake power respectively was plotted. The equivalence ratio is found to be within the range for the part load operation of the engine at 1500rpm for all fuels tested. A smoother tendency of the curve is observed for the additive blend of 3%. The equivalence ratio for 3% of additive blend is found to be minimum i.e. 0.52215, however the performance can be rated better when compared to other fuel samples. The BSFC for the 3% additive in biodiesel is reduced by 2.64% compared with neat PME. This is because of the combustion refinement in the presence of additive with better atomization and vis-a-vis lesser auto ignition temperature.

Brake Thermal Efficiency (η_{th})

Brake Thermal Efficiency (BTE) curve has better trend for PME with 3% 1, 4-Dioxane additive blend and the maximum η_{th} attained is 33.79% at an equivalence ratio of 0.52215. The variation of Brake Thermal Efficiency with respect to Brake Power of neat diesel, PME and additive blends are plotted. Brake Thermal Efficiency has increased with 1, 4-Dioxane additive percentage despite the reason that the heat value of additive is lesser than both diesel and biodiesel with economical mass consumption of additive in biodiesel (PME) and improvement in combustion propensity. 3percent 1, 4- Dioxane ad-

Figure 12. Fatty acid compositions of selected biodiesels

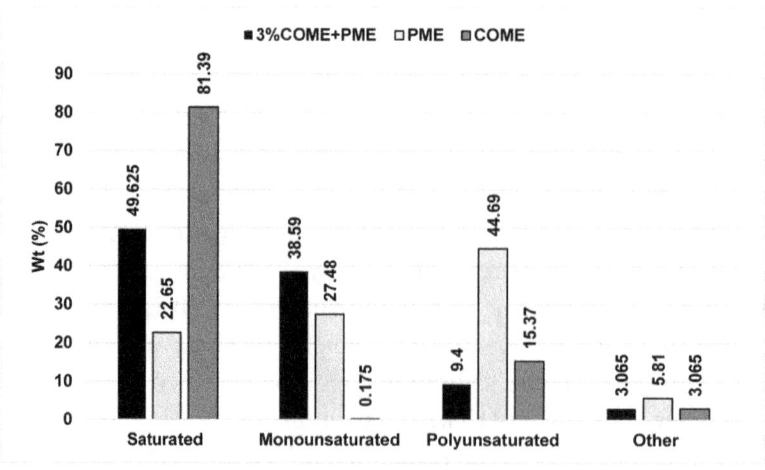

ditive blend achieved maximum Brake Thermal Efficiency of 33.79% at a Brake Power of 2.7 kW (i.e. maximum load on the engine) with an increase of 9.89% when compared with conventional diesel fuel at the same load. Thermal efficiency with PME has a sharp shoot up in the plot where as other additive percentages in biodiesel exhibited gradual rise in the same parameter. This is because of changes with the parameters like Cetane number, partial distillation of the combo fuel. However, at lower loads, the thermal efficiencies of the blends are higher than that of biodiesel and diesel.

Hydro Carbon (HC) and Carbon Monoxide (CO) Emissions

Studies on IDI engine indicates lower emissions in exhaust. Additive mixing further reduced the HC emission and CO emission to greater extent. 3percent 1,4-Dioxane in PME were compared with conventional diesel fuel at maximum load and these are decreased by 2ppm in case of HC, 40% for CO. This result can expected with the properties of additive which has got lower boiling point and auto ignition temperature. The above properties encourage combustion initiation more in advance than biodiesel, making the progress of the combustion very smoother in both the chambers. This combustion process reduces the heterogeneity and reduces pollutant formation leading to complete combustion.

Oxides of Nitrogen (NO_x)

This emission is reduced by 56 ppm (or by 18.42%) at high loads with the 3% 1,4-Dioxanel additive in PME compared with conventional diesel fuel as shown in This may because of colder combustion owing to higher latent heat of the additive and earlier evaporation which encourages low temperature combustion and hence lesser formation of NO_x as shown in the figure 13.

Figure 13. NOx emissions at varying loads

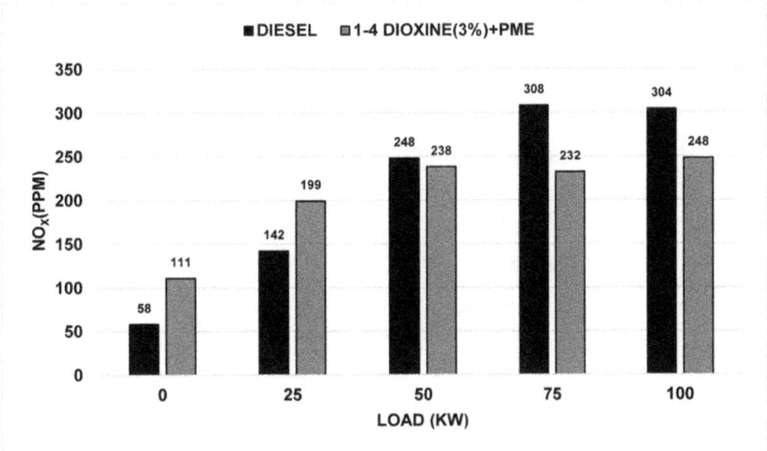

Based on the explanation given above, the author felt that the above fuel combination is the right choice for the replacement of conventional diesel fuel in IDI Engine chosen and tested in the laboratory. This conclusion is based on the real pressure plots obtained, the cumulative and net heat release rate plots derived and vibration plots compared. Since it is accepted that the vibration plots envisage greater insight in to the one degree of crank revolution in time domain, assessment has been made with narrow brand, log-log FFT plots with higher range of frequency and time wave forms. Knocking or any abnormality in combustion which obviously give set back to the power generation can be easily investigated by the micro level study of cylinder excitation with vibration. Finally, it can be summarized that the 3% 1,4-Dioxane additive in PME is the best alternative in view of the engine performance and reliability without any crank case oil dilution. It can be understood in a comprehensive way that this fuel combination systemizes the combustion in pre and main combustion chambers with effective increase in thermal efficiency at the loads arbitrarily chosen for the automobile engine operation without gear box and clutch assembly. Vibration curves are potential expressions referring to the combustion phenomena; figures shown above give ample evidence of non-polarized and polarized combustions. This data has been recorded at three strategic points on the engine and presented for comparison. There is non-polarization of combustion clearly indicated in the vibration signature recorded vertical on the cylinder head at two strategic loads when the new mixture 3% 1,4-Dioxane additive in PME is implemented in the diesel engine. The polarization in the vicinity of 2560 Hz in the case of diesel fuel is an indication for mild knocking in the case of diesel fuel which is well understood in the engine base vibration also at the referred loads. Dilution of 1,4- Dioxane vapor (lower boiling and auto ignition temperatures) in the atomized biodiesel after injection may be set to combustion before it reaches auto ignition temperature might have been the reason for changing the combustion philosophy reducing spontaneity and heterogeneity of combustion. From figure 14 there is an appreciable increase in thermal efficiency, and since the additive encourages earlier combustion, the premixed combustion is looking better leading to better torque conversion as noticed in the time wave forms recorded in the direction of piston movement on the cylinder head as shown in the figure 15

Figure 14. Comparison plot for brake thermal efficiency and brake specific fuel consumption for the selected fuels

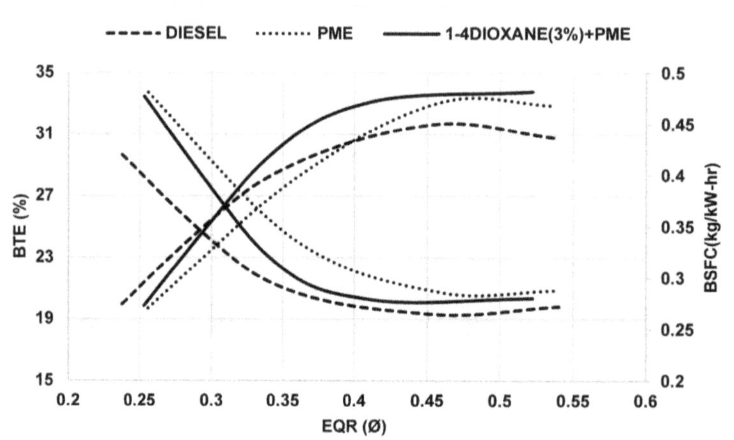

Figure 15. Time wave spectrum for diesel, neat palm methyl ester and neat palm methyl ester with 1,4-Dioxane

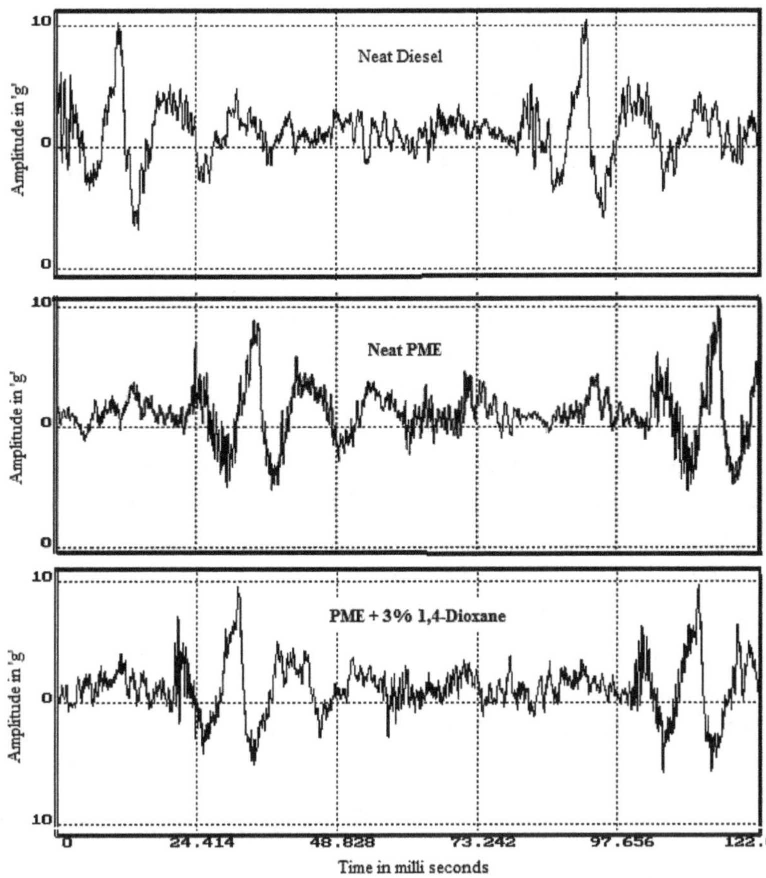

Methyl ester derived from beef tallow with ethyl alcohol enriched with 5% water is used as an additive. Water addition is meant to enhance the latent heat of ethyl alcohol to encourage low temperature combustion. The very introduction of ethyl alcohol by ***secondary injection*** at suction side (3 kg/cm^2) is to reduce the viscosity of the injected Beef Tallow Methyl Ester (BTME) after vapor entrainment and to save the escaping heat from the combustion mass of the fuel. Bajaj RE IDI engine (variable speed), single cylinder, and 4-stroke diesel is employed for testing. The power rating of the engine is 5kW with injection pressure 142 to 148 kg/cm^2. Supercharger with delivering capacity of 1.7 m^3/min (350W) is employed for boosting the air capacity to the engine. In between the engine and supercharger, there is a surge tank to store the air in the idle strokes. The surge tank capacity is six times the cylinder volume. Engine test rig is erected without the clutch and gear box and the speed and load on the engine are redefined to exclusively test the combustion aspect of the engine to investigate into the suitable alternate fuel totally replacing the diesel fuel. The fuels viz. Diesel fuel, BTME, BTME with 5% water added ethanol additive are tested to elicit information about the performance of the engine, emissions and vibration attitude of the engine. Vibration on the cylinder head is being measured to collate the relationship with the combustion propensity with different fuel samples. There is 66% of saturated fatty acid content in the BTME which ought to give better results fuel wise by replacing diesel fuel. It is observed that the saturated fatty acid content in the biodiesel derived from vegetable oil, and waste cooking oil are far lesser than the BTME and it is paradoxical the saturated fatty acids in the waste cooking oil are 18% and the remaining fatty acids are unsaturated fatty acids as estimated from chromatography.

2% hydrated ethyl alcohol in the Beef Tallow Methyl Ester (BTME) is justified suitable choice for the replacement. Hydrated ethyl alcohol maintains low temperature combustion which yields encouraging emissions. This conclusion is arrived at by studying various parameters like engine performance, emissions, smoke, combustion pressure signatures, derivatives of the combustion pressure plots, derived net and cumulative heat release rate plots, vibration signatures in the form of FFT, and time wave forms to finally understand the combustion comparing all these aspects.

It is a fact that the lower calorific value of the diesel fuel is higher than BTME and BTME with additive mentioned above. Despite this reason BSFC of the renewable fuels is higher than the diesel fuel combustion with the fuels formulated are showing up in most of the parameters. BSFC of diesel fuel is consumption is the lowest at all loads. For example, at maximum load and at the next lower load, the diesel fuel consumption (BSFC) is reduced by 2.71% and 3.63% respectively comparing with the fuel containing 2% of the hydrated ethyl alcohol in the methyl ester of beef tallow. When tried methyl ester into consideration and comparing it with the diesel fuel there is an increase of 5.80% and 7.90% at maximum load and the next lower load respectively. It can be inferred that 2% hydrated alcohol in BTME performed well comparing with neat BTME. This may be due to lesser heat transfer from the core of the combustion because of low temperature combustion. BSFC verses the equivalence ratio exhibited the performance of the above fuel in a lucid way with 2.72% & 3.71% increase in equivalence ratio at upper two loads and in the case of neat BTME the respective percentages are 6.02 and 8.26. In this aspect also 2% hydrated ethyl alcohol in BTME shown up beneficially when compared to neat diesel fuel. One can clearly visualize the better fuel consumption with respect to the diesel fuel in the mentioned plot. It is a well-known fact that the new engines also produce lesser equivalence ratios. Thermal efficiency increases by 9% & 8% respectively at the upper loads in the case of 2% hydrated ethyl alcohol additive in BTME. 5.15% & 2.96% are the hikes in thermal efficiencies when compare to diesel fuel and BTME fuels.

It can be observed that part load performance of the diesel engine is better than even when run with diesel fuel. There is substantial improvement with the 2% hydrated ethyl alcohol in BTME. This is a fact not withstanding that the heat value of the diesel fuel is greater than the new fuel used. It is also a fact that the diesel fuel has shown up at lower loads but the performance plummeted moderately at higher loads in comparison. But the thermal efficiency suffered at all loads with the engine run with diesel fuel in comparison., There is a 2.72% & 3.71% increase in Equivalence Ratio (EQR) for 2% additive in BTME and 6.02% & 8.26% EQR increase in the case of neat BTME. This is an indication of better combustion with 2% hydrated ethyl alcohol in BTME. In other cases of higher percentages of ethyl alcohol in BTME there is an increase in thermal efficiency at higher loads more than diesel because of the reason that there is a specific combustion regulation in limited directions decreasing the extent of heterogeneity. This combustion propensity is also being corroborated by the vibration analysis on the cylinder head. The narrow band analysis presented in log-log form in figures 16 (a, b, c) depict higher spectral densities for fuel diesel and biodiesel where as for the combo with coconut biodiesel the spectral density is lesser replete with harmonic frequencies. Same is observed in the case of time waves in the figures 17 (a, b, c). Reduction in NO indicates low temperature combustion as shown in the figure 18 in the case of combo fuel but the HC emission is compromised. There is better indication of net heat release rate and brake thermal efficiency in the case of combo fuel and of course, there is a visible increase in the specific fuel consumption in the case of combo fuel because of lower heat capacity of the alcohol mixed with water (figures 19 &20). Figures 21, 22, and 23 are correlated combustion pressure diagrams with time waves in the real time presentation which indicate the smother torque conversion in that duration.

Figure 16. (a) Log-log velocity spectrum for diesel fuel; (b) Log-log velocity spectrum for neat beef tallow biodiesel; (c) Log-log velocity spectrum for neat beef tallow biodiesel with 2% Ethanol

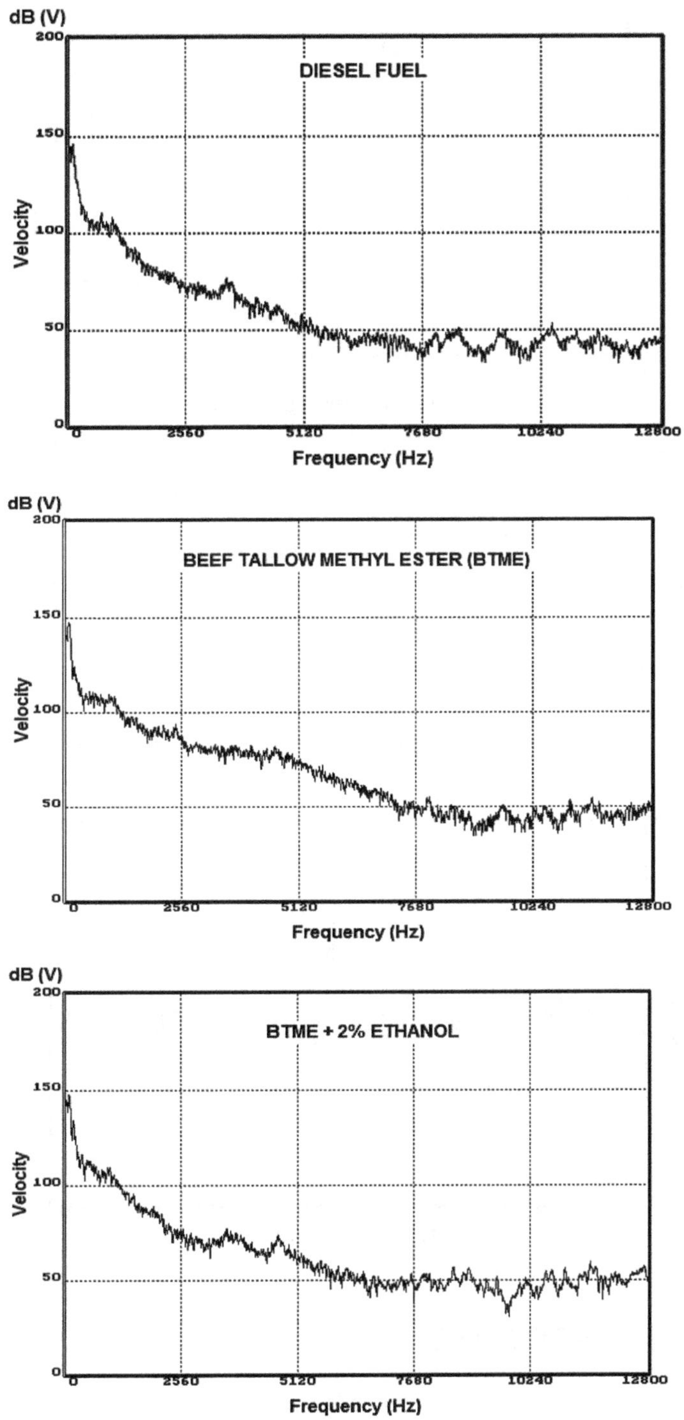

Figure 17. (a) Time wave spectrum for diesel fuel; (b) Time wave spectrum for neat beef tallow biodiesel; (c) Time wave spectrum for neat beef tallow biodiesel with 2% additive

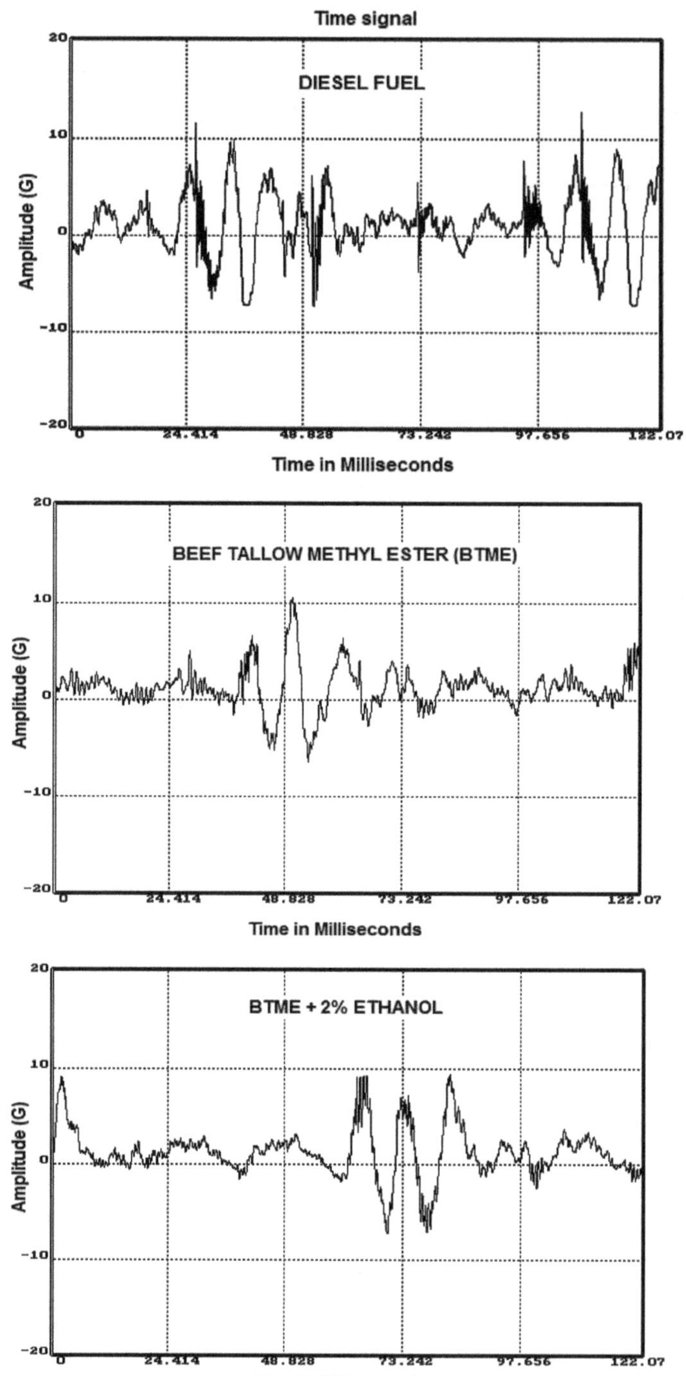

FUTURE RESEARCH DIRECTIONS

Figure 18. NO emissions at varying loads

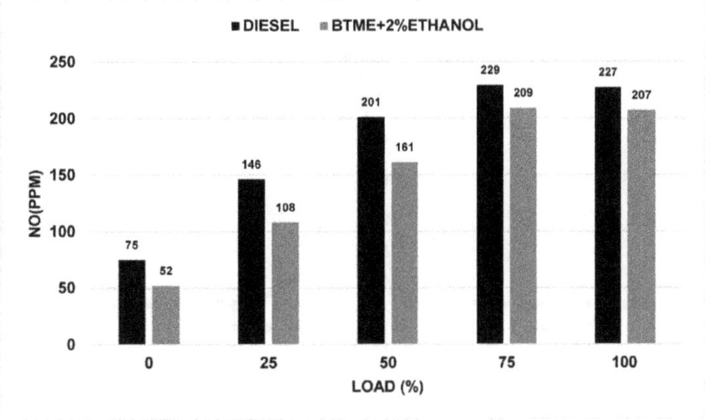

Figure 19. Net heat release rate at maximum load

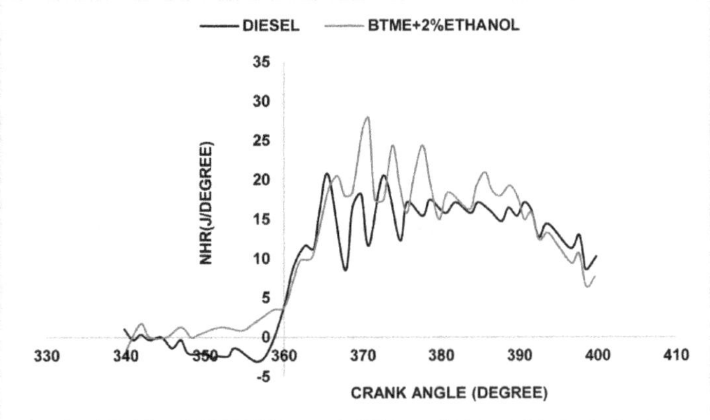

Figure 20. Comparison plot for brake thermal efficiency and brake specific fuel consumption for selected fuels

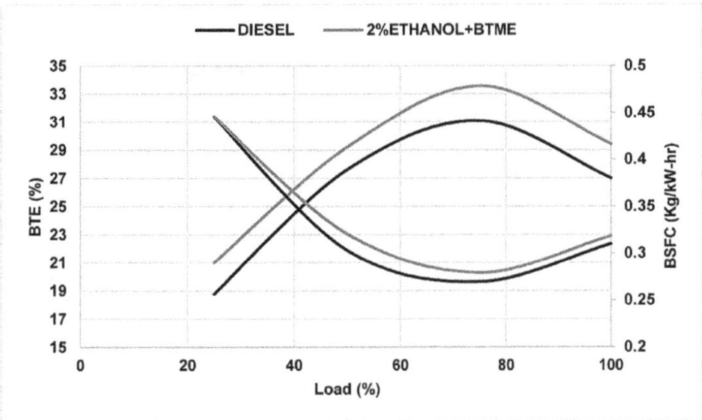

Figure 21. Combustion pressure and time wave interpretation for diesel fuel

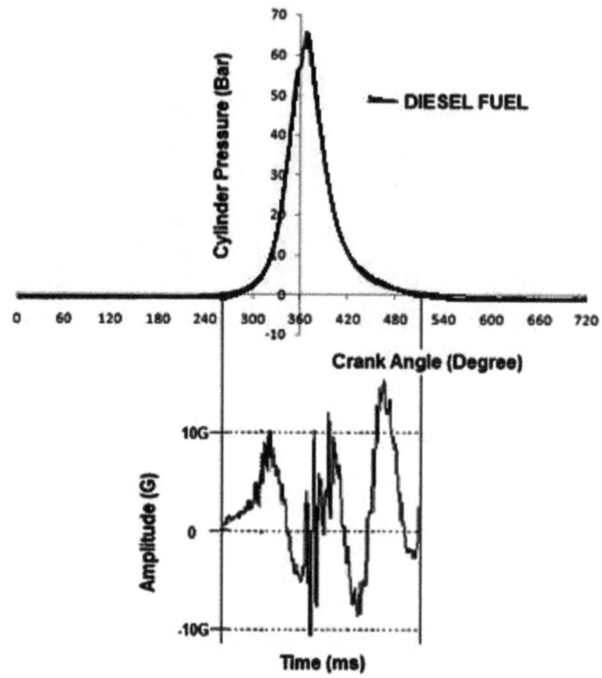

Figure 22. Combustion pressure and time wave interpretation for neat beef tallow biodiesel

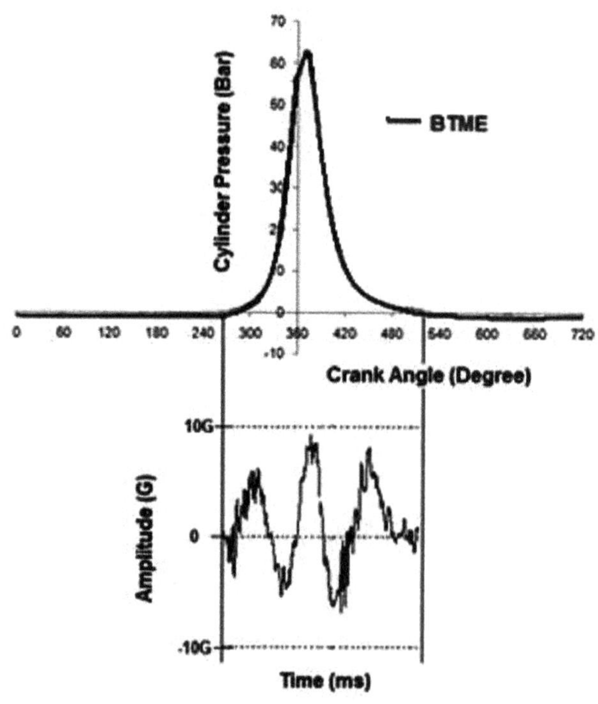

Figure 23. Combustion pressure and time wave interpretation for neat beef tallow biodiesel with additive

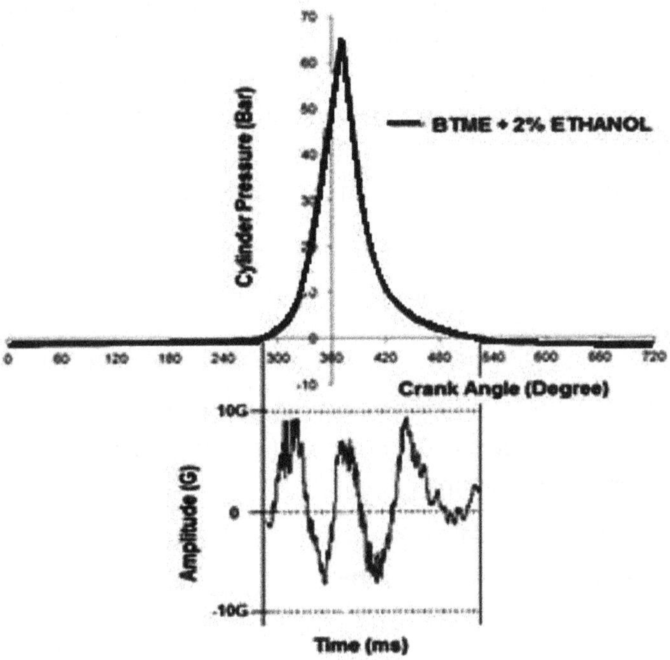

Common Rail Diesel injection concept is another trial that can be implemented with biodiesel application in the future scope of work. This also helps in stratified injection of fuel under consideration. Gear box and clutch operation can be attached to the engine to change the speed of the engine and this enables drawing of speed torque curves to finally arrive at the rated power and speed fixing of the engine in general. The measurement of combustion temperatures during the combustion period gives important information about the temperature-entropy plots of the burned fraction. This is possible if combustion photography is used as a tool to find out various parameters including the burn fraction verses emissions and it can be ascertained more information about the smoothness of combustion by finding the wavelets (time verses frequency) emanated from the engine cylinder during the combustion.

CONCLUSION

Several attempts have been made more with alcohols blending, benzene derivatives, using biodiesel as an additive, and secondary injection trials were used as fuels in the IDI engine. The comprehensive report indicates individual advantages with each trial. Comparing all the trials, coconut additive is yielding better results in the aspect of net heat release. Even though secondary injection involves additional expenditure of modifying the engine, it evades the concept of density separation and emulsion instability.

1,4- dioxane and coconut biodiesel blends produce saw tooth waves in their net heat release rate plots indicating better torque conversion in the diffused combustion zone. 1,4- dioxane possesses better physical properties like lower boiling point, least auto ignition temperature etc. which triggers the combustion ahead of the main fuel leading to better fuel air mixing in both the chambers.

Beef tallow biodiesel possesses higher Cetane number and hence leads to better premixed combustion. The time recorded is completely reversible indicating better torque conversion without shock on the flywheel. The engine is super charged in the case of beef tallow application and coconut biodiesel as an additive. Super charging of the engine might have provided excess oxygen for better combustion and cooling of the engine without exchanging the heat outside the engine with appreciable thermal efficiency increase in both the cases.

This research received no specific grant from any funding agency in the public, commercial, or not-for-profit sectors.

REFERENCES

Ashok Kumar Reddy, Y. (2013). *Study of Performance, Combustion, Emission and Vibration analysis of an Indirect Injection (IDI) Diesel Engine using preheated Jatropha Methyl Ester* (Unpublished doctoral dissertation). Andhra University, Visakhapatnam, India.

Kolakoti, A., & Appa Rao, B. V. (2017). Performance and emission analysis of a naturally aspirated and supercharged IDI diesel engine using palm methyl ester. *Biofuels*, 1–12. doi:10.1080/17597269.2017.1374770

Kolakoti, A., & Appa Rao, B. V. (2019). Effect of fatty acid composition on the performance and emission characteristics of an IDI supercharged engine using neat palm biodiesel and coconut biodiesel as an additive. *Biofuels*, *10*(5), 591–601. doi:10.1080/17597269.2017.1332293

Pentakota Santosh Babji. (2018). *Application of methyl ester of waste cooking oil as a replacement to diesel fuel in IDI engine with supercharging* (Unpublished doctoral dissertation). Andhra University, Visakhapatnam, India.

Prasada Raju, N. (2015). *Experimental Investigation on Suitability of Pongamia Methyl Ester (PME) and Isobutanol Additive Blends in Indirect Diesel Injection (IDI) Engine replacing Diesel Fuel* (Unpublished doctoral dissertation). Andhra University, Visakhapatnam, India.

Prasada Rao, K., & Appa Rao, B. V. (2017). Experimental Investigation on Cylinder Vibration analysis, Combustion, Emission and Performance of an IDI Engine. *International Journal of Manufacturing, Materials, and Mechanical Engineering*, *7*(1), 18–36. doi:10.4018/IJMMME.2017010102

Ramana Murty Naidu, S. C. V. (2015). *Feasibility Study on Indirect Injection Diesel Engine Replacing Diesel Fuel with Palm Methyl Ester And 1,4- Dioxane Additive Blends* (Unpublished doctoral dissertation). Andhra University, Visakhapatnam, India.

Sukumar Puhan, N. (2010). Effect of biodiesel unsaturated fatty acid on combustion characteristics of a DI compression ignition engine. *Biomass and Bioenergy*, *34*(8), 1079–1088. doi:10.1016/j.biombioe.2010.02.017

Vector Babu, Prasada Rao, & Appa Rao & Kolakoti. (2016). Experimental investigation on combustion, emissions, performance and cylinder vibration analysis of an IDI engine with RBME along with isopropanol as an additive. *Biofuels*, *8*(3), 307–321. doi:10.1080/17597269.2016.1226723

Vedulla, G. (2017). *Experimental Analysis of Supercharged Single Cylinder IDI Engine using Beef Tallow Methyl Ester with Hydrated Ethyl Alcohol as an Additive* (Unpublished doctoral dissertation). Andhra University, Visakhapatnam, India.

Chapter 3
Simulation Tool for Cable Design

Leonid Burstein
Independent Researcher, Israel

ABSTRACT

This chapter is an updated, enhanced, and expanded paper previously published in IJMMME. An attenuation-based mathematical model and computational tool for communication cable design is presented in the chapter. The relationship derived for the model creation connects attenuation with the following parameters of cable design and testing: copper wire, aluminum and insulation thicknesses, cable impedance, test frequency, and lay length. The expression found from numerous measurement data was used to simulate the Monte Carlo prediction of attenuation in the cable. To realize the simulations, a life script and regular MATLAB script programs were developed. The programs produce a special graphical user interface suitable for non-programmers, cable engineers, and technicians. Results of calculations obtained with the developed tool were verified by experiments and indicate that the tool can be used to design cables with optimal parameters.

INTRODUCTION

There are many fields where wired cables are used. The quality of the cable design and manufacture are the critical requirements for wired communications. In digital communications, the multicore and symmetrical cables should meet the requirements of the international standards IEC 61156-1, 5, 6, 7 (2002-2012) that determine their transmission characteristics and testing methods for frequencies up to 1200 MHz. The design of each component of the cable includes the correct selection of the following mechanical parameters of cable: diameter of the copper conductor, diameter of the insulated wire pair, lay length, individual wire material indices, and some other parameters. These parameters should conform to standardized requirements of the operational characteristics of the cable, which tests during manufacturing, and it is in the first way: the impedance, attenuation, frequency, near-end crosstalk, and return loss.

Design of the analog/digital cable parameters and cable technology were developed long ago and in detail, in terms of their strength and reliability, as described for example by US Navy Handbook (1976),

DOI: 10.4018/978-1-7998-4939-1.ch003

Velinski (1988, 1989), Costello (1997, 2003). In contrast, relation between physical-mechanical parameters of the cable and the testing of electrical cable characteristics have been developed much later and improved relatively recently with appearance of the International Standards IEC 61156 1-7 (2002 - 2009). This time, some modeling and simulation studies appeared for assuring quality of communication cables, Tranter et all (2003), Hyland (2012), and steel cords, Turkseven and Ertek (2003), Gerdemeli, Kurt, and Anil (2013). In the recent years, some practical and theoretical aspects of the cable design and its test characteristics were studied (Lin et al, 2018, Poltz, 2017, Li and Zou,2016), nevertheless, there are no studies that integrate the designed mechanical parameters and tested electrical characteristics of cables in one mathematical expression that describes the actual technological data.

In this study, we will assume that correctly designed and qualitative manufactured cables, with optimal wire and insulation diameters, lay length, aluminum thickness, impedance and frequency performance, should guarantee lower attenuation of the transmitted signal. These characteristics are not related to any unified theoretical relationship. Thus, the aid of the study is to find any empirical relation that describes the practical data and then use this relation for reliable cable design.

BACKGROUNDS, TEST DATA AND THEIR FITTING

The cable and test parameters that were assigned to search for an unifying relation are shown in.

Figure 1. Cable (a) and test (b) parameters selected to the relationship search.

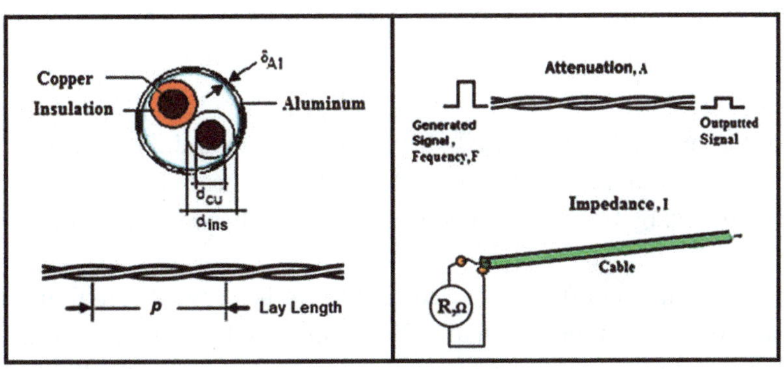

The attenuation was selected as the main parameter that should be kept and should be related with other design and test parameters. The general searching dependence reads

$$A = f\left(d_{cu}, d_{ins}, \delta_{Al}, I, F, p\right) \tag{1}$$

where d_{cu} is the copper wire diameter, mm; d_{ins} diameter of the insulator, mm; δ_{Al} is the aluminum thickness, μm; I – impedance value required in the cable testing, Ohm ; F – the frequency assigned for the cable test, MHz, ; p – cable lay length, mm; A – attenuation, dB per 100 m of the cable.

The data that connects parameters in eq.(1) were assembled in production line of Teldor Cables and System Ltd. (Appendix 1) for cables designed from four individually foil-shielded twisted pairs with tinned copper drain conductor, and overall braid-shielded and jacketed.

To fit these data, the least squares fitting technique was implemented. The objective function

$$S = \min\left\{\sum_i \left[A_i - f\left(d_{cu}, d_{ins}, \delta_{Al}, I, F, p, a\right)\right]^2\right\} \tag{2}$$

Where i is number of the data point; a is a vector of coefficients that should be defined from the data by the minimization of the objective function S.

To decrease the number of variables, diameters of copper and insulation were united in the ratio $d=d_{ins}/d_{cu}$ and eq. (2) reduces to

$$S = \min\left\{\sum_i \left[A_i - f\left(d, \delta_{Al}, I, F, p, \alpha\right)\right]^2\right\} \tag{3}$$

The variables in eq.(3) are of a different order of magnitude, e.g. cable lay length is 100 m and aluminum thickness 0.04 mm. Thus, the scaling should be used; data containing values of parameters of eq. (3) pretreated as follows:

$$X = \frac{x - x_{\min}}{x_{\max} - x_{\min}} \tag{4}$$

where x notes one of the original dimensional variables in eq.(3) and its minimal, min, and maximal, max, values; X is the same variables but dimensionless, after preprocessing with this equation.

Each variable x calculated by (4) changes between 0 to 1 and hence contribute equally to the seeking function f.

Main difficulty in defining a-coefficients of eq.(2) is that the view of function f is unknown a priori and should be selected so to obtain the minimal S. Wherein, it is desirable that fitting expression would not be complicated. Thus, further the following linear and nonlinear (in relay to the components of expression) functions are tested:

$$A_{fit} = a_0 + \sum_{j=1}^{5} a_j X_j \tag{5}$$

$$A_{fit} = a_0 + \sum_{j=1}^{5} a_j X_j + \sum_{j=1}^{4} \sum_{\substack{k=2 \\ k>j}}^{5} a_l X_j X_k \tag{6}$$

where index l is different for each X_jX_k and increases from 6 to 15 with increment k and j values.

For fitting data with these relations, the simple linear method of the minimization can be used. To realize this purpose, a special MATLAB program were created (Appendix 2); the program uses the regress command that processes multiple variable data and returns coefficients a, 95% confidence intervals for the coefficient estimates, residuals, and some statistical indices. The program includes:

- data reading from the Excel file;
- preprocessing data by eq.(4);
- data fit by estimating coefficients in eqs. (5, 6);
- attenuation calculation by the both defined equations and outputting coefficients with their upper and lower limits; displaying table with the mean and maximal errors, standard deviation and r-square.
- plot generation with two graphs – residuals and residual distribution.

One of the displayed tables and generated plot are given in Table 1 and Figure 2 respectively (see also Appendix 2).

Table 1. Basic statistics

	Mean_Error	Max_Error	St_Deviation	r_square
Equation 1	4.1350	13.325	3.5466	0.8378
Equation 2	3.7528	9.9062	2.2302	0.8957

Figure 2. Typical view of the residuals and its distribution for: equation (5) - left column with plots and equation (6) – right column.

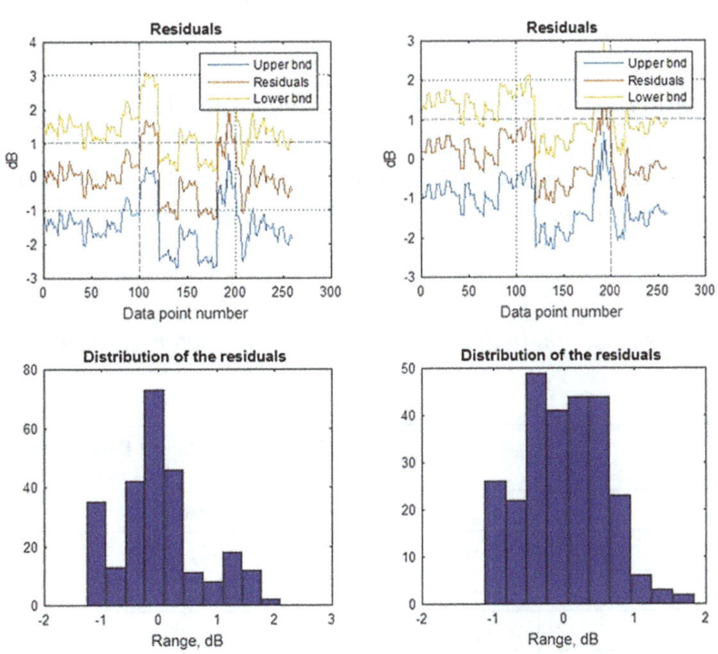

These results show that equation (6) leads to most little average errors and to better r-square, close to 0.9, thus this equation and will be used for the attenuation prediction by the assigned parameters. The coefficients defined for eq. (6) are a_0= 0.69326, a_1=-0.5933, a_2=-0.24276, a_3=- 0.5468, a_4=0.20271, a_5=0.077789, a_6=0.1269, a_7=0.53919, a_8=-0.2253, a_9=-0.028895, a_{10}=0.18077, a_{11}=-0.025033, a_{12}=0.016667, a_{13}=-0.045515, a_{14}=0.0051571, a_{15}=-0.010343. The minimal values of the *X*-parameters (eq.(5)) are $d_{\mathrm{cu,min}}$=0.4800, $d_{\mathrm{ins,min}}$ =1.0400, $\delta_{\mathrm{Al,min}}$=87.54, I_{min}=25, F_{min}=14.802, p_{min}= 48.083, and A_{min}=11.14; the ranges $\Delta x=x_{\mathrm{max}}-x_{\mathrm{min}}$ of these parameters are Δd_{cu}=0.1630, Δd_{ins}=0.5200, $\Delta\delta_{\mathrm{Al}}$=17.52, ΔI_{min} =15, ΔF=6.967, Δp=3.7445, and ΔA=7.224.

Each of the parameters in equation (6) varies (see data, Appendix 1) in certain ranges and has appropriate mean and acceptable error (stipulated by the standard deviation). Therefore, resulted attenuation value should be averaged from large number *A*-values defined by all possible combinations of the *d*, δ_{Al}, *I, F, p* parameters. The latter consists in simulating dependence (2) with random sampling of the input parameter distributions and performance of the necessary calculations (trials) for prediction of the output response. For such calculations, the Monte Carlo method can be applied (see for example Jain, 1973); further, we describe Monte-Carlo simulation technique and MATLAB consideration of the problem.

SIMULATION AND LIVE SCRIPT PROGRAM FOR ATTENUATION CALCULATIONS

The simulation process can be realized in following stages:

- determining the suitable distribution and 95% lower and upper mean uncertainties for each of parameters in eq.(6) ;
- random sampling according to chosen distribution for each of the parameters and appropriate attenuation calculations with final obtaining the mean and 98% error;
- comparison of the derived attenuation value with limited value; the latter should be taken from the standard for the actual cable.

The first stage is in fitting and deriving the best distribution for the *d*, δ_{Al}, *I, F* and *p* parameters. For this the n…fit commands (see Burstein, 2015) can be used; the n… here marks the distribution name abbreviation, for example, when n notes inif, the uniffit command should be used for uniform distribution. Such way defined distribution and some its parameters are listed in Table 2 for each component of the equation (6).

Table 2. Parameters of eqs.(5,6), their distributions, and parameter limits intended for Monte Carlo sampling.

Parameter	Distribution	Lower Parameter Limit	Upper Parameter Limit
Diameter ratio, $d_{\mathrm{ins}}/d_{\mathrm{cu}}$	Triangular, asymmetry	2.1667	2.4507
Aluminum thickness, δ_{Al}, µm	Uniform	87.54	106.06
Impedance, *I*, Ohm	Uniform	25	40
Frequency, *F*, MHz	Triangular, asymmetry	14.802	21.77
Lay length, *p*, mm	Triangular, asymmetry	40.083	51.828

On the second stage, for each of assigned parameters (Table 2) their values should be generated with some random number generator. In particular, the random MATLAB command can be used for uniform distribution, and this command together with the makedist command - for the triangular distribution. For each combination of the played parameters, the attenuation value should be calculated by eq. (6) with defined a- coefficients.

Finally, the histogram of the attenuation values and its required limit should be generated and compared; if 95% of the derived A values are less than assigned limit thus inputted set of parameters is suitable.

The live script (based on Burstein, 2020) program realizing the described scheme are presented in the Figure 3.

Figure 3. Live script program representing attenuations for different cable and test parameters

```
clear,close all
Dc=.565,Di=1.34,im=100.6,del=40,p=20.412,fre=48.28,sl=11.99;%Default Ini Data
a=[ 0.69326;-0.5933;-0.24276;-0.5468;0.20271;0.077789;0.1269;0.53919;...
-0.2253;-0.028895;0.18077;-0.025033;0.016667;-0.045515;0.0051571;-0.010343]';
Dat_maxmin=[1.613 1.501;4.155 4.155;2.500 2.500;0.980 0.980;3.334 3.334]./100;
Dat_min=[0.4800 1.0400 2.1667 87.54 25 14.802 48.083 11.14];
Dat_range=[0.1630 0.5200 0.29038 17.52 15 6.969 3.744468 7.224];
Di2Dc=Di/Dc;Dat0=[Di2Dc im del p fre];Dat=(Dat0-Dat_min(3:end-1))./Dat_range(3:end-1);
% Monte-Carlo
dU_min=Dat-Dat.*Dat_maxmin(:,2)';dU_max=Dat+Dat.*Dat_maxmin(:,1)';Dat_normdist=Dat;
  N=1000;ATT=zeros(1,N);
   for j=1:N
       for i=[1 4 5],Dat(i)=tria(dU_min(i),dU_max(i),Dat_normdist(i));end
       for i=2:3,Dat(i)=random('unif',dU_min(i),dU_max(i));end
         sum1=a(1)+sum(a(2:6).*Dat);
         sum2=a(7)*Dat(1)*Dat(2)+a(8)*Dat(1)*Dat(3)+a(9)*Dat(1)*Dat(4)+...
         a(10)*Dat(1)*Dat(5)+a(11)*Dat(2)*Dat(3)+a(12)*Dat(2)*Dat(4)+...
         a(13)*Dat(2)*Dat(5)+a(14)*Dat(3)*Dat(4)+a(15)*Dat(3)*Dat(5)+...
         a(16)*Dat(4)*Dat(5);
         ATT(j)=(sum1+sum2)*Dat_range(end)+Dat_min(end);
   end
        N_in=length(find(ATT<sl));percent_in=N_in/N*100;N_out=100-percent_in;
        [n,w]=hist(ATT);n_les_sl=n(w<sl);n_great_sl=n(w>=sl);
        if ~isempty(n_les_sl), bar(w(w<sl),n_les_sl),end
        hold on,axis manual
        if length(n_great_sl)>1; bar(w(w>=sl),n_great_sl,'r'),end
        plot([sl sl],[0 max(n)],'-r','Linewidth',7)
        hold off
        xlabel('Attenuation'),ylabel('At. Frequences'),grid
        title({['Attenuation is ' num2str(mean(ATT)) ['±',num2str(3*std(ATT))]];...
             'with 98% probability'})
        legend(['In range ',num2str(percent_in),' %'],...
            ['Out of range ',num2str(N_out),' %'],'location','best')

    function rtria=tria(lo,up,av)
        % generate random numbers by triangular distribution
        pd = makedist('Triangular','a',lo,'b',av,'c',up);
        rtria = random(pd);
    end
```

```
Dc = 0.5650
Di = 1.3400
im = 100.6000
del = 40
p = 20.4120
fre = 48.2800
```

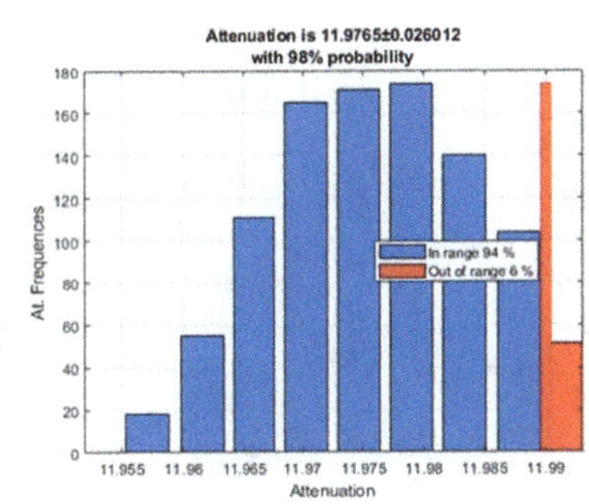

The live script outputs the inputted cable/test numbers and resulting plot on the right part of the Live Editor window. Notations above the plot are: Dc – copper wire diameter, mm; Di – diameter of insulation, mm; Im – required impedance value; del – thickness of the aluminum, µm; p – cable lay length, m; fre – test frequency, MHz . As for this live script example, here the 94% of attenuation values fall into the required area, so the cable/test parameters should be changed to meet the 98% requirement.

To use this simulation program there is acquaintance with MATLAB are required, that is not always convenient, especially for non-programmer engineers or technicians. Therefore, a special GUI was developed, more adapted for practical use.

SIMULATION TOOL

A program that implements the scheme discussed above and creates a GUI suitable for practical use is given in Appendix 3. The program generates special interface that permits input desirable parameter values with required attenuation limit, realize Monte Carlo simulation and represents histogram of the calculated attenuations with assigned limit line. The program displays defined mean *A* value and its standard deviation. Assumed number of trials in the simulation was 1000. Typical view the GUI with derived results are presented in the Figure 2. The v d_{ins}, d_{cu}, δ_{Al}, *l*, and *P* values together with the required impedance limit should be inputted, after pressing on the 'Show Attenuation' button the 'Calculation executes…' text appears in the GUI left panel. Then the following text and image are appeared: histogram with calculated value of the mean attenuation; 98% error of the defined *A*, and percentages of the *A*-values, which lower and higher of the limiting *A*-value (red vertical line). To run a new simulation, the 'Clear plots' button should be pressed; after this the new desirable parameters should be inputted and the 'Show Attenuation' button should be pressed again. The histogram bars with *A*-values exceeding required attenuation limit appear in red – see Figure 4.

Figure 4. Typical view of the simulation GUI for the attenuation calculations

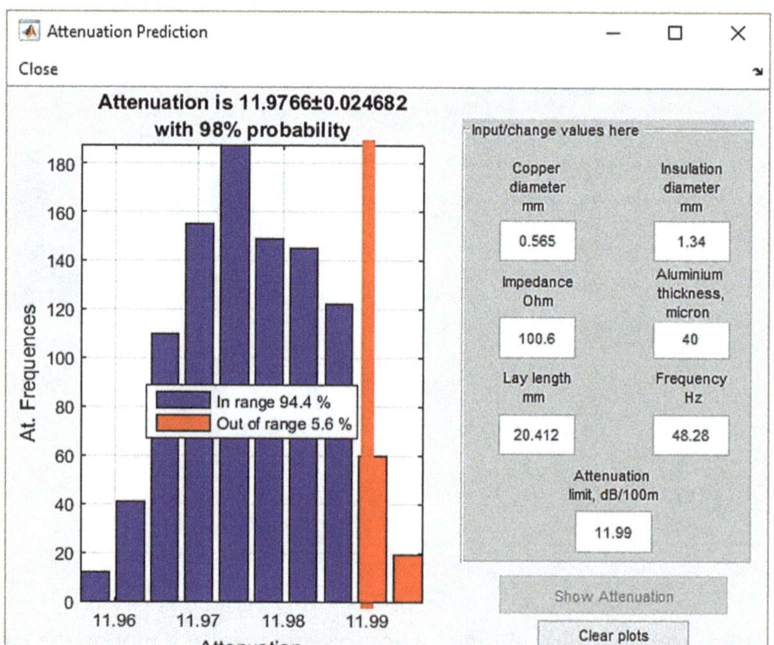

EXPERIMENTAL AND DISCUSSION

The described GUI that uses eq.(6) with defined coefficients was implemented to design cables of the 6, 6_A, 7, and 7_A categories (Figure 5); used parameters and attenuation limits required by standard represented in Table 3. Designed cables were manufactured and their attenuations was tested. Overall, 200 cables were testes for each of the seven sets of the cable parameters. Obtained attenuation means and their 98%-errors were compared with the corresponding predicted values.

Figure 5. Typical view of paired cables designed with the 'Attenuation Prediction' interface

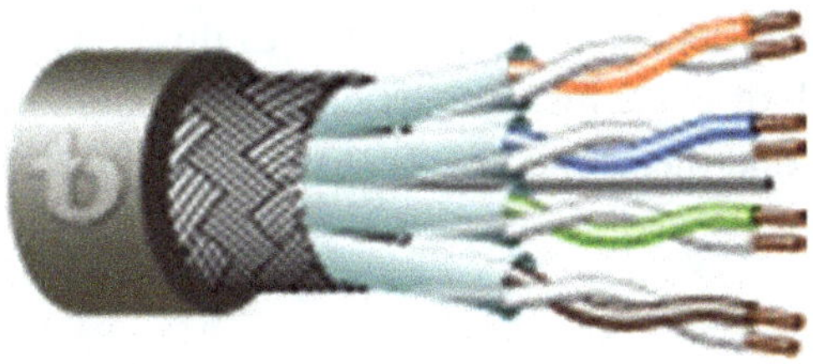

Table 3. Parameters and attenuation limits, A_{limit}, used in cable simulation design for further testing.

Category	6	6_A		7	7_A		
d_{cu}, mm	0.582	0.582	0.570	0.582	0.643	0.582	0.582
d_{ins}, mm	1.38	1.38	1.35	1.38	1.56	1.38	1.45
δ_{Al}, μm	25	25	25	40	40	40	40
p, mm	16.5	15.19	27	20.41	17.3	14.9	20.412
F, MHz	50	50	50	50	50	50	50
A_{limit}, dB/100m	13.75	13.36	13.36	13.256	13.013	13.013	13.013

The impedance value was assigned equal 100 Ohm in all cable categories. As can be seen from Table 4, measured (upper numbers in the cells) and calculated values of the attenuation are in very close agreement; in seven tested cables with different parameters, the discrepancy of the mean attenuation value is between 0% to 1.39%; the mean attenuation error is about 0.6%

Table 4. Attenuation (dB/100m) of the tested (upper line) and predicted by eq.(6) cables

Category	6	6_A		7	7_A		
Upper 98% (mean)	12.78 12.71	12.78 12.69	13.03 13.03	12.25 12.23	12.07 12.22	12.93 12.75	12.30 12.23
Mean value	12.74 12.83	12.74 12.61	12.98 12.93	12.20 12.20	12.02 12.19	12.89 12.85	12.26 12.17
Lower 98% (mean)	12.69 12.55	12.69 12.58	12.93 12.83	12.16 12.17	11.98 12.16	12.84 12.74	12.21 12.12
Error in mean A, %	0.70	1.02	0.36	0.00	1.39	0.31	0.73

For all tested cable sets, the obtained discrepancies lay in the predicted upper attenuation limits. The t-test also shows that tested and predicted mean attenuation values are not statistically different from each other. It indicates that defined equation (6) is accurate and permits correct cable design. Thus, the developed interface can be recommended for design of the high quality cables. Apparently, the applied method can be used also for other types of similarly constructed cables for which the processed data are available. For cables of categories 6_A and 7_A, respectively two and three cables were manufactured with different design values of the cable parameters; this allows you to choose the optimal cable parameters, based on the attenuation values better agreement. Thereby, developed tool based on the fitting relation (6) can be used in optimization problems when parameters of this equation, and hence cable parameters, can be defined in the framework of some restrictions on their combination. This is an actual and independent problem which lies beyond the scope of this study.

CONCLUSION

The original relationship for defining the attenuation of manufactured cables was derived and fitted to experimental data. The relationship connects some parameters of cable design and testing: copper wire, aluminum, and insulation thicknesses, cable impedance, test frequency, and lay length. The relation used for Monte-Carlo simulation of the cable attenuation prediction. To realize the simulations a number of MATLAB programs was developed; resulting computational program creates special graphical user interface intended for cable design. After assigning the desired cable/test parameters, this interface generates a plot with distribution of the attenuation values and required attenuation limit, and outputs defined mean attenuation, its 2% error, and numbers of values that get in and out of the limiting value. Data of calculations were verified by experiments and reveal good accordance with the actual data. The realized methodic together with developed program and interface can be used as compact tool for designing cables with optimal parameters.

ACKNOWLEDGMENT

The Research Foundation of the Kinneret Academic College on the Sea of Galilee has supported an early version of this study. This help is gratefully acknowledged. I thank my former student, today's engineer Asher Bohbot, who co-autosized in an earlier version of this chapter and also initiated this work in the past. I also wish to thank Teldor Cables and System Ltd., which kindly provided the cable data and enabled tests of the cables designed with the simulation tool created in this investigation.

REFERENCES

Burstein, L. (2015). *MATLAB in quality assurance sciences*. Amsterdam: Elsevier - Woodhead Publishing.

Burstein, L. (2020). *A MATLA Primer for Technical Programming for Materials Science and Engineering*. Amsterdam: Elsevier.

CEI-IEC 61156, 2002-12. Multicore and symmetrical pair/quad cables for digital communications, Parts1, 5, 6, and 7. International standard.

Costello, G. A. (1997). Theory of wire rope. Springer.

Gerdemeli, I., Kurt, S., & Anil, A. S. (2013). Fatigue Life Analysis of Wire Rope Strands with Finite Element Method. *Key Engineering Materials*, *572*(Sep), 513–516. doi:10.4028/www.scientific.net/KEM.572.513

Hyland, B. (2012). An Improved and Simple Cable Simulation Model. *Maxim Integrated Application Note*, *5141*, 1–12.

Jain, A. K. (1973). Monte Carlo simulation of crosstalk in communication tables. *WSC '73 Proceedings of the 6th conference on Winter simulation*, 844-857.

Li, B., & Zhou, Z. (2016). Test method for coupling attenuation of screened twisted pair cable based on mixed-mode S-parameters. In *Proceedings of 2016 Asia-Pacific International Symposium on Electromagnetic Compatibility (APEMC)*. Shenzhen, China: IEEE.

Lin, Y., Wang, K., Wu, J., Sun, O., & Chen, K. (2018). Using Deep Learning Algorithm to Modify the Design Parameters for Extending the Operation Life of the Flexible Cables. *Proceedings 67th International Cable and Connectivity Symposium (IWCS 2018). Providence, Rhode Island. Leadership, Innovation and Professional Development*, 139-150.

Poltz, J. (2017) Attenuation of screened twisted pairs. *Proceedings 66th International Cable and Connectivity Symposium (IWCS 2017) Orlando, Florida, Industry Leadership, Innovation and Professional Development*, 219-226.

Tranter, W.H., Shanmugan, K.S., Rappaport, T. S., & Kosbar, K.L. (2003). Modeling and simulation of waveform channels. In *Principles of Communication Systems Simulation with Wireless Applications* (pp. 529-582). Academic Press.

Turkseven, C. H., & Ertek, G. 2003, Simulation modeling for quality and productivity in Steel cord manufacturing. *Proceedings of the 2003 Winter Simulation Conference*, 1225-1229. 10.1109/WSC.2003.1261554

US Navy, . (1976). *Wire-rope analysis and design data* (Vol. 2). Washington, DC: Naval See Systems Command.

Velinsky, S. A. (1988). Design and mechanics of multi-lay wire strands. *J. Mech. Trans. Automat. Des. Trans. ASME*, *110*, 152–160.

Velinsky, S. A. (1989). On the design of wire rope. *J. Mech. Transm. Automat. Des. Trans. ASME*, *111*, 382–388.

KEY TERMS AND DEFINITIONS

Attenuation: A decrease in electric signal value during its propagation through wires, measured at a specific cable length.

Cable: Is two or more coupled insulated wires wrapped in one insulating shield and located in a tubular jacket.

Cable Design: Here means searching for the best cable parameters, namely the best copper wire diameter, aluminum and insulation thickness, cable impedance, test frequency and gasket length, that provides the required attenuation value limit.

Fitting: Is finding the accurate mathematical expression that best matches the observed/test data.

Frequency: The number of waves that have passed a fixed length at a given time; the cable should provide a certain frequency range to the broadcast signal; therefore cables are tested at the required frequencies.

Graphical User Interface: Developed window/s with graphic items (menu, buttons, boxes, etc.) that allows users interact with computer when solving his problem.

Impedance: Complex resistance between two nodes of the circuit (e.g., two points of a section of the wire-insulator).

Insulator: A material that prevents the passage of electric current through it.

Lay Length: A distance of one full revolution of two twisted and isolated wires.

Live Script: An interactive program that presents together codes, texts, obtained values and graphs on the same page of a special editor.

Monte-Carlo Simulation: A computational program/algorithm that uses multiple random sampling to obtain the most probably value of a parameter of interest and its possible deviation.

Simulation Tool: An interface for non-programmer users allowing analyze, design, and manufacturing product with optimal parameters; the tool based on special programs providing modelling and working with the simulation tool.

APPENDIX 1

Nomenclature

a – fitting coefficients;
A – attenuation, dB;
d_{cu} - the copper wire diameter, mm;
d_{ins} - diameter of the insulation, mm;
d= d_{ins}/ d_{cu} - insulation to wire diameter ration;
f - generalized attenuation function to fit;
F – the frequency assigned for the cable test, MHz;
I – impedance value required in the cable testing, Ohm ;
p – cable lay length, mm;
r-squared – coefficient of determination used to evaluate of the fit quality;
S – generalized objective function;
x - one of the fitting parameters, dimensional;
X – scaled value of the one of the fitting parameters, dimensionless;
δ_{Al} – thickness of the aluminum, μm;
Δ – range of the parameter, dimensional;

Subscripts

Al – aluminum;
cu – copper;
i – data point number;
ins – insulator;
l – fitting coefficient indices;
j, *k* – indices of the adjusted parameters
min – minimal value;
max – maximal value.

APPENDIX 2

Cable Geometry and Some Other Tests Data Used in Fitting (Partly)

These data should be written into the EXCEL file named CableData2.xlsx for further fitting by the program represented in Appendix 2. The total number of data points used in this study was equal to 260.

Table 5.

d_{cu},mm	d_{ins},mm	I,Ω	δ_{Al},μm	p,mm	F,MHz	A,dB
0.5820	1.3800	104.77	40	20.41	48.28	12.17
0.5820	1.3800	101.05	40	20.41	49.12	12.28
0.5820	1.3800	100.10	40	20.41	49.95	12.38
0.5820	1.3800	104.08	40	20.41	50.78	12.50
0.5820	1.3800	102.19	40	20.41	51.61	12.59
0.5820	1.3800	98.97	40	14.98	48.28	12.57
0.5820	1.3800	98.75	40	14.98	49.12	12.68
0.5820	1.3800	97.68	40	14.98	49.95	12.80
0.5820	1.3800	96.41	40	14.98	50.78	12.92
0.5820	1.3800	97.76	40	14.98	51.61	13.01
0.5820	1.3800	101.15	40	19.57	48.28	12.17
0.5820	1.3800	101.88	40	19.57	49.12	12.28
0.5820	1.3800	101.31	40	19.57	49.95	12.38
0.5820	1.3800	100.04	40	19.57	50.78	12.50
0.5820	1.3800	98.21	40	19.57	51.61	12.59
0.5820	1.3800	102.66	40	17.54	48.28	12.57
0.5820	1.3800	100.66	40	17.54	49.12	12.68
0.5820	1.3800	100.34	40	17.54	49.95	12.80
0.5820	1.3800	100.25	40	17.54	50.78	12.92
0.5820	1.3800	99.15	40	17.54	51.61	13.01
0.5820	1.3800	99.13	37	20.41	49.00	12.45
0.5820	1.3800	99.51	37	20.41	49.50	12.51
0.5820	1.3800	99.50	37	20.41	50.00	12.58
0.5820	1.3800	99.29	37	20.41	50.50	12.65
0.5820	1.3800	100.18	37	20.41	51.00	12.71
0.5820	1.3800	98.22	37	14.98	49.00	12.85
0.5820	1.3800	100.24	37	14.98	49.50	12.92
0.5820	1.3800	98.66	37	14.98	50.00	12.99
0.5820	1.3800	96.88	37	14.98	50.50	13.05
0.5820	1.3800	96.54	37	14.98	51.00	13.11
0.5820	1.3800	100.22	37	19.57	49.00	12.51
0.5820	1.3800	100.77	37	19.57	49.50	12.58
0.5820	1.3800	100.22	37	19.57	50.00	12.64
0.5820	1.3800	100.55	37	19.57	50.50	12.71
0.5820	1.3800	100.18	37	19.57	51.00	12.78
0.5820	1.3800	100.74	37	17.54	49.00	12.46
0.5820	1.3800	100.82	37	17.54	49.50	12.52
0.5820	1.3800	100.78	37	17.54	50.00	12.59
0.5820	1.3800	101.81	37	17.54	50.50	12.66

APPENDIX 3

Computational Program for Cable Data Fit

```
function [Coeff1,Coeff2]=CableFit
% calculates fit coefficients:
% in equations 1
% a=a1+a2*D+a3*eps+a4*del+a5*P+a6*F
% and in equation 2
% a=a1+a2*D+a3*eps+a4*del+a5*P+a6*F+a7*D*eps+a8*D*del+...
%   a9*D*P+a10*D*f+a11*eps*del+a12*eps*P+a13*eps*F+a14*del*P+...
%   a15*del*F+a16*P*F
% Output: coefficients with their boundaries for eq 1 and eq 2
% to run: >>[Coeff1,Coeff2]=CableFit
%===============================
% Leonid Burstein, Feb 06, 2016
% Revised: Dec 02, 2019
%===============================
clear
% reading data from 'CableData.xlsx'
filename = 'CableData2.xlsx';
sheet = 1;Datarange = 'A2:G261';
Dat0 = xlsread(filename, sheet, Datarange);
Dat0=[Dat0(:,2)./Dat0(:,1) Dat0(:,3:end)];
close all
%============== Scaling
Dat_max=max(Dat0);Dat_min=min(Dat0);
Dat_range=(Dat_max-Dat_min);
Dat=(Dat0-ones(length(Dat0),1)*Dat_min)./(ones(length(Dat0),1)*Dat_range);%ndm

%============== fitting, coefficients estimation
X1=[ones(length(Dat0),1) Dat(:,1:end-1)];% for equation 1
X2=[ones(length(Dat0),1) Dat(:,1:end-1) Dat(:,1).*Dat(:,2) Dat(:,1).*Dat(:,3)
...
Dat(:,1).*Dat(:,4) Dat(:,1).*Dat(:,5) Dat(:,2).*Dat(:,3) Dat(:,2).*Dat(:,4)
...
Dat(:,2).*Dat(:,5) Dat(:,3).*Dat(:,4) Dat(:,3).*Dat(:,5) Dat(:,4).*Dat(:,5)];%
for equation 2
[a1,bint1,r1,rint1,stats1] = regress(Dat(:,end),X1);% equation 1
[a2,bint2,r2,rint2,stats2] = regress(Dat(:,end),X2);% equation 2
att1=ones(length(X1),1)*a1'.*X1;% attenuation by eq 1, dimensionless
att2=ones(length(X2),1)*a2'.*X2;% attenuation by eq 2, dimensionless
```

```
r1=Dat_range(end)*r1;% residuals by eq 1, dimensional
r2=Dat_range(end)*r2;% residuals by eq 2, dimensional
rint1=Dat_range(end)*rint1;% residual confidence bounds, eq 1, dimensional
att_fit1(:,1)=Dat_range(end)*sum(att1')+Dat_min(end);% attenuation by eq 1,
dimensional
Error1=abs(r1)./Dat0(:,end)*100;% error in Att estimation, eq 1, percentage
rint2=Dat_range(end)*rint2;% residual confidence bounds, eq 2, dimensional
att_fit2(:,1)=Dat_range(end)*sum(att2')+Dat_min(end);%attenuation by eq 2, di-
mensional
Error2=abs(r2)./Dat0(:,end)*100;% error in Att estimation, eq 2, percentage
Mean_Error1=mean(Error1);Max_Error1=max(Error1);
Mean_Error2=mean(Error2);Max_Error2=max(Error2);
format shortG
Coeff1=[a1 bint1];%defined coefficients a1
Coeff2=[a2 bint2];%defined coefficients a2
r1_square=stats1(1);
r2_square=stats2(1);
resid1=[rint1(:,1) r1 rint1(:,2)];%residuals with boundaries, eq. 1
resid2=[rint2(:,1) r2 rint2(:,2)];%residuals with boundaries, eq. 2

% tables with statistics and errors
Equation={'Equation 1';'Equation 2'};
Mean_Error=[Mean_Error1;Mean_Error2];Max_Error=[Max_Error1;Max_Error2];
r_square=[r1_square;r2_square];St_Deviation=[std(Error1);std(Error2)];
Table_1_Basic_statistics = table(Mean_Error,Max_Error,St_Deviation,...
    r_square,'RowNames',Equation)
Attenuation=Dat0(:,end); A_fitted_eq_1=att_fit1;Error_eq_1=Error1;
A_fitted_eq_2=att_fit2;Error_eq_2=Error2;
Table_2_Attenuation = table(Attenuation,A_fitted_eq_1,Error_eq_1,...
    A_fitted_eq_2,Error_eq_2)

% plot generation
subplot(2,2,1)
plot(1:length(Dat),resid1),grid
legend('Upper bnd','Residuals','Lower bnd','location','best')
title('Residuals'),ylabel('dB'),xlabel('Data point number')
subplot(2,2,2)
plot(1:length(Dat),resid2),grid
legend('Upper bnd','Residuals','Lower bnd','location','best')
title('Residuals'),ylabel('dB'),xlabel('Data point number')
subplot(2,2,3)
hist(r1),xlabel('Range, dB')
title('Distribution of the residuals')
```

```
subplot(2,2,4)
hist(r2),xlabel('Range, dB')
title('Distribution of the residuals')
```

After inputting in the Command Window the command

```
>>[Coeff1,Coeff2]=CableFit
```

The figure with residuals and its distribution (Figure 2, subchapter 2), the table with basic statistics (Table 1, subchapter 2) and the following data are displayed:

Table 6a. Attenuation =

Attenuation	A_fitted_eq_1	Error_eq_1	A_fitted_eq_2	Error_eq_2
12.17	11.967	1.6692	11.993	1.4532
12.28	12.363	0.6727	12.088	1.5597
12.38	12.539	1.2847	12.19	1.5334
12.50	12.325	1.3989	12.328	1.3765
12.59	12.576	0.1112	12.423	1.3304
12.57	12.368	1.6087	12.077	3.9258
12.68	12.486	1.5267	12.191	3.8548
12.80	12.672	0.9972	12.293	3.961
12.92	12.874	0.3551	12.388	4.115
13.01	12.868	1.088	12.526	3.724
12.17	12.244	0.6118	12	1.3998
12.28	12.288	0.0641	12.114	1.3517
12.38	12.434	0.4380	12.221	1.2868
12.50	12.636	1.0879	12.319	1.4461
12.59	12.882	2.3201	12.408	1.4459
12.57	12.103	3.7143	12.056	4.08
… (shortened).				

Table 6b. Coeff1 =

0.76792	0.71151	0.82433
-0.62903	-0.6813	-0.57677
-0.19202	-0.27826	-0.10578
-0.033387	-0.061916	-0.0048581
0.010359	-0.032266	0.052984
0.063039	0.023581	0.1025

Table 6c. Coeff2 =

0.69336	0.51278	0.87393
-0.59339	-0.80811	-0.37867
-0.24273	-0.53643	0.050962
-0.54688	-0.71082	-0.38294
0.2025	-0.013391	0.41839
0.077819	-0.065831	0.22147
0.1269	-0.13059	0.38439
0.53927	0.40883	0.66971
-0.22518	-0.39808	-0.052284
-0.028871	-0.15837	0.10063
0.1807	-0.015382	0.37678
-0.024965	-0.35074	0.30081
0.016571	-0.2287	0.26185
-0.045364	-0.12177	0.03104
-0.067372	0.077691	0.0051595
-0.010325	-0.12354	0.10289

APPENDIX 4

Program to Attenuation Simulation and Generating Tool (Gui) for Cable Design

```
function CableSimulation
% Monte-Carlo cable attenuation
% GUI generation
% includes fitting coefficients and minimal, maximal and range
% values of the cable parameters
```

```
% the program includes call to logo.jpg-file with Teldor
% company and Kinneret college graphic symbols
% to run: >>CableSimulation
%=====================================
% Burstein L, Feb 06, 2016
% Revised: Dec 02, 2019
%=====================================
clear,
close all
%figure and menu
f=figure('MenuBar','None','Resize','on',...
    'NumberTitle','off','Name','Attenuation Prediction');
fm=uimenu(f,'Label','Close');
uimenu(fm,'Label','Close Figure','Callback','close')
% for frames/panel
clr=[.8 .8 .8];
uipanel('Title','Input/change values here','Uni','Norm',...
        'Bac',clr,'Pos',[.6 .17 .38 .78]);
Dc=.582;Di=1.38;im=100.6;del=40;p=20.412;freq=48.28;lsl=13.6;%Default Ini Data
ax1 = axes('Position',[0.1 0.1 0.45 0.8],'color',[.8 .8 .8]);
C = imread('logo.jpg');image([.1 .45],[.1 .8],C)%f
flag=1;
set(ax1,'XTick',[],'YTick',[]),box on
%Input Fields
ps1=[.65 .7 .1 .07];ps2=[.85 .7 .1 .07];ps3=[.65 .53 .1 .07];
ps4=[.85 .53 .1 .07];ps5=[.65 .36 .1 .07];ps6=[.85 .36 .1 .07];
ps7=[.75 .19 .1 .07];
ps1t=[.65 .78 .1 .1];ps2t=[.85 .78 .1 .1];ps3t=[.65 .61 .1 .07];
pos4t=[.85 .595 .1 .1];ps5t=[.65 .44 .1 .07];ps6t=[.85 .44 .1 .07];
ps7t=[.74 .27 .12 .07];psbtn=[.65 .08 .3 .07];psclc=[.7 .02 .2 .05];
%Strings for the edit windows
str1=sprintf('Wire diameter\n     mm');str2=sprintf(' Isolation diameter\n
mm');
str3=sprintf('Impedance\n   ndm');str4=sprintf('Aluminium thickness\n
mm');
str5='Lay';str6=sprintf('Frequency\n    Hz');
str7='Attenuation limit, ndm';str8='Show Attenuation';
str9='Clear plots';
% GUI commands
DC=uicontrol(f,'Sty','edit','Uni','Norm','Pos',ps1,'Str',num2str(Dc));
uicontrol(f,'sty','tex','Uni','Norm','Pos',ps1t,'Str',str1,'Bac',clr)
DI=uicontrol(f,'sty','edit','Uni','Norm','Pos',ps2,'Str',num2str(Di));
uicontrol(f,'sty','tex','Uni','Norm','Pos',ps2t,'Str',str2,'Bac',clr)
```

```
IMP=uicontrol(f,'sty','edit','Uni','Norm','Pos',ps3,'Str',num2str(im));
uicontrol(f,'sty','tex','Uni','Norm','Pos',ps3t,'Str',str3,'Bac',clr)
DEL=uicontrol(f,'sty','edit','Uni','Norm','Pos',ps4,'Str',num2str(del));
uicontrol(f,'sty','tex','Uni','Norm','Pos',pos4t,'Str',str4,'Bac',clr)
P=uicontrol(f,'sty','edit','Uni','Norm','Pos',ps5,'Str',num2str(p));
uicontrol(f,'sty','tex','Uni','Norm','Pos',ps5t,'Str',str5,'Bac',clr)
FRE=uicontrol(f,'sty','edit','Uni','Norm','Pos',ps6,'Str',num2str(freq));
uicontrol(f,'sty','tex','Uni','Norm','Pos',ps6t,'Str',str6,'Bac',clr)
LSL=uicontrol(f,'sty','edit','Uni','Norm','Pos',ps7,'Str',num2str(lsl));
uicontrol(f,'sty','tex','Uni','Norm','Pos',ps7t,'Str',str7,'Bac',clr)
% start button
pbt1=uicontrol(f,'Str',str8,'Uni','Norm','Pos',psbtn,'Cal',@Callback_calcul);
%clear button
uicontrol(f,'Uni','Norm','Pos',psclc,'Str',str9,'Cal',@clear_axis);
function clear_axis(~,~)
        cla(ax1),set(ax1,'color',[.8 .8 .8])
        set(ax1,'XTick',[]),set(ax1,'YTick',[])
        xlabel(ax1,' '),ylabel(ax1,' '),title(ax1,'')
        if flag~=1,legend(ax1,'');end
        set(pbt1,'Enable','on');
end
function Callback_calcul(~,~)
        set(pbt1,'Enable','off');
a=[ 0.69326;-0.5933;-0.24276;-0.5468;0.20271;0.077789;0.1269;0.53919;...
-0.2253;-0.028895;0.18077;-0.025033;0.016667;-0.045515;0.0051571;-0.010343]';
Dat_maxmin=[1.613 1.501;4.155 4.155;2.500 2.500;0.980 0.980;3.334 3.334]./100;
Dat_min=[0.4800 1.0400 2.1667 87.54 25 14.802 48.083 11.14];
Dat_range=[0.1630 0.5200 0.29038 17.52 15 6.969 3.744468 7.224];
 Dc=str2double(get(DC,'Str'));Di=str2double(get(DI,'Str'));
 im=str2double(get(IMP,'Str'));del=str2double(get(DEL,'Str'));
 Pn=str2double(get(P,'Str'));fre=str2double(get(FRE,'Str'));
 sl=str2double(get(LSL,'Str'));
Di2Dc=Di/Dc;
Dat0=[Di2Dc im del Pn fre];
Dat=(Dat0-Dat_min(3:end-1))./Dat_range(3:end-1);
% Monte-Carlo
        dU_min=Dat-Dat.*Dat_maxmin(:,2)';dU_max=Dat+Dat.*Dat_maxmin(:,1)';
        Dat_normdist=Dat;
        N=1000;
        ATT=zeros(1,N);
        if flag~=1,txt_x=.22;txt_y=.5;else txt_x=.19;txt_y=.5;end
        text(txt_x,txt_y,'Calculations execute...','Bac',[.8 .8 .8])
        flag=0;
```

```
        pause(0.1)
    for j=1:N
        % normal distribution
        %for i=1:length(Dat),Dat(i)=random('norm',Dat_normdist(i),dU(i));end
        % triangular distribution
        for i=[1 4 5]
            Dat(i)=tria(dU_min(i),dU_max(i),Dat_normdist(i));
        end
        % uniform distribution
        for i=2:3,Dat(i)=random('unif',dU_min(i),dU_max(i));end
        %for i=1:length(Dat),Dat(i)=random('unif',dU_min(i),dU_max(i));end
        sum1=a(1)+sum(a(2:6).*Dat);
        sum2=a(7)*Dat(1)*Dat(2)+a(8)*Dat(1)*Dat(3)+a(9)*Dat(1)*Dat(4)+...
        a(10)*Dat(1)*Dat(5)+a(11)*Dat(2)*Dat(3)+a(12)*Dat(2)*Dat(4)+...
        a(13)*Dat(2)*Dat(5)+a(14)*Dat(3)*Dat(4)+a(15)*Dat(3)*Dat(5)+...
        a(16)*Dat(4)*Dat(5);
        ATT(j)=(sum1+sum2)*Dat_range(end)+Dat_min(end);
    end
      %[ATT_calc mean(ATT)]%for comparing the direct and the stat calculations
        N_in=length(find(ATT<sl));
        percent_in=N_in/N*100;N_out=100-percent_in;
        [n,w]=hist(ATT);
        n_les_sl=n(w<sl);
        n_great_sl=n(w>=sl);
        if ~isempty(n_les_sl), bar(w(w<sl),n_les_sl),end
        hold on,axis manual
        if length(n_great_sl)>1; bar(w(w>=sl),n_great_sl,'r'),end
        plot([sl sl],[0 max(n)],'-r','Linewidth',7)
        hold off
        xlabel('Attenuation'),ylabel('At. Frequences'),grid
        title({['Attenuation is ' num2str(mean(ATT))
['±',num2str(3*std(ATT))]];...
             'with 98% probability'})
        legend(['In range ',num2str(percent_in),' %'],...
            ['Out of range ',num2str(N_out),' %'],'location','best')
        axis tight
end
    function rtria=tria(lo,up,av)
        % generate random numbers by triangular distribution
        pd = makedist('Triangular','a',lo,'b',av,'c',up);
        rtria = random(pd);
    end
end
```

After inputting the >>CableSimulation in the Command Window, the following graphical user interface (GUI) generated.

Figure 6. View of the graphical user interface generated by the CableSimulation program.

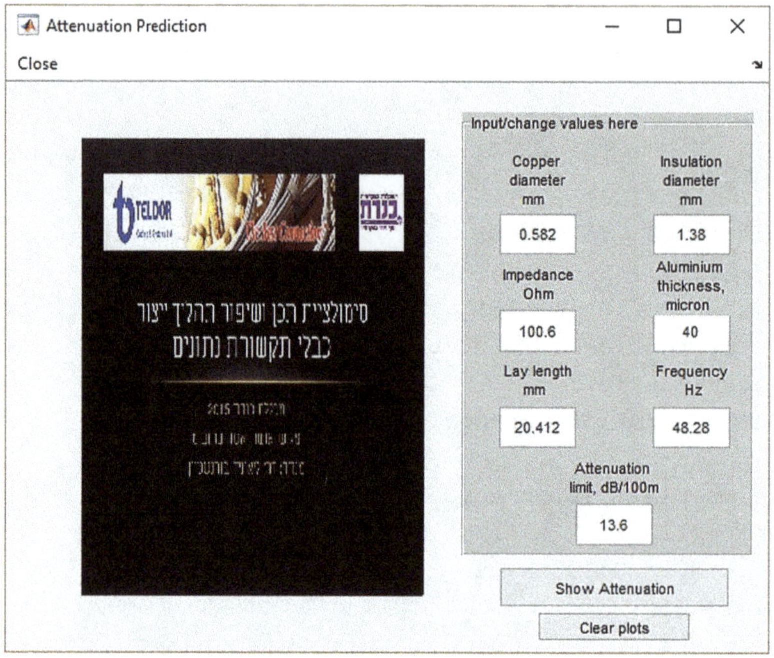

Chapter 4
Biodegradable Acoustic Proficiency in Sound Absorption Capacity After Water Treatment and Testing by Impendence Tube Method:
Hot Water and Saline Treatment to Test Acoustic Property

Tushar Kanta Mohapatra
KIIT University, India

Isham Panigrahi
KIIT University, India

Suchismita Satapathy
https://orcid.org/0000-0002-4805-1793
KIIT University, India

Debesh Mishra
https://orcid.org/0000-0003-2665-1365
KIIT University, India

ABSTRACT

The present research investigation aims to fabricate a new class of fiber reinforced composite material by using biodegradable materials with epoxy as the strengthening agent. In order to explore the possibilities of using the new class of composite material in required application areas, the research work is carried out mainly in the field of the acoustic absorption properties of these bio fibers as an alternate building material. Also, the utilization of these materials as sound insulation will also provide a good solution to the waste management.

DOI: 10.4018/978-1-7998-4939-1.ch004

INTRODUCTION

Nowadays sustainability has become a great urgency for the researchers to use environment friendly building materials. The cellulosic fibers that originate from various sources like wood, plants and waste retains a great capability for the replacement of synthetic, glass and asbestos fibers as reinforcements in composites. If we consider the cost then the natural fibers have an added advantage over the synthetic fibers as they are relatively less costly. Moreover since the natural fibers are biodegradable they have additional advantage of reducing CO2 emissions from the building materials production. Furthermore one big plus point of these building materials is that they are created from the wastes itself so in this manner the waste management also takes place side by side. So from the above mentioned qualities of these biofibers they have been chosen as the material for the research. The reinforcing phase of a composite constitutes mainly of two types i.e. fiber reinforcement or particle reinforcement. So if the fiber is obtained from natural sources like plants then the composites can be called as natural fiber reinforced composite. A lot of research work has already been done by several researchers to explore these fibers for applications in various fields including acoustics as one of the major field of investigation. The natural fibers itself provides a lot of advantages like they are available in plenty, they are light in weight, environment friendly and also biodegradable.

The incident sound striking a room surface yields certain sound energy which involves reflected sound, absorbed sound and transmitted sound. The sound reflectors prevent sound transmission by forming a solid impervious barrier. On the other hand most good sound absorbers readily transmit the sound. So the sound absorbers are generally porous, lightweight material.

Thus it can be said that the acoustic material are those that are specifically produced for the purpose of providing high values of absorption.

The sound absorbing material absorbs most of the sound energy that strikes upon them and reflect very little. Therefore, these materials are quite important in controlling noise. For this purpose there exists a wide range of sound absorbing materials which provides absorption properties that depends on the frequencies, composition, thickness, density, porosity etc. However the material that have a great value of sound absorption coefficient are usually porous.

A porous absorbing material is a solid that has cavities or interstices so that sound waves are able to enter through them and these sound waves are converted to heat energy. Thickness plays an important role in sound absorption. The thicker the material the greater is its sound absorption coefficient. Most of the porous sound absorbing material that are commercially available are fibrous. These materials are composed of a set of continuous filament that trap the air between them. They are produced in rolls with different thermal, acoustical and mechanical properties.

LITERATURE REVIEW

Sung et.al have explained commercially available sound absorption materials could be generally divided into two categories: resonant sound absorption materials and porous sound absorption materials .Rey et.al (2007),Kozumi (2002) have explained natural fibers like bamboo fibres as accoustic, Zulkifh et.al have (2008) chosed multilayer coir fibre for accoustic, Asby et.al. (2003) have selected metal foam as accoustics in producing sound absorbing material.Hau et.al have explained that Metal foam has a porous index between 75 to 95 percent.

Thilagavathi et al. (2010) have studied the use of the biodegradable nonwovens for automobile interior components which normally uses glass and other manufactured fibers that are difficult to recycle. Rwawiire et al. (2017) have investigated that barkcloth which is a cellulose nonwoven natural fiber had a low thermal conductivity and good sound absorption properties. It displayed that by using multilayers of barkcloth the sound absorption coefficient is improved. Fouladi et al. (2011) have investigated the acoustic absorption of coir fiber for natural fresh samples and industrial prepared fibers mixed with binders. They also presented that modelling the industrial fiber is vital and inevitable since natural coir fiber is enhanced for commercial use. A novel method was developed by Hudaa and Yang (2009) to fabricate composites from mechanically split cornhusks (MSHs).They have also shown that these materials displayed good sound absorption properties due to the presence of micro air pockets in husks and also their cost is substantially low as compared with other natural fibers. Khidir et al. (2014) have studied the sound absorption properties of a single layer date palm fiber based on the flow resistivity value using differential tube. Navacerrada et al. (2014) have studied the usefulness of fique fiber which is used for the manufacture of packaging and cordages. They also studied that by addition of various natural fibers like fique a good quality of construction material can be developed. Ganesan and Karthik (2016) have studied the effect of blend proportion of kapok and milkweed fibers with cotton, fabric GSM. They found that the sound reduction increased with increase in blend proportion of kapok and milkweed fibers. Liu et al. (2015) have investigated that at lower frequency range the kapok fiber nonwoven fabrics had superior acoustical properties. Xianga et al. (2013) have showed that the kapok fiber has excellent acoustical damping performance due to its natural hollow structure and the sound absorption coefficient of kapok fibrous assemblies were significantly affected by bulk density, thickness and arrangement of kapok fibers, also the results demonstrated that kapok fiber has a great scope to be used as an environment friendly sound absorption material. Berardi and Iannace (2015) have done a comprehensive study on the use of several natural fibers that are a valid alternative to traditional synthetic ones for sound absorption treatments at a reduced cost too. Jayamani et al. (2014) have studied the development of epoxy/luffa composites and the investigation of their mechanical and acoustical properties were done. Moreover the fibers went on an alkalization treatment and its effects on the mechanical and sound absorption properties of the composites were measured by the utilization of universal testing machine and two-microphone transfer function impedance tube methods. Dong and Yan (2012) have researched that natural fibers with their composites had a superior capability of noise reduction as compared with the synthetic fibers and their composites. Fatima and Mohanty (2018) have studied the properties regarding the acoustic and flammability of jute fibers along with its composites. The fire resistant tests prove that they possess a great quality to suppress the fire. Koizumi et al. (2002) have studied on the use of natural bamboo fiber as a sound absorbing material. They also found that the bamboo fiber retain acoustic properties identical to those of the glass wool fibers. Kabir et al. (2012) have studied for the advanced composite applications the various surface treatments that is done to the natural fibers. And as a result of which a drastic change is seen regarding the improvement of mechanical strength and dimensional stability of resultant composites. Prabhakaran et al. (2014) have investigated that flax is a natural fiber which has a good mechanical properties and they also studied their sound absorption and vibration damping properties and found that the sound absorption and vibration damping property of flax fiber was much higher as compared to the glass fiber. White and Ansell (1983) have studied that the annual crop fibers that have a rich cellulose content are quite cheap and are rapidly renewable source of fibers with the potential for polymer reinforcement and are found to be very useful as a core building material for structural board products. Sanadi et al. (1986) have used the polyester composites reinforced with continuous unidirectional sun hemp fibers to study the tensile and impact behavior. They showed that sunhemp fibers has a great potential as reinforcing fillers in plastics in order to produce inexpensive materials with a high toughness. Putra et al. (2013) have investigated on finding a sustainable and eco-friendly material for sound absorber. They showed that sugarcane wasted fibers displayed a good acoustic properties which is even comparable with that of the classical synthetic absorbers. Ballagh (1996) have investigated on the use of wool as sound absorbing material and founded that wool is a great material for sound absorption since they prove to be a great use as duct linings because of their low flow resistivity and also achieve a higher absorption coefficients for control of room reverberation. Tang et al. (2018) have investigated on the use of corn husk as a sound absorption material and founded that it had a good acoustic absorption which is great for noise reduction. It also had an advantage of being extremely thin and light weight compared with other commonly used porous materials. Rao et al. (2010) have studied on the use of new natural fiber called vakka by using it as the fiber and mixing with the matrix polyester resin. It was founded that not only the sound absorption and tensile strength property had increased but also it had a unique feature of being dielectric and by that way it can be used for electrical insulation applications. Ku et al. (2011) have investigated mainly on the tensile properties of the natural fibers reinforced polymer composites. They founded that the matrix fiber bonding plays a very vital role in the tensile strength of composite and also established that the halpin-Tsai equation which is the most effective way of predicting the Young's modulus of the composite having various kinds of natural fibers. Ho et al. (2012) have studied mainly on the various types of manufacturing processes that should be used to manufacture the biodegradable composite taking into several factors like type of material, mechanical and thermal properties of the fiber and matrices etc. So they showed that by using the proper fibers, matrices, additives and production methods the manufactured composite would have great tensile strength, sound absorption properties and the other desired characteristics that would make it a suitable material. Mittal et al. (2016) have investigated extensively about the use of the natural fiber composites and found that there is a need for the improvement of the surface properties of the natural fibers by using various methods like physical and chemical treatments, addition and coupling agent's etc. in order to manufacture the composites having the desired properties.

RESEARCH METHODOLOGY

In the present research work, the composites are created utilizing the straightforward hand-lay-up procedure which is considered as the most seasoned and most economical open trim technique for composite manufacture. The pre-prerequisites of this procedure are less and the manufacture steps are peaceful basic. The accompanying are the general steps utilized as a part of this examination for get ready epoxy-resin composites of various materials.

1. 1st clean all the fiber properly and put under the sun for removing the water particle. Then put in NaOH solution to dehydrate.
2. Mix both the resin (L-12) and hardener (K-6) at a ratio of 10:1 for producing mixture for producing material.
3. Then cut all the fiber in a small size i.e. 2mm to 5mm and put them in different bag.
4. After that fiber are added to this solution of epoxy resin and hardener in hand layup method.
5. Then make a round shape mold which diameter is equal to the diameter of impedance tube.

6. Before pour the mixture in the mold add some realizing agent lay all over the mold surface. Which helps to easily removal of material from the mold after get harder.
7. Then left all the material of 24 hour for better hardener with a hydraulic pressure over this. After the completion of 24 hours the material are removed from the mold for checking and then it is used in test for sound absorbing.

In this experiment many biodegradable materials has been used for acoustic testing like sugarcane, bamboo, coconut, rice straw, and husk. So these forms the fiber material. These are the materials that are waste materials which are available on earth. These prove to be a great material for acoustic testing since they have low cost and also are eco-friendly.

An impedance tube method is one of the most cost effective methods for measuring sound absorption. It mainly consists of a speaker, tube, microphones and sample holder. The sample of the material to be tested is fixed on one end of the tube with the help of sample holder while the speaker is present on the other end of the tube. A particular frequency of sound is produced by the help of software or signal generator. This signals travels down the tube where it encounters the material where some part of the sound is absorbed while some is reflected. In order to get the value of the reflected sound signal the microphones are placed at specific distances within the tube. And the signal is analyzed with the help of LabVIEW.

RESULT AND DISCUSSION

From the collected data of the above mentioned samples with three types of impedance tube material and with the varying test conditions two different types of graphs is obtained. The first graphs shows the SAC vs. frequency plot of different samples in different tubes in different conditions. The second graph is the NRC vs. Types of material used in varying conditions and varying tubes during is wet testing (dry,cold, hot and saline water) with the same three types of impedance tube materials. All the sound testing is done at a frequency range of 500 Hz to 5000 Hz.

After that in order to calculate the sound absorption and noise reduction coefficient all the data obtained is transferred into a tabular format. At first all the materials result is analyzed individually then a detail comparison is done in order to obtain the best material as well as best sound absobing medium for sound absorption. The incident sound striking a room surface yields certain sound energy which involves reflected sound, absorbed sound and transmitted sound. The sound reflectors prevent sound transmission by forming a solid impervious barrier. On the other hand most good sound absorbers readily transmit the sound. So the sound absorbers are generally porous, lightweight material.

Thus it can be said that the acoustic material are those that are specifically produced for the purpose of providing high values of absorption.

The sound absorbing material absorbs most of the sound energy that strikes upon them and reflect very little. Therefore, these materials are quite important in controlling noise. For this purpose there exists a wide range of sound absorbing materials which provides absorption properties that depends on the frequencies, composition, thickness, density, porosity etc. However the material that have a great value of sound absorption coefficient are usually porous.

A porous absorbing material is a solid that has cavities or interstices so that sound waves are able to enter through them and these sound waves are converted to heat energy. Thickness plays an important role in sound absorption. The thicker the material the greater is its sound absorption coefficient. Most of the porous sound absorbing material that are commercially available are fibrous. These materials are composed of a set of continuous filament that trap the air between them. They are produced in rolls with different thermal, acoustical and mechanical properties.

Before knowing the reason for choosing the difference impedance tube materials it is necessary to understand the details about sound absorption process. Sound absorption is defined as a process in which a material takes in sound energy when sound waves strikes upon them. Amount of the sound energy absorbed or reflected is dependent on the sound absorption coefficient (α). So it is a dimensionless quantity whose value ranges from 0 to 1. It indicates how much of the sound will get absorbed by that material. Mathematically, it can be expressed as:

$$\alpha = 1 - \frac{I_r}{I_i}$$

Where, I_r is one sided intensity of reflected sound and I_i is one sided intensity of incident sound.

So the value of α=0 indicates that the entire sound gets reflected and the value of α=1 indicates that the entire sound gets absorbed.

Now in order to calculate the sound absorption coefficient of a material we take the help of impedance tube method. Impedance tube method has more advantage when compared to other method because of its compact, low cost and fast result. In this method sound is created inside the impedance tube and made to incident on the acoustic material. From the reflected waves some of the characteristic of the acoustic materials can be calculated. Impedance tube method is well suitable for the small sample size. It allows to make quick and easy yet perfect measurements of absorption coefficients. The tube should be straight and it's inside surface should be smooth, nonporous, and free from dust to maintain low sound attenuation. The current investigations shows the potency of the various natural acoustic material regarding their sound absorption capability. For this study five different materials are taken namely the fibers of sugarcane, coconut, bamboo, rice straw and husk. These materials are tested at similar environmental conditions but with four stages. First the dry testing is conducted with the three different types of impedance tube materials like stainless steel, glass and PVC. The next phase is wet testing (cold, hot and saline water) with the same three types of impedance tube materials. All the sound testing is done at a frequency range of 500 Hz to 5000 Hz.

After that in order to calculate the sound absorption and noise reduction coefficient all the data obtained is transferred into a tabular format. At first all the materials result is analyzed individually then a detail comparison is done in order to obtain the best material for sound absorption.

Table 1 shows that for the dry samples in Glass impedance tube bamboo has the highest value of SAC at 3000 Hz and rice straw has the overall lowest SAC at 4500 Hz. From Table.2 it can be concluded that for the dry samples in PVC impedance tube husk has the highest value of SAC at 2500 Hz and sugarcane has the overall lowest SAC at 2500 Hz. It can be concluded From Table.3 that for the dry samples in Stainless Steel impedance tube bamboo has the highest value of SAC at 3500 Hz and rice straw has the overall lowest SAC at 4000 Hz.

When the same materials are treated with Cold water, found that that for the cold water treated samples in Glass impedance tube coconut has the highest value of SAC at 3000 Hz and sugarcane has the overall lowest SAC at 5000 Hz. from Table. 4.Table.5 and Table.6 shows that for the cold water treated

Table 1. SAC for dry sample in Glass Tube

Sl.No.	Frequency	Sugarcane	Bamboo	Coconut	Rice Straw	Husk
1.	500	0.23921061	0.446754	0.419853	0.42144	0.412149
2.	1000	0.55557044	0.693459	0.429274	0.568155	0.662534
3.	1500	0.26172072	0.629382	0.507097	0.602449	0.516167
4.	2000	0.4041228	0.618392	0.455777	0.597101	0.590575
5.	2500	0.37859996	0.623946	0.515538	0.376318	0.597216
6.	3000	0.50085352	0.808383	0.525891	0.543734	0.794363
7.	3500	0.40095731	0.749248	0.660841	0.536861	0.728193
8.	4000	0.31449845	0.659834	0.494504	0.344215	0.612484
9.	4500	0.26648495	0.573945	0.540558	0.20237	0.520219
10.	5000	0.33684347	0.729658	0.520184	0.472345	0.698483

Table 2. SAC for dry sample in PVC Tube

Sl.No.	Frequency	Sugarcane	Bamboo	Coconut	Rice Straw	Husk
1.	500	0.104995	0.3599	0.024605	0.337299	0.398677
2.	1000	0.030793	0.341715	0.067784	0.37445	0.5305
3.	1500	0.013276	0.326623	0.137297	0.278324	0.503577
4.	2000	0.013048	0.31512	0.137969	0.350919	0.447544
5.	2500	0.01181	0.282069	0.041195	0.384593	0.593951
6.	3000	0.012721	0.36545	0.056818	0.415206	0.423279
7.	3500	0.014132	0.180538	0.254142	0.172044	0.370121
8.	4000	0.029762	0.332003	0.101388	0.291192	0.507034
9.	4500	0.014545	0.310831	0.043566	0.459573	0.427259
10.	5000	0.028063	0.328112	0.150619	0.236521	0.429922

Table 3. SAC for dry sample in Stainless Steel Tube

Sl.No.	Frequency	Sugarcane	Bamboo	Coconut	Rice Straw	Husk
1.	500	0.412458	0.744945	0.651388	0.475567	0.649586
2.	1000	0.752636	0.364096	0.491743	0.516748	0.512831
3.	1500	0.660227	0.173123	0.519454	0.454227	0.534407
4.	2000	0.687131	0.718093	0.54332	0.506551	0.452985
5.	2500	0.58993	0.758329	0.431587	0.428097	0.690121
6.	3000	0.413817	0.696227	0.329154	0.385105	0.636649
7.	3500	0.373046	0.842649	0.341986	0.317973	0.581078
8.	4000	0.633421	0.644747	0.020886	0.000492	0.29229
9.	4500	0.653567	0.629517	0.191576	0.159434	0.527102
10.	5000	0.612826	0.258765	0.358809	0.034792	0.324439

Table 4. SAC for cold water treated sample in Glass Tube

Sl.No.	Frequency	Sugarcane	Bamboo	Coconut	Rice Straw	Husk
1.	500	0.029828	0.069681	0.049277	0.037918	0.08092
2.	1000	0.772711	0.653348	0.692477	0.326399	0.817535
3.	1500	0.615208	0.493929	0.771687	0.605473	0.473536
4.	2000	0.821568	0.438061	0.687087	0.712823	0.736475
5.	2500	0.743692	0.699613	0.661785	0.645585	0.61775
6.	3000	0.794363	0.632282	0.835805	0.656733	0.811175
7.	3500	0.604256	0.765546	0.741314	0.733878	0.83324
8.	4000	0.683789	0.514665	0.77255	0.587518	0.393648
9.	4500	0.766424	0.607771	0.615544	0.58476	0.170553
10.	5000	0.672382	0.791639	0.384051	0.446694	0.82125

Table 5. SAC for cold water treated sample in PVC Tube

Sl.No.	Frequency	Sugarcane	Bamboo	Coconut	Rice Straw	Husk
1.	500	0.594465	0.73047	0.693555	0.610595	0.685401
2.	1000	0.527593	0.759141	0.702981	0.864677	0.661314
3.	1500	0.655318	0.861651	0.79962	0.86552	0.670964
4.	2000	0.677388	0.740561	0.820501	0.917251	0.893483
5.	2500	0.739338	0.842459	0.51791	0.904768	0.714218
6.	3000	0.854459	0.92791	0.839804	0.857048	0.829537
7.	3500	0.810418	0.876263	0.865461	0.851369	0.816997
8.	4000	0.857099	0.867619	0.868189	0.765484	0.79516
9.	4500	0.873869	0.803858	0.90413	0.840745	0.885922
10.	5000	0.874217	0.672302	0.776173	0.75753	0.718008

Table 6. SAC for cold water treated sample in Stainless Steel Tube

Sl.No.	Frequency	Sugarcane	Bamboo	Coconut	Rice Straw	Husk
1.	500	0.01976	0.179831	0.036424	0.421789	0.586887
2.	1000	0.532332	0.552206	0.4798	0.619764	0.579361
3.	1500	0.067238	0.327797	0.341438	0.759378	0.367753
4.	2000	0.472512	0.40188	0.452985	0.553609	0.462926
5.	2500	0.746199	0.252886	0.711722	0.763147	0.511257
6.	3000	0.178148	0.479472	0.741108	0.707105	0.669771
7.	3500	0.691224	0.458467	0.623378	0.720548	0.454278
8.	4000	0.581494	0.245071	0.55222	0.594291	0.040509
9.	4500	0.651924	0.193925	0.002822	0.612977	0.654857
10.	5000	0.564482	0.271342	0.494758	0.296213	0.257232

Table 7. SAC for hot water treated sample in Glass Tube

Sl.No.	Frequency	Sugarcane	Bamboo	Coconut	Rice Straw	Husk
1.	500	0.025463	0.027872	0.263495	0.018469	0.411089
2.	1000	0.853587	0.789979	0.924451	0.321301	0.898824
3.	1500	0.856535	0.595891	0.617565	0.717529	0.295948
4.	2000	0.502538	0.754509	0.849977	0.607023	0.267241
5.	2500	0.801369	0.562537	0.757792	0.774903	0.696464
6.	3000	0.847897	0.359943	0.842638	0.852886	0.840929
7.	3500	0.867729	0.799778	0.841574	0.880422	0.801444
8.	4000	0.856504	0.705551	0.812979	0.858547	0.687932
9.	4500	0.775349	0.755555	0.805222	0.780389	0.80334
10.	5000	0.862067	0.751137	0.893839	0.862157	0.869538

Table 8. SAC for hot water treated sample in PVC Tube

Sl.No.	Frequency	Sugarcane	Bamboo	Coconut	Rice Straw	Husk
1.	500	0.087357	0.309513	0.430717	0.177496	0.115364
2.	1000	0.694092	0.639828	0.649518	0.691522	0.737318
3.	1500	0.800725	0.760419	0.606425	0.615735	0.778021
4.	2000	0.886105	0.861288	0.76412	0.915304	0.909286
5.	2500	0.736864	0.713792	0.685729	0.737674	0.638348
6.	3000	0.76789	0.764468	0.827694	0.871791	0.705629
7.	3500	0.744915	0.655746	0.675044	0.699988	0.633159
8.	4000	0.674527	0.583482	0.794591	0.65901	0.826501
9.	4500	0.77505	0.703501	0.799754	0.764279	0.583439
10.	5000	0.773215	0.856607	0.741907	0.855262	0.655094

Table 9. SAC for hot water treated sample in Stainless Steel Tube

Sl.No.	Frequency	Sugarcane	Bamboo	Coconut	Rice Straw	Husk
1.	500	0.009039	0.045326	0.028817	0.192437	0.207661
2.	1000	0.032202	0.052135	0.065083	0.046313	0.402071
3.	1500	0.416718	0.518669	0.449331	0.61913	0.403604
4.	2000	0.020055	0.687307	0.285951	0.725745	0.576566
5.	2500	0.181856	0.640442	0.599733	0.527623	0.576058
6.	3000	0.718177	0.73121	0.489368	0.471435	0.435376
7.	3500	0.298049	0.670686	0.060397	0.126599	0.760499
8.	4000	0.329064	0.082373	0.126358	0.037371	0.085839
9.	4500	0.369193	0.015027	0.388434	0.035078	0.299283
10.	5000	0.064419	0.522705	0.423498	0.440549	0.299427

samples in PVC impedance tube bamboo has the highest value of SAC at 3000 Hz and coconut has the overall lowest SAC at 2500 Hz and for the cold water treated samples in Stainless Steel impedance tube rice straw has the highest value of SAC at 2500 Hz and coconut has the overall lowest SAC at 4500 Hz.

The same sample when hot water treatment is done, Found in Table 7, Table 8 and Table 9. for The hot water treated samples in Glass impedance tube coconut has the highest value of SAC at 1000 Hz and rice straw has the overall lowest SAC at 500 Hz., for the hot water treated samples in PVC impedance tube rice straw has the highest value of SAC at 2000 Hz and sugarcane has the overall lowest SAC at 500 Hz.and that for the hot water treated samples in Stainless Steel impedance tube husk has the highest value of SAC at 3500 Hz and sugarcane has the overall lowest SAC at 500 Hz.

When the same samples are treated with Saline water, Table 10,11 and 12 shows that, for the saline water treated samples in Glass impedance tube husk has the highest value of SAC at 1000 Hz and husk has the overall lowest SAC at 500 Hz., for the saline water treated samples in PVC impedance tube rice straw has the highest value of SAC at 1500 Hz and husk has the overall lowest SAC at 500 Hz. and for

Table 10. SAC for saline water treated sample in Glass Tube

Sl.No.	Frequency	Sugarcane	Bamboo	Coconut	Rice Straw	Husk
1.	500	0.371105	0.35211	0.447607	0.032544	0.021432
2.	1000	0.665381	0.812667	0.822219	0.80789	0.860751
3.	1500	0.651228	0.604704	0.585336	0.459549	0.844392
4.	2000	0.765701	0.586452	0.506297	0.676666	0.555279
5.	2500	0.667123	0.597765	0.595666	0.753685	0.510381
6.	3000	0.78915	0.719165	0.78861	0.730088	0.837288
7.	3500	0.768579	0.406	0.793923	0.62242	0.676397
8.	4000	0.464513	0.544104	0.668655	0.826844	0.452969
9.	4500	0.455495	0.704736	0.83133	0.387432	0.413358
10.	5000	0.792719	0.793528	0.667026	0.580037	0.773998

Table 11. SAC for saline water treated sample in PVC Tube

Sl.No.	Frequency	Sugarcane	Bamboo	Coconut	Rice Straw	Husk
1.	500	0.196108	0.302466	0.291254	0.073309	0.018667
2.	1000	0.682421	0.681748	0.627442	0.651372	0.526077
3.	1500	0.710207	0.729339	0.866243	0.869389	0.64269
4.	2000	0.852821	0.563535	0.788642	0.826537	0.825112
5.	2500	0.602523	0.727737	0.694728	0.797509	0.678778
6.	3000	0.505461	0.716072	0.740643	0.788337	0.827914
7.	3500	0.635959	0.809435	0.774229	0.74133	0.462137
8.	4000	0.862534	0.407924	0.70828	0.810897	0.659974
9.	4500	0.82074	0.587413	0.669178	0.762826	0.746712
10.	5000	0.631972	0.656483	0.556421	0.712138	0.778503

Table 12. SAC for saline water treated sample in Stainless Steel Tube

Sl.No.	Frequency	Sugarcane	Bamboo	Coconut	Rice Straw	Husk
1.	500	0.239912	0.165506	0.169435	0.113775	0.012987
2.	1000	0.439954	0.172997	0.247364	0.32705	0.394607
3.	1500	0.164204	0.120488	0.173997	0.042214	0.547784
4.	2000	0.7122	0.450434	0.188757	0.693464	0.40329
5.	2500	0.614439	0.549638	0.018857	0.460165	0.714548
6.	3000	0.572074	0.697795	0.591574	0.449485	0.551986
7.	3500	0.684582	0.673343	0.508226	0.684684	0.704609
8.	4000	0.05016	0.054854	0.43336	0.30759	0.565165
9.	4500	0.449556	0.158962	0.055022	0.733811	0.039184
10.	5000	0.33366	0.067487	0.538212	0.061246	0.306551

Table 13. Comparative analysis between SAC found in all materials in dry,wet(hot,cold and saline water treatment)

Materials	Maximum No of Times Found
Sugarcane	1
Bamboo	2
Coconut	2
Rice Straw	4
Husk	3

the saline water treated samples in Stainless Steel impedance tube rice straw has the highest value of SAC at 4500 Hz and husk has the overall lowest SAC at 500 Hz.

From Table 13. it can be inferred that Rice Straw comes out to be the best acoustic material. Considering the overall test conditions the actual SAC value for Rice Straw is 0.80 .

From all the above tests we have found that rice straw is the best building material. And also as it is cheap and easily available it is used as cattle feeding material and also it can be used as an alternate building material.

Noise reduction coefficient is found by finding average of Sound absobtion coefficient.

Comparative analysis for the best building material has been done by comparing Noise reduction coefficient.

From Fig1 to 4, we can say that the PVC impedance tube material comes out to be the best material for the samples that are treated with the saline water.

So comparing the data from Fig. 1 to Fig. 4., it can be inferred that the NRC values increases tremendously after treatment with water (cold, hot and saline) as compared to its value when the sample is dry. So it can be said that the NRC values of most of the samples increases after water absorption.

Figure 1.

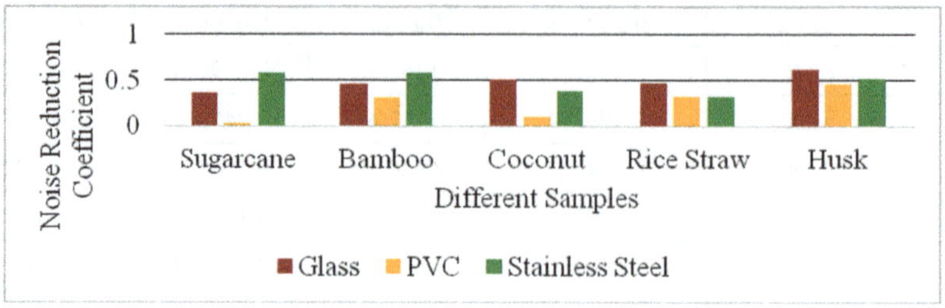

Figure 2. Comparative analysis of different samples in cold water treated condition

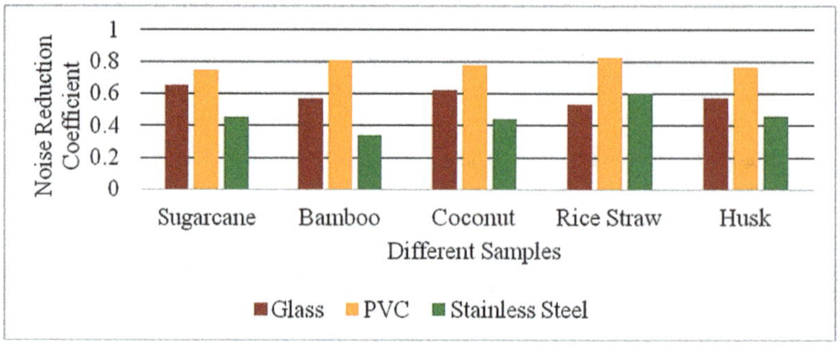

Figure 3. Comparative analysis of different samples in dry condition

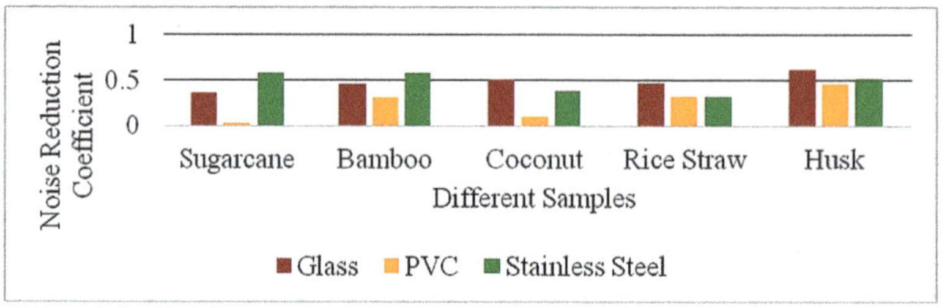

Figure 4. Comparative analysis of different samples in hot water treated condition

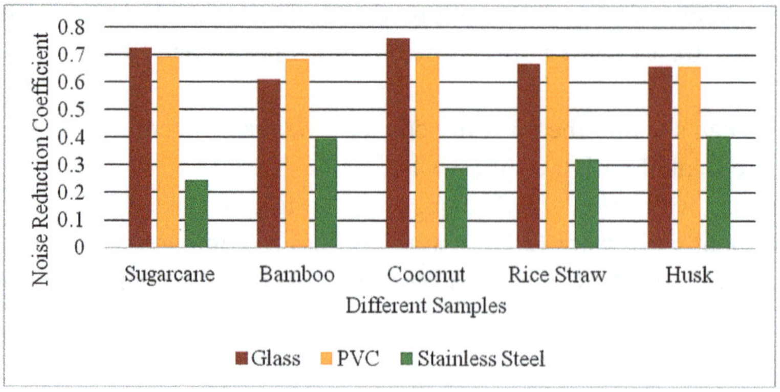

CONCLUSION

Noise control in home appliances has to be economic, simple and easily implemented, since there is global competition among manufacturers. New eco-friendly materials for noise control are being explored and their acoustical properties are good to using a sound absorbing material. The production of these materials generally has a lower environmental impact than conventional ones though a proper analysis. Effective creation of epoxy composites strengthened with normal strands is conceivable through straightforward hand-lay-up procedure. From the impedance tube test technique it is found that as the epoxy strengthened in the test is expanded the physical properties and the acoustic absorption coefficient additionally increased. Suggesting that these composites filaments have great acoustic protection properties. Further, with increment in the thickness of the composites, the acoustic absorption coefficient likewise expanded.

REFERENCES

Ashby, M. F., & Lu, T. (2003). Metal Foams: A Survey. *Science in China. Series B, Chemistry*, *46*, 521–532. doi:10.1360/02yb0203

Ballagh, K. O. (1996). Acoustical properties of wool. *Applied Acoustics*, *48*(2), 101–120. doi:10.1016/0003-682X(95)00042-8

Berardi, U., & Iannace, G. (2015). Acoustic characterization of natural fibers for sound absorption applications. *Building and Environment*, *94*, 840–852. doi:10.1016/j.buildenv.2015.05.029

R. Del Rey, J. Alba, & A. S. V. Sanch (Eds.). (2007). Proposal of an Empirical Model for Absorbent Acoustical Materials Based in Kenaf. *Proceedings of the 19th International Congress on Acoustics*, 3361-3366.

Dong, Y. W., & Yan, L. (2012). Sound absorption performance of natural fibers and their composites. *Science China. Technological Sciences*, *55*(8), 2278–2283. doi:10.100711431-012-4943-1

Fouladi, M. H., Ayub, M., & Nor, M. J. M. (2011). Analysis of coir fiber acoustical characteristics. *Applied Acoustics*, *72*(1), 35–42. doi:10.1016/j.apacoust.2010.09.007

Ganesan, P., & Karthik, T. (2016). Development of acoustic nonwoven materials from kapok and milkweed fibers. *Journal of the Textile Institute*, *107*(4), 477–482. doi:10.1080/00405000.2015.1045251

Hao, G. L., Xu, Q. P., Han, F. S., Li, W. D., & Bai, S. M. (2009). Processing and Damping Behaviour of Porous Copper. *Powder Metallurgy*, *52*(2), 145–150. doi:10.1179/003258908X356885

Ho, M. P., Wanga, H., Lee, J. H., Ho, C. K., Lau, K. T., Leng, J., & Hui, D. (2012). Critical factors on manufacturing processes of natural fiber composites. *Composites. Part B, Engineering*, *43*(8), 3549–3562. doi:10.1016/j.compositesb.2011.10.001

Hudaa, S., & Yang, Y. (2009). A novel approach of manufacturing light-weight composites with polypropylene web and mechanically split cornhusk. *Industrial Crops and Products*, *30*(1), 17–23. doi:10.1016/j.indcrop.2008.12.007

Jayamani, E., Hamdan, S., Rahman, M. R., Heng, S. K., & Bakri, M. K. B. (2014). Processing and Characterization of Epoxy/Luffa Composites: Investigation on Chemical Treatment of Fibers on Mechanical and Acoustical Properties. *BioResources*, *9*(3), 5542–5546. doi:10.15376/biores.9.3.5542-5556

Kabir, M. M., Wang, H., Lau, K. T., & Cardona, F. (2012). Chemical treatments on plant-based natural fiber reinforced polymer composites: An overview. *Composites. Part B, Engineering*, *43*(7), 2883–2892. doi:10.1016/j.compositesb.2012.04.053

Khidir, E. A., Ali, M. O. M., Ali, M. M., Tahir, M. F. M., & Zulkifli, R. (2014). Analysis of Sound Absorption of Date Palm Fibers Based on Flow Resistivity. *Applied Mechanics and Materials*, *471*, 285–290. doi:10.4028/www.scientific.net/AMM.471.285

T. Koizumi (Ed.). (2002). The Development of Sound Absorbing Materials Using Natural Bamboo Fibers and Their Acoustic Properties. *Proceedings of the Inter-Noise*.

Koizumi, T., Tsujiuchi, N., & Adachi, A. (2002). The Development of Sound Absorbing Materials Using Natural Bamboo Fibers. *High Performance Structures and Composites*, *59*, 157–166.

Ku, H., Wang, H., Pattarachaiyakoop, N., & Trada, M. (2011). A review on the tensile properties of natural fiber reinforced polymer composites. *Composites. Part B, Engineering*, *42*(4), 856–873. doi:10.1016/j.compositesb.2011.01.010

Liu, X., Yan, X., Li, L., & Zhang, H. (2015). Sound-Absorption Properties of Kapok Fiber Nonwoven Fabrics at Low Frequency. *Journal of Natural Fibers*, *12*(4), 311–322. doi:10.1080/15440478.2014.919891

Mittal, V., Saini, R., & Sinha, S. (2016). Natural fiber-mediated epoxy composites A review. *Composites. Part B, Engineering*, *99*, 425–435. doi:10.1016/j.compositesb.2016.06.051

Navacerrada, M. A., Diaz, C., & Fernández, P. (2014). Characterization of a Material Based on Short Natural Fique Fibers. *BioResources*, *9*(2), 3480–3496. doi:10.15376/biores.9.2.3480-3496

Prabhakaran, S., Krishnaraj, V., Kumar, M. S., & Zitoune, R. (2014). Sound and Vibration Damping Properties of Flax Fiber Reinforced Composites. *Procedia Engineering*, *97*, 573–581. doi:10.1016/j.proeng.2014.12.285

Putra, A., Abdullah, Y., Efendy, H., Farid, W. M., Ayob, M. R., & Py, M. S. (2013). Utilizing Sugarcane Wasted Fibers As A Sustainable Acoustic Absorber. *Procedia Engineering*, *53*, 632–638. doi:10.1016/j.proeng.2013.02.081

Putra, A., Or, K. H., Selamat, M. Z., Nora, M. J. M., Hassanb, M. H., & Prasetiyo, I. (2018). Sound absorption of extracted pineapple-leaf fibers. *Applied Acoustics*, *136*, 9–15. doi:10.1016/j.apacoust.2018.01.029

Rao, K. M. M., Rao, K. M., & Prasad, A. V. R. (2010). Fabrication and testing of natural fiber composites: Vakka, sisal, bamboo and banana. *Materials & Design*, *31*(1), 508–513. doi:10.1016/j.matdes.2009.06.023

Rwawiire, S., Tomkova, B., Militky, J., Hes, L., & Kale, B. M. (2017). Acoustic and thermal properties of a cellulose nonwoven natural fabric (barkcloth). *Applied Acoustics*, *116*, 177–183. doi:10.1016/j.apacoust.2016.09.027

Sanadi, A. R., Prasad, S. V., & Rohatgi, P. K. (1986). Sunhemp fiber-reinforced polyester. *Journal of Materials Science*, *21*(12), 4299–4304. doi:10.1007/BF01106545

Tang, X., Zhang, X., Zhang, H., Zhuang, X., & Yan, X. (2018). Corn husk for noise reduction: Robust acoustic absorption and reduced thickness. *Applied Acoustics*, *134*, 60–68. doi:10.1016/j.apacoust.2018.01.012

Thilagavathi, G., Pradeep, E., Kannaian, T., & Sasikala, L. (2010). Development of Natural Fiber Nonwovens for Application as Car Interiors for Noise Control. *Journal of Industrial Textiles*, *39*(3), 276–278. doi:10.1177/1528083709347124

White, N. M., & Ansell, M. P. (1983). Straw-reinforced polyester composites. *Journal of Materials Science*, *18*(5), 1549–1556. doi:10.1007/BF01111977

Xianga, H. F., Wanga, D., Liua, H. C., Zhaoa, N., & Xu, J. (2013). Investigation on Sound Absorption Properties of Kapok Fibers. *Chinese Journal of Polymer Science*, *31*(3), 521–529. doi:10.100710118-013-1241-8

Zhai, W., Yu, X., Song, X., Ang, L. Y. L., Cui, F. S., Lee, H. P., & Li, T. (2018). Microstructure-based experimental and numerical investigations on the sound absorption property ofopen-cell metallic foams manufactured by a template replication technique. *Materials & Design*, *137*, 108–116. doi:10.1016/j.matdes.2017.10.016

Zhao, X. D., Yu, Y. J., & Wu, Y. J. (2016). Improving low-frequency sound absorption of microperforated panel absorbers by using mechanical impedance plate combined with helmholtz resonators. *Applied Acoustics*, *114*, 92–98. doi:10.1016/j.apacoust.2016.07.013

Zulkifh, R., Nor, M. J. M., Tahir, M. F. M., Ismail, A. R., & Nuawi, M. Z. (2008). Acoustic Properties of Multi-Layer Coir Fibres Sound Absorption Panel. *Journal of Applied Sciences (Faisalabad)*, *8*(20), 3709–3714. doi:10.3923/jas.2008.3709.3714

KEY TERMS AND DEFINITIONS

Acoustics: Acoustic materials play an important role in noise control solutions.

Biodegradable: Biodegradation is the breakdown of organic matter by microorganisms, such as bacteria and fungi. Biodegradable Plastics refers to a material that maintains its mechanical strength during practical use but break down into low-weight compounds and non-toxic byproducts after their use.

Composite Material: A composite material (also called a composition material or shortened to composite, which is the common name) is a material made from two or more constituent materials with significantly different physical or chemical properties that, when combined, produce a material with characteristics different from the individual components.

Impedance Tube: The method of testing materials using impedance tube is called as normal incident sound absorption or transmission loss testing because of the normally incident sound waves on to the surface of test sample in the tube.

Sound Absorption: Acoustic absorption refers to the process by which a material, structure, or object takes in sound energy when sound waves are encountered, as opposed to reflecting the energy. Part of the absorbed energy is transformed into heat and part is transmitted through the absorbing body.

Water Absorption: The effect of water absorption on the mechanical properties of flax reinforced bio epoxy composites has been studied by the immersion of the samples in water at room temperature. It shows that water absorption increases with an increase in fibre weight fraction due to a higher cellulose content.

Chapter 5
Evaluating the Influence of Energy Consumption on Process Plan Attributes:
Process Plans Indexing Based on Electrical Demand and Energy Consumption

Omar Ahmed Al-Shebeeb
IMSE Department, West Virginia University, USA & Mechanical Engineering Department, University of Diyala, Iraq

Bhaskaran Gopalakrishnan
IMSE Department, West Virginia University, USA

Dayakar Devaru
JSS Science and Technology University, India

ABSTRACT

This chapter presents an idea of including the energy cost in the total cost of manufacturing. The main objective of this work is to index and evaluate the process plan based on the energy consumption of different machining processes while considering machining cost to index the process plans based on energy consumption and to select the best process plan based on machining cost. The second portion is the energy evaluation part, in which the process planner, after selecting machining parameters and considering several alternative process plans, analyzes and compares the energy cost and the total machining cost for the part. MPSEL program was used to develop alternative process plans. VISUALMACH program was used to calculate process plan attributes. Different process plans were compared based on the costs of energy, machining, material, and total cost. The obtained results from the proposed methodology showed a significant contribution toward the effectiveness of energy management function within enterprises to meet targets and goals on energy and emissions reduction.

DOI: 10.4018/978-1-7998-4939-1.ch005

NOMENCLATURE

Table 1.

CAPP	Computer Aided Process Planning
NC	Numerical Control
MPSEL	Machining Parameter Selection System
VISUAL MACH	Visual Basic Machining
BWM	Power Consumption for Base Work Material
DWM	Power Consumption for Design Work Material
CNC	Computer Numerical Control
VSD	Variable Speed Drive
HP	Horse Power
EPLAN	Energy Indexing and Energy Evaluation
DFMT	Design for Manufacturing Techniques
CAPP	Computer Aided Process Planning
MP	Machining Parameters
AM	Additive Manufacturing
RS	Rough Set
HMEA	Hybrid Multi-objective Evolutionary Algorithm
CBR	Cased-Based Reasoning
MOSEL	Machining Parameters Selection
PPAC	Process Plan Attributes Calculator

INTRODUCTION

In metal cutting and manufacturing industry, the electrical energy cost is always treated as an indirect cost and only the total machining cost, which includes tooling cost, machine and labor cost, is attributed to the cost of the part. The cost of electrical energy has increased over time and industry realizes the importance of energy cost invested in manufacturing. A part that can be produced based upon varying machining or metal cutting process plans can vary not only in its machining cost but also in energy cost. Based on the process plan developed, the cost of energy, which includes electrical energy cost as well as the electrical demand cost should be used in the selection of process plan. Process plans are selected based upon the production rate, machining and tooling cost and not based upon the energy consumed in the machining process. Enhanced energy indexing and evaluation of process plans should be made available to the process planner and manufacturing engineers and managers to contribute towards cost reduction and enhance the sustainability initiatives.

In metal cutting and manufacturing, electricity is treated as an indirect cost. The machine tools are driven by electrical motors which consume significant electrical energy. The machine tools, irrespective of their age and level of sophistication, utilize electrical motors for spindle movement, table movement, and alignment of part and machine elements. The spindle motor consumes the most energy as it positions the part or rotates it for precision manufacturing. The actual power and energy consumed will depend upon the load on the motor, often influenced by the material hardness and the intensity of metal removal dependent upon the depth of cut.

The domain of computer aided process planning, whether it is generative or variant, involves the specification of steps in producing the part with details on process parameters. Significant variations are possible in terms of the steps used in process planning due to the variety of process parameters that are available to utilize. The process plans, in their attempt to fulfill the product design attributes, focus on achieving it with the least possible cost of manufacturing the part. Adding the consideration of energy cost into the mix creates an interesting and important sustainability consideration in product manufacturing. Hence, the objective of this research is to analyze the process plans and process parameters used in the machining process and to compare and contrast the process plans in terms of machining cost, tooling cost, and energy cost.

The objectives of the process plan may differ variously due to following reasons:

1. Manufacture of a new product
2. Greater stability in the attainment of desired quality level
3. Increase in product volume
4. Increased utilization of facility resources
5. Cost reduction

The process plan typically includes the following information:

1. Type of process
2. Type of machine and cutting tool used
3. Sequence of processes
4. Cutting fluid used
5. Cutting parameters involved in each process
6. Type of jigs and fixtures
7. Labor cost and machining cost
8. Machining time in each process and tooling time in each process.

LITERATURE REVIEW

In the domain of computer aided process planning, various methodologies have been developed for process planning which focus on cost reduction and optimization (Gupta & Gopalakrishnan, 2010). (Fuh, Chang, & Melkanoff, 1992) developed a process planning methodology based on the characteristics of the work material surface. A real time computer aided process planning (CAPP) system as a support tool for economic product design was developed by (Park & Khoshnevis, 1993). The reduction of the idle time of machines and other resources in process planning was analyzed by (Petuelli & Müller, 1995). (Akturk & Avci, 1996) & (Leonesio, Tosatti, Pellegrinelli, & Valente, 2013) developed a mathematical model for the operation of CNC machine tool which included system characterization, cutting conditions and tool life relationship and related constraints to optimize cutting and idle time by considering tool consumption and non-machining time cost components. (Fuh et al., 1992) & (Mourtzis, Doukas, Lalas,

& Papakostas, 2015) developed a methodology and a prototype system in which the product design and the manufacturing planning were integrated.

(Grzesik & Kwiatkowska, 1997) worked on the beneficial effect of controlling the cutting contact area of tool with respect to energy reduction. Their result showed an improvement in energy reduction with an increase in the feed rate. They conducted experimental tests to measure the cutting force, cutting energy, and the temperature at tool chip interface together with chip identification form. (Gopalakrishnan, Plummer, & Alkadi, 2001) discussed about the opportunities and energy savings in the glass manufacturing facilities. Manufacturing energy conservation evaluation should include a well-structured team effort to create energy awareness in terms of monitoring, collecting and organizing energy cost and consumption data; identifying, analyzing and implementing energy conservation opportunities; and monitoring results (McCoy & Douglass, 2000). (Tsai, Chen, Leu, Chang, & Lin, 2013) A mathematical programming methodology to analyze green manufacturing initiatives has led to improvements in green manufacturing technology investments. (Baumers, Dickens, Tuck, & Hague, 2016) The utilization of additive manufacturing techniques in terms of productivity enhancement and potential attainment of different levels of system productivity has been examined. (Dixon, McGowan, Onysko, & Scheer, 2010) The consideration of US energy conservation policies and their impact on national economy and future prospects is described. (Lauwers et al., 2014) The definition and classification of hybrid manufacturing processes and principles that impact energy sources and tools have been found to have a significant effect on process performance. (Balogun, Edem, Adekunle, & Mativenga, 2016) The evaluation of machine efficiency and the relationship to shear energy in machining processes can be linked to enhanced productivity. The importance of integrating the process planning and scheduling functions has been emphasized for productivity gains (Zhang, Tang, Peng, Tao, & Jia, 2016) The importance of integrating the process planning and scheduling functions has been emphasized for productivity gains.

(O. A. Al-Shebeeb, 2019) studied the synergistic integration of DFMT with CAPP and MP to produce desirable results in terms of production cost, production rate, satisfaction of design and system parameters, and most importantly the functionality of the product. Algorithm was developed, discussed, and tested in this dissertation. This algorithm has sequenced elements to investigate and analyze each DFMT. This analysis identifies several potential process plans, from which the plan with the lowest projected cost and highest production rate is selected and constructed. The results showed that the developed process plans used inside this work showed significant influence on cost reduction and production rates. These findings suggest that further beneficial outcomes and variety might be obtained by applying this algorithm. (J. Jiang et al., 2019) enhanced the production line of AM by reducing production time, materials consumption, and energy consumption. They presented new support generation strategy by optimizing two main factors (print path and print orientation) to enhance some AM outputs (production time, raw materials consumption, and energy consumption). After applying several experiments of this strategy, the results indicated that this strategy can reduce the production time, raw materials consumption, and energy consumption significantly. (Jin et al., 2019) investigated environmental performance factors and makespan influence on energy consumption by machine tool. These factors, such as total power consumption, integrated process planning, and scheduling could enhance. Several programming models have been utilized to perform a proposed algorithm for investigating this study. The outcomes of this work explained that the energy consumption can be influenced by variety of automation levels and workloads of machine tools. (O. Al-Shebeeb & Gopalakrishnan, 2016) studied the effectiveness of Computer Aided Process Planning (CAPP) on manufacturing outputs (production rate and production cost). This study has been performed by using CAPP software MPSEL and VISUAL MACH to create and calculate the outputs of several process plans and appropriate machining parameters, then make comparison between these process plans toward their manufacturing outputs (manufacturing cost and throughput). The acquired results showed that one process plan has a significant influence than other and the reason for that has been explained. (Dai et al., 2015) analyzed the relationship between process planning and scheduling in term of energy-efficient optimization. Mathematical model has been developed to integrate process planning with scheduling by engaging the mathematical model with energy consumption and makespan, the optimal solution between energy consumption and scheduling was adopted and explored by using modified generic algorithm. The final results were collected from a case study and validated with other method, this method can enhance the energy efficiency of manufacturing processes in process planning. (Kanneganti et al., 2017) investigated the energy assessment by creating a methodology and reporting format suitable for ISO 50001. This methodology connected Sections 4.4.3 (Energy Review) to 4.4.6 (Objectives, Targets and Action Plans) in ISO 50001 to reduce energy aspect for energy assessment (time and other recourse and simplify the implementation of ISO 50001. (Krog & Sperling, 2019) tested the strategy of energy planning in Denmark by using Danish approach and investigating the history of strategic energy planning. They developed framework to evaluate strategic energy planning, and it has been concluded that strategic energy planning could be integrated into Danish approach. (Gong et al., 2019) presented mathematical model to address the gap between scheduling and process planning. Hybrid Multi-objective Evolutionary Algorithm (HMEA) was developed to evaluate this work. Several multidimensional encoding operators were created to assist in this research for obtaining the best quality of population and enhancing the pace of the algorithm. With a specific legalising way and intensive computational experiments were applied on HMEA, improvements were attained in this work. (Z. Jiang et al., 2019) presented a way to improve the quality of manufacturing products by creating a hybrid method from combining Rough Set (RS) and Cased-Based Reasoning (CBR) methods. The results indicated this algorithm improved the quality of manufactured products positively and applied in prototype system. The adopted hybrid method showed its effectiveness to increase the speed of using process planning in manufacturing.

MPSEL and VISUAL MACH Computer Programs

MPSEL is an acronym for Machining Parameter Selection System (Gopalakrishnan, 1989; Gopalakrishnan & Pandiarajan, 1991). It is an expert system developed to select different machining parameters like machine tools, cutting tools, cutting fluids and to indicate the different cutting conditions. The different cutting conditions include, thermal shock, high tool chip friction etc. The expert system considers the following machine shop environment in which engine lathe, NC lathe, turret lathe, single spindle automat, cylindrical grinding machine, surface grinding machine, horizontal milling machine, vertical milling machine, NC mill, Turret drilling machine, vertical drilling machine and radial drilling machine are available. The processes that are considered in this system are, turning, facing, milling, grinding, drilling, reaming, boring, tapping and threading. Selection of machines and cutting tools is a time consuming process as it involves many parameters and is subject to operational constraints such as machine capacity, tool work combination, and surface finish requirement. MPSEL is designed to select the appropriate machines and cutting tools for a machining process based on the design of the part. This is done by consulting with the expert system, which has expert knowledge, built in about the processes. This minimizes the time and effort required for the selection of cutting tools, cutting fluids and machine tools.

MACH is a spreadsheet application that can be used for calculating the machining cost and tool cost for a particular machining operation (Gopalakrishnan, 1989; Gopalakrishnan & Pandiarajan, 1991). The total cost of machining is calculated for each cutting pass and they are integrated to arrive at the total cost to produce a single part using the selected machining process parameters. Machine setup time cost, work loading, unloading, and tool changing cost are also included in the machining cost. The spreadsheet includes calculation of cost for turning, face or end milling, drilling, surface grinding, reaming and tapping. The cost of each machining process per pass is obtained by using the formula discussed in the following chapter. It also gives detailed information about the machining time components, tool cost components and the machining cost components. These details are more useful in the analysis of the selected machining parameters in terms of machining cost. Similar to MPSEL, the MACH spread sheet program was also run for each and every process for calculating the machining time, machining cost and tooling cost. In each process, it was run for every cutting pass. The cutting parameters used were selected from industry recommended values for depth of cut, feed rate and cutting speed. The constraints of machines such as machine capacity, tooling capacity and tool life were also considered in coming up with the cutting parameters. New software interface, Process Plan Attributes Calculator (PPAC) created by the researcher is constructed through coded spreadsheets, using Visual Basic for Applications (VBA®). This interface was used because it is easier and faster to use than VISUALMACH. The user can change any element from inputs directly in one step (O. A. Al-Shebeeb, 2019).

ENERGY INDEXING OF PROCESS PLANS

Energy indexing of process plans would identify the electrical energy and demand cost associated with the spindle motors utilized by various processes at the individual steps in the process planning framework. The process plans that are evaluated are PP1, PP2, PP3, PP4, PP5, and PP6 as shown in tables 2, 3, 4, 5, 6, and 7.

Table 2. Represents PP1

Process Plan 1 - PP1	
Work Material used:	Austenitic Stainless Steel (*Mcmaster.Com*, n.d.)
Density:	= 0.280 lbs/in3
Hardness:	= 42 Rockwell
Heat Treatment:	= Annealed
Melting Point:	= 2500ºF

Table 3. Shows PP2

Process Plan 2 – PP2	
Work Material used:	Aluminum Alloy (*Mcmaster.Com*, n.d.)
Density:	=0.0978 lbs/in^3
Hardness:	=150 Brinell
Heat Treatment:	=Low=elastic as cast
Melting Point:	=1000°F

Table 4. Represents PP3

Process Plan 3 – PP3	
Work Material used:	Aluminum Alloy

Table 5. Shows PP4

Process Plan 4 – PP4	
Work Material used:	Austenitic Stainless Steel

Table 6. Represents PP5

Process Plan 5 – PP5	
Work Material used:	Austenitic Stainless Steel

Table 7. Represents PP6

Process Plan 6 – PP6	
Work Material used:	Aluminum Alloy

The product and process parameters that affect energy usage are considered in indexing of the process plans and have been outlined herein.

1. Stock removal volume
2. Work material selection
3. Type of machine used
4. Type of machine category
5. Machine attributes
6. Machine load capacity

As metal cutting machines age, they tend to consume more energy due to internal power losses and due to the decrease in machine efficiency. Energy consumed in machining depends upon the type of machine, machining process, cutting tools, work material and cutting conditions. Energy consumed also depends upon the work size and the capacity and functional specification of the machine used in the machining process. Recently, metal cutting machine tools have been equipped with energy saving devices on the motors such as variable speed drives and transmission energy loss prevention.

RAW MATERIAL STOCK VOLUME REMOVAL

Based on (Creese & Adithan, 1992), the statistics shows that for a typical machining operation, 62% of the original volume of the raw material stock is removed. This leads to significant energy required to remove the excess material in the raw material stock. A reduction in the amount of raw material stock volume, commensurate with the requirements of the machining process is an important contribution to energy reduction. More the volume of the raw material stock, more will be the number of passes required to obtain the final part geometry. At the end of each pass, the spindle motor is idle to enable part set up and possible machining parameters set up. Idling motors consume significant energy and this can be avoided by lessening the number of passes through the reduction in raw material stock volume. Moreover, any increase in stock volume also leads to decrease in tool life, which leads to tool nose radius failure, leading to more power consumption.

The machine idle running time and handling time is estimated from (Creese & Adithan, 1992). The machining time allowance is 9% of the total machining time and 15% of the total machining time is considered for handling time. The machine idle running power requirement is assumed to be 30% of the total machining power requirement as a conservative value (Koenigsberger, 2013).

The base stock volume of the material can be calculated by considering the machining allowance that is left for the final dimension of the part to be manufactured. Generally, 62% of the initial volume material is removed (Creese & Adithan, 1992). Since the stock material available in the market, ranges in a wide variety, as a conservative value, the base stock volume can be considered to be 1.3 times the final volume of the work material before machining. Energy ratio based on the stock volume removal is calculated as follows:

$$Stock\ volume\ ratio = \frac{\left(Actual\ stock\ volume - Base\ stock\ volume\right)}{Actual\ stock\ volume}$$

$$Stock\ energy\ index = \left[\frac{\left(Total\ power\ consumed\ in\ Actual\ stock - Total\ power\ consumed\ in\ Base\ stock\right)}{Total\ power\ consumption\ in\ actual\ stock}\right]$$

The following portion will explain the principle of raw material stock material volume based energy indexing score for the process plan # PP2, which is presented in the appendix 7 for the turning process for a 50 HP NC lathe, for the part shown in fig 4.1 and 4.2. Insert figures here.

The specific cutting power requirement for aluminum alloy is 0.3 HP/(in^3 min) (*Mcmaster.Com*, n.d.)

Volume of stock material = 10" x 3.14 x 2.5" x 2.5"
= 196.35 in^3
Final part volume = 8" x 3.14 x 2.25" x 2.25"
= 127.23 in^3
Stock removal volume= Volume of stock material – final part volume
= 196.35 – 127.23

= 69.12 in^3
Material removal rate (MRR) = 12 x feed x depth of cut x cutting speed
= 12 x 0.01in x 1000 sfpm x 0.25 in
= 30.00 in^3/min
Actual cutting time = Stock removal volume/MRR
= 69.12/30
= 2.304 min
Actual power consumption = (2.304 x 0.3 x 30 x 0.746 x 1/60)
+ (0.3 x 50 x 0.24 x 2.304 x 0.746 x 1/60).......[based on equation 18]
= 0.36 kWh
Base volume = 1.3 x (8" x 3.14 x 2.25" x 2.25")
= 165.4 in^3
Stock energy index volume = Base volume – Final volume
= 165.4 – 127.23
= 38.17 in^3
Base volume cutting time =Stock energy index volume/MRR
= 38.17/ 30
= 1.27 min
Base power consumption = (1.27 x 0.3 x 30 x 0.746 x 1/60)
+ (0.3 x 50 x 0.24 x 1.27 x 0.746 x 1/60). [based on equation 18]
= 0.2 kWh
Stock energy index = [(0.36 – 0.2)/0.36]*100
= 44%
= 4 (Rounded to 4)

It is evident that higher the value of the stock energy index, more is the level of energy consumed.

WORK MATERIAL

The work material influences the energy consumption, specifically in terms of its properties and characteristics. Machinability of materials is an aspect often discussed by practitioners in manufacturing. Based on the literature search, it can be surmised that (Schey, 2000) the machinability index for the materials can be based upon cutting force, temperature developed in machining, chip formation or power consumed in machining. Hence, the energy index can be developed based on the type of work selected in order to fulfill the functionality of the product design.

Specific cutting energy of a material is defined as the energy consumed in removing a unit volume of the same material (Schey, 2000). Cutting energy P_c is the force multiplied by the distance l over which the force acts. Since the volume of work material removed is defined as the product of width, length and depth of the work material removed, the specific cutting energy can be written as:

$$P_p = (P_c \text{ x } l)/ (h \text{ x } W_w \text{ x } l)$$

Where,

P_p = Unit power required in specific process (It varies for turning, milling, drilling etc.)
P_c = Cutting energy
h = Length of work material
W_w = Width of work material
l = Length of work material

The power required at the machine spindle can be defined as P_p x MRR. With MRR being constant, the power required in the machining process will vary proportional to work material specific cutting energy. For purposes of energy indexing, the following methodology has been developed. The base work material has been defined as the material that has been selected as an alternative to the design work material in an effort to reduce energy consumption.

Power consumption for base work material (BWM) = M/c time x MRR x P_{pbwm}

Power consumption for design work material (DWM) = M/c time x MRR x P_{pdwm}

Work material ratio = [Power consumed in (DWM – BWM)]/[Power consumed in DWM].

For energy indexing example, based on process plan PP2, the design work material selected is an aluminum alloy, which has a specific cutting energy of P_t of 0.3. It has been determined that the magnesium alloy can be used instead as the base work material.

Power consumed in actual work material = 2.304 x 30 x 0.3 x 0.746 x 1/60
= 0.258 kWh
Power consumed in base material = 2.304 x 30 x 0.2 x 0.746 x 1/60
= 0.172 kWh
Work material index = [(0.258 – 0.172)/0.258]x 100 = 33%
=3 (rounded)

It should be noted that the energy index of a process plan based on work material selection is likely to be higher with work materials with lower specific cutting energy. However design effort needs to be focused on determining alternative materials that would satisfy the design functionality.

MACHINES INFORMATION

Type of Machine Used in the Process

The machine selected and used in the process affects the energy consumed in the machining process. In cutting parameters selection, depth of cut is first determined, followed by feed rate and then cutting speed in the increasing order of importance for the tool life. In the case of conventional machine tools with non-Computer Numerical Controls (CNC), after each pass the operation has to be stopped for enabling change in cutting speed, which is altered due to changes in product geometry. Moreover, on such machines, the exact cutting speed that is desired may not be possible to set as the cutting speed potentially could be set only at predetermined values within ranges. This causes excessive time wasted between passes as well as shift in cost to higher values due to the cutting speed being set at non optimal values. In the energy indexing context, such conventional machine tools are rated lower than the CNC machine tools that provide part programming and desirable variability in setting of the cutting speeds. The CNC machine tools would be rated higher on the energy indexing scale while the semi-automatic machine tools would be rated within the ranges of the two types of machine tools described herein. More the time for machining, more time is spent on machine idling or cutting metal, and the CNC machines have a high production rate and hence avoid this problem of excessive energy consumption.

They ratings of the machine tools for purposes of energy indexing are:

1. Conventional machines with a rank of 5,
2. Semi-automatic or special purpose automated machines with a rank of 3 and
3. CNC machines with a rank of 1.

Since NC lathe is used in the turning process for the process plan PP2, the energy index based on machine used will be 1 for this process plan.

Type of Machine Category

The number of spindles on the machine as well as the potential for multiple tools to operate on the part at the same time significantly reduces the product machining time, specifically the idle time required to alter machining parameters. Hence, multi-spindle and turret tool machines are rated higher with respect to energy indexing as opposed to single spindle machines with a single tool machining system. The multi-spindle machines are intended for performing many machining processes with the single set up of the work piece, thus reducing the total machining time. The multi-spindle machines also have the capacity of machining more than one work piece at a time. Some of the spindles are engaged in machining while some are used in loading and unloading of work pieces. For the energy-indexing example, based on process plan PP2, in the type of machine category, the machine used is single spindle NC lathe. Although the machine has effective tooling capabilities, since it does not have multiple spindles, for energy indexing purposes, the machine is rated at 3.

Machine Attributes

According to (Koenigsberger, 2013), the motor of a machine is continuously operating, but the loading and unloading of the motor varies highly within a machining cycle. Most of the machines have a spindle motor that is driven by AC current. Due to changing load on the spindle motor of the machines, a variable speed drive (VSD) on the motor would reduce the energy consumption in a significant manner. For purposes of energy indexing, a machine that has a VSD on the spindle motor or a DC current driven is rated as 1and a machine not having these characteristics is rated as 3. For the energy-indexing example, based on process plan PP2, the machine selected is assumed as one with a VSD on the spindle motor, hence given a rating of 1.

Machine Load Capacity

Figure 1.

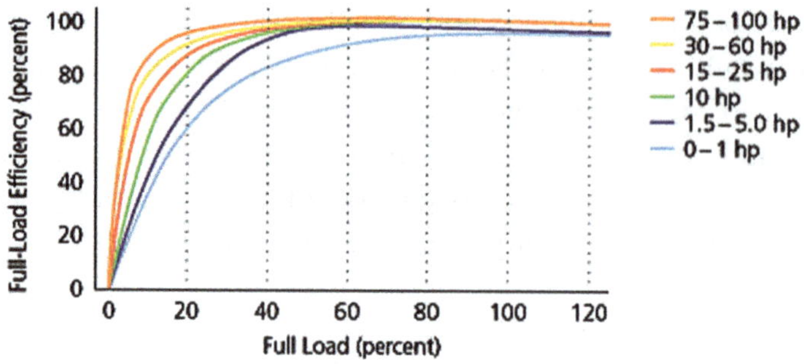

The machine load, determined by the load on the spindle motor, indicates the intensity of the metal removal rate. A large value for the metal removal rate will indicate a relatively large value for the motor load. In most cases, the spindle motor is severely under loaded. As shown on figure 1, at very low motor loads, the efficiency of the motor decreases considerably. Hence, higher input power to the motor is required to deliver the same output power. For purposes of energy indexing, the rating of the machine or motor load will be:

1. 1 - 10% load – Rank 10 (Which occurs seldom)
2. 11 - 20% load – Rank 9
3. 21 - 30% load – Rank 8 …and so on,
4. 91 – 100% load – Rank 1.

For the energy indexing example, which is based on process plan PP2, the following calculation explains the machine load capacity index:

MRR = 30 in^3

HP required at the spindle = 0.3 x 30
= 9

HP available at the machine = 50 HP (assuming 85% as machine efficiency)

Machine load capacity index = [9 / (50 x 0.85)] x 100
= 22%.
= 8 {since, only 22% loaded, the energy index is 8}

The overall indexing for the six process plans is summarized in Table 8. It is interesting to note that the viability of the process plans based on energy can be evaluated based on individual factors or in an

Table 8. Indexing of process plans based on process parameters and machine

Parameter	Index for Process Plan					
	PP1	PP2	PP3	PP4	PP5	PP6
Stock	4	4	4	4	4	4
Work Material	9	3	3	9	9	3
Machine Type	1	1	5	5	5	5
Machine category	3	3	3	3	3	3
Machine Attributes	1	1	3	3	3	3
Machine Load Capacity	4	8	9	7	7	9
Overall average	5	5	7	8	8	7

overall sense. It is also possible to weight some factors more than others, if it is commensurate with the objectives of the process planner and the organization.

ENERGY EVALUATION

A software program, in a spreadsheet format, has been developed (EPLAN) to assist the process planner in evaluating the energy effectiveness of the process plans. EPLAN can also be used to calculate the energy cost, machining cost and the total cost for each operation in the process plan. The total cost will be comprised of the energy cost, machining cost and tooling cost. The total machining cost includes the machine cost and labor overhead cost and tooling cost. The total energy cost includes the energy usage cost and the demand cost. The following tables and graphs will illustrate the evaluation of the process plans, in terms of the cost parameters.

- The machine and labor overhead cost for conventional machines is $30/hour
- The machine and labor overhead cost for non-conventional machines is $40/hr
- The electrical demand cost is assumed to be $10/kW
- The electrical energy cost is assumed to be $0.05/kWh
- The lot size considered is 50 parts per lot.

ANALYSIS OF PROCESS PLANS AND PARAMETERS

The following tables explain about the variation in the total cost based on the energy cost, machining cost with respect to the six different process plans generated. It clearly shows that a process plan might be desirable with respect to machining cost, but it may have high energy cost, as in the case of PP1 and PP6. The process plan PP2 has not only the lower manufacturing cost but also lower energy cost when compared to process plan PP1. Table 9 illustrates the variation in energy cost and total cost among process plans for various types of work materials.

Table 9. Energy cost analysis for different process plans

Work Material Type	Process Plan	Material Cost	Machining Cost	Energy Cost	Total Cost	EC as % of Total Cost
Austenitic Stainless Steel	PP1	$199.57	$11.13	$7.94	$218.64	3.63%
Aluminum Alloy	PP2	$34.75	$8.60	$5.43	$48.78	11.13%
Aluminum Alloy	PP3	$34.75	$48.28	$1.78	$84.81	2.10%
Austenitic Stainless Steel	PP4	$199.57	$55.46	$3.63	$258.66	1.40%
Austenitic Stainless Steel	PP5	$199.57	$54.39	$3.62	$257.58	1.41%
Aluminum Alloy	PP6	$34.75	$41.70	$1.78	$78.23	2.28%

The Table 10 shows the variation in the energy cost, in terms of electrical demand cost, subject to variations in the cost of electrical demand. It is evident that the value for the electrical demand rate can influence the effectiveness of a process plan.

Table 10. Energy cost analysis for different demand rates

Demand Rate ($)	Total Energy Cost ($)					
	PP1	PP2	PP3	PP4	PP5	PP6
5	$4.11	$2.76	$0.93	$1.72	$1.72	$0.93
6	$4.88	$3.30	$1.10	$2.01	$2.01	$1.10
7	$5.64	$3.83	$1.27	$2.30	$2.29	$1.27
8	$6.41	$4.36	$1.44	$2.59	$2.58	$1.44
9	$7.18	$4.90	$1.61	$2.87	$2.86	$1.61
10	$7.94	$5.43	$1.78	$3.16	$3.15	$1.78

For PP4 in Table 11, the machine HP 10, 15, 15 indicates that the engine lathe (10 HP), turret drilling machine (15 HP) and the vertical milling machine (15 HP) are associated with the machining related to

Table 11. Energy cost analysis for variation in Energy cost based on machine HP

Process Plan #	M/c's HP	Machining Cost (MCC) $	Energy Cost $	Total M/c Cost (TMC) $	EC as % of MCC
PP4	10,15,15	$55.46	$3.16	$58.62	5.39%
PP4	15,20,20	$55.46	$3.56	$59.02	6.03%
PP4	20,25,25	$55.46	$3.96	$59.42	6.66%
PP4	25,30,30	$55.46	$4.36	$59.82	7.29%
PP4	35,40,40	$55.46	$5.16	$60.62	8.51%
PP4	45,50,50	$55.46	$5.97	$61.43	9.72%
PP4	55,75,75	$55.46	$7.97	$63.43	12.57%

the process plan. The process plan PP4 was evaluated with varying power of the machines associated with the manufacturing, in terms of energy and manufacturing cost.

METHODOLOGY FOR EVALUATING PROCESS PLANS BASED ON ENERGY AND MANUFACTURING COST

The methodology is depicted on figure 2. The process planner or any other user in the organization will be able to use EPLAN to evaluate process plan based on cost and will also be able to alter variables to obtain updated cost information that would allow for enhanced evaluation.

Figure 2.

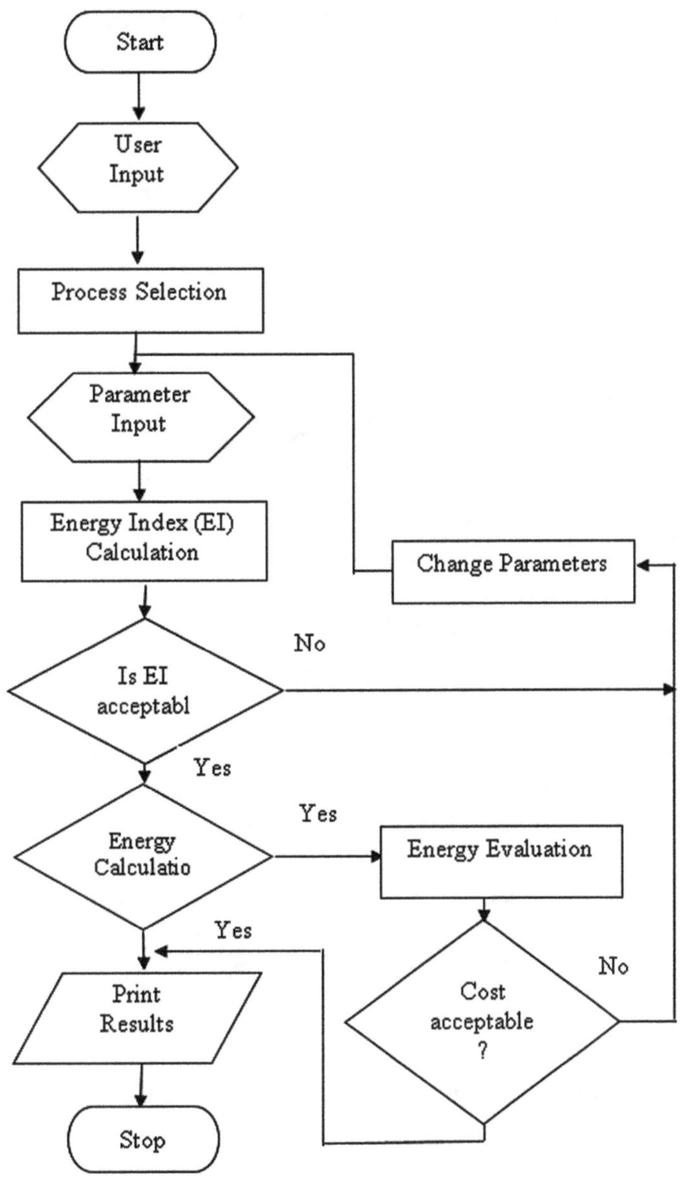

PROGRAMS FOR ENERGY INDEXING AND ENERGY EVALUATION

EPLAN was developed using VisualBasic®. The software contains databases of numerous materials in the form of alloys as well as catering to commonly used machining processes. The material characteristics in terms of hardness and specific cutting power are incorporated. A typical data input screen of EPLAN has been shown in figure 3. The EPLAN database also contains commonly used machine tools, both conventional and sophisticated numerically controlled machines, along with their attributes.

Figure 3.

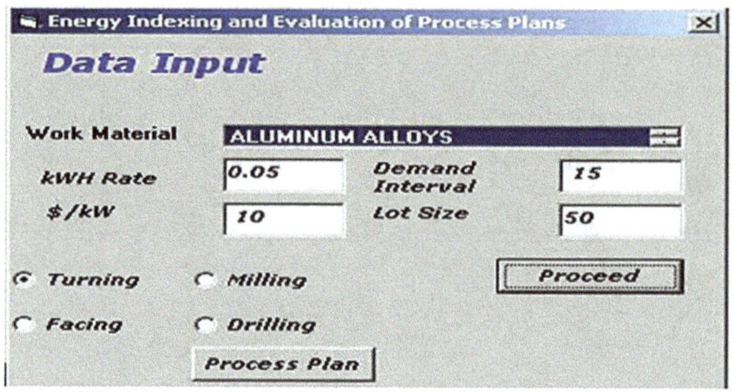

Figure 4.

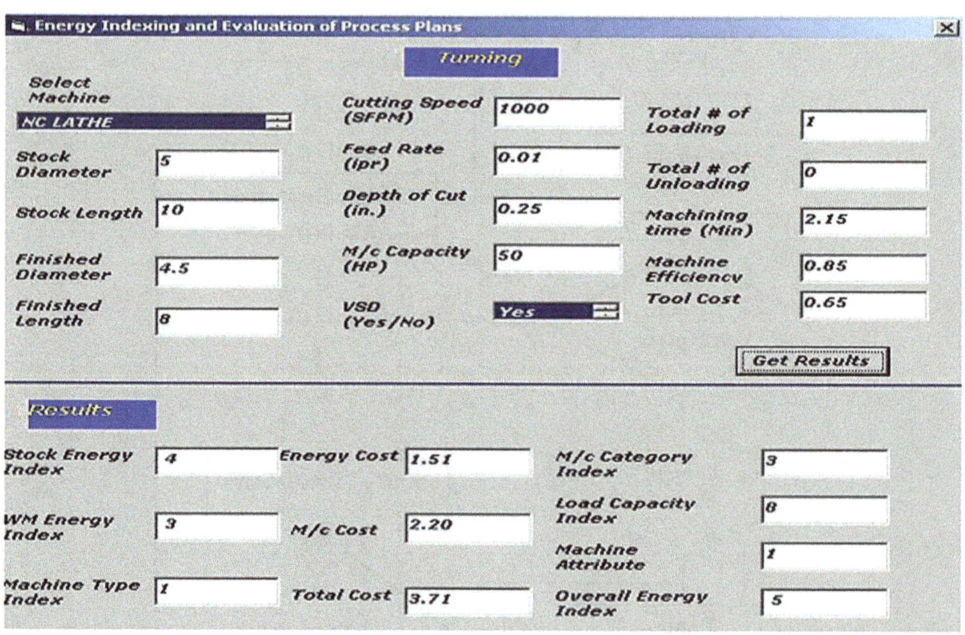

The figure 4 shows the input and output from analyzing the turning operation on a process plan.

DISCUSSION

The evaluation of process planning in the manufacture of discrete metal parts should be done in terms of energy efficiency as well as cost and production rate. This effort will play a large role in ensuring that sustainability aspects are nurtured and developed in this domain. The consideration of energy efficiency in process planning will enable additional cost savings and lead to a more desirable option in the selection of a process plan.

CONCLUSION

Energy indexing of process plans is essential for enabling enterprises to contribute to sustainability and reduce their carbon footprint. Reduced energy cost in manufacturing products increases profits and leads to enhanced global competitiveness and job growth. The energy indexing methodology will work well within the context of concurrent or simultaneous engineering by providing alternatives that can be an ingredient within the dialogue between the design, manufacturing, and other downstream functions. The energy indexing aspects can contribute towards the effectiveness of energy management function within enterprises and contribute towards meeting targets and goals on energy and emissions reduction.

ACKNOWLEDGMENT

The work was partly supported by the Industrial Assessment Center at West Virginia University, funded by the US Department of Energy. Special thanks for Mr. S. Rangaswamy for his participation in the early journal paper of this work.

REFERENCES

Akturk, M. S., & Avci, S. (1996). An integrated process planning approach for CNC machine tools. *International Journal of Advanced Manufacturing Technology*, *12*(3), 221–229. doi:10.1007/BF01351201

Al-Shebeeb, O., & Gopalakrishnan, B. (2016). *Computer Aided Process Planning Approach for Cost Reduction and Increase in Throughput*. Academic Press.

Al-Shebeeb, O. A. (2019). *Analysis of the Integration of DFM Techniques and Effective Machining Parameter Selection in Metal Parts Manufacturing*. Academic Press.

Balogun, V. A., Edem, I. F., Adekunle, A. A., & Mativenga, P. T. (2016). Specific energy based evaluation of machining efficiency. *Journal of Cleaner Production*.

Baumers, M., Dickens, P., Tuck, C., & Hague, R. (2016). The cost of additive manufacturing: Machine productivity, economies of scale and technology-push. *Technological Forecasting and Social Change*, *102*, 193–201.

Creese, R., & Adithan, M. (1992). *Estimating and costing for the metal manufacturing industries*. CRC Press.

Dai, M., Tang, D., Xu, Y., & Li, W. (2015). Energy-aware integrated process planning and scheduling for job shops. *Proceedings of the Institution of Mechanical Engineers, Part B: Journal of Engineering Manufacture, 229*(1_suppl), 13–26. 10.1177/0954405414553069

Dixon, R. K., McGowan, E., Onysko, G., & Scheer, R. M. (2010). US energy conservation and efficiency policies: Challenges and opportunities. *Energy Policy, 38*(11), 6398–6408. doi:10.1016/j.enpol.2010.01.038

Fuh, J., Chang, C., & Melkanoff, M. (1992). A logic-based integrated manufacturing planning system. *Computers in Engineering, 1*, 391–391.

Gong, G., Deng, Q., Chiong, R., Gong, X., Huang, H., & Han, W. (2019). Remanufacturing-oriented process planning and scheduling: Mathematical modelling and evolutionary optimisation. *International Journal of Production Research*, 1–19. doi:10.1080/00207543.2019.1634848

Gopalakrishnan, B. (1989). Computer integrated machining parameter selection in a job shop using expert systems. *Journal of Mechanical Working Technology, 20*, 163–170. doi:10.1016/0378-3804(89)90027-2

Gopalakrishnan, B., & Pandiarajan, V. (1991). Materials and manufacturing processes selection system for product designs in concurrent engineering. *Journal of Materials Processing Technology, 28*(1), 93–103. doi:10.1016/0924-0136(91)90209-W

Gopalakrishnan, B., Plummer, R., & Alkadi, N. (2001). Analysis of energy conservation opportunities in glass manufacturing facilities. *Energy Engineering, 98*(6), 27–49. doi:10.1080/01998590109509330

Grzesik, W., & Kwiatkowska, E. (1997). An energy approach to chip-breaking when machining with grooved tool inserts. *International Journal of Machine Tools & Manufacture, 37*(5), 569–577. doi:10.1016/S0890-6955(96)00076-4

Gupta, D., & Gopalakrishnan, B. (2010). Energy sensitive machining parameter optimisation. *International Journal of Industrial and Systems Engineering, 5*(4), 405–423. doi:10.1504/IJISE.2010.032964

Jiang, J., Xu, X., & Stringer, J. (2019). Optimization of process planning for reducing material waste in extrusion based additive manufacturing. *Robotics and Computer-integrated Manufacturing, 59*, 317–325. doi:10.1016/j.rcim.2019.05.007

Jiang, Z., Jiang, Y., Wang, Y., Zhang, H., Cao, H., & Tian, G. (2019). A hybrid approach of rough set and case-based reasoning to remanufacturing process planning. *Journal of Intelligent Manufacturing, 30*(1), 19–32. doi:10.100710845-016-1231-0

Jin, L., Zhang, C., & Fei, X. (2019). Realizing energy savings in integrated process planning and scheduling. *Processes (Basel, Switzerland), 7*(3), 120. doi:10.3390/pr7030120

Kanneganti, H., Gopalakrishnan, B., Crowe, E., Al-Shebeeb, O., Yelamanchi, T., Nimbarte, A., ... Abolhassani, A. (2017). Specification of energy assessment methodologies to satisfy ISO 50001 energy management standard. *Sustainable Energy Technologies and Assessments, 23*, 121–135. doi:10.1016/j.seta.2017.09.003

Koenigsberger, F. (2013). *Design principles of metal-cutting machine tools*. Elsevier.

Krog, L., & Sperling, K. (2019). A comprehensive framework for strategic energy planning based on Danish and international insights. *Energy Strategy Reviews*, *24*, 83–93. doi:10.1016/j.esr.2019.02.005

Lauwers, B., Klocke, F., Klink, A., Tekkaya, A. E., Neugebauer, R., & Mcintosh, D. (2014). Hybrid processes in manufacturing. *CIRP Annals-Manufacturing Technology*, *63*(2), 561–583. doi:10.1016/j.cirp.2014.05.003

Leonesio, M., Tosatti, L. M., Pellegrinelli, S., & Valente, A. (2013). An integrated approach to support the joint design of machine tools and process planning. *CIRP Journal of Manufacturing Science and Technology*, *6*(3), 181–186. doi:10.1016/j.cirpj.2013.03.002

McCoy, G., & Douglass, J. G. (2000). *Energy management for motor driven systems*. Department of Energy's Motor Challenge Program.

Mcmaster.com. (n.d.). www.mcmaster.com

Mourtzis, D., Doukas, M., Lalas, C., & Papakostas, N. (2015). Cloud-based integrated shop-floor planning and control of manufacturing operations for mass customisation. *Procedia CIRP*, *33*, 9–16.

Park, J. Y., & Khoshnevis, B. (1993). A real-time computer-aided process planning system as a support tool for economic product design. *Journal of Manufacturing Systems*, *12*(2), 181–193. doi:10.1016/0278-6125(93)90017-N

Petuelli, G., & Müller, U. (1995). Integration of metal cutting tool resharpening into manufacturing process planning. *Assembly and Task Planning, 1995. Proceedings., IEEE International Symposium On*, 305-310. 10.1109/ISATP.1995.518787

Schey, J. A. (2000). *Introduction to manufacturing processes*. New York: McGraw-Hill.

Tsai, W., Chen, H., Leu, J., Chang, Y., & Lin, T. W. (2013). A product-mix decision model using green manufacturing technologies under activity-based costing. *Journal of Cleaner Production*, *57*, 178–187. doi:10.1016/j.jclepro.2013.04.011

Zhang, Z., Tang, R., Peng, T., Tao, L., & Jia, S. (2016). A method for minimizing the energy consumption of machining system: Integration of process planning and scheduling. *Journal of Cleaner Production*.

Chapter 6
Effect of Boundary Conditions and Taper Patterns on Geometrically Nonlinear Frequency Response of Axially Graded Beams on Elastic Foundation

Hareram Lohar
Jadavpur University, India

Anirban Mitra
Jadavpur University, India

Sarmila Sahoo
https://orcid.org/0000-0001-7000-3758
Heritage Institute of Technology, India

ABSTRACT

Forced vibration analysis is performed on a tapered axially functionally graded beam resting on elastic foundation under externally applied harmonic excitations to present the effect of boundary conditions and taper patterns on the frequency response. The elastic foundation is modelled in the present analysis as Winkler foundation. A displacement based semi-analytical method is adopted for mathematical formulation and the derivation of governing equations is carried out following Hamilton's principle. Von Karman nonlinear strain-displacement relation employed to incorporate geometric nonlinearity. Broyden method is adopted to solve the nonlinear set of equations. Frequency response curves are plotted in non-dimensional frequency-amplitude plane to represent nonlinear forced vibration characteristic of the system. New benchmark results are also provided for different combination of system parameters (i.e., excitation amplitudes, foundation stiffness values, material models, taper patterns, and flexural boundary conditions). Operational deflection shapes (ODS) are also presented.

DOI: 10.4018/978-1-7998-4939-1.ch006

INTRODUCTION

In the recent years it has emerged that pure homogeneous materials, due to certain insufficiency of their properties, are limited in terms of applications in the more technologically advanced domains, such as aerospace industries. New class of advanced materials, namely, composites and functionally graded materials (FGM) have come to the forefront in case of these applications. FGMs are characterized by continuous variation in properties along the spatial directions. Two or more constituent materials are mixed functionally and continuously to obtain FGMs (Koizumi, 1997). These materials provide unique thermo-mechanical properties which are different from the parent materials. It has excellent thermal resistance, low density and high toughness. FGM is also free from interface problems which are severe in case of laminated composite (Suresh and Mortensen, 1998; Nguyen et al., 2013). Due to excellent properties and increasing demand in various areas, a vast body of researchers is devoted to explore their behavior in different ways.

FG beams can be graded in thickness direction, axial direction and simultaneously along the two directions as well. The thickness directional gradations are most common and have been explored quite comprehensively over the last two decades. Various methodologies like analytical method, semi-analytical method and numerical methods are employed to study the static, dynamic and buckling behavior (Kodali et al., 2008; Yaghoobi and Fereidoon, 2010; Hemmatnezhad et al., 2013; Su et al., 2013; Pradhan and Chakraverty, 2014; Ebrahimi and Salari, 2015; Chen et al., 2015; Jin and Wang, 2015; Jia et al, 2015; Ebrahimi and Zia, 2015) of beams which are graded in the thickness direction. On the other hand, recent years has seen quite a few research works on axially graded beams.

Shahba et al. (2011) extended the free vibration and stability analysis to axially graded Timoshenko beams with tapered profile. Huang and Luo (2011) put forward a simple new technique for calculation of critical buckling load of axially arbitrarily inhomogeneous beams with various end conditions and continuous elastic restraint. Shahba and Rajasekaran (2012) investigated non-uniform Euler–Bernoulli beams with axial inhomogeneity by solving the equations of motion, governing free vibration and stability behaviour of the system. Li et al. (2013) derived closed form characteristic equations for axially functionally graded (exponential gradation) beams with various end conditions, such as clamped, pinned, guided, free etc., in order to analyse the influence of gradient on frequency spectrum. Huang et al. (2013) proposed a new approach for exact calculation of natural frequencies of vibration of an AFG non-uniform Timoshenko beam having both ends clamped. Euler Bernoulli theory alongside modiðed couple stress theory was utilised by Akgöz and Civalek (2013) to present free vibration behaviour of non-homogenous (along the axis) and non-uniform fixed-free microbeams. A finite element analysis was performed by Kien (2013) to capture the large amplitude free vibration characteristics of axially functionally graded non-uniform cross-section Timoshenko beams. Sarkar and Ganguli (2014) developed closed form polynomial solutions for a certain class of axially functionally graded uniform Timoshenko beams with fixed–fixed boundary condition, whose governing equations are expressed as coupled differential equations with varying coefficients. Kumar et al. (2015) employed appropriate energy principles to study free vibration behaviour under loaded and no-load condition of axially graded slender beams with different taper profiles. In this paper authors considered two separate boundary conditions, clamped-clamped and clamped-free. Zeighampour and Beni (2015) used strain gradient theory to investigate the vibration of axially functionally graded material (AFGM) nanobeam. Ghayesh and Farokhi (2018) numerically analysed nonlinear bending and free vibration behaviour of coupled AFG tapered beams. Rajasekaran and Khaniki (2018) performed a FE based static and dynamic analysis of AFG non-uniform small-scale beams through nonlocal strain gradient theory. Sınır et al. (2018) conducted nonlinear free and forced vibration analyses of AFG Euler-Bernoulli tapered beams. Huang et al. (2018) studied the free vibration analysis of a AFG spinning thick beam to investigate the effect of AFG material on whirling frequencies and to determine the critical speed.

The theory of beams on elastic foundation one of the cornerstones of engineering mechanics and is applicable to a wide range of practical problem in both the civil and mechanical engineering fields. It has application in geotechnics, bio-mechanics, road, railroad and marine engineering, since it constitutes the best practical idealization for many problems. The importance of the problem is underscored by large number of research articles which are in existence in the literature about beams on elastic foundation. Its beginning can be traced back to 1867 classical textbook by Emil Winkler. However, throughout the years the interest in this particular research domain has remained strong and a survey of existing literature reveals a plethora of recent work related to this field. Simsek and Cansız (2012) worked on elastically connected double-functionally graded beam system (DFGBS) which carrying a moving harmonic load at a constant speed and studied the dynamic responses of the system by using Euler–Bernoulli beam theory. Yas and Samadi (2012) obtained equations of motion of single-walled carbon nanotube reinforced nanocomposite beams supported on elastic foundation through Hamilton's principle and generalized differential quadrature method was employed to solve these equations to generate natural frequencies and buckling loads of the system corresponding to different boundaries. Simsek and Reddy (2013) used the modiðed couple stress theory (MCST) to represent a uniðed higher order beam theory for buckling of a functionally graded (FG) micro-beam, embedded in elastic Pasternak medium. Murin et al. (2013) considered a homogenized FGM (polynomial spatial variation of material properties) beam resting on longitudinal variable (Winkler) elastic foundation and conducted a modal analysis to calculate the eigenvalues (natural frequencies) and corresponding eigenvectors (modeshapes) of the system. Kanani et al. (2014) investigated the large amplitude free and forced vibration of FG beam resting on nonlinear elastic foundation containing shearing layer and cubic nonlinearity. Akgöz and Civalek (2015) performed a study based on microstructure dependent trigonometric beam model and Navier's solution technique to investigate the static bending behaviour of embedded FG microbeams, simulated by Winkler foundation. Niknam and Aghdam (2015) made an attempt to investigate the effect of nonlinear elastic foundation and nonlocality in nonlinear free vibration analysis and buckling of functionally graded nanobeam. Tossapanon and Wattanasakulpong (2016) investigated buckling and dynamic behaviour of FG sandwich beams mounted on two parameters elastic foundation. Deng et al. (2017) performed free vibration and stability analysis of double-FG Timoshenko beam system resting on Winkler-Pasternak elastic foundation. Arefi and Zenkour (2017a, 2017b) analysed wave propagation characteristics of FG magneto-electro-elastic nanobeam resting on Visco-Pasternak foundation. Lohar et al. (2016a, 2016b, 2016c, 2018a, 2018b, 2019) recently performed a series of studies where free vibration characteristics of axially functionally graded tapered beams supported on elastic foundations was investigated and detailed results corresponding to various boundary conditions, material property variations and taper profiles were presented through backbone curves and mode shape diagrams.

From the existing literature, it is observed that free vibration studies are quite abundant, while limited numbers of papers are available for forced vibration analysis of AFG beams on elastic foundation. At this point it is necessary to mention that non-uniform beams, i.e. beams with continuous or abrupt changes in cross-sectional profile along the length of the beam are quite common in engineering applications. A few examples of engineering structures that can be modelled as non-prismatic beams are helicopter blades and turbine, bridges and aircraft structures. The reason behind such extensive application is mainly due to two particular specifications, namely, the capability to optimize weight and strength and the ability to meet aesthetic architectural needs. Hence, in the present analysis, large amplitude forced vibration analysis of an axially functionally graded (AFG) tapered beam resting on elastic foundation is performed using Broyden method, which is a multidimensional secant method used for numerically solving a system of nonlinear equations. Inclusion of the elastic foundation creates an added level of complexity. The results of the present study are furnished in terms of response curves in the normalized amplitude-frequency domain for different combinations of classical end conditions, axial material property variation model and foundation stiffness. To the best of authors' knowledge, such frequency response curves are not available in the existing literature.

MATHEMATICAL FORMULATION

Forced vibration analysis is conducted with an objective to find out the response of the system, in terms of displacement amplitude, under externally applied time varying excitations. Presently, large amplitude forced vibration analysis of AFG beams resting on elastic foundation subjected to transverse harmonic excitation is performed. In the current study, only steady-state response is presented and frequency of response of the undamped system is assumed to be equal to that of the external excitation. An indirect approach is adopted for solving the problem, where it is reduced to a static scenario by assuming that under maximum amplitude of excitation, i.e., when the system suffers maximum deformation, the dynamic system fulfils force equilibrium conditions. This supposition converts the problem of solving for the dynamic system into an equivalent static situation, where the excitation frequency and amplitude of the harmonic excitation are the parameters that control system response.

Figure 1. AFG beam supported on elastic foundation represented as series of linear springs

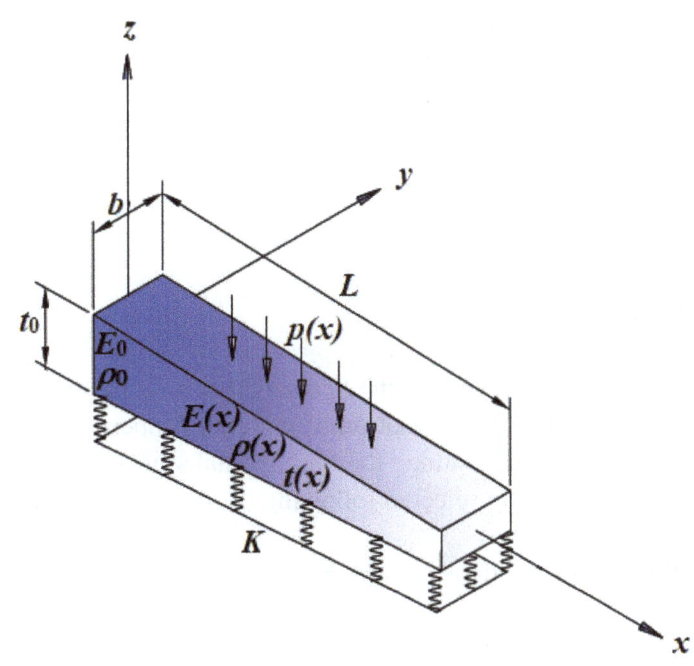

Figure 2. Three different taper profiles (a) linear taper (b) parabolic taper (c) exponential taper

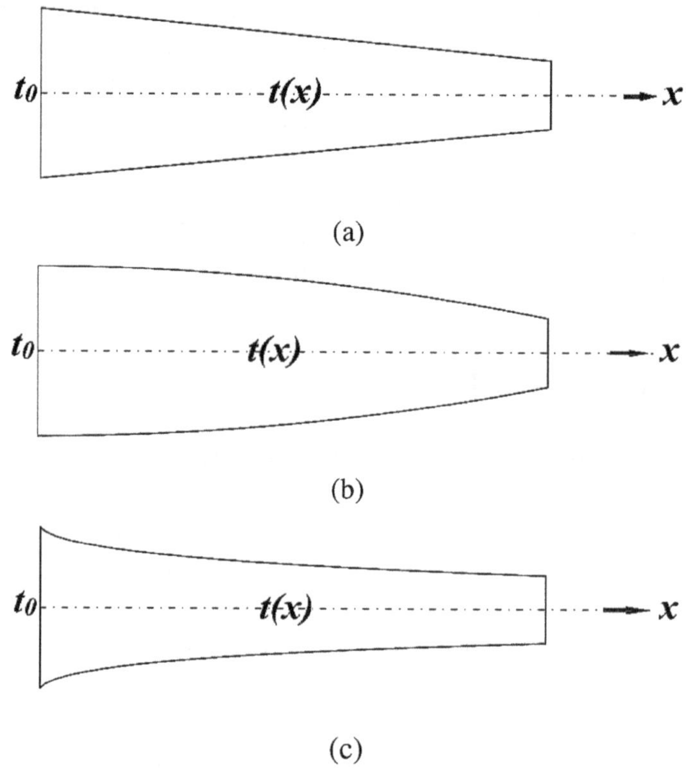

The present system under consideration is a tapered AFG beam resting on a linear elastic foundation as shown in Figure 1. The length of the beam is expressed by the parameter L, while the cross-sectional dimensions are noted as width b and variable thickness $t(x)$. The variation of thickness is considered along the length direction following three different taper profiles, which are linear taper, parabolic taper and exponential taper respectively (Figure 2), and expressed as,

Linear taper: $t(x) = t_0\,(1 - \alpha x)$

Parabolic taper: $t(x) = t_0\,(1 - \alpha x^2)$

Exponential taper: $t(x) = t_0\,(-\alpha x^{1/2})$

Here, t_0 represents the root side thickness of the beam as shown in Figure 1 and α is the parameter which control the degree of taper profile. Gradation of material properties i.e. modulus of elasticity, $E(x)$, and mass density, $\rho(x)$ in the beam is along the longitudinal axis. For that purpose, in the present study three different material models are selected following the expression given below,

Material 1: Elastic modulus, $E(x) = E_0$, density, $\rho(x) = \rho_0$

Material 2: Elastic modulus, $E(x) = E_0\,(1+x)$, density, $\rho(x) = \rho_0\,(1+x+x^2)$

Material 3: Elastic modulus, $E(x) = E_0 (e^x)$, density, $\rho(x) = \rho_0 (e^x)$

Here, E_0 and ρ_0 represent the corresponding value of the material properties at the root side of the beam, as shown in Figure 1. Poisson's ratio (μ) is taken as constant. It is to be noted from the above expressions that material 1 is the case of homogeneous beam, in which material properties are constant throughout the entire beam geometry. The foundation stiffness is taken as K and the effect of the foundation on the system is incorporated in terms of series of linear springs with stiffness coefficient, K (Figure 1) same as that of the foundation. Furthermore, three different classical flexural boundary conditions are considered and these are clamped-clamped (CC), clamped-simply supported (CS) and simply supported-simply supported (SS) boundaries. So, it is quite clear that two classical end conditions of the beam, which are Clamped (C) and Simply supported (S) are taken into account. The beam is subjected to a transverse uniformly distributed time varying harmonic excitation, $p(x,\tau)$, expressed by, $p= \overline{p}(x) e^{j\omega\tau}$, here, ω represents the frequency of excitation, $j = \sqrt{-1}$, $\overline{p}$ denotes the intensity of the harmonic excitation per unit length of the beam and τ is the time.

Energy Expression

In the present paper, the mathematical formulation is carried out taking the beam to follow Euler – Bernoulli hypothesis, as a result of which the effect of rotary inertia and shear deformation is neglected. It should be mentioned here that according to the above hypothesis plane cross-section of the beam remains plane and normal to the deformed axis once deformation has taken place. To introduce geometric nonlinearity into the formulation, Von Karman type nonlinear strain-displacement relations are considered along with Euler-Bernoulli theory.

$$\varepsilon_x^b = -z\frac{d^2w}{dx^2} \tag{1a}$$

$$\varepsilon_x^s = \frac{du}{dx} + \frac{1}{2}\left(\frac{dw}{dx}\right)^2 \tag{1b}$$

In the overall expression of axial strain due to bending (ε_x^b) and stretching (ε_x^s), terms corresponding to linear bending strain as well as linear in-plane strain and non-linear in-plane strains are present. Equation (1b) altogether represents Von Karman type of nonlinear strain displacement relation. The nonlinear in-plane term is associated with large deformation. Here, w and u represent transverse and in-plane displacement fields, respectively, which are defined at the mid-plane of the beam and are functions of axial coordinate.

The present study is based on energy method. Hence, the strain-displacement expressions provided in equation (1) helps derive the energy expressions. As already mentioned with regards to strain-displacement relations (Equation (1)), both bending and stretching effect of the beam is considered to incorporate large displacement behaviour. Thus, strain energy components corresponding to both bending and stretching are present in the overall strain energy expression. Beside this, strain energy stored in the system due to deformation of elastic foundation (equivalently represented by deformation of cluster of springs) is also taken into consideration. Therefore, total strain energy (U) of the system is presented as

$$U = U_b + U_m + U_f \tag{2}$$

Here, U_b and U_m are total strain energies caused by bending deformation and stretching, respectively, while, U_f is the strain energy stored due to deformation of the elastic foundation i.e., in terms of formulation this part of the strain energy is taken care of by the deflection of the springs replacing the foundation. Substituting the relevant strain-displacement expressions, total strain energy of the system can be expressed as,

$$\begin{aligned} U &= \frac{1}{2}\int_0^L \left(\frac{d^2w}{dx^2}\right)^2 E(x)I(x)\,dx \\ &+ \frac{1}{2}\int_0^L \left[\left(\frac{du}{dx}\right)^2 + \frac{1}{4}\left(\frac{dw}{dx}\right)^4 + \left(\frac{dw}{dx}\right)^2\left(\frac{du}{dx}\right)\right] E(x)A(x)\,dx + \frac{1}{2}\int_0^L Kw^2 dx \end{aligned} \tag{3}$$

Here, $A(x)$ represents the cross-sectional area of the beam and $I(x)$ is the area moment of inertial of the cross-section of the beam about the mid-plane.

The total potential energy (V) due to externally applied transverse harmonic excitation, $p(x,\tau)$, can be given by,

$$V = \int_0^L p(x,\tau)\,w\,dx \tag{4}$$

The present formulation can be applied for other type of transverse excitations as well by changing the loading pattern in the equation (4). Hence, the present study is certainly not limited to only uniformly distributed excitation.

The kinetic energy (T) of the system can be given by,

$$T = \frac{1}{2}\int_0^L \left\{(\dot{w})^2 + (\dot{u})^2\right\}\rho(x)A(x)dx \tag{5}$$

Here, the overdots denote differentiation with respect to time (τ).

Forced Vibration Analysis

The present analysis is a semi-analytical displacement based method that employs appropriate energy method to derive the governing equations. In the present formulation of geometrically nonlinear forced vibration, Hamilton's principle (mathematically expressed in equation (6)) is applied to obtain the governing set of equation.

Table 1. Start functions for various boundary conditions

Displacement Field	Boundary Conditions	Start Function
w	CC	$\phi_1=[(x/L)\{1-(x/L)\}]^2$
	CS	$\phi_1=(x/L)^2\{2(x/L)^2-5(x/L)+3\}$
	SS	$\phi_1=sin(\pi x/L)$
u	CC	$\psi_1=(x/L)\{1-(x/L)\}$
	CS	
	SS	

$$\delta\left(\int_{\tau_1}^{\tau_2}\left(T-U-V\right)d\tau\right)=0 \tag{6}$$

Here, δ is the variational operator. The expressions for these energy functionals (U, V and T) are furnished in the previous section (Equations (3), (4) and (5) which are determined from the assumed dynamic displacement fields, which are given by,

$$w\left(x,\tau\right)=\sum_{i=1}^{nw}d_i\varphi_i\left(x\right)e^{i\omega\tau} \tag{7a}$$

$$u\left(x,\tau\right)=\sum_{i=nw+1}^{nw+nu}d_i\psi_i\left(x\right)e^{i\omega\tau} \tag{7b}$$

The transverse (w) and in-plane (u) displacement fields are defined at the neutral axis of the beam and are assumed to be separable in space and time. As shown in Equation (7) they are approximately represented by linear combinations of a set of unknown parameters (d_i) and orthogonal admissible functions (φ_i and ψ_i). It is quite clear that φ_i is associated with the transverse displacement (w), whereas ψ_i defines the in-plane displacements (u). Being admissible functions, φ_i and ψ_i possess certain properties such as, continuity and differentiability within the domain. Also satisfaction the flexural and membrane boundary conditions of the system is another major criterion. In order to generate a complete set of orthogonal admissible functions, first of all a start function satisfying the boundary condition is chosen. For transverse displacement, beam deflection functions, derived from static deflection shape of the beam, corresponding to the boundary condition of the system are taken as the start function. The starting functions for stretching of the beam (u) come from the in-plane boundary conditions. For the present analysis, the in-plane boundary conditions are considered as immovable, i.e., no in-plane displacements are allowed at the boundaries. These functions are tabulated in the Table 1 for different flexural and in-plane boundary conditions. Once the start functions are selected, generation of higher order functions is accomplished by Gram-Schmidt orthogonalization (Kumar et al., 2015) scheme. The number of functions for w and u are denoted by nw and nu, respectively. For the present mathematical formulation, it is assumed that number of functions for w and u are equal i.e., $nw = nu$. In expression (7) ω is the response frequency of the system and according to the basic assumption of current analysis response frequency is equal to the excitation frequency. Substitution of the equation (3), (4) and (5), as well as the dynamic displacement fields (7), in equation (6) the governing equation of the geometrically nonlinear forced vibration problem in the following matrix form.

$$-\omega^2 [M]\{d\} + [K]\{d\} = \{f\} \tag{8}$$

$[K]$, $[M]$ and $\{f\}$ are stiffness matrix, mass matrix and load vector, respectively. $\{d\}$ is the vector of unknown coefficient. The form and the elements of the above mentioned matrices and vector are provided in Appendix.

Equation (8) represents the governing set of nonlinear equations for the dynamic system. But the basic assumption of the present study reduces the problem to an equivalent static case. At the peak excitation amplitude value, it is assumed that the system satisfies the force equilibrium condition and thus the problem can be considered as a static one where the system response is dependent on the excitation frequency and the amplitude of the harmonic excitation. So, at each excitation frequency for a given harmonic excitation amplitude the corresponding unknown coefficients are to be determined. However, determination of these unknown parameters is not straightforward and cannot be directly implemented through simple matrix inversion followed by pre-multiplication with the load vector as shown in equation (9), as the stiffness matrix ($[K]$) itself is a function of the unknown parameters. A closer look at the stiffness matrix elements presented in the previous section clearly shows that certain terms indeed contain the unknown coefficients (d_i). This coupling arises due to the consideration in-plane stretching term in the strain-displacement relation at the mathematical formulation stage.

$$\{d\} = \left[[K] - \omega^2 [M]\right]^{-1} \{f\} \tag{9}$$

To solve the set of nonlinear governing equations an iterative procedure with an initial guess value of the unknowns must be employed. In the present work, a multidimensional quasi-Newton method, known as Broyden's method that utilizes the Jacobian matrix and subsequent updation/correction of the initial Jacobian, is used for the numerical solution of the nonlinear system equations. Determination of the Jacobian matrix is usually an extensive and difficult computational task. Even for sufficiently simple functions, for which the partial derivatives can be analytically found out, the extent of computations necessary may well be excessive. In case of majority of practical problems, however, complicated functions are encountered and numerical determination of an approximation to the Jacobian matrix is inevitable. In the present scenario, an initial Jacobian of the system is computed based on the initial guess and subsequently, only updates, ensuring the minimum change to the Jacobian from the previous iteration, are effected, instead of determination of the whole Jacobian matrix in each step of iterations. So, in this case, the partial derivatives are not estimated or evaluated directly, but corrections to the approximate Jacobian matrix are computed from the function values. The function is constituted by rearranging the terms of the governing equation (8) as follows,

$$F(d) = \left[[K] - \omega^2 [M]\right]\{d\} - \{f\} \tag{10}$$

So, the objective of the Broyden method is to solve for the roots of the set of equations, $F(d) = 0$. The codes for implementation of Broyden method is developed in Matlab following the algorithm provided by Press et al [2005]. Once the values of the unknown parameters are solved for a given excitation amplitude and frequency, the displacement fields can be computed following equation (7). Finally, the maximum deflection (w_{max}) value corresponding to a frequency step is searched out from within the transverse displacement field (w).

Determination of Backbone Curves

Nonlinear behaviour of the system can be easily predicted by plotting backbone curves. They are basically zero amplitude frequency response of the system. In this regard, the study of backbone curves is important aspect of forced vibration analysis, which further necessitates a free vibration study to obtain it. However, the details free vibration study is conducted in two distinct steps. In the initial step, the deflected configuration under static loading is presented through a static analysis and prediction of the maximum deflection can be done. In the later step, a subsequent free vibration study is performed on the previously obtained deflected configuration to determine the loaded natural frequencies of the system. Consequently, loaded natural frequencies vs. maximum deflection plots are presented in non-dimensional plane to obtain backbone curves.

Static Analysis

To formulate static analysis principle of minimum total potential energy is used, mathematically expressed as,

$$\delta(U + V) = 0 \tag{11}$$

Here, the expression of U and V remain as unchanged to those presented in equations (3 and 4). The assumed displacement field for the energy functionals in static analysis are only function of the axial coordinate and are made time independent by erasing the temporal part ($e^{j\omega\tau}$) in equations (7). The sets of orthogonal admissible functions (φ_i and ψ_i) remain same as of the force vibration analysis. Now, carefully substituting assumed displacement fields into energy functionals and hence derived energy functionals into the energy principle, the set of governing equations for static analysis is obtained in the form given below,

$$[K]\,\{d\} = \{f\} \tag{12}$$

The form and elements of $[K]$ and $\{f\}$ are same as of previously explained forced vibration study. The derived set of equations are nonlinear in nature and are solved using a numerical scheme following direct substitution method to obtain the deflected configuration of the system.

Free Vibration Analysis

To formulate free vibration problem Hamilton's principle, as expressed in equation (6), is utilized. The expressions of T and U are remaining as unchanged as expressed in previous section in equations (3 and 4), whereas, the term V (Equation (5)) no longer be present and reduced to zero. This is due to the assumption of the present formulation, in which free vibration analysis is performed on the statistically deflected configuration of the beam whose static solution is already known. The dynamic displacement fields have the identical as that of forced vibration analysis, as expressed in equation (7). However, here, ω represent the natural frequency of the vibratory system. Also, d_is are the new set of unknown coefficients which are different from static analysis. The sets of orthogonal admissible functions (φ_i and ψ_i) remain same as of static and as well as the force vibration analysis. After careful substitution of assumed dynamic displacement field into energy functionals and applying energy principle, the set of governing equations for the free vibration study is obtained as,

$$[K]\{d\} - \omega^2 [M]\{d\} = 0 \tag{13}$$

The above expression indicates a standard eigenvalue problem and its solution is obtained by using Matlab's intrinsic solver. The square root of the eigenvalues provides the loaded natural frequencies of the deflected configuration. The variation of the loaded natural frequencies and corresponding maximum amplitudes in a dimensionless plane represents the backbone curve of the system.

RESULTS AND DISCUSSIONS

The present work investigates the influence of various classical flexural boundary conditions as well as material property variations and stiffness of the elastic foundation on the large amplitude forced vibration behavior of axially functionally graded taper beams under harmonic excitation. For the current analysis only uniformly distributed excitation pattern has been considered. The intensity of the harmonic excitation, $\overline{p}$, is made dimensionless using the expression, $\overline{p}^* = \overline{p}\,(L^4/E_0 I_0 t_0)$. In the present study, four different values for $\overline{p}^*$ are selected, which are 30, 40, 50 and 60 respectively.But it is possible to replace the given expression with mathematical form of any other type of excitation pattern as well. Three different classical flexural boundary conditions are considered i.e. clamped-clamped (CC), clamped-simply supported (CS) and simply supported-simply supported (SS). For the non-uniform beam geometry, length (L), width (b) and root side thickness (t_0) are considered as 1.0m, 0.02m and 0.005m, respectively. Three different taper profiles, named as linear taper, parabolic taper and exponential taper are considered. For controlling the taper profiles four different values of taper parameter (α) are selected for each types of taper pattern. To compere the generated results accurately thickness at the extreme end of beam kept as unchanged for all three types of taper profile and, accordingly, the α values are selected in each cases. For linear and parabolic pattern, the selected values are 0, 0.2, 0.4 and 0.6 whereas for exponential taper these values are calculated as 0, 0.223144, 0.510826 and 0.916291. It should be noted that $\alpha = 0$ resembles a uniform beam. Figure 3 shows the variation of the thickness along the length for each taper profiles in a non-dimensional plane. From the figures it is prominent that $\alpha = 0$ shows no variation in thickness for all taper patterns. Taper parameter, $\alpha = 0.6$ indicates maximum variation of thickness for linear and parabolic profiles whereas, for exponential profile it is at $\alpha = 0.916291$.

Figure 3. Variation of thickness for (a) linear taper (b) parabolic taper and (c) exponential taper for different taper parameters

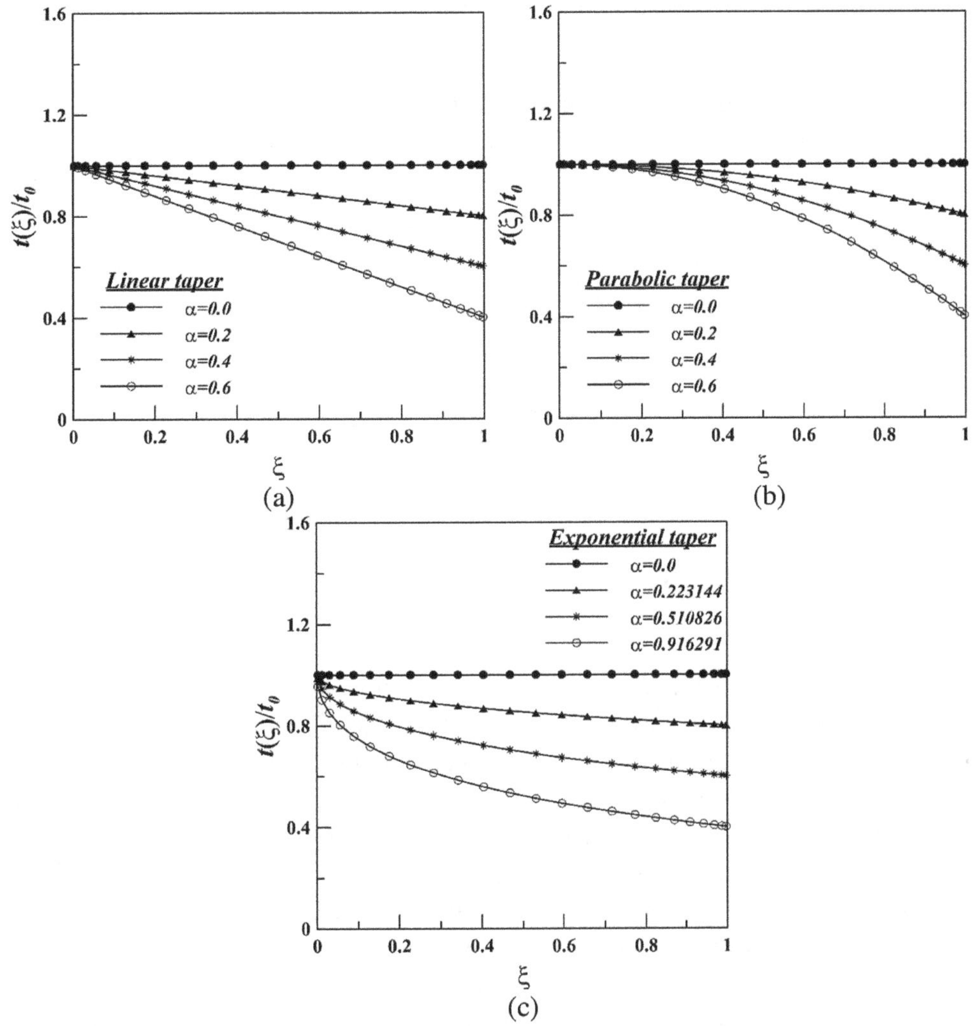

Material property variations along the length of the beam are taken into account by considering gradation of elastic modulus and density. Three different models for variation of these two material parameters are presented in the paper. The material properties at root side (x=0) are considered as, E_0 = 70 GPa and ρ_0 = 7850 kg/m^3. Figure 4 plots the variation of the elastic modulus and density along the length of the beam in normalized plane for different material models. It is prominent from the figure that material 1 shows no variation of material properties, which clearly indicates the case of homogeneous beam, whereas, material 3 shows maximum variation of material properties. Poisson ratio (μ) value throughout the entire analysis is taken as 0.3. The other aspect of complexity of the current system is the elastic foundation upon which the system rests. As already mentioned, the foundation has been incorporated into the mathematical formulation by a series of parallel massless linear springs with equivalent spring stiffness (K). Four different values of dimensionless stiffness [$K_f = K\{(L/t_0)^3/E_0 b\}$] are taken for the present study and these values are 0, 10, 20, 30 and 40 respectively. Here, K represents the dimensional

(N/m) form of stiffness value. The situation with '0' spring stiffness corresponds to a case where the foundation is absent, i.e., the beam is not supported by the foundation at all. This case has also been included to provide a better comparison for the response with and without the foundation.

Figure 4. Gradation of (a) elastic modulus ($E(\xi)/E_0$) and (b) density ($\rho(\xi)/\rho_0$) in the axial direction (ξ) for different material model.

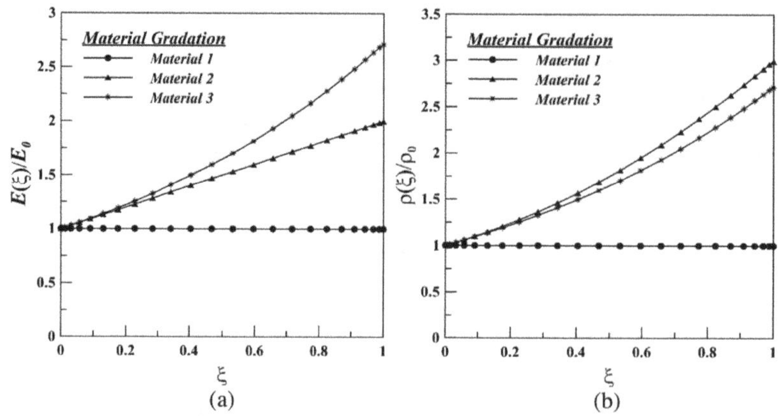

The present study is conducted in a normalised computational domain that varies from 0 to 1. The physical domain in x coordinate is converted to the computational domain by dividing the coordinate by the beam length (L), i.e., $\xi = x/L$. In this computational domain Gausss points (ng) are generated that work as the reference points with respect to which all the numerical computations are performed. Hence, a necessary convergence test must be performed for selecting these gauss points. Similar study is also needed to decide the number of higher order functions for w and u (nw and nu) to be generated. Because, those number significantly influence the accuracy of the analysis outcome. The convergence study is carried out for a clamped beam resting on an elastic foundation of stiffness, $K = 10^3$ N/m and the profile of the beam is parabolic in nature with taper parameter (α) of 0.2. Material model for this particular study is considered as material 2, as shown in Table 1. In order to conduct the test, the maximum normalised deflection of the system is tracked against the variation of relevant parameters at a given excitation amplitude (10 N/m) and excitation frequency value (31.85 Hertz). The results of the convergence study are presented in Figure 5. Normalized maximum deflection (w_{max}/t_p) of the system is plotted against number of Gauss points (ng) is plotted in Figure 5(a). A choice of 24 Gauss points for the present work is made from this study. The positions of these Gauss points within the computational domain are indicated in Table 2. It is evident that these points are equally spread out and they are densely spaced at the two extremities, while sparsely spaced towards the middle. The second part of the convergence study fixes the number of functions to be generated. The previously mentioned system is analyzed for variation in number of orthogonal functions assuming that the number of functions for w and u are equal ($nw = nu$). The convergence results are furnished in Figure 5(b) and a choice of 8 functions are made.

Figure 5. Convergence Study for (a) Number of Gauss point and (b) Number of orthogonal functions

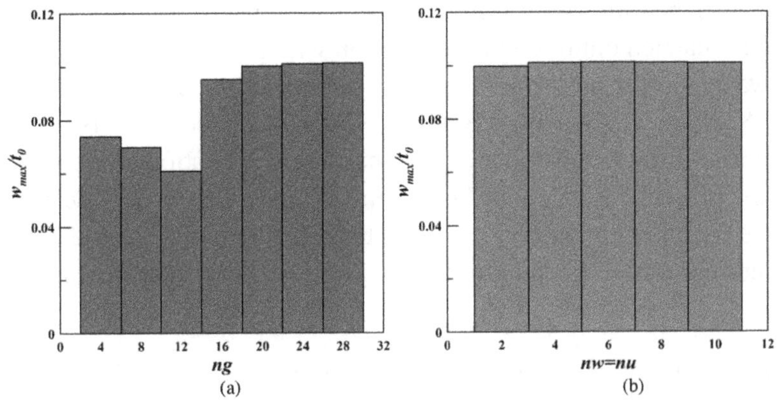

Table 2. Position of Gauss points within computational domain

Gauss Point Number	Position in Normalised Coordinate (ξ)
1	0.0024
2	0.0126
3	0.0309
4	0.0568
5	0.0900
6	0.1299
7	0.1760
8	0.2273
9	0.2831
10	0.3425
11	0.4044
12	0.4680
13	0.5320
14	0.5956
15	0.6575
16	0.7169
17	0.7727
18	0.8240
19	0.8701
20	0.9100
21	0.9432
22	0.9691
23	0.9874
24	0.9976

Once the above-mentioned parameter values corresponding to the numerical scheme are ascertained, the next task is to establish the validity of the present methodology and solution technique. The validation of the current method is carried out through comparison with previously established results. However, the authors are not aware of any benchmark results that provide frequency response curves for the system under consideration. So, comparison with a system having reduced complexity is undertaken. A simple clamped homogeneous uniform beam under a point harmonic excitation at the mid-span is considered (Ribeiro, 2004). In the present formulation the system is simulated by considering a foundation stiffness $K = 0$ N/m, while the taper parameter is reduced to 0. The generated results for frequency response curve as shown in Figure 6 exhibit good agreement with result published by Ribeiro (Ribeiro, 2004).

Figure 6. Validation plot for large amplitude forced vibration response of clamped uniform beam under point load (0.134N) at mid-span of the beam

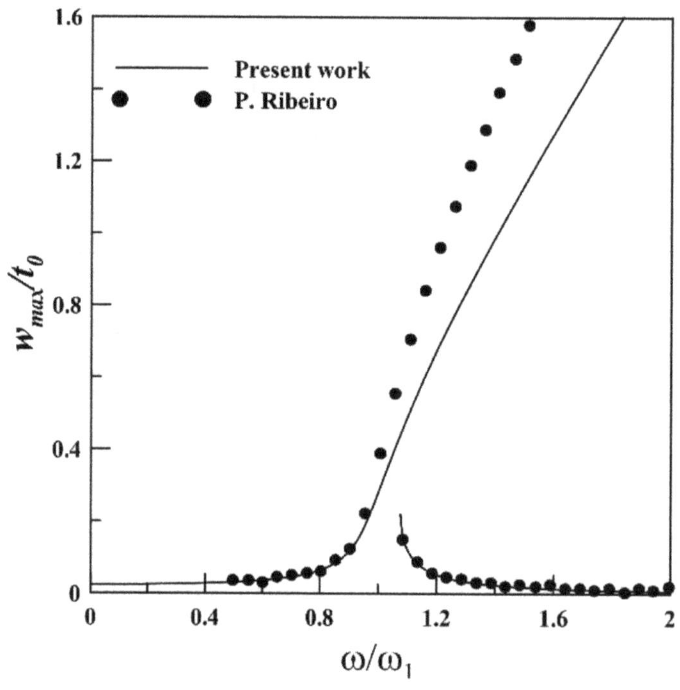

Numerical results for large amplitude forced vibration analysis are presented in terms of normalised frequency response plots, where the normalised excitation frequency (ω_f/ω_1) is the abscissa, while ordinate represents normalised maximum deflection. Maximum amplitude of vibration is normalized by root thickness (t_0) of the beam. On the other hand, normalization of the forcing frequency is carried out by dividing with the fundamental frequency (ω_1) of the system, which are tabulated which are tabulated in Table 3-4 for different combination of system parameter. It should be mentioned here that the natural frequencies (ω_1) of the system are determined following a separate geometrically nonlinear free vibration analysis. Results are generated for different combinations of excitation amplitude, foundation stiffness, flexural boundaries, material models and taper patterns, which are resented in Figures 7-15.

Figures 7-15 are characterized by two distinct zones as the response curve obtained is discontinuous in nature. One branch of the response curve increases monotonically with increase in excitation frequency. It is observed that this branch of the response curve starts from the vertical *y*-axis i.e., at zero excitation frequency. Needless to say that this point refers to a pure static scenario where $\omega = 0$ and the governing equation reduces to, $[K]\{d\} = \{f\}$. This increasing branch of response is obtained through a forward sweep in Broyden method. The other branch of response curve is usually characterised by decrease in response amplitude with increase in excitation frequency, at least initially. This branch has lower values of amplitude, as compared to the first branch, and it remains completely below the increasing branch. A backward sweep is necessary for determining this part of the response by Broyden method. From the figures it is also evident that the response curves tilt towards the right of the vertical indicating hardening type nonlinearity. For this type of nonlinear systems, increase in nonlinear frequency values is observed with rise in amplitude of vibration. In the case of systems under consideration, stretching effect associated with large deflection produces an additional stiffening of the system, thereby causing an increase in the frequency values. Some of the response curves exhibit multi response zone, which is another typical nonlinear phenomenon. It is observed that beyond certain excitation frequency two response amplitudes corresponding to the previously mentioned distinct zones of response are exhibited. The branches of the response curve in these two zones represent stable solutions. Theoretically, a third and unstable steady-state solution also exists in the multi-valued interval of the response but the current methodology is unable to capture that response.

The effect of excitation amplitude on the frequency response of AFG beam is shown in Figure 7. Three plots for three different boundary conditions (CC, CS and SS) are presented corresponding to linear taper beam and material 2. Foundation stiffness and taper parameter have been fixed at 10 and 0.4, respectively, whereas the non-dimensional excitation amplitude is varied from 30 to 60 in steps. The normalizing factor

Table 3. Values of dimensionless fundamental natural frequency $\left(\Omega_1 = \omega_1 L^2 \sqrt{\rho_0 A_0 / E_0 I_0}\right)$ *for different combinations of taper pattern, foundation stiffness, material model and boundary conditions.*

Taper Pattern	Taper Parameter	Stiffness	Ω_1								
			Material 1			Material 2			Material 3		
			CC	CS	SS	CC	CS	SS	CC	CS	SS
Linear	0.4	00	17.57	12.97	07.73	15.77	11.07	07.15	17.35	12.08	07.81
		10	21.49	18.05	14.54	18.24	14.27	11.65	19.78	15.27	12.30
		20	24.79	21.99	19.04	20.42	16.87	14.84	21.94	17.91	15.55
		30	27.70	25.32	22.66	22.38	19.12	17.45	23.90	20.20	18.22
Parabolic	0.4	00	18.28	13.77	08.58	16.42	11.82	07.93	18.07	12.90	08.66
		10	21.71	18.19	14.50	18.58	14.59	11.80	20.18	15.66	12.51
		20	24.67	21.73	18.62	20.51	16.91	14.68	22.09	18.00	15.42
		30	27.30	24.77	21.98	22.28	18.94	17.07	23.85	20.06	17.87
Exponential	0.510826	00	16.39	11.79	06.86	14.75	10.05	06.33	16.23	10.95	06.89
		10	21.02	17.78	14.97	17.70	13.86	11.66	19.14	14.77	12.25
		20	24.80	22.21	19.75	20.22	16.81	15.21	21.65	17.78	15.88
		30	28.08	25.88	23.68	22.46	19.32	18.06	23.90	20.34	18.82

Figure 7. Effects of the excitation amplitude on frequency response for different boundary conditions: (a) CC (b) CS and (c) SS

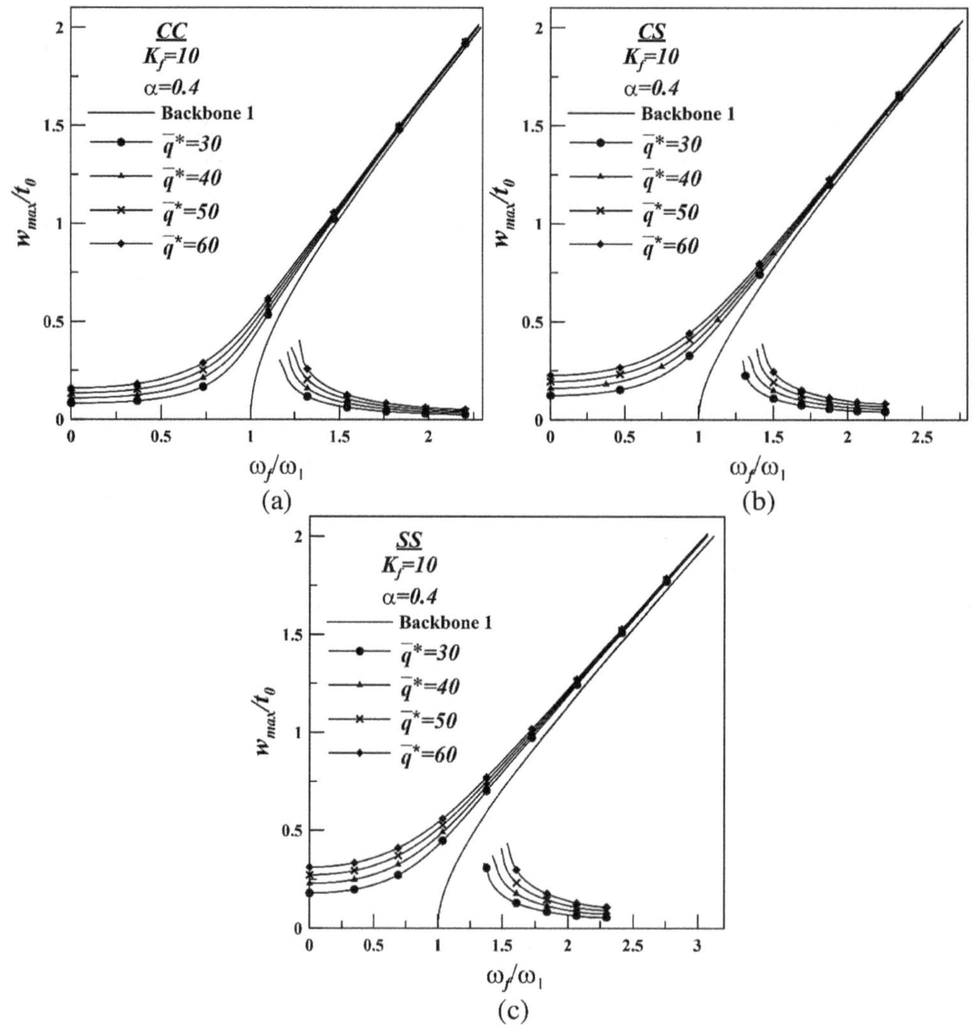

(fundamental frequency) for excitation frequency for the three boundary conditions are different and these values are furnished in Table 3. Backbone curves corresponding to the system are also incorporated into the frequency response plots. These backbone curves are generated from the large amplitude free vibration analysis performed in a separate section. It is observed from Figure 7 that at a given excitation frequency, the response amplitude of beam with SS boundary conditions (Figure 7(c)) is the highest, whereas for beam with CC boundary conditions (Figure 7(a)) it is the lowest. It is also observed that the plots are tilted towards the right of vertical in all three cases. This behaviour can be attributed to the stretching effect associated with large deflection resulting in additional stiffening of the system. It is also observed that the two segments of the frequency response envelop the backbone curve of the system. As the excitation amplitude is continuously decreased the amplitude of the response also decreases and at very low excitation amplitude the response curve is likely to merge with the backbone curve. In fact, this behaviour corroborates the statement that the backbone curve is actually system response under zero amplitude excitation.

The effect of foundation stiffness on the frequency response of AFG beam is shown in Figures 8-14 for different combination system parameters. Figures 8-10 represent the frequency response in the vicinity of fundamental mode of vibration of linearly tapered beam on elastic foundation with three different material models (Material 1, Material 2 and Material 3). In each figure, there are three sets of plots for three different boundary condition (CC, CS and SS) and in each plot four response curves are depicted corresponding to various spring stiffness values which varies from 0 to 30. The taper parameter is kept as constant, at 0.4 to plot the results with an objective to study the influences of the foundation stiffness. Similar plots for parabolic tapered beams are provided in Figures 11-12, whereas response curves for exponential AFG beams are shown in Figures 13-14. Results corresponding to Material 1, which corresponds to a homogeneous material, is included for linearly tapered beams only. For the other two taper patterns (parabolic and exponential), results are provided only for Materials 2 and 3. In these sets of plots, with change in the foundation stiffness, four different backbone curves would have been obtained

Figure 8. Frequency response of AFG linear taper beam (Material 1) for different boundary conditions: (a) CC (b) CS and (c) SS

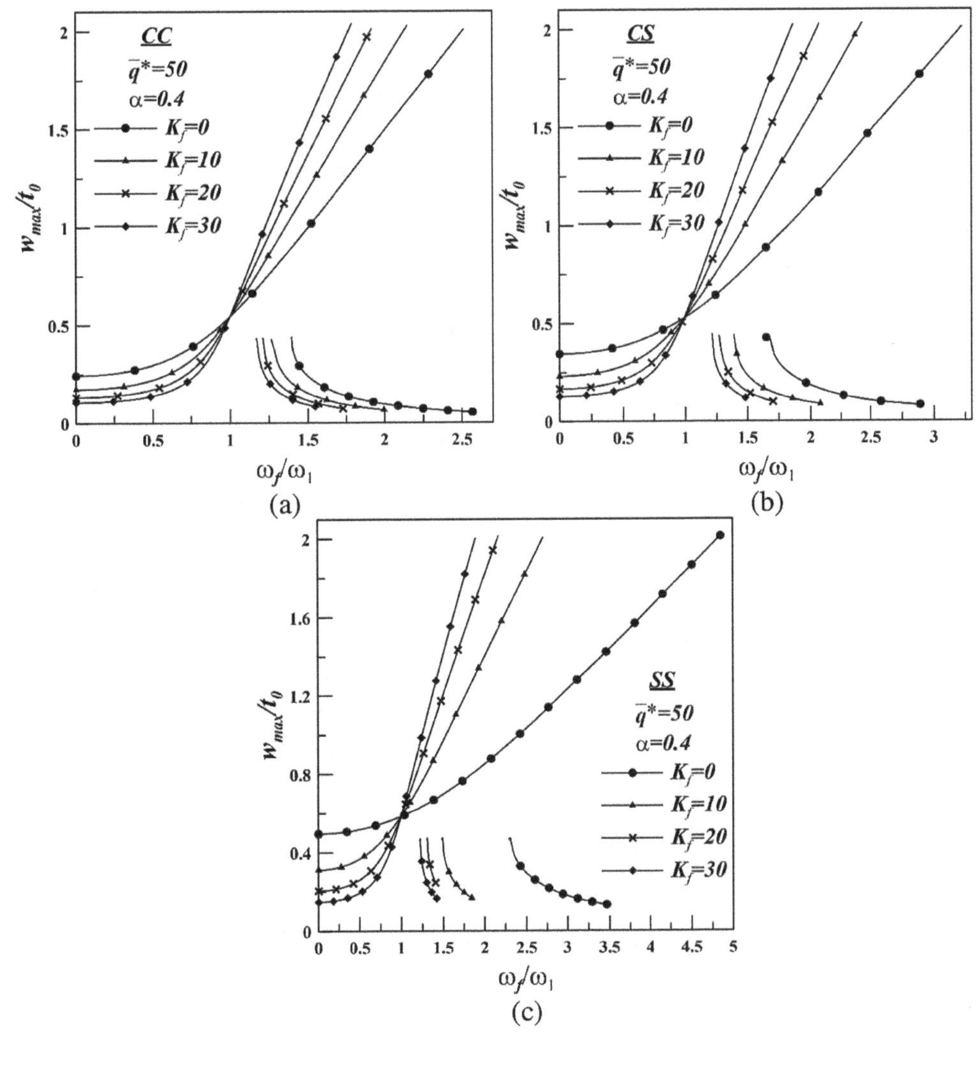

and putting them all together in the same figure would have made it cluttered. Hence, a conscious decision not to incorporate these system backbones is taken in order to maintain the clarity of the figures. In the plot, the excitation frequencies are normalized by dividing with fundamental frequencies (ω_1) which are tabulated in Table 3. It is observed from Figures 8-14, in all the cases, with increase of foundation stiffness the response amplitude decreases in the low excitation frequency zone. But the trend is completely opposite in the higher frequency zone. There appears to be cross-over point in the response behaviour around $\omega_f/\omega_1 = 1.00$ in each case. Another important observation is that the response curves in forward sweep of SS beam are more diverging in the non-dimensional plane from each other than the other two boundaries. From the above discussion it may be concluded that nonlinearity involved in case of SS beam is higher than the other two taper patterns. For CC and CS beam, the difference between the response curves are found to be small. For linear taper pattern, it is quite clear that the change in response curves for different material model is hardly found for a particular case of boundary condition,

Figure 9. Frequency response of AFG linear taper beam (Material 2) for different boundary conditions: (a) CC (b) CS and (c) SS

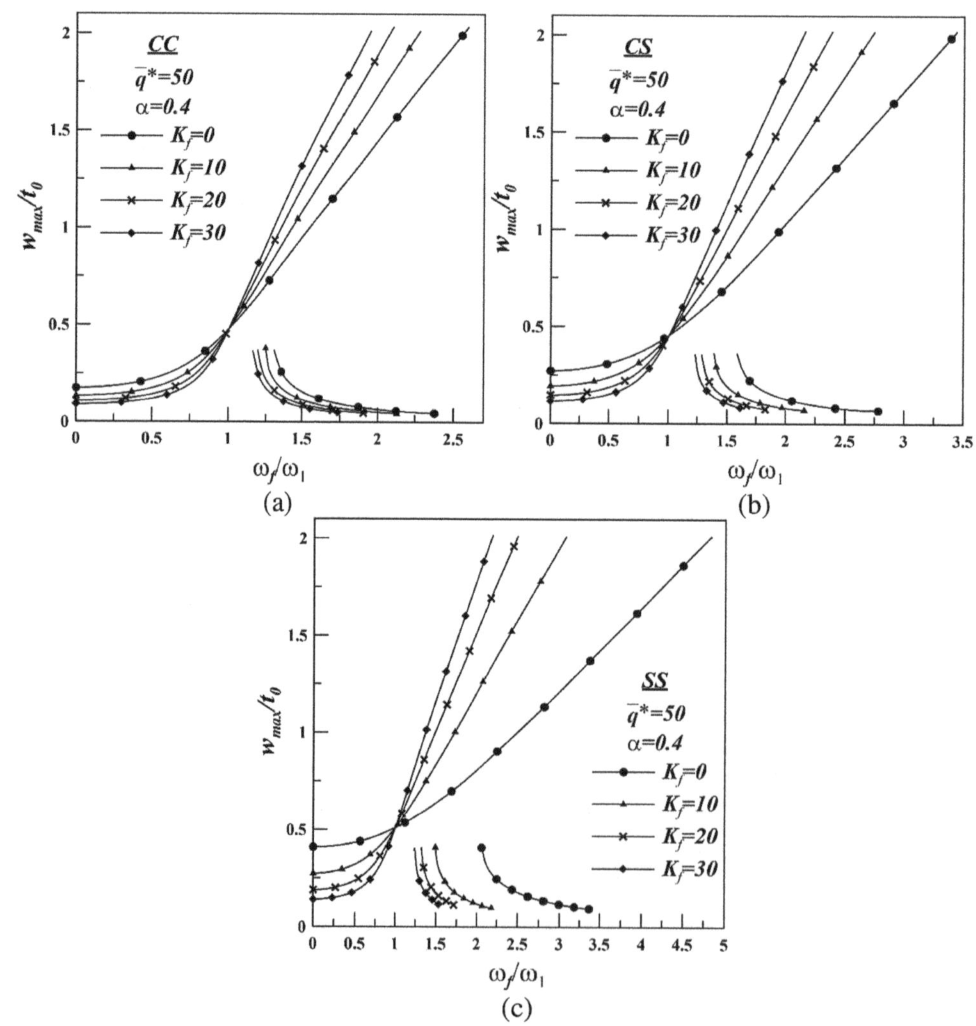

Figure 10. Frequency response of AFG linear taper beam (Material 3) for different boundary conditions: (a) CC (b) CS and (c) SS

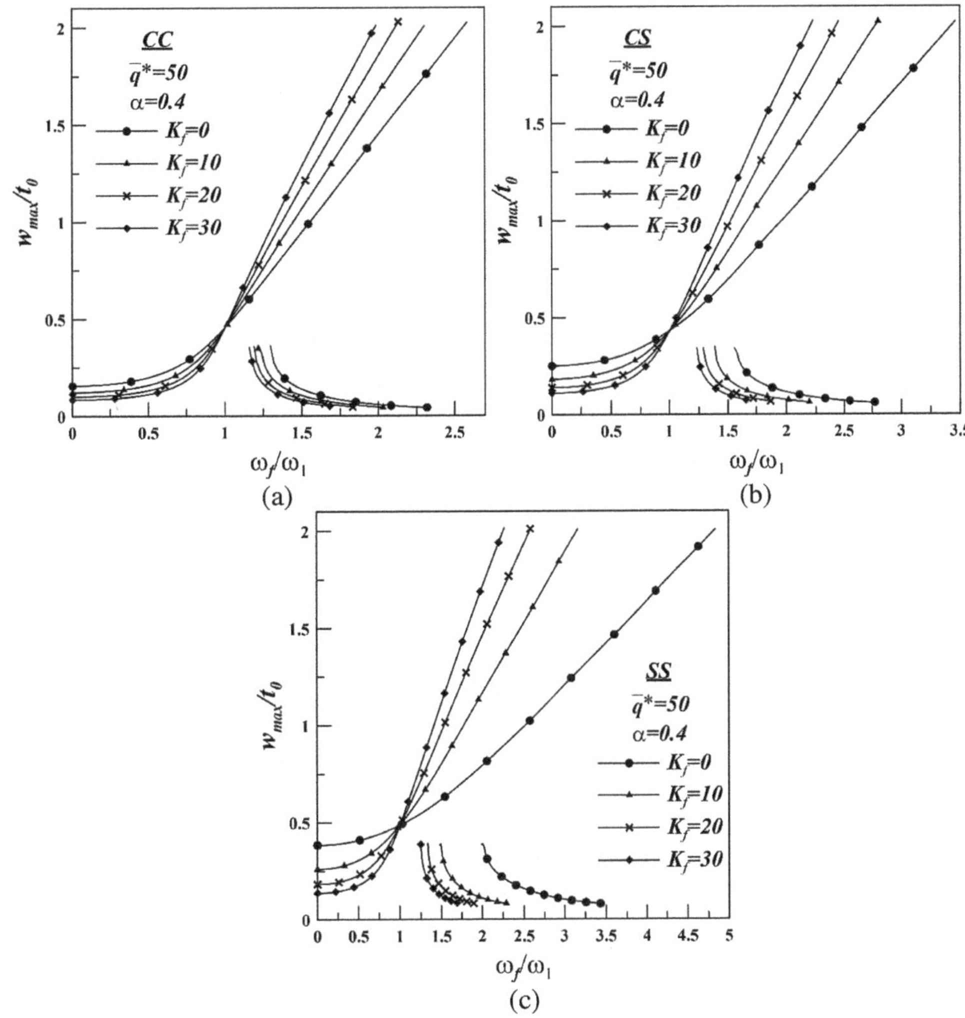

as shown in Figures 8-10. For other type of taper pattern i.e. parabolic and exponential taper same trend can be found. It is also observed from the figures that for particular boundary condition and material model, in exponential taper the response curves are more openly clustered than the other two (linear and parabolic) for different foundation stiffness values.

Figure 15 shows the effect of taper parameter on forced vibration response of AFG beam for different boundaries. Three different taper patterns are considered in which CC boundary condition is taken. For linear and parabolic taper pattern the taper parameter values are varied from 0 to 0.6, whereas, in the case of exponential taper pattern value are varied from 0 to 0.912691. It is to be noted that taper parameter 0.0 represents the case of uniform beam. The excitation amplitude and foundation stiffness have been fixed at 50 and 10, respectively. In the plot, the excitation frequencies are normalized by dividing with fundamental frequencies (ω_1) which are tabulated in Table 4. From the Figure 15, it is observed that with the increase of taper parameter values, the amplitude of the response is increasing

Figure 11. Frequency response of AFG parabolic taper beam (Material 2) for different boundary conditions: (a) CC (b) CS and (c) SS

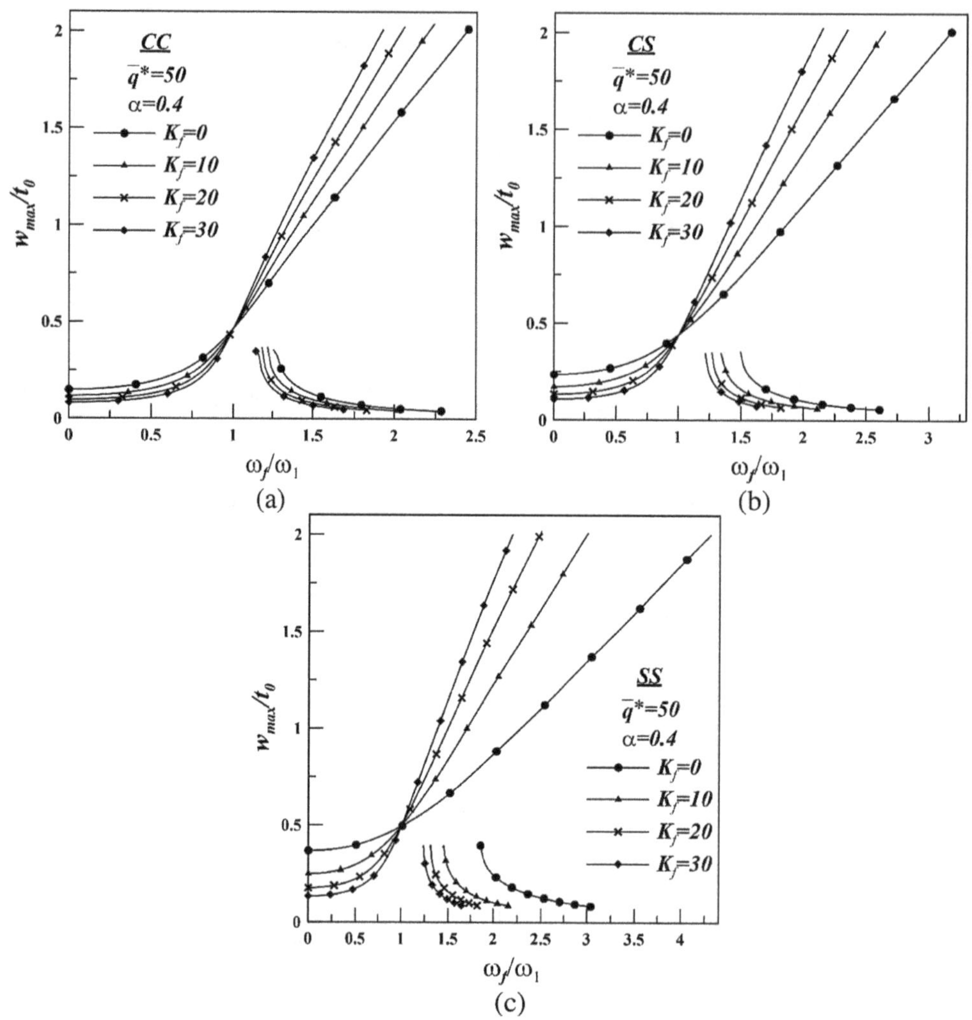

in nature in the low frequency domain, whereas, the trend is reversed at higher frequency range. This type of trend is found due to removal of material with increasing taper which further contributes to the reduction of beam stiffness. In these sets of figures, the backbone curves for the four individual cases are not incorporated for the sake of better clarity.

The shape attained by the system while vibrating under the influence of external excitation is often referred to as operational deflection shape (ODS). For the present analysis, evolution of the ODS as one moves along the branches of the frequency response curves is studied carefully. In other words, the effects of excitation frequency on the ODS is investigated by analysing a clamped parabolic tapered AFG beam with taper parameter of 0.2 under the action uniformly distributed excitation with amplitude of 10 N/m. The beam is resting on a foundation with stiffness of 10^3 N/m and the material model 2 is considered. Figure 16(a) shows the frequency-response curve along with some representative points. The ODS corresponding to these representative points are shown in Figure 16(b). The points on the response

Figure 12. Frequency response of AFG parabolic taper beam (Material 3) for different boundary conditions: (a) CC (b) CS and (c) SS

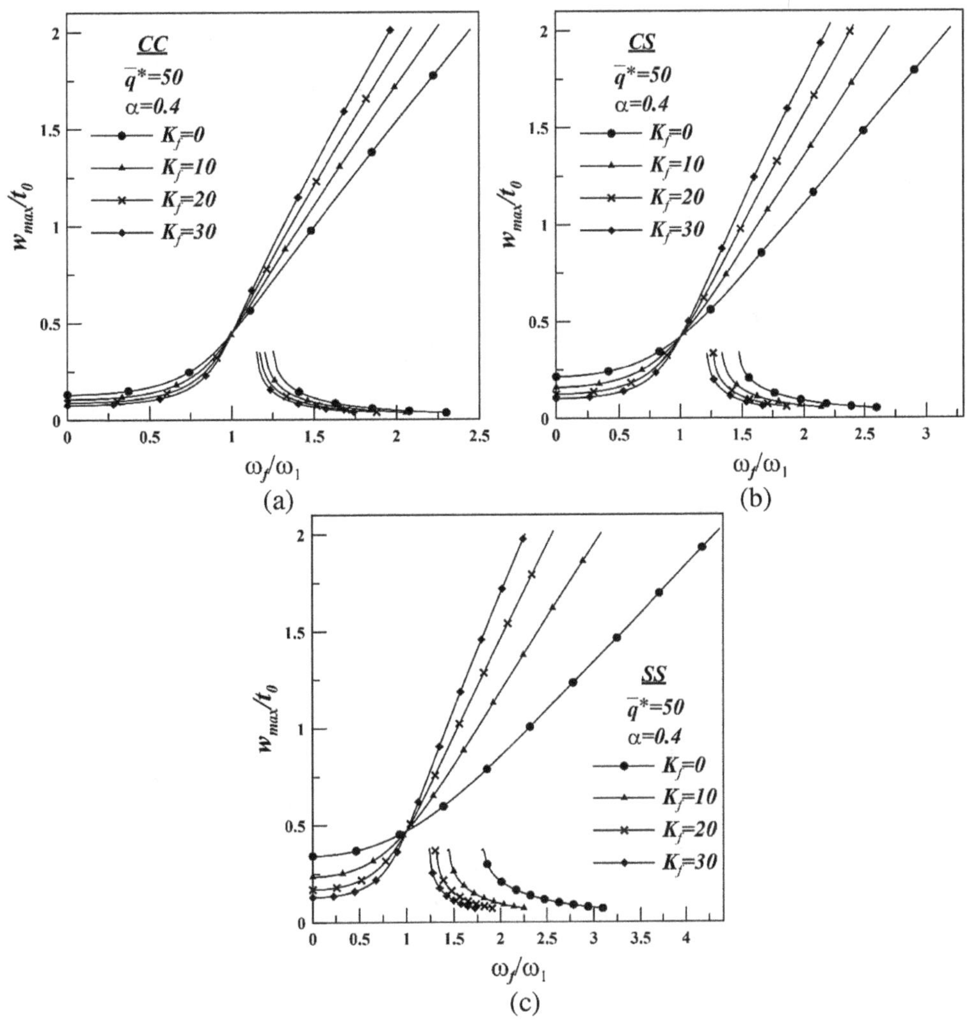

curve at which the ODS are shown correspond to different frequency-amplitude combinations. The start of the multi-response zone is designated by point 'a' on the higher branch of the response. Point 'a1' is obtained by moving vertically down from point 'a' and it corresponds to the same excitation frequency level as 'a'. Point 'b' represents the case when the system attains normalized response amplitude of 2.0. Point with the same frequency as point 'b' but on the lower branch is represented by point 'b1'. The ODS of different representative points seems similar with different maximum response amplitude. However, one major difference that can be noted is that the response at the lower branch is out of phase with the response in the increasing branch.

Figure 13. Frequency response of AFG exponential taper beam (Material 2) for different boundary conditions: (a) CC (b) CS and (c) SS

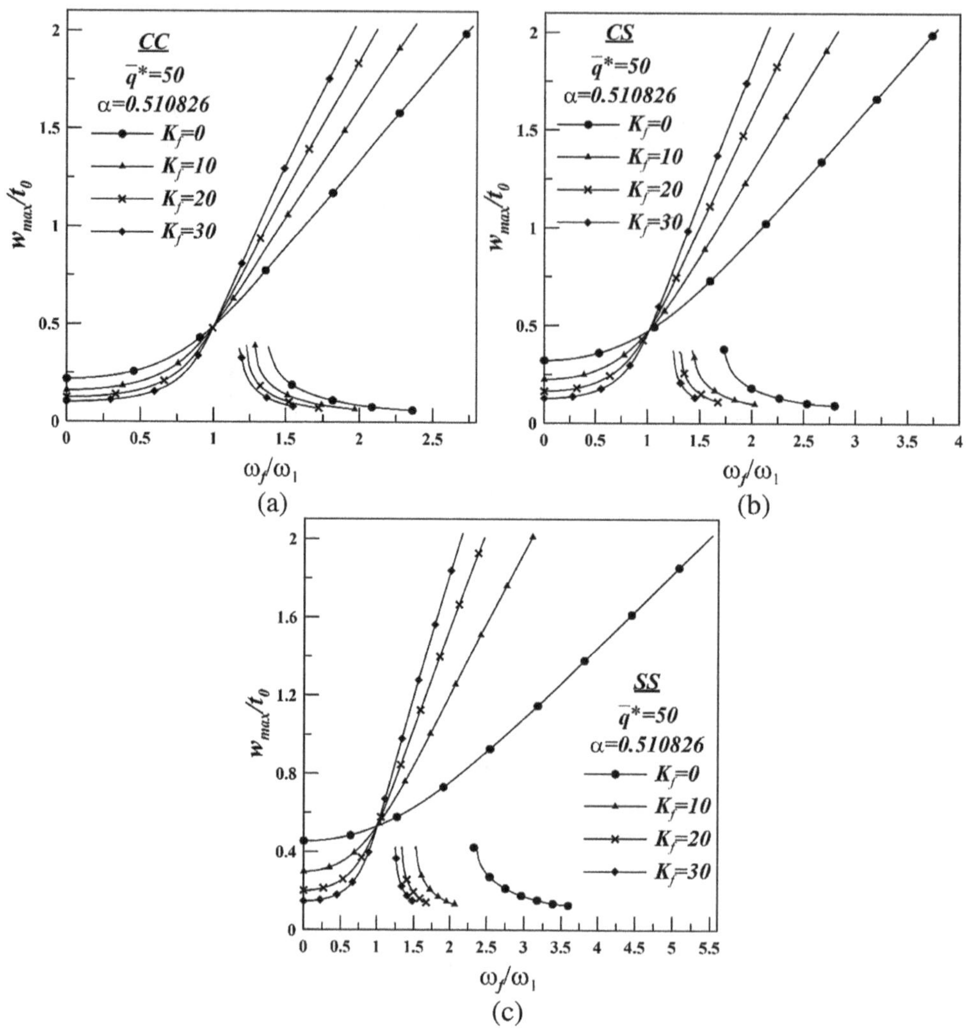

CONCLUSION

In the present analysis, large amplitude force vibration analysis is performed on an axially functionally graded tapered thin beam which is resting on elastic foundation with different end condition. A displacement based semi-analytical method along with energy principles is employed for the mathematical formulation and the governing differential equation of the system is developed through application of Hamilton's principle. Due to consideration of nonlinear strain-displacement relations the set of governing equations also exhibits nonlinear features. To solve for the set of nonlinear equations Broyden's method (quasi-Newton method) is applied and numerical solutions in terms of transverse and in-plane displacement fields are obtained. A convergence study is performed to determine the values for various parameters related to the numerical scheme. Established result from existing literature is used to provide validation for the present method and solution procedure. The geometric nonlinear forced vibration char-

Figure 14. Frequency response of AFG exponential taper beam (Material 3) for different boundary conditions: (a) CC (b) CS and (c) SS

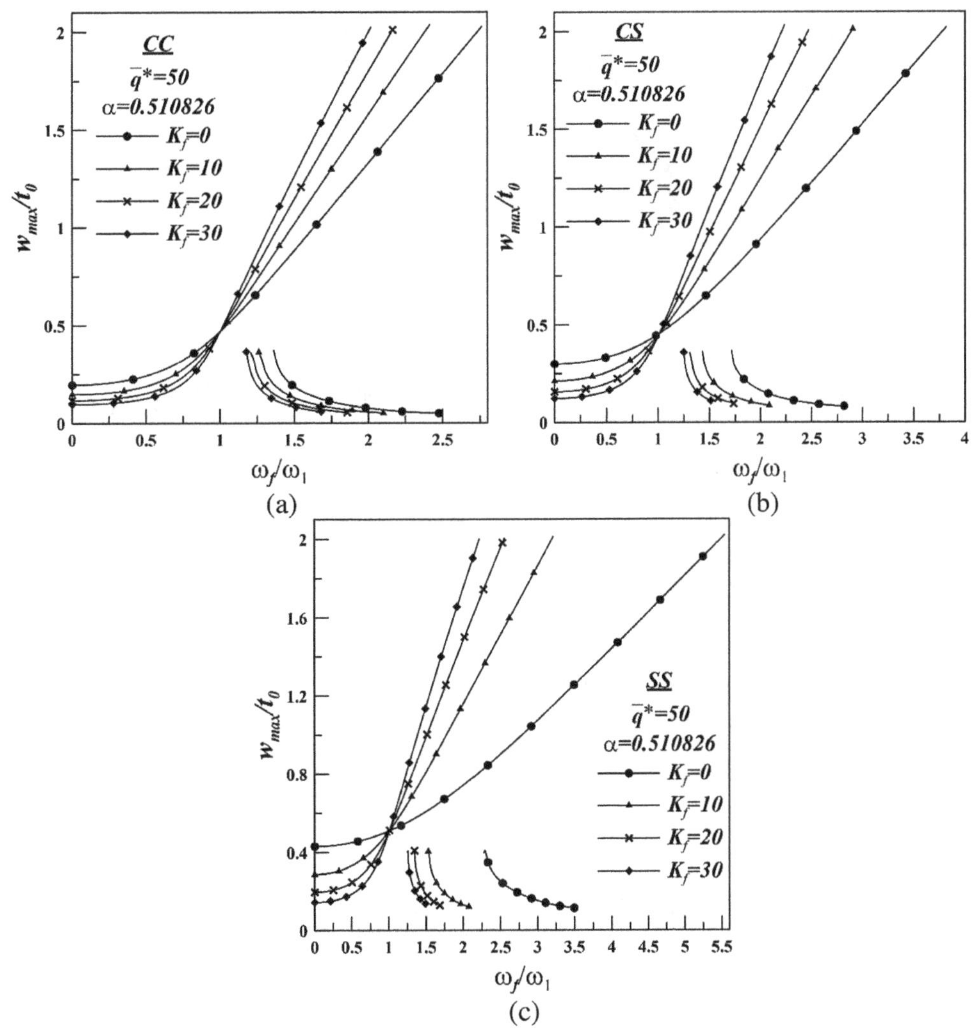

Table 4. Values of dimensionless fundamental natural frequency $\left(\Omega_1 = \omega_1 L^2 \sqrt{\rho_0 A_0 / E_0 I_0}\right)$ *for different combinations of taper pattern considering particular material model, boundary condition and foundation stiffness.*

Taper Pattern	Taper Parameter	Material 2	Taper Pattern	Taper Parameter	Material 2	Taper Pattern	Taper Parameter	Material 2
		CC			CC			CC
		Ω_1			Ω_1			Ω_1
Linear	0	22.01	Parabolic	0	22.01	Exponential	0	22.01
	0.2	20.11		0.2	20.32		0.223144	19.8
	0.4	18.24		0.4	18.58		0.510826	17.7
	0.6	16.47		0.6	16.78		0.916291	16.07

Figure 15. Effects of the taper parameter on frequency response for CC beam for (a) linear taper (b) parabolic taper and (c) exponential taper

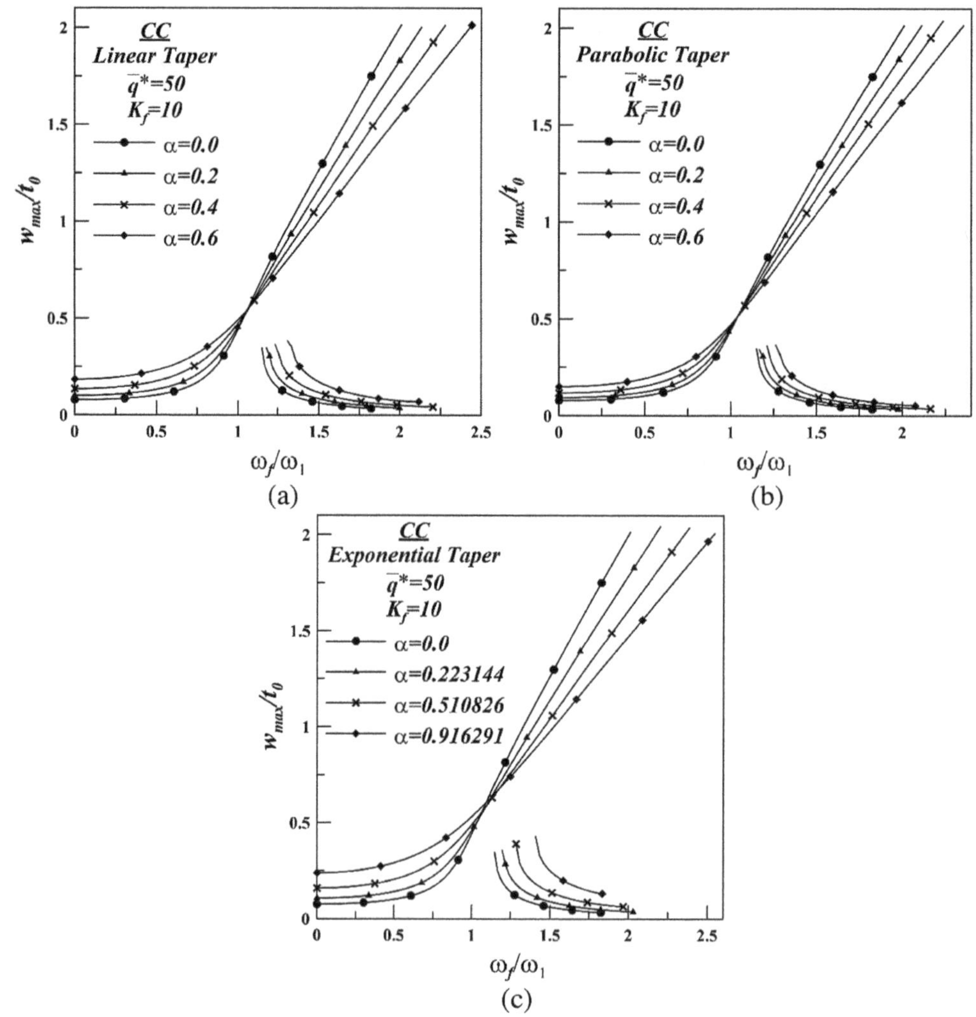

Figure 16. Operational deflection shape (ODS) plots for non-uniform AFG beam with clamped boundary conditions under uniformly distributed excitation: (a) response curve showing representative points and (b) operational deflection shapes corresponding to the representative points.

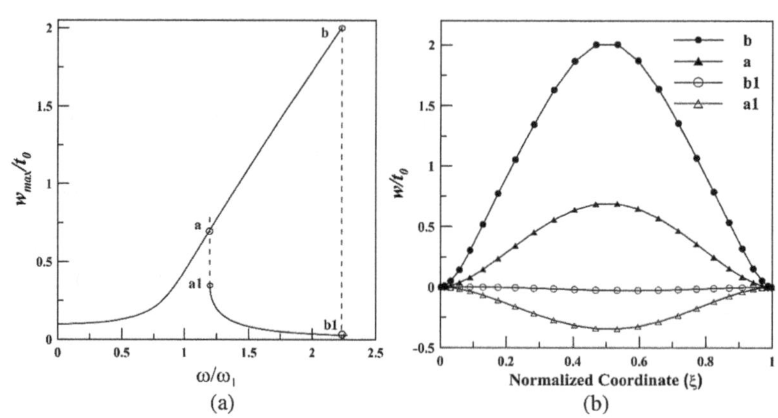

acteristic of the system is represented through frequency response curves in non-dimensional excitation frequency-maximum response amplitude diagrams. New results, capable of acting as benchmark results, are provided for a different combination of excitation amplitude, flexural boundary conditions, material models, foundation stiffness values and taper profiles. However, the general nature of the formulation and solution technique ensures that other type of taper pattern, material gradation as well as excitation pattern can be handled with minimum modifications.

REFERENCES

Akgöz, B., & Civalek, Ö. (2013). Free vibration analysis of axially functionally graded tapered Bernoulli–Euler microbeams based on the modiðed couple stress theory. *Composite Structures*, *98*, 314–322. doi:10.1016/j.compstruct.2012.11.020

Akgöz, B., & Civalek, Ö. (2015). Bending analysis of FG microbeams resting on winkler elastic foundation via strain gradient elasticity. *Composite Structures*, *134*, 294–301. doi:10.1016/j.compstruct.2015.08.095

Arefi, M., & Zenkour, A. M. (2017). Wave propagation analysis of a functionally graded magneto-electro-elastic nanobeam rest on Visco-Pasternak foundation. *Mechanics Research Communications*, *79*, 51–62. doi:10.1016/j.mechrescom.2017.01.004

Arefi, M., & Zenkour, A. M. (2017). Analysis of wave propagation in a functionally graded nanobeam resting on visco-Pasternak's foundation. *Theoretical & Applied Mechanics Letters*, *7*(3), 145–151. doi:10.1016/j.taml.2017.05.003

Chen, D., Yang, J., & Kitipornchai, S. (2015). Elastic buckling and static bending of shear deformable functionally graded porous beam. *Composite Structures*, *133*, 54–61. doi:10.1016/j.compstruct.2015.07.052

Deng, H., Chen, K. D., Cheng, W., & Zhao, S. G. (2017). Vibration and buckling analysis of double-functionally graded Timoshenko beam system on Winkler-Pasternak elastic foundation. *Composite Structures*, *160*, 152–168. doi:10.1016/j.compstruct.2016.10.027

Ebrahimi, F., & Salari, E. (2015). Thermal buckling and free vibration analysis of size dependent Timoshenko FG nanobeams in thermal environments. *Composite Structures*, *128*, 363–380. doi:10.1016/j.compstruct.2015.03.023

Ebrahimi, F., & Zia, M. (2015). Large amplitude nonlinear vibration analysis of functionally graded Timoshenko beams with porosities. *Acta Astronautica*, *116*, 117–125. doi:10.1016/j.actaastro.2015.06.014

Ghayesh, M. H., & Farokhi, H. (2018). Bending and vibration analyses of coupled axially functionally graded tapered beams. *Nonlinear Dynamics*, *91*(1), 17–28. doi:10.100711071-017-3783-8

Hemmatnezhad, M., Ansari, R., & Rahimi, G. H. (2013). Large-amplitude free vibrations of functionally graded beams by means of a finite element formulation. *Applied Mathematical Modelling*, *37*(18-19), 8495–8504. doi:10.1016/j.apm.2013.03.055

Huang, Y., & Luo, Q. (2011). A simple method to determine the critical buckling loads for axially inhomogeneous beams with elastic restraint. *Computers & Mathematics with Applications (Oxford, England)*, *61*(9), 2510–2517. doi:10.1016/j.camwa.2011.02.037

Huang, Y., Wang, T., Zhao, Y., & Wang, P. (2018). Effect of axially functionally graded material on whirling frequencies and critical speeds of a spinning Timoshenko beam. *Composite Structures*, *192*, 355–367. doi:10.1016/j.compstruct.2018.02.039

Huang, Y., Yang, L., & Luo, Q. (2013). Free vibration of axially functionally graded Timoshenko beams with non-uniform cross-section. *Composites. Part B, Engineering*, *45*(1), 1493–1498. doi:10.1016/j.compositesb.2012.09.015

Jia, X. L., Ke, L. L., Feng, C. B., Yang, J., & Kitipornchai, S. (2015). Size effect on the free vibration of geometrically nonlinear functionally graded micro-beams under electrical actuation and temperature change. *Composite Structures*, *133*, 1137–1148. doi:10.1016/j.compstruct.2015.08.044

Jin, C., & Wang, X. (2015). Accurate free vibration analysis of Euler functionally graded beams by the weak form quadrature element method. *Composite Structures*, *125*, 41–50. doi:10.1016/j.compstruct.2015.01.039

Kadoli, R., Akhtar, K., & Ganesan, N. (2008). Static analysis of functionally graded beams using higher order shear deformation theory. *Applied Mathematical Modelling*, *32*(12), 2509–2525. doi:10.1016/j.apm.2007.09.015

Kanani, A. S., Niknam, H., Ohadi, A. R., & Aghdam, M. M. (2014). Effect of nonlinear elastic foundation on large amplitude free and forced vibration of functionally graded beam. *Composite Structures*, *115*, 60–68. doi:10.1016/j.compstruct.2014.04.003

Kien, N. (2013). Large displacement response of tapered cantilever beams made of axially functionally graded material. *Composites. Part B, Engineering*, *55*, 298–305. doi:10.1016/j.compositesb.2013.06.024

Koizumi, M. (1997). FGM activities in Japan. *Composites. Part B, Engineering*, *28B*(1-2), 1–4. doi:10.1016/S1359-8368(96)00016-9

Kumar, S., Mitra, A., & Roy, H. (2015) Geometrically nonlinear free vibration analysis of axially functionally graded taper beams. *Engineering Science and Technology, an International Journal, 18,* 579-593.

Li, X. F., Kang, Y. A., & Wu, J. X. (2013). Exact frequency equations of free vibration of exponentially functionally graded beams. *Applied Acoustics*, *74*(3), 413–420. doi:10.1016/j.apacoust.2012.08.003

Lohar, H., Mitra, A., & Sahoo, S. (2016) Free vibration analysis of axially functionally graded linearly taper beam on elastic foundation. *IOP Conf. Series: Materials Science and Engineering, 149.* 10.1088/1757-899X/149/1/012130

Lohar, H., Mitra, A., & Sahoo, S. (2016). Geometric nonlinear free vibration of axially functionally graded non-uniform beams supported on elastic foundation. *Curved and Layered Structures*, *3*(1), 223–239. doi:10.1515/cls-2016-0018

Lohar, H., Mitra, A., & Sahoo, S. (2016). Natural frequency and mode shapes of exponential tapered AFG beams on elastic foundation. *International Frontier Science Letters*, *9*, 9–25. doi:10.18052/www.scipress.com/IFSL.9.9

Lohar, H., Mitra, A., & Sahoo, S. (2018). Large amplitude forced vibration analysis of an axially functionally graded tapered beam resting on elastic foundation. *Materials Today: Proceedings*, *5*(2), 5303–5312. doi:10.1016/j.matpr.2017.12.114

Lohar, H., Mitra, A., & Sahoo, S. (2018). Free vibration of initially deflected axially functionally graded non-uniform Timoshenko beams on elastic foundation. *Romanian Journal of Acoustics and Vibration*, *15*(2), 75–89.

Lohar, H., Mitra, A., & Sahoo, S. (2019). Nonlinear response of axially functionally graded Timoshenko beams on elastic foundation under harmonic excitation. *Curved and Layered Structures*, *6*(1), 90–104. doi:10.1515/cls-2019-0008

Murin, J., Aminbaghai, M., Kutis, V., & Hrabovsky, J. (2013). Modal analysis of the FGM beams with effect of axial force under longitudinal variable elastic winkler foundation. *Engineering Structures*, *49*, 234–247. doi:10.1016/j.engstruct.2012.09.039

Nguyen, T., Vo, T., & Thai, H. (2013). Static and free vibration of axially loaded functionally graded beams based on the first-order shear deformation theory. *Composites. Part B, Engineering*, *55*, 147–157. doi:10.1016/j.compositesb.2013.06.011

Niknam, H., & Aghdam, M. M. (2015). A semi analytical approach for large amplitude free vibration and buckling of nonlocal FG beams resting on elastic foundation. *Composite Structures*, *119*, 452–462. doi:10.1016/j.compstruct.2014.09.023

Pradhan, K. K., & Chakraverty, S. (2014). Effects of different shear deformation theories on free vibration of functionally graded beams. *International Journal of Mechanical Sciences*, *82*, 149–160. doi:10.1016/j.ijmecsci.2014.03.014

Press, W. H., Teukolsky, S. A., Vetterling, W. T., & Flannery, B. P. (2005). *Numerical Recipes in Fortran 77* (2nd ed.). Cambridge, MA: Press Syndicate.

Ribeiro, P. (2004). Non-linear forced vibrations of thin/thick beams & plates by the finite element and shooting methods. *Computers & Structures*, *82*(17-19), 1413–1423. doi:10.1016/j.compstruc.2004.03.037

Sarkar, K., & Ganguli, R. (2014). Closed-form solutions for axially functionally graded timoshenko beams having uniform cross-section and ðxed–ðxed boundary condition. *Composites. Part B, Engineering*, *58*, 361–370. doi:10.1016/j.compositesb.2013.10.077

Shahba, A., Attarnejad, R., Marvi, T. M., & Hajilar, S. (2011). Free vibration and stability analysis of axially functionally graded tapered timoshenko beams with classical and non-classical boundary conditions. *Composites. Part B, Engineering*, *42*(4), 801–808. doi:10.1016/j.compositesb.2011.01.017

Shahba, A., & Rajasekaran, S. (2012). Free vibration and stability of tapered euler–bernoulli beams made of axially functionally graded materials. *Applied Mathematical Modelling*, *36*(7), 3094–3111. doi:10.1016/j.apm.2011.09.073

Simsek, M., & Cansız, S. (2012). Dynamics of elastically connected double-functionally graded beam systems with different boundary conditions under action of a moving harmonic load. *Composite Structures*, *94*(9), 2861–2878. doi:10.1016/j.compstruct.2012.03.016

Simsek, M., & Reddy, J. N. (2013). A unified higher order beam theory for buckling of a functionally graded microbeam embedded in elastic medium using modified couple stress theory. *Composite Structures, 101*, 47–58. doi:10.1016/j.compstruct.2013.01.017

Su, H., Banerjee, J. R., & Cheung, C. W. (2013). Dynamic stiffness formulation and free vibration analysis of functionally graded beams. *Composite Structures, 106*, 854–862. doi:10.1016/j.compstruct.2013.06.029

Suresh, S., & Mortensen, A. (1998). *Fundamentals of Functionally Graded Materials*. London: IOM Communications Limited.

Yaghoobi, H., & Fereidoon, A. (2010). Influence of neutral surface position on deflection of functionally graded beam under uniformly distributed load. *World Applied Sciences Journal, 10*, 337–341.

Yas, M. H., & Samadi, N. (2012). Free vibrations & buckling analysis of carbon nanotube-reinforced composite Timoshenko beams on elastic foundation. *International Journal of Pressure Vessels and Piping, 98*, 119–128. doi:10.1016/j.ijpvp.2012.07.012

Zeighampour, H., & Beni, Y. T. (2015). Free vibration analysis of axially functionally graded nanobeam with radius varies along the length based on strain gradient theory. *Applied Mathematical Modelling, 39*(18), 5354–5369. doi:10.1016/j.apm.2015.01.015

APPENDIX 1: NOMENCLATURE

A_0: cross-sectional area of the beam at the root
d_i: unknown coefficients for dynamic analysis
E_0: elastic modulus of the beam material at the root
$\{f\}$: load vector
I_0: moment of inertia of the beam at the root
K: stiffness of the elastic foundation
$[K]$: stiffness matrix
L: length of the beam
$[M]$: mass matrix
nw: number of constituent functions for w
nu: number of constituent functions for u
ng: number of Gauss points
p: harmonic excitation for uniformly distributed type loading
t_0: thickness of the beam at the root
T: kinetic energy of the system
u: displacement field in x-axis
U: strain energy stored in the system
V: potential energy of the external forces
w: displacement field in z-axis
w_{max}: maximum deflection of the beam
α: taper parameter
δ: variational operator
ε_x^b : axial strains due to bending
ε_x^s : axial strains due to stretching
ρ_0: density of the beam material at the root
τ : time coordinate
ω: frequency of excitation
ω_1: fundamental frequency of system
ξ : normalized axial coordinate
ψ_i : set of orthogonal functions for u
ϕ_i : set of orthogonal functions for w

APPENDIX 2

The form of the stiffness matrix $[K]$, mass matrix $[M]$ and force $\{f\}$ vector are given as below,

$$[K]=\begin{bmatrix}K_{11} & K_{12}\\ K_{21} & K_{22}\end{bmatrix},\ [M]=\begin{bmatrix}M_{11} & M_{12}\\ M_{21} & M_{22}\end{bmatrix},\ \{f\}=\{f_{11} \quad f_{12}\}^{T}$$

Elements of stiffness matrix $[K]$, mass matrix $[M]$ and force vector $\{f\}$ are,

$$[K_{11}]=\frac{1}{L^3}\sum_{j=1}^{nw}\sum_{i=1}^{nw}\int_0^1 \frac{d^2\phi_i}{d\xi^2}\frac{d^2\phi_j}{d\xi^2}E(\xi)I(\xi)d\xi$$
$$+\frac{1}{2L^3}\sum_{j=1}^{nw}\sum_{i=1}^{nw}\int_0^1\left(\sum_{i=1}^{nw}d_i\frac{d\phi_i}{d\xi}\right)^2\frac{d\phi_i}{d\xi}\frac{d\phi_j}{d\xi}E(\xi)A(\xi)d\xi$$
$$+\frac{1}{L}\sum_{j=1}^{nw}\sum_{i=1}^{nw}\int_0^1\left(\sum_{i=1+nw}^{nw+nu}d_i\frac{d\psi_{i-nw}}{d\xi}\right)^2\frac{d\phi_i}{d\xi}\frac{d\phi_j}{d\xi}E(\xi)A(\xi)d\xi$$
$$+KL\sum_{j=1}^{nw}\sum_{i=1}^{nw}\int_0^1\phi_i\phi_j d\xi$$

$$[K_{12}]=0$$

$$[K_{21}]=\frac{1}{2L^2}\sum_{j=nw+1}^{nw+nu}\sum_{i=1}^{nw}\int_0^1\left(\sum_{i=1}^{nw}d_i\frac{d\phi_i}{d\xi}\right)^2\frac{d\phi_i}{d\xi}\frac{d\psi_{j-nw}}{d\xi}E(\xi)A(\xi)d\xi$$

$$[K_{22}]=\frac{1}{L}\sum_{j=nw+1}^{nw+nu}\sum_{i=nw+1}^{nw+nu}\int_0^1\frac{d\psi_{i-nw}}{d\xi}\frac{d\psi_{j-nw}}{d\xi}E(\xi)A(\xi)d\xi$$

$$[M_{11}]=L\sum_{j=1}^{nw}\sum_{i=1}^{nw}\int_0^1\phi_i\phi_j\rho(\xi)A(\xi)d\xi$$

$$[M_{12}]=0,\ [M_{21}]=0$$

$$[M_{22}]=L\sum_{j=nw+1}^{nw+nu}\sum_{i=nw+1}^{nw+nu}\int_0^1\psi_{i-nw}\psi_{j-nw}\rho(\xi)A(\xi)d\xi$$

$$\{f_{11}\}=L\sum_{j=1}^{nw}\int_0^1 p(\xi)\phi_j d\xi$$

$$\{f_{12}\}=0$$

Chapter 7
Static and Dynamic Analysis of Deformable Fractal Surface in Contact With Rigid Flat

Tamonash Jana
https://orcid.org/0000-0002-4673-5534
Jadavpur University, India

Anirban Mitra
Jadavpur University, India

Prasanta Sahoo
https://orcid.org/0000-0002-1538-0646
Jadavpur University, India

ABSTRACT

The chapter consists of static and dynamic analyses of a fractal rough surface in contact with a rigid flat. The fractal surface is constructed using modified Weierstrass-Mandelbrot function. A rigid flat surface touches the topmost point of the rough surface, which moves towards the rough surface and deforms it. Different contact parameters (e.g., contact force, contact area, contact stress, etc. for varying fractal and material properties are obtained through finite element based static analysis. A parameter denoting the degree of nonlinearity of the contact system is extracted from the force-displacement plot of the surface. This parameter is utilized to explain the dynamic behaviour of the fractal surface which vibrates under the influence of the externally excited rigid flat surface. The dynamic analysis of the contact system is carried out by modelling the contact interface as a single degree of freedom (SDOF) spring-mass-damper system. The dynamic behavior of the system is investigated in terms of frequency response curves, time-displacement plots, and phase plots.

DOI: 10.4018/978-1-7998-4939-1.ch007

INTRODUCTION

Analysis of solid rough contact surface has immense importance in the present scenario because in MEMS and other micro-scale devices, the effect of the surface topography on the contact interactions becomes crucial (Dickrell et al., 2007; Rezvanian et al., 2007). To predict and control the solid surface interactions, clear understanding of the contact behavior is essential under static as well as under dynamic conditions. Fractal geometry first introduced by Mandelbrot (Mandelbrot, 1967; Mandelbrot & Blumen, 1989) is extensively utilized to characterize the realistic rough surface due to due to its scale-dependent, non-stationary, self-similar nature. Majumdar and Tien (1990), Majumdar and Bhushan (1990), and Ling (1990) formulated general methods to apply fractal geometry to model rough surface. Ausloos and Berman (1985) and Berry and Lewis (1980) furnished the applicability of the modified Weierstrass–Mandelbrot equation to represent rough fractal surface. The method to determine the value of the fractal dimension of physical rough surface was presented by Dubuc et al. (1989). Yan and Komvopolos (1998) investigated the elastic-plastic contact behavior between two approaching fractal rough surfaces through analytical approach. Several researchers carried out investigations on the normal contact stiffness of fractal rough surfaces under different conditions (Buczkowski et al., 2014; Pohrt & Popov, 2012; Xiao et al., 2019; Zhang et al., 2017). Since finite element analysis softwares have evolved as an extremely efficient tool for multiphysics analysis, it is also utilized to carry out contact analysis of fractal rough surfaces by several researchers (Chatterjee & Sahoo, 2013; Hyun et al., 2004; Pei et al., 2005; Sahoo & Ghosh, 2007).

Vibration at solid contact interface has direct influence to fatigue and wear which leads to damage of surfaces of several machine elements with solid surface to surface contact like gear, cam-follower, rolling element bearings etc. For the sake of controlling the vibration at the contact interface, the nature of the dynamic behaviour should be analyzed and understood. From the literature it has been observed that, when it comes to dynamic contact interactions between rough surfaces, Hertzian theory has been employed most extensively for modelling. First significant theoretical work on contact dynamics was by Nayak (1972), who developed some of the theoretical groundwork necessary for detailed physical explanations of experimentally observed phenomena in vibratory point contact by modelling a single-degree-of-freedom dynamic system and obtained analytical solutions by using a single term harmonic balancing method (HBM) (Nayfeh & Mook, 1995). Hess, Soom and Kim (Hess & Soom, 1991, 1993; Hess, Soom, & Kim, 1992) studied nonlinear vibrations dynamic contact situation theoretically and experimentally and quantified the reduction in mean contact compression, contact area and mean frictional force due to the nonlinear Hertzian stiffness. The authors obtained solutions for both the average and instantaneous contact deflections. Perret-Liaudet investigated a sphere-plane (Hertzian) contact problem for sub-harmonic and super-harmonic resonance of order two using MMS and expanding the nonlinearity by third order Taylor series (Perret-Liaudet, 1997, 1998). The condition for contact loss was taken into consideration for the first time in this work. Sabot et al. (1998) studied the non-linear vibrations of a sphere–plane contact excited by a normal load equal to the sum of a static load and a harmonic load and predicted the contact natural frequency, frequency contents and softening behavior using numerical and analytical methods Perret-Liaudet and Sabot (1999) analyzed the sphere-plane model of Hertzian contact by numerical methods and confirmed the significance of clearance at the contact on super-harmonic and sub-harmonic resonances, which lead to vibro-impact responses. Rigaud and Perret-Liaudet (Perret-Liaudet & Rigaud, 2003; Rigaud & Perret-Liaudet, 2003) investigated dynamic behaviour of a preloaded double sphere–plane Hertzian contact under sinusoidal excitation and random normal excitation and solution was obtained using the shooting method in conjunction with parametric continuation technique. The authors also analyzed the dynamic response of an impacting Hertzian contact subjected to an order-2 subharmonic excitation and an order-2 superharmonic excitation (Perret-Liaudet & Rigaud, 2006; Perret-Liaudet & Rigaud, 2007). Ma et al. (2007) studied the sphere-plane contact model with possible contact loss experimentally and analytically using a multi-term harmonic balancing method. Tian and Xie (2008) analyzed the dynamic contact stiffness at the interface between a vibrating rigid sphere and a semi-infinite transversely isotropic viscoelastic solid with a harmonic force superimposed onto a static compressive force. Xiao et al. (2011) used exact method, MMS and HBM to determine the natural frequency of a mass interacting with a Hertzian contact stiffness. Recently Jana et al. (2019) carried out dynamic analysis of functionally graded single asperity hemispherical and semi-cylindrical contact.

When it comes to dynamic analysis of multi-asperity surface contact, there exists a very few number of works in the literature. However, Xiao et al. (2015) modelled an elastic fractal surface vibrating while in contact with a rigid flat surface. They analyzed its force-displacement relation and studied the free and forced vibration responses of the system. Jana et al. (2017) investigated the influence of surface and material properties on the dynamic behaviour of an isotropic rough fractal surface. Recently, Huang et al. (2019) studied a vibrating coupled two-degree-of-freedom system involving rough fractal surface. Bifurcation and chaotic response was observed for the vibrating contact system under certain conditions.

From the literature survey it has been found that dynamic fractal surface contact with rigid flat surface is contained to an elastic domain. However, realistic materials are not infinitely elastic and exhibit definite yield point. Post elastic dynamic behaviour of rough surface in contact with rigid flat remains largely unexplored. Hence, in the present paper the effect of variation of material properties such as tangent modulus (assuming bilinear model) and yield strength values of the contacting rough surfaces on forced vibration responses of the system is analyzed. Forced response of the system corresponding to variation of fractal parameters is also studied.

The current work focuses on static and dynamic analysis of rough fractal surfaces. To be more specific, it investigates the influence of different fractal and material parameters on the static as well as dynamic behaviour of the contact interface. A three-dimensional rough fractal surface is generated using a modified two-variable Weierstrass-Mandelbrot function (Yan & Komvopoulos, 1998). Different contact parameters are obtained through static analysis using commercial FEA package ANSYS. The rough fractal surface is assumed to come in contact with a rigid flat surface. External harmonic excitation in the form of displacement is imparted on the rigid flat. For the present system, the equation of motion is represented by a forced Helmholtz–Duffing equation whose characteristics are dependent on the power value of the force-displacement relationship. Solution of the above mentioned equation leads to frequency response curves of the system. Superharmonic responses are obtained for the system and it is analyzed for the variation of the nonlinearity exponent value.

FRACTAL SURFACE MODELLING

3D fractal surface topography (Ling, 1990) can be utilized to realistically model rough surfaces. In the present paper the fractal surface is generated following modified two-variable Weierstrass-Mandelbrot function (Yan & Komvopoulos, 1998) expressed mathematically as follows.

$$z(x,y)=L\left(\frac{G}{L}\right)^{D-2}\left(\frac{\ln\gamma}{M}\right)^{1/2}\sum_{\mu=0}^{M}\sum_{f=0}^{f_{\max}}\gamma^{(D-3)f}\times\left\{\cos\varphi-\cos\left[\frac{2\pi\gamma^{f}\left(x^{2}+y^{2}\right)^{1/2}}{L_{0}}\times\cos\left(\tan^{-1}\left(\frac{y}{x}\right)-\frac{\pi\mu}{M}\right)+\varphi\right]\right\} \tag{1}$$

where, L and G are the sample length and height scaling parameter independent of frequency within the scale range, wherein fractal power law behavior exists. D, M, γ and ϕ are fractal dimension, number of superimposed ridges to construct the surface, frequency density and random phase angle, respectively. For a real physical rough surface the value of fractal dimension (D) is found to be between 2.0 and 3.0 (Yan & Komvopoulos, 1998) Frequency index, represented by f, has a lower limit at zero for a truncated series of the height function and the upper limit is given by,

$$f\,\mathrm{int}\left[\frac{\log\left(L/L_{s}\right)}{\log\gamma}\right]_{\max} \tag{2}$$

Here, int[.] represents the maximum integer value of the number within the bracket and L_s is the cut off length, which is in the order of about six lattice distances. The surface height function provided by equation (1) has the following properties –

1. continuous
2. non-differentiable
3. scale-invariant (within the range determined by the upper and lower wavelengths used in the truncated series)
4. asymptotically self-affine

The surface heights $z(x,y)$ at different points (x,y) are evaluated using a MATLAB code with the value of φ generated by the random number generator function of MATLAB. The values of sample length and cut-off length are set to be 1×10^{-6} m and 1.5×10^{-7} m, respectively, whereas, values of the parameters M and γ are taken as 1 and 1.5, respectively. Figure 1(a) shows the constructed fractal surface simulated for $D = 2.4$, $G = 1.36\times10^{-11}$ m in MATLAB with 50×50 grid. It should be mentioned here that low value of G and high value of D correspond to smoother and dense surface profile and vice-versa.

FINITE ELEMENT MODELLING

The surface heights $z(x,y)$ as per the supplied x and y values generated in MATLAB are imported to ANSYS 14.5 as keypoints by defining array parameter. To construct a cubic block on which the rough surface is to lay, additional keypoints are created and these keypoints are joined to construct lines. Subsequently, areas are formed with the lines generated from the keypoints. Finally a cubic solid block of dimension 1μm × 1μm × 1μm is generated whose top surface is the fractal rough surface profile constructed from modified two-variable Weierstrass-Mandelbrot function. The rough surface on the top is set as the contact surface. A rigid flat surface, which is identified as the target surface, is constructed to touch the contact surface from top. Finite element modeling of the contact interface is achieved using three different types of elements, namely, solid element, contact element and target element. A higher order 3D, 10-node solid element, known as SOLID187, is used to mesh the rough deformable solid body. The target surface is meshed with TARGE170 element which comes in contact with the CONTA174 elements overlaying the SOLID187 elements. Figure 1(b) shows the meshed model of fractal surface having the same G and D values.

Figure 1. Weierstrass-Mandelbrot surface for D=2.4, G = 1.36×10^-11 m simulated in (a) MATLAB with 50×50 grid and (b) meshed model of the 8×8 fractal surface in ANSYS

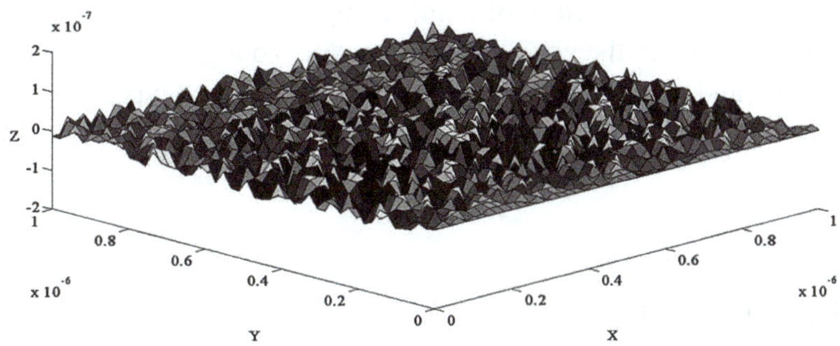

(a)

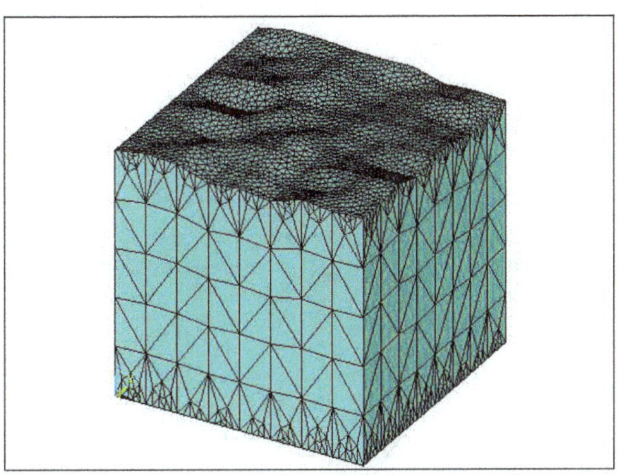

(b)

At the center of the rigid target surface a pilot node is chosen and the external load/displacement is applied at this node. Constraints are applied to the rigid flat to allow motion in z direction only and to restrict motion along the remaining two coordinate axes (x and y). All the degrees of freedom in all the directions at the nodes at the bottom of the solid elements (i.e. all the nodes at $z = 0$) are constrained. The contact with the rigid plane is realized using surface-to-surface contact elements that use the augmented Lagrangian method (Zhong, 1993). In the present method a downward (towards negative z direction) displacement is imposed on the rigid surface incrementally with 10 to 100 sub-steps to come in contact and deform the deformable rough surface. After the solution is obtained, different contact parameters e.g. contact force, contact area, contact stress etc. at the contact surface are recorded. The non-dimensional displacement ($\delta^*=\delta/L$) vs. non-dimensional load $\left(P^* = P / \left(A_0 E'\right)\right)$ curves are fitted in a power law relationship in the form,

$$P^* = k(\delta^*)^n \tag{3}$$

where, k is nonlinear contact stiffness which is a constant parameter with positive value. In this case, a bulk stiffness is considered for the whole surface instead of the individual stiffness of the asperities. The parameter n can be considered as a measure of nonlinearity of the system and is termed as 'nonlinearity exponent'. E', A_0, P, δ and L are the equivalent composite Young's modulus, nominal contact area, applied load, displacement and sample length, respectively. The equivalent composite Young's modulus is expressed as $E' = E / \left\{1 - \nu^2\right\}$, where, E is the elastic modulus and ν is the Poisson's ratio of the deformable rough surface.

MESH CONVERGENCE STUDY

Prior to performing any finite element analysis, it is essential to perform a mesh convergence study in order to determine the optimum mesh density. The mesh convergence study is performed in terms of

Figure 2. Mesh convergence plot in terms of contact force corresponding to non-dimensional rigid flat displacement (δ/L) of 0.03

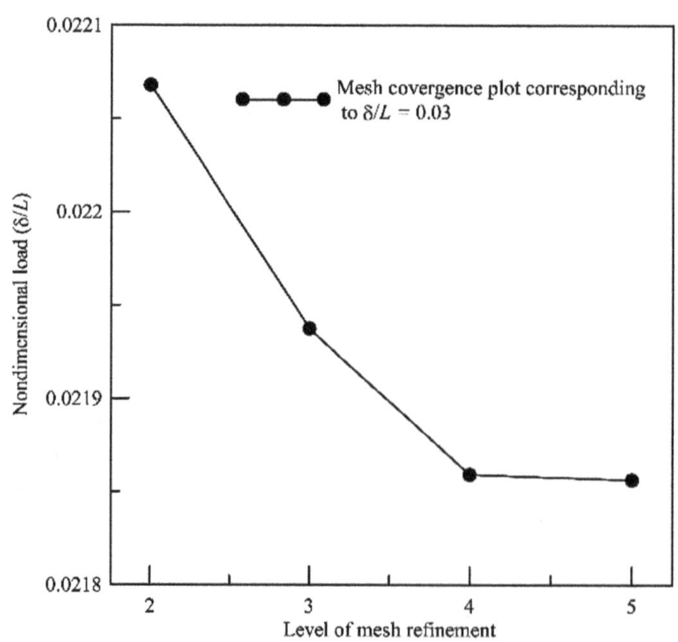

contact force. Fractal surface with $D = 2.5$ and $G = 1.36\times10^{-11}$ m is adopted for the mesh convergence study. The material of the fractal surface is taken as linearly elastic with elastic modulus of 200 GPa and Poisson's ratio of 0.3. Static finite element analysis is performed following the methodology mentioned in the previous section. The value of the contact reaction force corresponding to $\delta^* = 0.03$ is obtained for FE models with increasing mesh density. The mesh configuration is made denser in steps until the difference of the value of contact force of two successive steps is less than 0.1%. Figure 2 furnishes the mesh convergence plot. The final mesh configuration corresponding to $D = 2.5$ and $G = 1.36\times10^{-11}$ m comprises 27897 SOLID187 elements, 3712 CONTA174 elements and 10 TARGE170 elements.

VALIDATION STUDY

Validation of the finite element modelling of contact between a deformable rough surface and a rigid flat is obtained by comparison with experimental results provided by Buzio et al. (2003) who studied the load deflection behavior between atomic force microscope (AFM) probes and self-affine fractal carbon films. The authors presented the experimental results for multiple asperity contact between nanostructured carbon and AFM probe. The AFM imaging provides fractal dimension $D = 2.3$ and fractal roughness $G = 1.55\times10^{-10}$ m along with the nominal contact area $A_0 = 4.8\times10^{-12}\text{m}^2$. The material properties for the system were taken as, equivalent composite Young's modulus (E') = 0.88 GPa, tangent modulus

Figure 3. Comparison of non-dimensional load-displacement results from present analysis with the results of Buzio et al. (2003) and Chatterjee and Sahoo (2014)

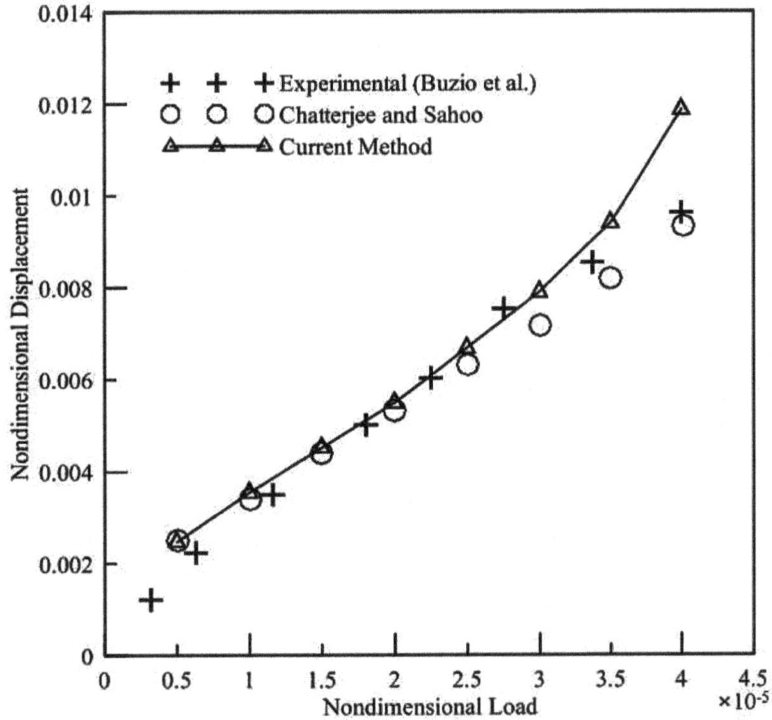

(E_t) = 0, Poisson's ratio (ν) = 0.3 and yield strength (Y) = 16.07 MPa. Keeping the material and fractal properties same, the force-displacement analysis is carried out in ANSYS. SOLID185, TARGE170 and CONTA174 elements are used to construct the solid, target and contact elements respectively during the finite element analysis. Downward force (towards negative z direction) is applied on the pilot node causing downward displacement of the rigid flat as well as the deformable rough surface due to deformation. The result as shown in Figure 3 is in favorable agreement with the experimental result as well as with the result by Chatterjee and Sahoo (2014)

FINITE ELEMENT ANALYSIS RESULTS

Different aspects of behaviour of fractal surface are analysed through finite element analysis. The values of fractal dimension (D) considered are 2.3, 2.4 and 2.5 whereas three values of G, 1.36×10^{-10} m, 1.36×10^{-11} m, 1.36×10^{-12} m and 1.36×10^{-13} m, are taken into account. The material is taken as infinitely linear elastic i.e. no yielding occurs up to infinite stress. The elastic modulus and the Poisson's ratio of the material is taken as 200 GPa and 0.3 respectively. The variation of nondimensional contact force ($P/E'L^2$) vs. nondimensional deflection (δ/L) behaviour corresponding to varying D and G values are shown in Figure 4(a) and Figure 4(b) respectively. It can be observed from the Figures that for lower value of D and higher value of G, i.e. for rougher surface topography, the resulting contact force is lower for a given value of rigid flat displacement (δ).

Fig 5(a) and 5(b) illustrate nondimensional rigid flat displacement vs. nondimensional contact area (A/L^2) plots for the same linear elastic fractal surface with varying D and G values. The Figures suggest that rougher surface (lower D and higher G) causes lower contact area and vice-versa. However, the trend is found to be opposite for nondimensional average contact pressure (P/AE') where rougher surface results in higher average contact pressure (Figure 6).

The effect of fractal parameters on the stress distribution at the rough surface is also analyzed. Figure 7 shows the contour plots of Von Mises stress (σ_{vm}) at the fractal surface for fixed value of G (1.36×10^{-11} m) and D values of 2.3 and 2.5 corresponding to $\delta/L = 0.025$. Figure 8 presents the same for fixed D (2.4) and two values of G (1.36×10^{-12} m and 1.36×10^{-10} m). It can be noted that the contact spots of fractal surfaces with lower value D and higher value G are induced with much higher values of Von Mises stress. This is because, for the rougher surface, the load is supported by smaller contact area hence resulting in higher stress. On the other hand, for smoother surface, load is distributed in larger contact areas.

The previous plots are presented under the assumption of infinite linear elasticity, which is not practical. To investigate the dynamic contact behaviour in post elastic domain, it is required to define the yield strength along with tangent modulus. Hence, keeping the surface properties fixed at $D = 2.4$ and $G = 1.36\times10^{-11}$ m, the material is considered to yield and behave according to a bilinear model after yielding. Effect of variation of tangent modulus with a fixed yield limit on the force-displacement behaviour is also an interesting aspect of study. Yield strength is considered to be fixed at 250 MPa and three value of the tangent modulus is taken as 100 GPa, 60 GPa and 10 GPa. The elastic modulus and Poisson's ratio is set fixed at the previously mentioned (infinitely linear elastic) values.

Similar analysis is also applied to obtain the force-displacement relationship to different yield strength values. The yield strength is varied as 250 MPa, 560.8 MPa, 911.5 MPa, 1265.3 MPa and 1619 MPa keeping the tangent modulus fixed at 10 GPa. These yield strength values cover a range of steels used in engineering applications (Chatterjee & Sahoo, 2013; Jackson & Green, 2005). Figure 9 illustrates

Figure 4. Plots of non-dimensional displacement (δ/L) versus non-dimensional contact force (P/ E'L²) of rough fractal surfaces for varying (a) fractal dimension (D) and (b) fractal roughness (G)

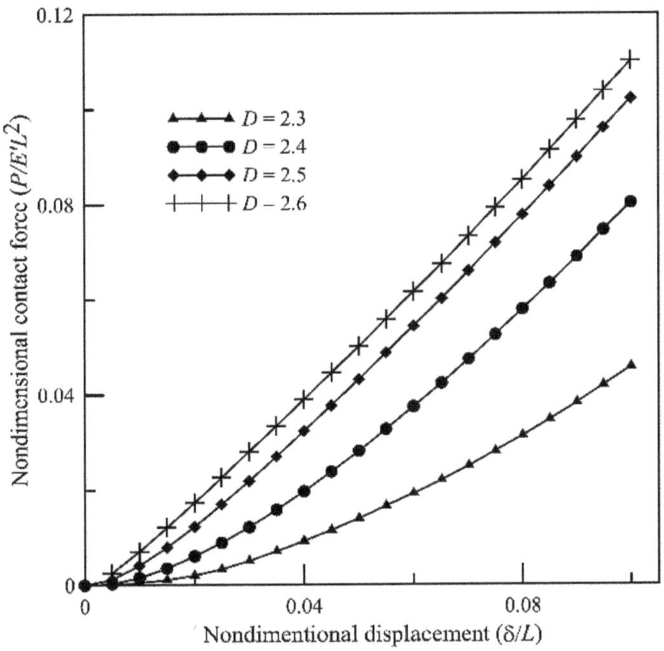

(a)

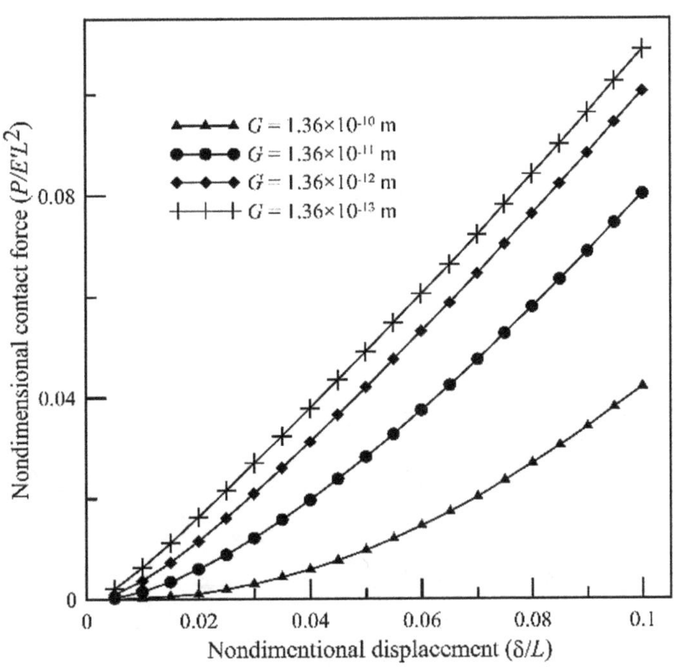

(b)

Figure 5. Plots of non-dimensional displacement (δ/L) versus non-dimensional contact area (A/L^2)of rough surfaces for varying (a) fractal dimension (D) and (b) fractal roughness (G)

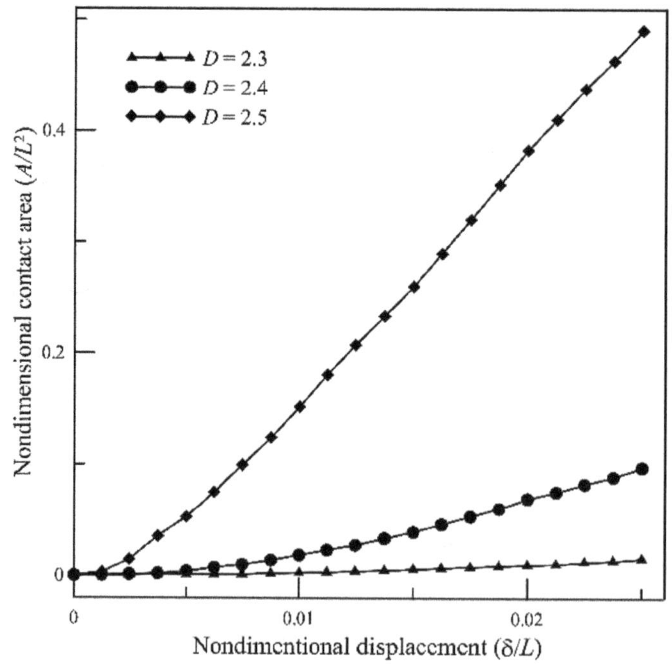

(a)

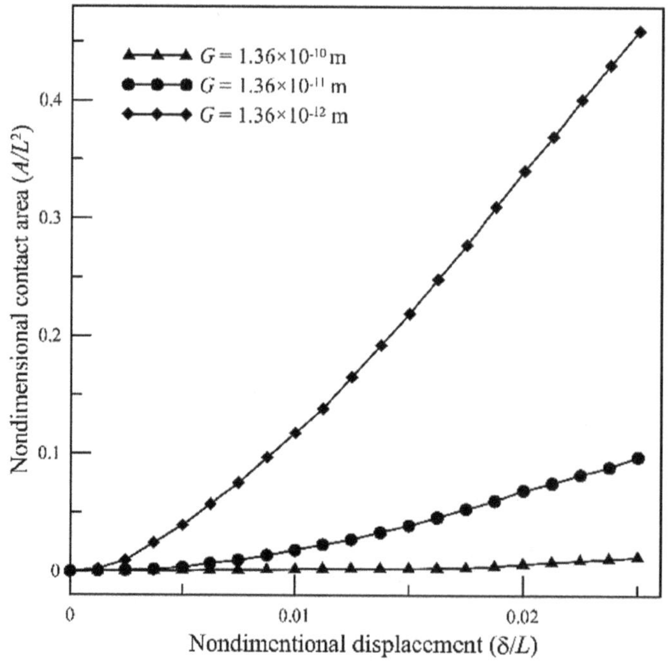

(b)

Figure 6. Plots of non-dimensional displacement (δ/L) versus non-dimensional average contact pressure (P/AE') of rough surfaces for varying (a) fractal dimension (D) and (b) fractal roughness (G)

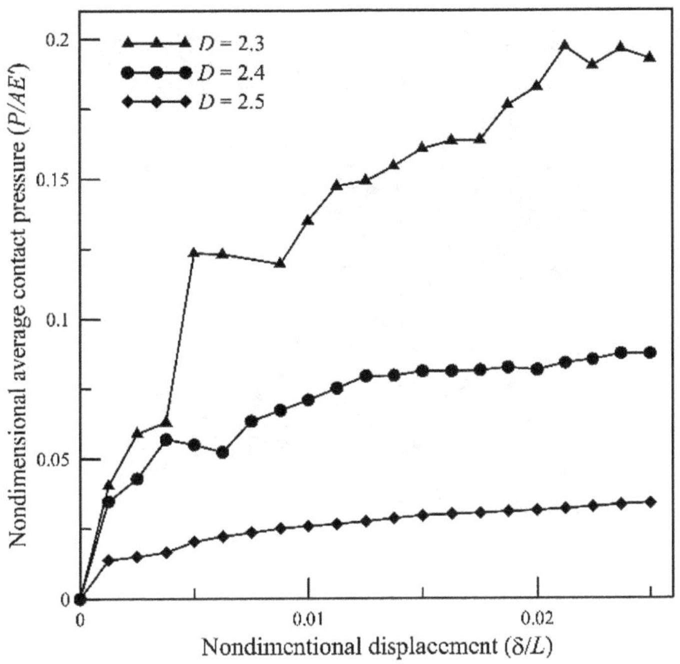

(a)

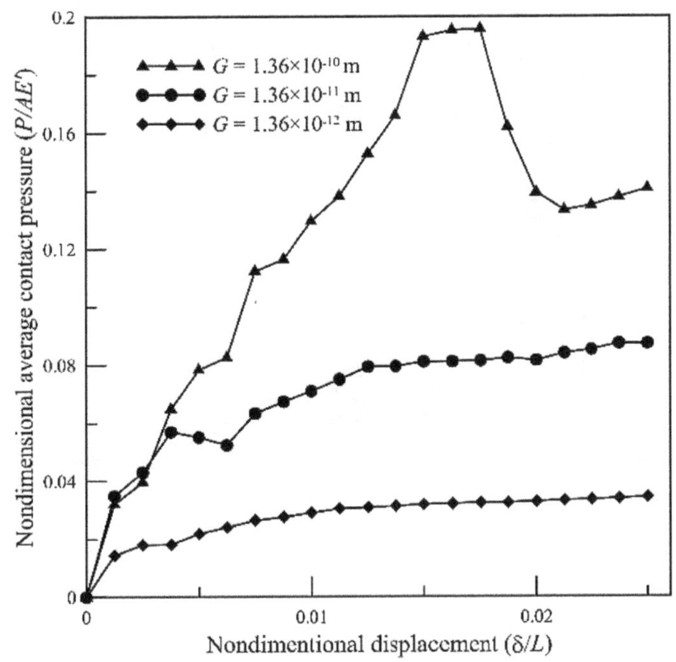

(b)

Figure 7. Contour plots of Von Mises stress (σ_{vm}) of the top surface of linear elastic fractal surface with $G = 1.36\times10^{-11}$ m, (a) D = 2.3 and (b) D = 2.5, corresponding to $\delta/L = 0.025$

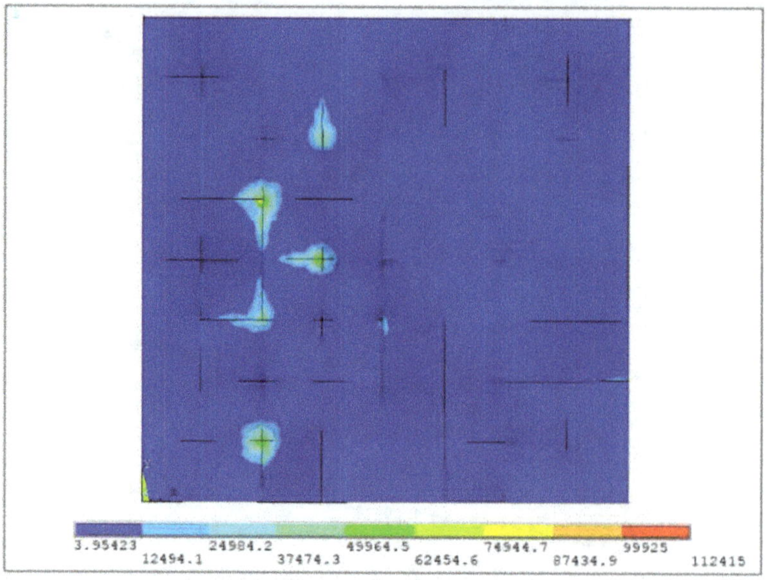

(a)

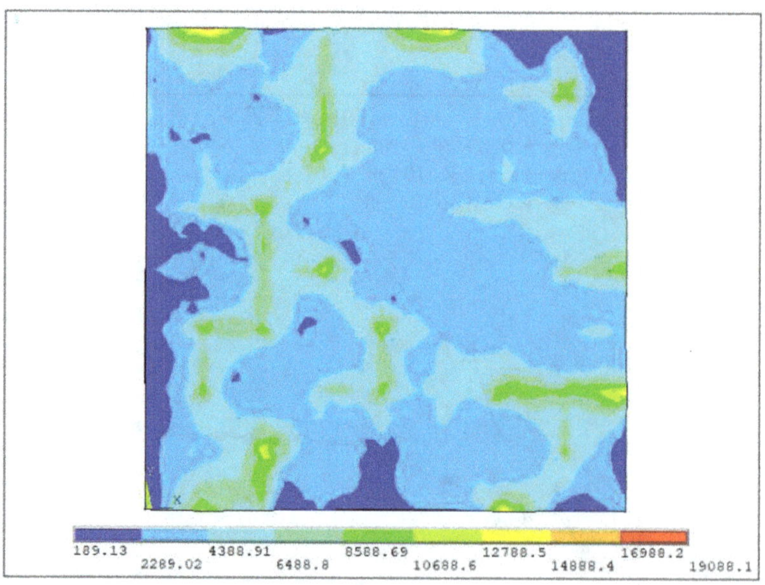

(b)

nondimensional rigid flat displacement vs. nondimensional contact force plots for varying tangent modulus (E_t) and yield strength (Y) values. Nondimensional average contact pressure is plotted against nondimensional rigid flat displacement for varying E_t and Y values in Figure 10. The Figures shows

Figure 8. Contour plots of Von Mises stress (σ_{vm}) of the top surface of linear elastic fractal surface with $D = 2.4$, (a) $G = 1.36{\times}10^{-12}$ m and (b) $G = 1.36{\times}10^{-10}$ m, corresponding to $\delta/L = 0.025$.

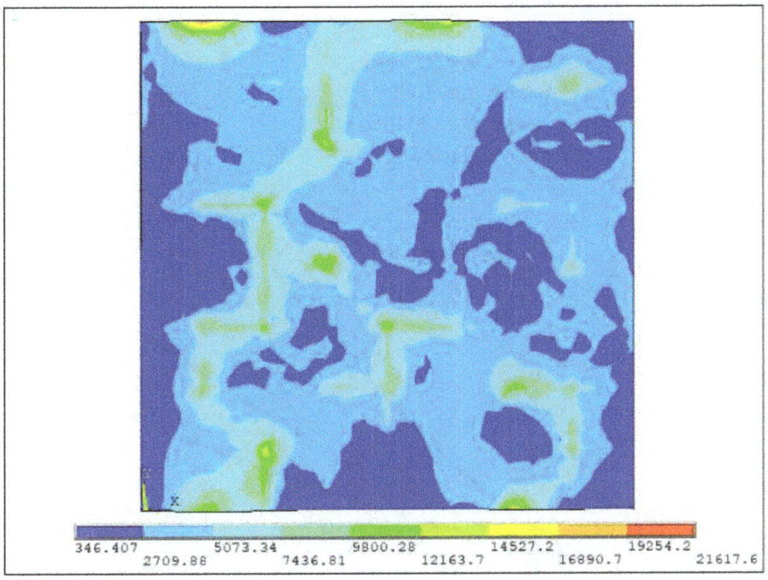

(a)

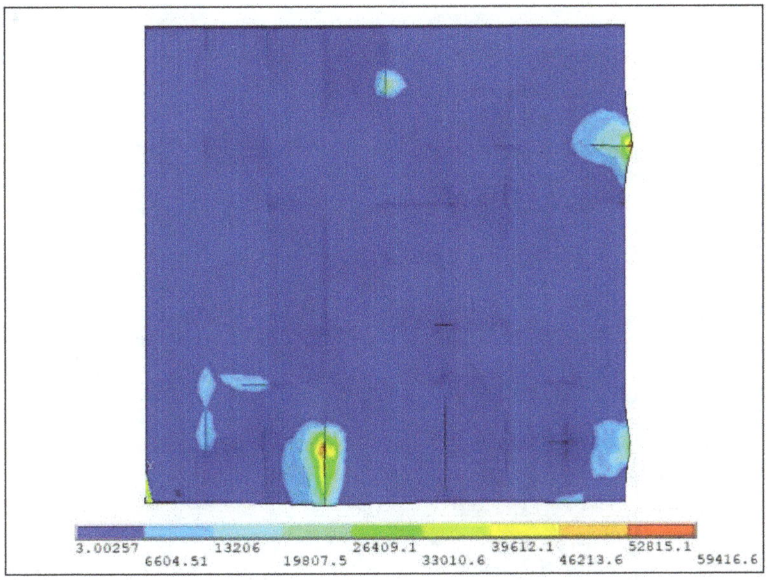

(b)

that higher tangent modulus and yield strength is associated with higher contact force as well as higher average contact pressure.

Figure 9. Plots of nondimensional displacement (δ/L) versus non-dimensional contact force ($P/E'L^2$) of rough fractal surfaces for varying (a) tangent modulus (E_t) and (b) yield strength (Y)

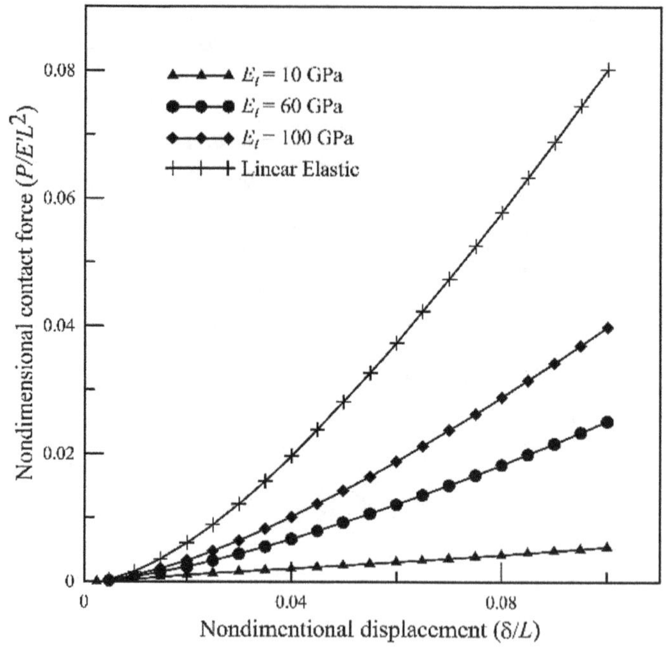

(a)

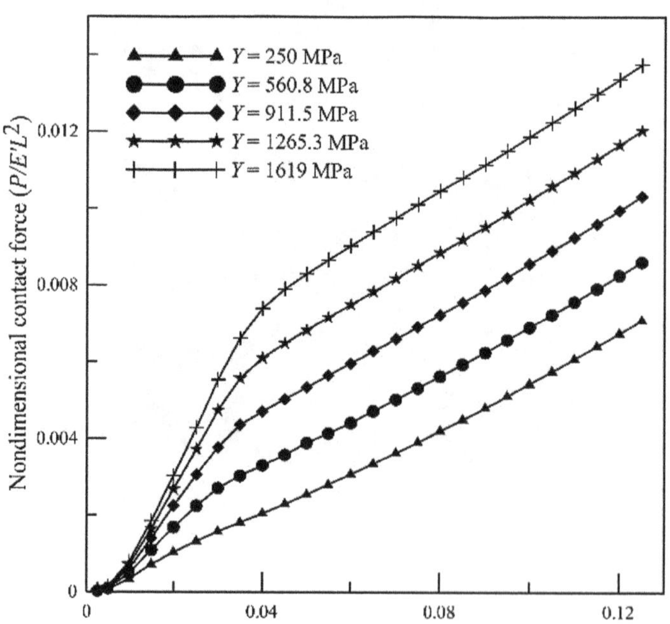

(b)

Figure 10. Plots of non-dimensional displacement (δ/L) versus non-dimensional average contact pressure (P/AE') of rough surfaces for varying (a) tangent modulus (E_t) and (b) yield strength (Y)

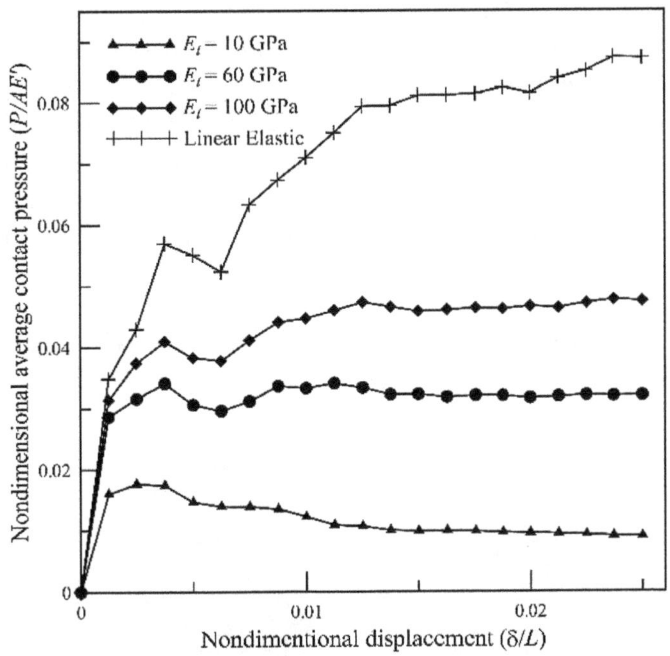

(a)

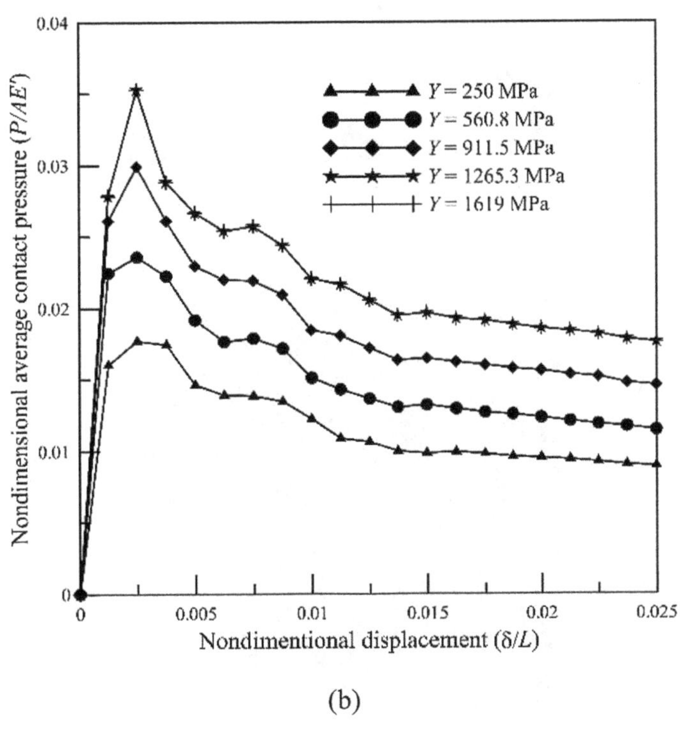

(b)

Figure 11 presents the contour plots of Von Mises stress (σ_{vm}) at the top surface of an elastic-plastic fractal surface (D = 2.4 and G = 1.36×10^{-11} m) with fixed yield strength (Y) 250MPa and and varying tangent modulus (E_t) of 10 and 100 GPa corresponding to nondimensional rigid flat displacement of 0.025. Higher Von Mises stress is induced in the fractal surface with higher tangent modulus. Similar nature is observed for the variation in yield strength, i.e. higher value of yield strength results in higher Von Mises stress at the rough surface. In order to get an idea of the state of the material located at the rough surface, stress state contour plot is presented which compares the Von Mises stress at any point on the surface with the yield strength. It provides the visualization of the information whether the material at the surface has yielded or not. Figure 13 presents the stress state contour plots for the variation of

Figure 11. Contour plots of Von Mises stress (σ_{vm}) at the top surface of the elastic-plastic fractal surface with (a) E_t = 10 GPa and (b) E_t = 100 GPa, corresponding to δ/L = 0.025

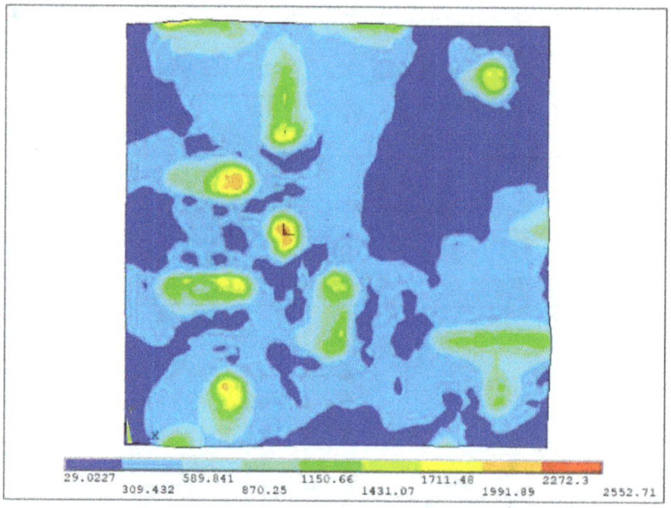

(a)

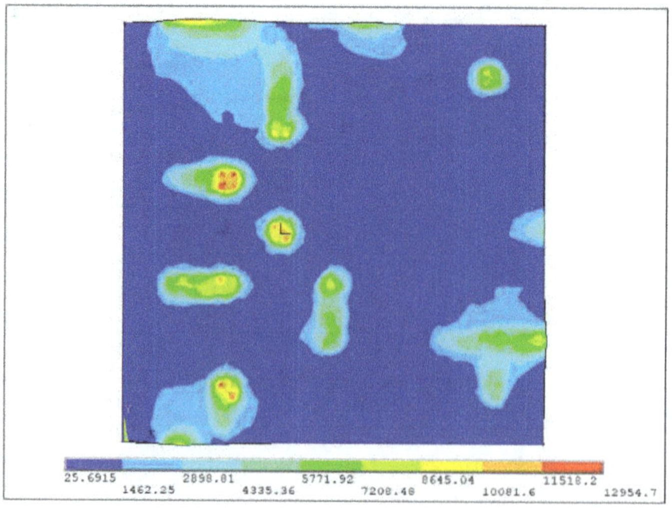

(b)

tangent modulus keeping the yield strength fixed at 250 MPa. It shows that for higher value of tangent modulus, more material at the surface gets yielded. Figure 14 presents similar stress state contour plots for two yield strength values i.e. 250 MPa and 1619 MPa, respectively, keeping the tangent modulus fixed at 10 GPa. The Figure shows that higher yield strength causes smaller yielded zone at the surface.

Figure 12. Contour plots of Von Mises stress (σ_{vm}) of the top surface of elastic-plastic fractal surface with (a) Y = 250 MPa and (b) Y = 1619 MPa, corresponding to δ/L = 0.025

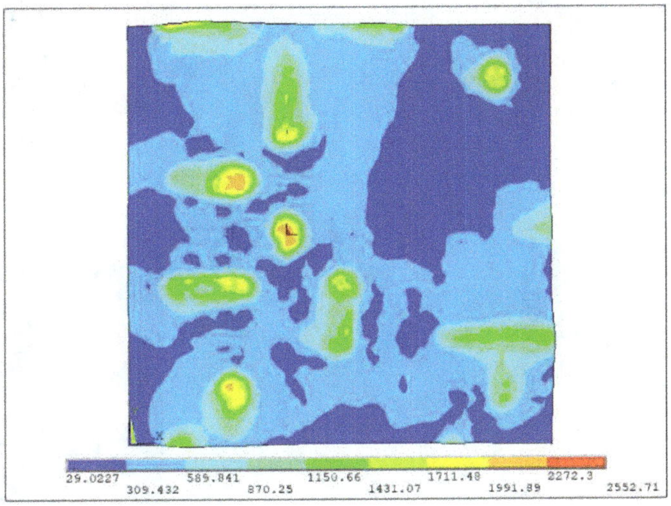

(a)

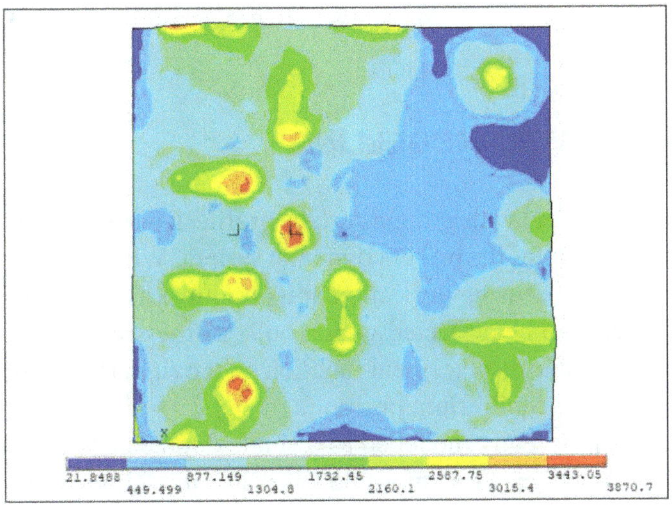

(b)

Figure 13. Stress state contour plots of the elastic-plastic fractal surface with (a) E_t = 10 GPa and (b) E_t = 100 GPa, corresponding to δ/L = 0.0125

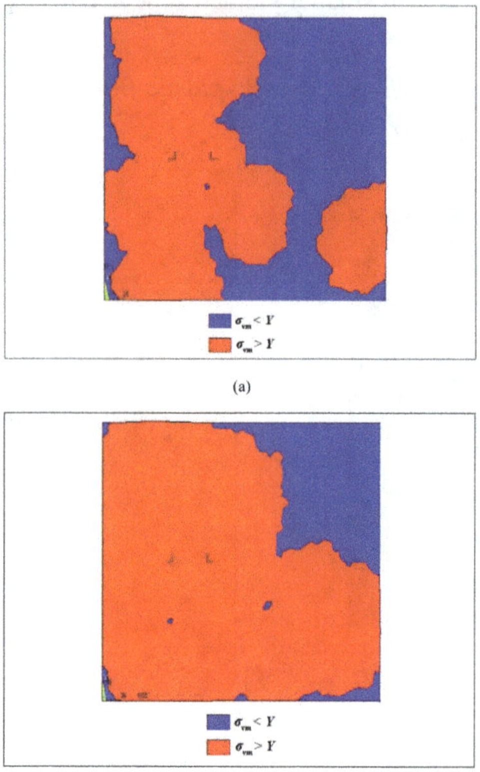

(a)

(b)

Figure 14. Stress state contour plots of the elastic-plastic fractal surface with (a) Y = 250 MPa and (b) Y = 1619 MPa, corresponding to δ/L = 0.0125.

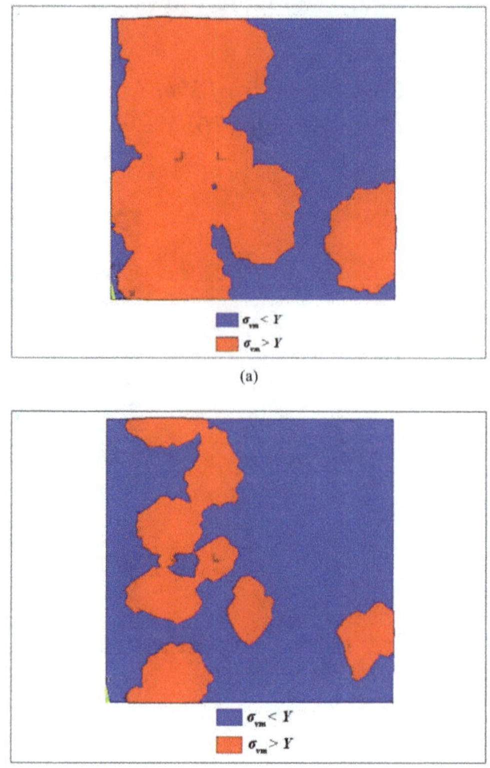

(a)

(b)

DYNAMIC CONTACT SYSTEM MODELLING

A deformable solid body (mass = *m*) with a rough surface profile at one extremity and a rigid flat surface capable of vertical motion constitutes the present dynamic system. While in motion, the rigid flat comes in contact with the rough surface causing deflection at its multiple asperities (Figure 15(a)). However, the rough surface is considered to be deflected with a bulk stiffness *k* for the entire surface. External excitation in the form of a harmonic displacement function on the rigid flat causes the deformable rough surface to vibrate about the static equilibrium position.

This dynamic system is equivalently modelled as a single degree of freedom (SDOF) spring-mass-damper system (Figure 15(b)) attached to a movable base. The model is considered to be valid for low frequency behaviour only (Komvopoulos & Ye, 2000). The mass block is considered to be equivalent to the deformable body in terms of mass, i.e., the block has the same mass as that of the dough solid. The spring stiffness is taken as the bulk stiffness of the surface. It should be pointed out that the spring is non-linear in nature, which implies that the restoring force in the system is a nonlinear function of the spring deformation. On the other hand, damping model is considered to be linear and viscous, with linear damping co-efficient *c*. The deflected position of the block due to its weight is considered as the static equilibrium position. The spring undergoes a deflection of z_s due to the weight of the block and the static equilibrium position of the system is denoted by o as shown in Figure 15(b). The system, under the influence of a harmonic external excitation, executes oscillation about this equilibrium position and $z(t)$ indicates the displacement of the system with respect to the equilibrium.

Figure 15. (a) Deformable rough surface in contact with rigid flat surface and (b) SDOF dynamic model representation

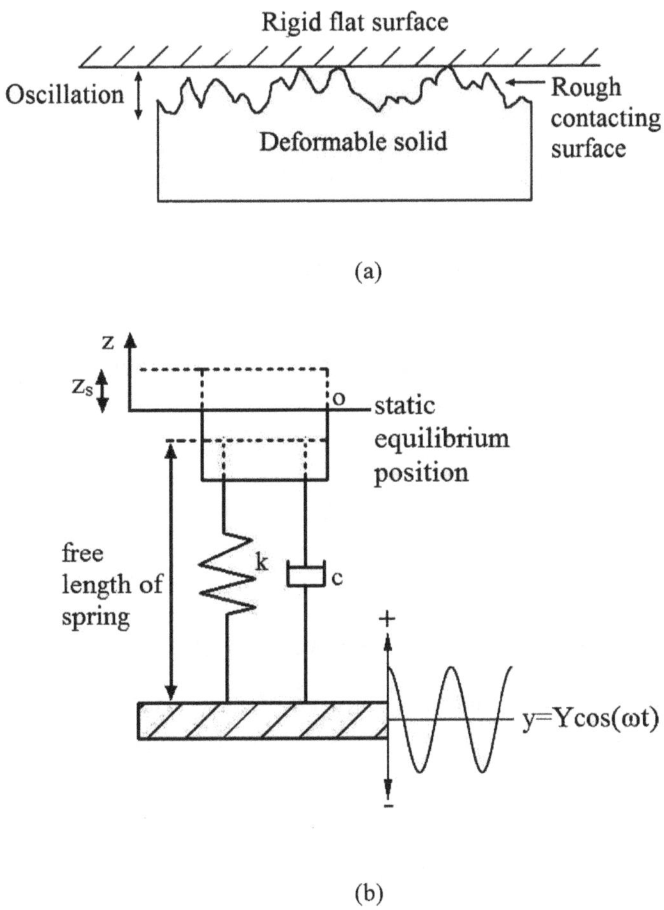

Since the contact interface is modelled by the nonlinear spring, the force displacement relationship for the nonlinear spring follows the same power law as equation (3).

$$F_s = k(z + z_s)^n \tag{4}$$

Here, F_s is the restoring force and k is the nonlinear stiffness of the spring. External harmonic excitation is applied as displacement on the base that the spring-mass-damper system is resting on. The mathematical form of the harmonic excitation is given as,

$$y = Y\cos(\omega t) \tag{5}$$

Hence,

$$\dot{y} = Y\omega\sin(\omega t) \tag{6}$$

where, y(t) denotes the displacement of the base. Considering both the motions of the mass and the base, the governing equation of the system can be written as,

$$m\ddot{z} + c\left(\dot{z} - \dot{y}\right) + k\left[\left(z + z_s\right)^n - y^n\right] - mg = 0 \tag{7}$$

Putting the values of y and $\dot{y}$ from equation (5) and (6) we obtain

$$m\ddot{z} + c\dot{z} + k\left(z + z_z\right)^n - mg = k\left[Y\cos\left(\omega t\right)\right]^n + c\omega Y\sin\left(\omega t\right) \tag{8}$$

The equation of motion can be expressed in normalized form using the following non-dimensional variables.

$$u = \frac{z}{z_s},\ \Omega = \frac{\omega}{\omega_s},\ \tau = \omega_s t,\ \zeta = \frac{c}{2m\omega_s} \text{ and } Z = \frac{Y}{z_s}$$

Natural frequency at the static equilibrium position, $\omega_s = \sqrt{\dfrac{nkz_s^{n-1}}{m}}$ is utilized to normalize the external excitation frequency. u, Ω, τ, Z and ζ are normalized displacement, frequency, time, excitation amplitude and damping coefficient, respectively. After normalization the dimensionless equation of motion becomes

$$u'' + 2\zeta u' + \frac{1}{n}\left[\left(u+1\right)^n - 1\right] = \frac{1}{n}Z^n\cos^n\left(\Omega\tau\right) + 2\zeta Z\Omega\sin\left(\Omega\tau\right) \tag{9}$$

where, $'$ denotes differentiation w.r.t. non-dimensional time. The expression $(u+1)^n$ is expanded into a third order Taylor series expansion and hence the following equation is obtained.

$$u'' + 2\zeta u' + u + \alpha_2 u^2 + \alpha_3 u^3 = \frac{1}{n}Z^n\cos^n\left(\Omega\tau\right) + 2\zeta Z\Omega\sin\left(\Omega\tau\right) \tag{10}$$

where, $\alpha_2 = \dfrac{\left(n-1\right)}{2}$ and $\alpha_3 = \dfrac{\left(n-1\right)\left(n-2\right)}{6}$.

The governing differential equation describes linearly damped forced vibration of the rough surface in contact with a harmonically excited rigid flat surface. Since the restoring force part of the left hand side of equation (10) contains a quadratic and a cubic term, it can be classified as a Duffing-Helmholtz type equation with natural frequency (ω_n) value of unity. The above mentioned equation is solved numerically using ODE solver of MATLAB to obtain the frequency response curve for the vibrating rough surface.

RESULTS OF DYNAMIC CONTACT ANALYSIS

The values of the nonlinear stiffness (k) and the nonlinearity exponent (n) are calculated by fitting the nondimensional force-displacement plots (Figure 4) in the form of equation (3) corresponding to varying fractal and material properties. The nonlinear stiffness (k) and the nonlinearity exponent (n) for different fractal dimension (D) and fractal roughness (G) values are shown in Table 1. Since the nonlinearity exponent (n) is the measure of nonlinearity, it can be seen that the nonlinearity increases for rougher surface topography i.e. higher G value and lower D value and vice versa. Following the same way, from the plots in Figure 9, the values of k and n corresponding to varying tangent modulus and yield strength values are furnished in Table 2. It can be observed that for higher tangent modulus (E_t) value the surface exhibits higher nonlinearity. On the other hand, the nonlinearity exponent is found to decrease with the increase of yield stress (Y). In all the previously mentioned cases, the value of n is found to be greater than unity, whereas, for the yield strength values of 560.8 MPa, 911.5 MPa, 1265.3 MPa, and 1619 MPa, the n value is found to be less than unity.

Table 1. Values of nonlinear stiffness (k) and nonlinearity exponent (n) for different fractal dimension (D) and fractal roughness (G) values

Fractal Roughness (G)	Fractal Dimension (D)	Nonlinear Stiffness (k)	Nonlinearity Exponent (n)
1.36×10^{-11}m	2.3	2.5751	1.74723
1.36×10^{-11}m	2.4	2.7927	1.5372
1.36×10^{-11}m	2.5	1.9254	1.2719
1.36×10^{-11}m	2.6	1.5275	1.1421
1.36×10^{-10}m	2.4	5.4703	2.1092
1.36×10^{-12} m	2.4	1.9939	1.2934
1.36×10^{-13} m	2.4	1.5909	1.1638

Frequency response curve is an effective way to represent dynamic behaviour of a system. The comparison of the frequency response curves for the surfaces with $G = 1.36\times10^{-11}$ m and fractal dimension (D) varying from 2.6 to 2.3 are presented in Figure 16(a) and 16(b). In Figure 15(a), frequency response curves for D values 2.4, 2.5 and 2.6 are plotted for the same level of excitation ($Z = 0.0198$), while in Figure 15(b) frequency response curves for D values 2.3 and 2.4 are presented corresponding to for $Z =$ 0.036. Similarly the comparison of the frequency response curves for the surfaces with $D = 2.4$ m and G varying from 1.36×10^{-11} m to 1.36×10^{-13} m are presented in Figure 17. The amplitude of excitation is

Table 2. Values of nonlinear stiffness (k) and nonlinearity exponent (n) for different tangent modulus and yield strength values.

Yield Strength (MPa)	Tangent Modulus (GPa)	Nonlinear Stiffness (k)	Nonlinearity Exponent (n)
N/A	Linear Elastic	2.7927	1.5372
250	100	1.2666	1.4989
250	60	0.71652	1.453
250	10	0.06957	1.1057
560.8	10	0.05301	0.8816
911.5	10	0.05112	0.7709
1265.3	10	0.05416	0.7174
1619	10	0.05944	0.6914

taken as $Z = 0.0192$. It can be clearly noted that for higher value of nonlinearity i.e. for rougher surface topography, the amplitude of response decreases for same level of excitation. Frequency response curves for the deformable surfaces having yield strength 250 MPa and varying tangent modulus values as 10, 60 and 100 GPa are presented in Figure 18 and surfaces with tangent modulus fixed at 10 GPa and varying yield strength values as 560.8 MPa, 911.5 MPa, 1265.3 MPa and 1619 MPa are showed in Figure 19. The excitation level is kept as $Z = 0.02$ and $Z = 0.000295$ for the two cases, respectively. Hence it could be observed from Figure 18 and 19 that lower tangent modulus and higher yield strength value results in higher response amplitude. When it comes to the nature of the frequency response i.e. whether the system is of softening or hardening type it can be observed that for the value of nonlinearity exponent (n) less than 1, the system is found to be hardening in nature and softening for nonlinearity exponent value greater than 1 as indicated by Figure 20.

For the present system, amplitude peaks are observed at excitation frequency values $\omega_n/2$, $\omega_n/3$ etc. i.e. the system exhibits superharmonic responses. In Figure 21(a) to 21(e) superharmonic response is shown for the variation of the nonlinearity exponent (n) from 1.2 to 2.0 keeping the excitation amplitude fixed at $Z = 0.02$. It can be seen that superharmonic response can be noted at normalized excitation frequency values of 0.5 ($\omega_n/2$), 0.33 ($\omega_n/2$), 0.25 ($\omega_n/2$). In Figure 22 the magnified view of the responses inside the rectangle of Figure 21(c) shows the superharmonic responses at frequencies $\omega_n/3$, $\omega_n/4$, $\omega_n/5$, $\omega_n/6$, and so on. For the increase of the nonlinearity exponent, as previously mentioned, the response amplitude decreases at natural frequency as well as at superharmonic frequencies. Interestingly, the response amplitude at $\omega_n/2$ is found to be higher than at the natural frequency value ω_n for the situation $n = 1.8$ (Figure 21(d)). But above a certain value of n, the amplitude at the natural frequency becomes fixed and does not change with increase of the nonlinearity exponent. Also it has been observed that the value of the amplitude it gets fixed at, depends upon the excitation amplitude and increases when the later increases.

For initial displacement of 0.5, the phase plots and time-displacement plots for n values 0.6, 1.0, 1.5, and 2.0 are presented in Figure 23 to 26.From the phase plot and the time-displacement plot (in non-dimensional plane) of the system a new phenomenon is observed for different nonlinearity exponent values. It can be observed from Figure 23 to Figure 26 that for higher value of nonlinearity exponent (n), the maximum negative displacement value of the vibrating system increases. Hence, it can be said that

Figure 16. Comparison between frequency response curves for surfaces having D values (a) 2.4 to 2.6 for Z = 0.0198 and (b) 2.4 and 2.3 for Z = 0.036

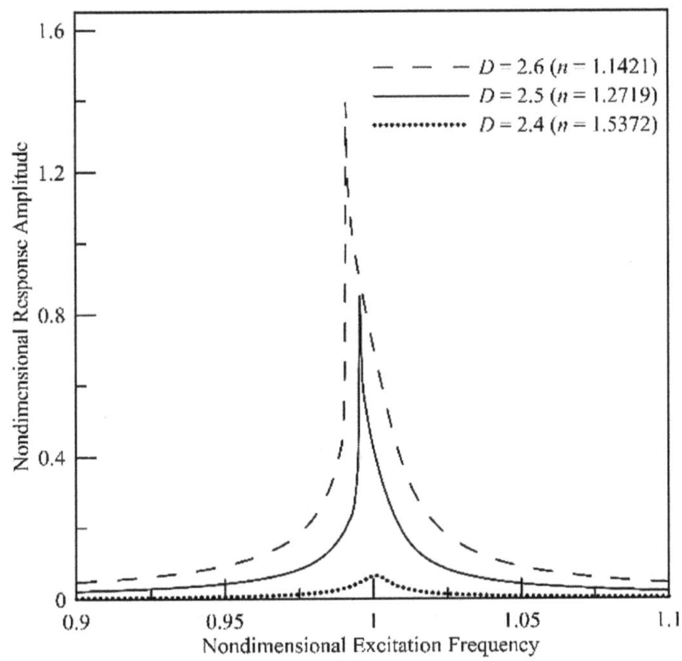

(a)

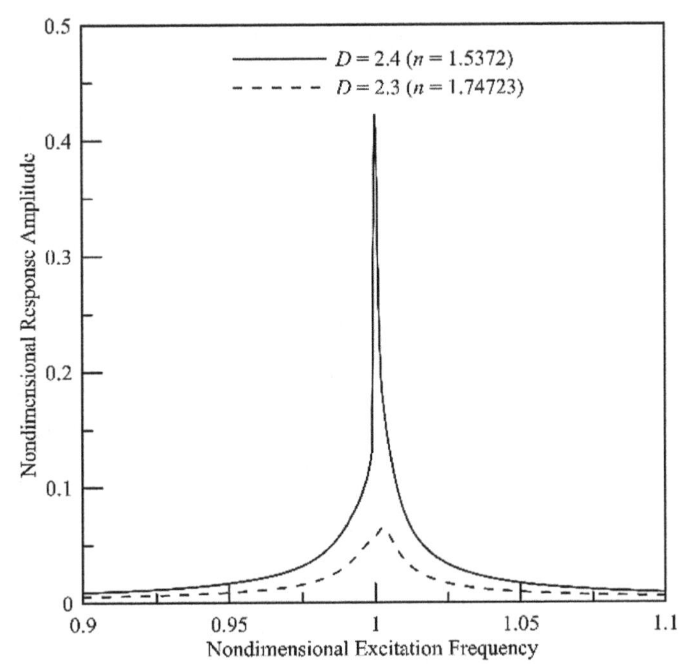

(b)

Figure 17. Comparison between frequency response curves for surfaces having G values varying from 1.36×10^{-11} m to 1.36×10^{-13} m for Z = 0.0192

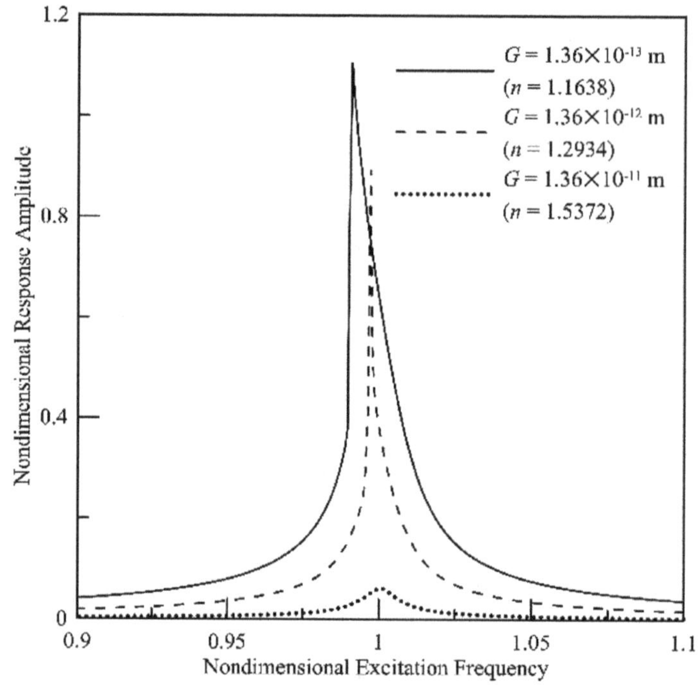

Figure 18. Frequency response curves for rough surface material with yield stress 250 MPa and varying tangent modulus (E_t) 10, 60 and 100 GPa for Z = 0.02

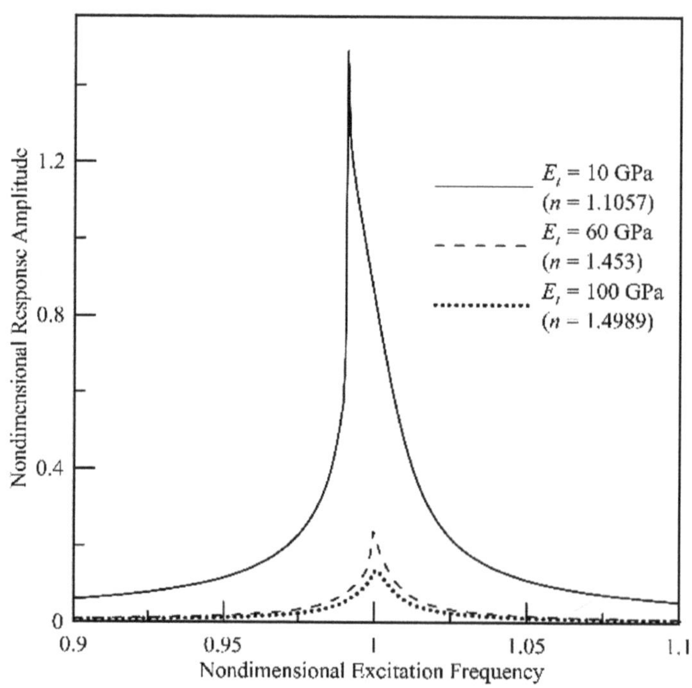

Figure 19. Frequency response curves for rough surface material with tangent modulus 10 GPa and varying yield strength (Y) 560.8 MPa, 911.5 MPa, 1265.3 MPa and 1619 MPa for Z = 0.000295

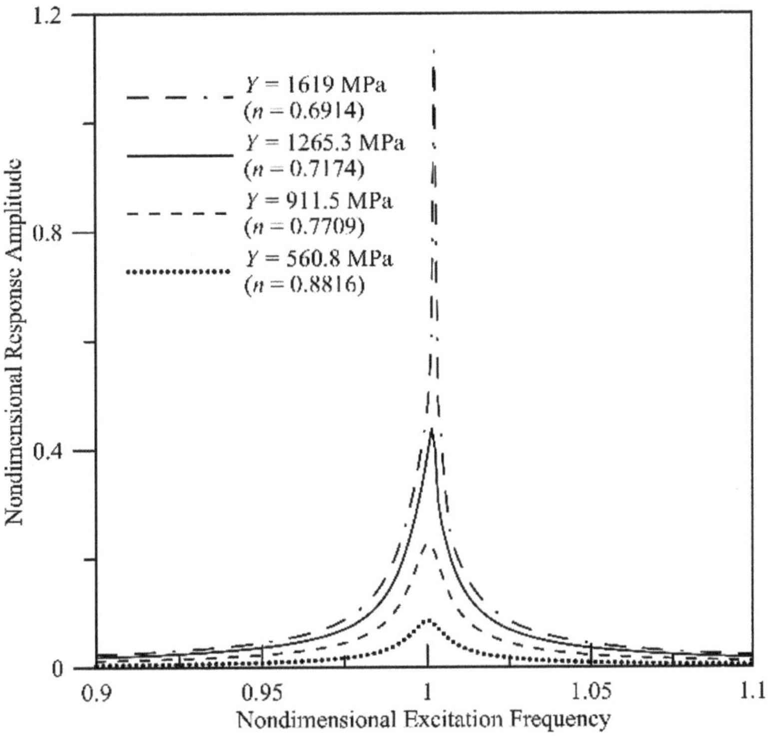

Figure 20. Comparison of frequency response curves with n values 0.98 and 1.02 for Z = 0.028

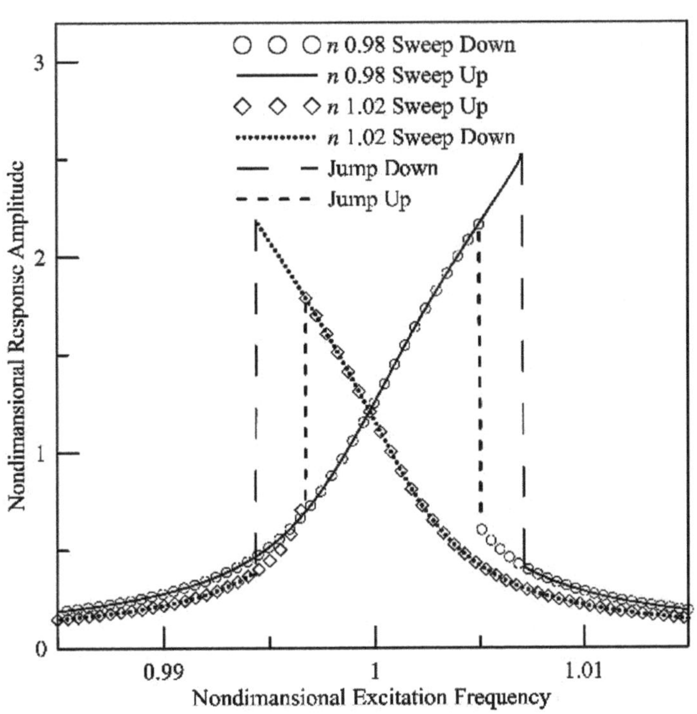

Figure 21. Superharmonic response for the variation of nonlinearity exponent from 1.2 to 2.0 (a-e)

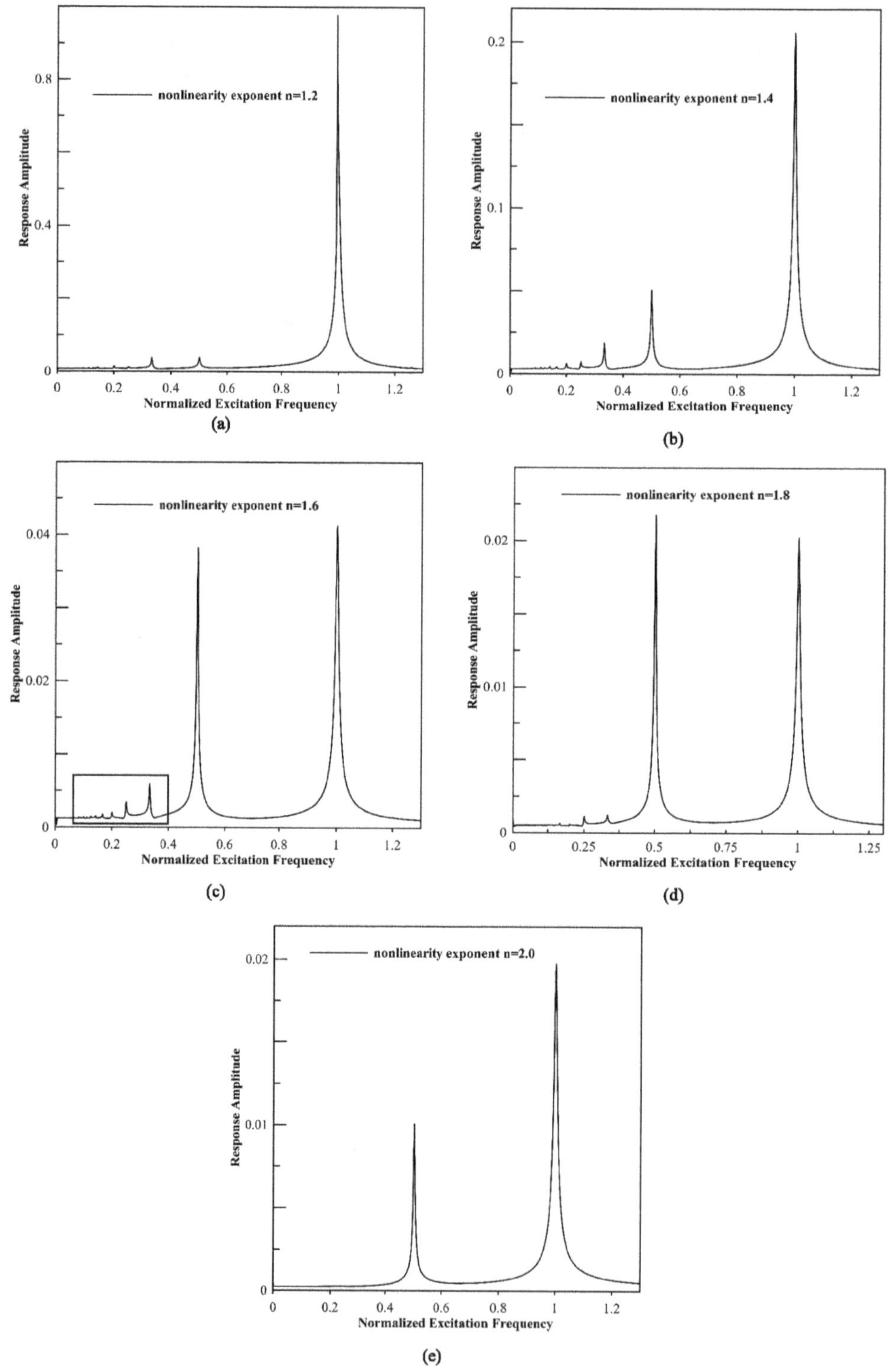

Figure 22. Magnified view of the superharmonic responses for nonlinearity exponent n = 1.6

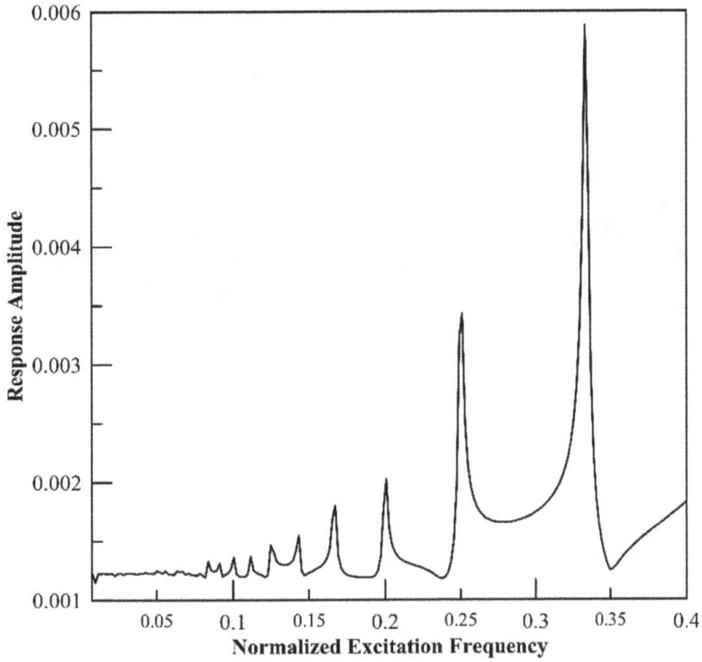

Figure 23. (a) Phase plot and (b) time-displacement plot of vibrating rough surface with n = 0.6

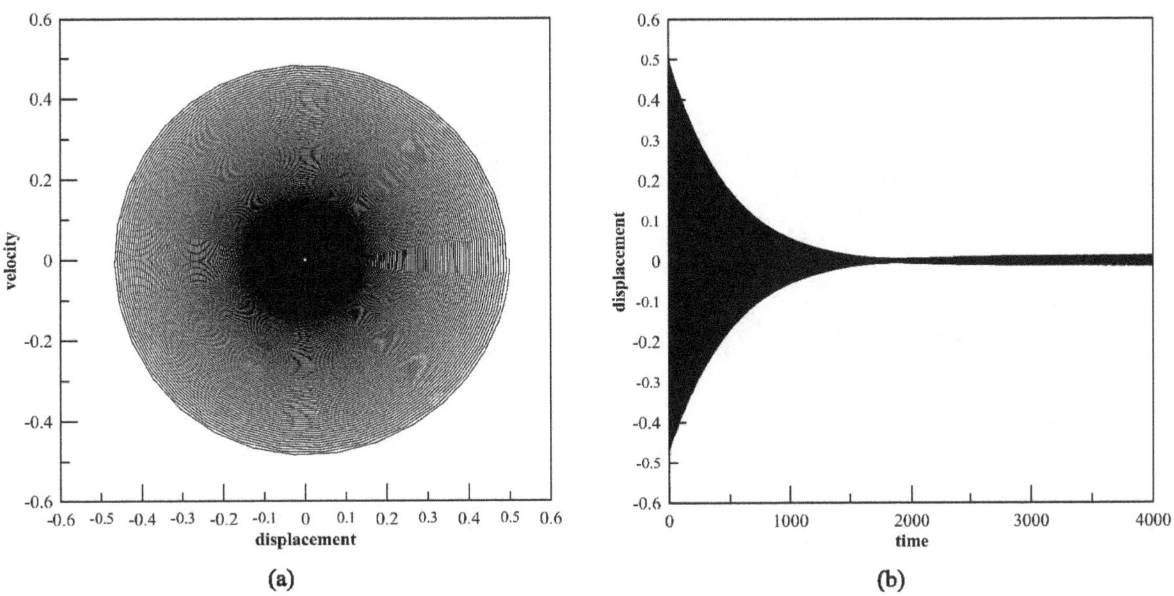

Figure 24. (a) Phase plot and (b) time-displacement plot of vibrating rough surface with n = 1.0

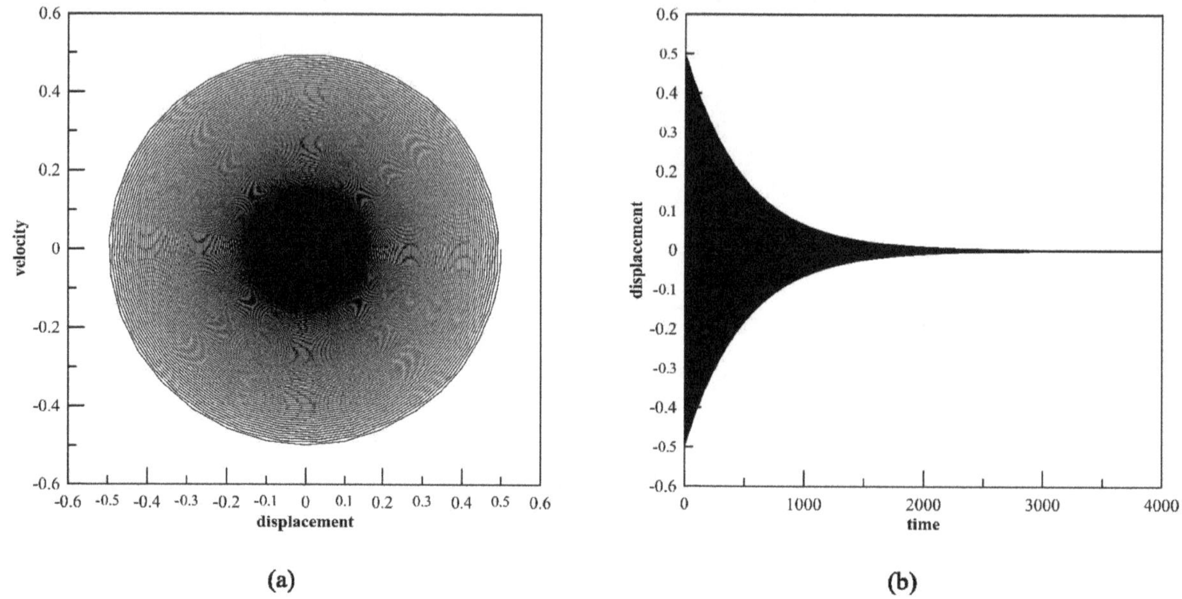

Figure 26. (a) Phase plot and (b) time-displacement plot of vibrating rough surface with n = 2.0

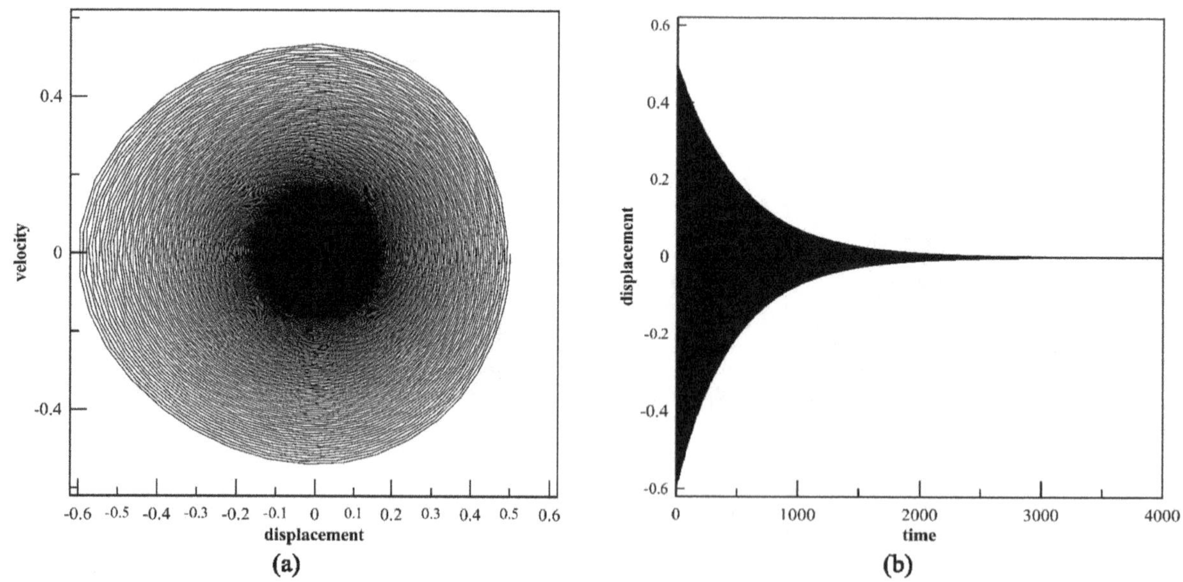

Figure 25. (a) Phase plot and (b) time-displacement plot of vibrating rough surface with n = 1.5

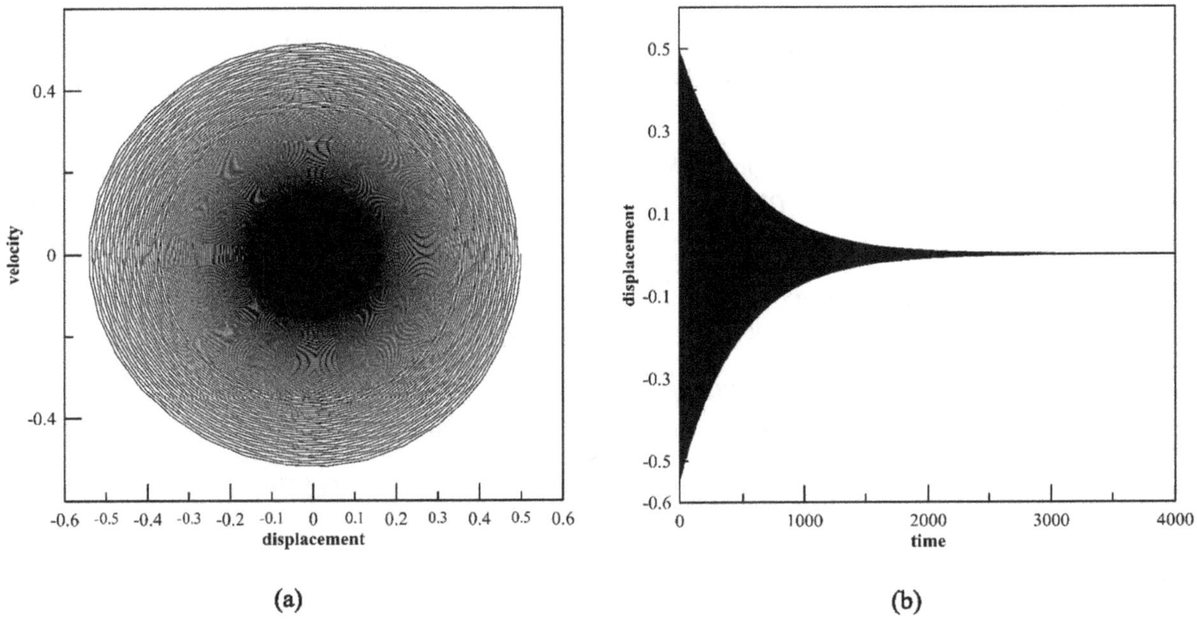

the maximum negative displacement of the vibrating system increases with rougher surface topography, higher tangent modulus, lower yield stress value and vice-versa.

CONCLUSION

In the current paper, fractal surface is generated from two-variable Weierstrass-Mandelbrot function in MATLAB and the coordinate points are imported to ANSYS as keypoints to construct the FE model of the rough fractal surface. Finite element base static analysis is performed for the fractal surface in contact with a rigid flat surface. From the results of the FE analysis, effect of variation in fractal parameters and material properties on different contact parameters e.g. contact force, contact area, contact stress etc. are investigated. In addition, dynamic analysis of the contact system is performed under harmonic excitation, which is imparted on the rigid flat. The dynamic system is represented by a single degree of freedom spring-mass-damper system with harmonic excitation on the base the spring-mass-damper is resting on. The parameter nonlinearity exponent is obtained from the force-displacement relationship of the deformable rough surface which is used to determine the dynamic characteristics of the system. The effect of the variation of fractal parameters and the material properties are also analyzed. It is observed that for higher nonlinearity the response amplitude decreases but above certain value of nonlinearity exponent the amplitude becomes constant. From phase plot and time-displacement plot it is noted that higher nonlinearity causes higher negative displacement of the vibrating system.

REFERENCES

Ausloos, M., & Berman, D. H. (1985). A Multivariate Weierstrass-Mandelbrot Function. *Proceedings of the Royal Society A: Mathematical, Physical and Engineering Sciences, 400*(1819), 331–350. 10.1098/rspa.1985.0083

Berry, M. V., & Lewis, Z. V. (1980). On the Weierstrass-Mandelbrot Fractal Function. *Proceedings of the Royal Society A: Mathematical, Physical and Engineering Sciences, 370*(1743), 459–484. 10.1098/rspa.1980.0044

Buczkowski, R., Kleiber, M., & Starzynski, G. (2014). Normal contact stiffness of fractal rough surfaces. *Archives of Mechanics, 66*(6), 411–428. doi:10.24423/AOM.1286

Buzio, R., Boragno, C., & Valbusa, U. (2003). Contact mechanics and friction of fractal surfaces probed by atomic force microscopy. *Wear, 254*(9), 917–923. doi:10.1016/S0043-1648(03)00247-3

Chatterjee, B., & Sahoo, P. (2013). Finite Element based Parametric Study of Elastic-plastic Contact of Fractal Surfaces. *Procedia Engineering, 64*, 552–561. doi:10.1016/j.proeng.2013.09.130

Chatterjee, B., & Sahoo, P. (2014). Finite Element Based Contact Analysis of Fractal Surfaces – Effect of Varying Elastic Modulus. *Procedia Engineering, 90*, 116–122. doi:10.1016/j.proeng.2014.11.823

Dickrell, D. J., Dugger, M. T., Hamilton, M. A., & Sawyer, W. G. (2007). Direct contact-area computation for MEMS using real topographic surface data. *Journal of Microelectromechanical Systems, 16*(5), 1263–1268. doi:10.1109/JMEMS.2007.901120

Dubuc, B., Zucker, S. W., Tricot, C., Quiniou, J. F., & Wehbi, D. (1989). Evaluating the Fractal Dimension of Surfaces. *Proceedings of the Royal Society A: Mathematical, Physical and Engineering Sciences, 425*(1868), 113–127. 10.1098/rspa.1989.0101

Hess, D. P., & Soom, A. (1991). Normal Vibrations and Friction Under Harmonic Loads: Part I—Hertzian Contacts. *Journal of Tribology, 113*(1), 80–86. doi:10.1115/1.2920607

Hess, D. P., & Soom, A. (1993). Normal vibrations and friction at a hertzian contact under random excitation: Perturbation solution. *Journal of Sound and Vibration, 164*(2), 317–326. doi:10.1006/jsvi.1993.1216

Hess, D. P., Soom, A., & Kim, C. H. (1992). Normal vibrations and friction at a Hertzian contact under random excitation: Theory and experiments. *Journal of Sound and Vibration, 153*(3), 491–508. doi:10.1016/0022-460X(92)90378-B

Huang, Z., Peng, X., Li, G., & Hao, L. (2019). Response of a Two-Degree-of-Freedom Vibration System with Rough Contact Interfaces. *Shock and Vibration, 2019*, 1–13. doi:10.1155/2019/1691582

Hyun, S., Pei, L., Molinari, J. F., & Robbins, M. O. (2004). Finite-element analysis of contact between elastic self-affine surfaces. *Physical Review E: Statistical Physics, Plasmas, Fluids, and Related Interdisciplinary Topics, 70*(2), 12. doi:10.1103/PhysRevE.70.026117 PMID:15447555

Jackson, R. L., & Green, I. (2005). A Finite Element Study of Elasto-Plastic Hemispherical Contact Against a Rigid Flat. *Journal of Tribology, 127*(2), 343–354. doi:10.1115/1.1866166

Jana, T., Mitra, A., & Sahoo, P. (2017). Dynamic contact interactions of fractal surfaces. *Applied Surface Science*, *392*, 872–882. doi:10.1016/j.apsusc.2016.09.025

Jana, T., Mitra, A., & Sahoo, P. (2019). Dynamic analysis of elastically and plastically graded spherical and cylindrical contact. *Proceedings of the Institution of Mechanical Engineers, Part J: Journal of Engineering Tribology*. 10.1177/1350650119841756

Komvopoulos, K., & Ye, N. (2000). Three-Dimensional Contact Analysis of Elastic-Plastic Layered Media With Fractal Surface Topographies. *Journal of Tribology*, *123*(3), 632–640. doi:10.1115/1.1327583

Ling, F. F. (1990). Fractals, engineering surfaces and tribology. *Wear*, *136*(1), 141–156. doi:10.1016/0043-1648(90)90077-N

Ma, Q. L., Kahraman, A., Perret-Liaudet, J., & Rigaud, E. (2007). An investigation of steady-state dynamic response of a sphere-plane contact interface with contact loss. *Journal of Applied Mechanics, Transactions ASME*, *74*(2), 249–255. 10.1115/1.2190230

Majumdar, A., & Bhushan, B. (1990). Role of fractal geometry in roughness characterization and contact mechanics of surfaces. *Journal of Tribology*, *112*(2), 205–216. doi:10.1115/1.2920243

Majumdar, A., & Tien, C. L. (1990). Fractal characterization and simulation of rough surfaces. *Wear*, *136*(2), 313–327. doi:10.1016/0043-1648(90)90154-3

Mandelbrot, B. (1967). How long is the coast of Britain? Statistical self-similarity and fractional dimension. *Science*, *156*(3775), 636–638. doi:10.1126cience.156.3775.636 PMID:17837158

Mandelbrot, B. B., & Blumen, A. (1989). Fractal Geometry: What is it, and What Does it do? [and Discussion]. *Proceedings of the Royal Society A: Mathematical, Physical and Engineering Sciences*, *423*(1864), 3–16. 10.1098/rspa.1989.0038

Nayfeh, A. H., & Mook, D. T. (1995). *Nonlinear Oscillations*. Wiley; doi:10.1002/9783527617586

Pei, L., Hyun, S., Molinari, J. F., & Robbins, M. O. (2005). Finite element modeling of elasto-plastic contact between rough surfaces. *Journal of the Mechanics and Physics of Solids*, *53*(11), 2385–2409. doi:10.1016/j.jmps.2005.06.008

Perret-Liaudet, J. (1997). Subharmonic resonance of order 2 on a sphere-plane contact. *Comptes Rendus de l'Academie Des Sciences Series IIB Mechanics Physics Chemistry Astronomy*, *325*(8), 443–448. doi:10.1016/S1251-8069(97)89445-6

Perret-Liaudet, J. (1998). Superharmonic resonance of order two on a sphere-plane contact. *Comptes Rendus de l'Academie Des Sciences Series IIB Mechanics Physics Astronomy*, *326*(12), 787–792. doi:10.1016/S1251-8069(99)80028-3

Perret-Liaudet, J., & Rigaud, E. (2003). Experiments and numerical results on non-linear vibrations of an impacting Hertzian contact. Part 2: Random excitation. *Journal of Sound and Vibration*, *265*(2), 309–327. doi:10.1016/S0022-460X(02)01267-1

Perret-Liaudet, J., & Rigaud, E. (2006). Response of an impacting Hertzian contact to an order-2 subharmonic excitation : Theory and experiments. *Journal of Sound and Vibration*, *296*(1–2), 319–333. doi:10.1016/j.jsv.2006.03.004

Perret-Liaudet, J., & Rigaud, E. (2007). Superharmonic resonance of order 2 for an impacting Hertzian contact oscillator: Theory and experiments. *Journal of Computational and Nonlinear Dynamics*, *2*(2), 190–196. doi:10.1115/1.2447549

Perret-Liaudet, J., & Sabot, J. (1999). Vibro-impacts induced by non-linear resonances in hertzian contacts. In *Dynamics of Vibro-Impact Systems* (pp. 251–260). Springer Berlin Heidelberg. doi:10.1007/978-3-642-60114-9_28

Pohrt, R., & Popov, V. L. (2012). Normal contact stiffness of elastic solids with fractal rough surfaces. *Physical Review Letters*, *108*(10), 104301. doi:10.1103/PhysRevLett.108.104301 PMID:22463410

Ranganath Nayak, P. (1972). Contact vibrations. *Journal of Sound and Vibration*, *22*(3), 297–322. doi:10.1016/0022-460X(72)90168-X

Rezvanian, O., Zikry, M. A., Brown, C., & Krim, J. (2007). Surface roughness, asperity contact and gold RF MEMS switch behavior. *Journal of Micromechanics and Microengineering*, *17*(10), 2006–2015. doi:10.1088/0960-1317/17/10/012

Rigaud, E., & Perret-Liaudet, J. (2003). Experiments and numerical results on non-linear vibrations of an impacting Hertzian contact. Part 1: Harmonic excitation. *Journal of Sound and Vibration*, *265*(2), 289–307. doi:10.1016/S0022-460X(02)01262-2

Sabot, J., Krempf, P., & Janolin, C. (1998). Non-linear vibrations of a sphere-plane contact excited by a normal load. *Journal of Sound and Vibration*, *214*(2), 359–375. doi:10.1006/jsvi.1998.1582

Sahoo, P., & Ghosh, N. (2007). Finite element contact analysis of fractal surfaces. *Journal of Physics. D, Applied Physics*, *40*(14), 4245–4252. doi:10.1088/0022-3727/40/14/021

Tian, J., & Xie, Z. (2008). Dynamic contact stiffness of vibrating rigid sphere contacting semi-infinite transversely isotropic viscoelastic solid. *Guti Lixue Xuebao*, *21*(6), 580–588. doi:10.100710338-008-0870-x

Xiao, H., Brennan, M. J., & Shao, Y. (2011). On the undamped free vibration of a mass interacting with a Hertzian contact stiffness. *Mechanics Research Communications*, *38*(8), 560–564. doi:10.1016/j.mechrescom.2011.07.012

Xiao, H., Shao, Y., & Brennan, M. J. (2015). On the contact stiffness and nonlinear vibration of an elastic body with a rough surface in contact with a rigid flat surface. *European Journal of Mechanics. A, Solids*, *49*, 321–328. doi:10.1016/j.euromechsol.2014.08.005

Xiao, H., Sun, Y., & Chen, Z. (2019). Fractal modeling of normal contact stiffness for rough surface contact considering the elastic–plastic deformation. *Journal of the Brazilian Society of Mechanical Sciences and Engineering*, *41*(1), 11. doi:10.100740430-018-1513-x

Yan, W., & Komvopoulos, K. (1998). Contact analysis of elastic-plastic fractal surfaces. *Journal of Applied Physics*, *84*(7), 3617–3624. doi:10.1063/1.368536

Zhang, D., Xia, Y., Scarpa, F., Hong, J., & Ma, Y. (2017). Interfacial contact stiffness of fractal rough surfaces. *Scientific Reports*, *7*(1), 12874. doi:10.103841598-017-13314-2 PMID:28993704

Zhong, Z.-H. (1993). *Finite element procedures for contact-impact problems*. Oxford University Press.

APPENDIX: NOMENCLATURE

A: Actual contact area
A_0: Nominal contact area
c: Viscous damping coefficient
D: Fractal dimension
E: Elastic modulus of deformable surface
E': Composite elastic modulus of equivalent rough surface
F_s: Restoring force of spring
f: Frequency index
f_{max}: Maximum frequency index
G: Fractal roughness
g: Gravitational acceleration
k: Nonlinear stiffness of spring/bulk stiffness of deformable surface
L: Sample length of fractal surface
L_s: Cut off length of fractal surface
M: Number of superimposed ridges to construct the fractal surface
m: Mass of the block
n: Nonlinearity exponent. Exponent value in the power law force-displacement relationship equation $P^*=k(\delta^*)^n$
P: Applied normal load on rough surface
P^*: Normalized applied normal load, $P^* = \frac{P}{A_0 E'}$
t: Time
u: Normalized vertical displacement
y: Vertical displacement of the base, the spring-mass-damper is resting on
Y: Base excitation amplitude (Harmonic)
Z: Normalized base excitation amplitude ($Z=Y/z_s$)
z: Vertical displacement of block
z_s: Static vertical displacement of block due to its own weight
γ: Frequency density
δ: Displacement of contacting rough surface
δ^*: Normalized displacement of contacting rough surface, $\delta^*=\delta/L$
ϕ: Random phase angle of Weierstrass-Mandelbrot function
ν: Poisson's ratio
ζ: Damping ratio
σ_{vm}: Von Mises stress
τ: Normalized time, $\tau = t\omega_s$
ω: Harmonic excitation frequency
ω_s: Natural frequency of system at the static equilibrium position
ω_n: Natural frequency in the non-dimensional equation of motion
Ω: Normalized harmonic excitation frequency

Chapter 8
Material Selection Considerations for Lightweight Automotive Bodies

Herman Tang
https://orcid.org/0000-0001-6609-7851
Eastern Michigan University, USA

ABSTRACT

Lightweight is a trend of new vehicle development, driven by government regulations, environmental concerns, and customer needs. A major effort in the automotive industry is on light weighting vehicle bodies. This chapter reviews the various materials, their characteristics, weight reduction potentials, and costs for light weighting vehicle bodies. The chapter also exams the joining technologies on their principles, applications for the lightweight materials, and influencing factors for choosing a joining process. Furthermore, this chapter discusses the development trends of material selection and joining technology applications.

INTRODUCTION

Demands of Weight Reduction

The better fuel economy of passenger vehicles is required by the US government and customers. The Corporate Average Fuel Economy (CAFE) standards are regulations in the United States. In 2011, CAFE required the improved fuel economy to 54.5 mpg for the cars and light-duty trucks in US by 2025 model year. The target 54.5 mpg is still under discussion and debates as EPA and NHTSA proposed new Safer Affordable Fuel-Efficient (SAFE) Vehicle Rules to replace the original CAFE for mode year 2021-2026 (DOT 2018). The CAFE/SAFE standards have been a main driving force for the automakers to seek various ways to improve the fuel consumption of new vehicles.

DOI: 10.4018/978-1-7998-4939-1.ch008

Light-weighting is an important path and approach to achieve a better fuel economy for automakers to meet these standards. It is reported that reducing a vehicle's mass by 10% can improve fuel economy by 6-8% (DOE, 2015). In addition, for new electric cars, their weight is extremely important to the battery power demand. A study showed that a 10% reduction in vehicle weight permits nearly a 10% smaller in battery size (Smock, 2010). Another study showed that 100kg weight reduction from a 1593-kg electric vehicle might improve the total cost of ownership by €500 over a 4-year holding period (Redelbach, 2012).

A midsize passenger car weights about 3400 lb (1542 kg). Many industrial experts consider a 450 lb (204 kg), or about 13%, reduction of vehicle weight is necessary by 2025. To reduce the total weight of a vehicle, its body structure is often considered a main target for a 30% reduction. The other segments of a passage vehicle have similar targets. For example, powertrain units may be expected to be reduced by 25%. Other units, such as chassis, interior, closures, and glazing, may be targeted for 20, 14, 8, and 3%, respectively.

Environmental Concern

The emission of CO_2, sometimes called Greenhouse gas (GHG), is a main environmental concern. The transportation is a large contributor to GHG emissions. Figure 1 shows a clear correlation between the fuel economy and CO_2 emission in the last four decades (EPA, 2019). Therefore, the fuel economy improvement directly benefits the GHG reduction.

Figure 1. EPA report on vehicle fuel economy and CO_2 emission

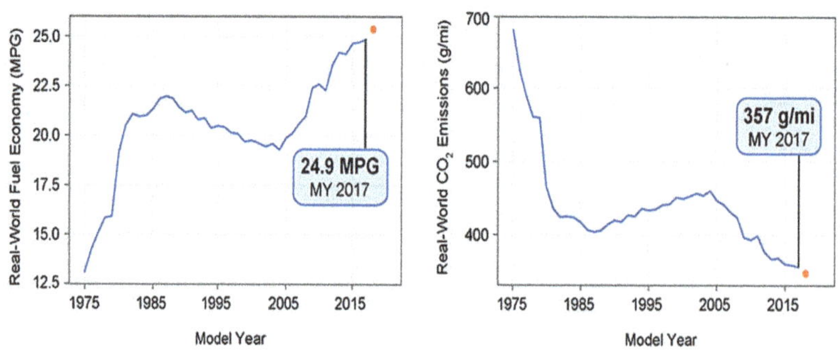

The CO_2 emission from vehicle fuel combustion can be calculated by

$$CO_2 = \frac{Gas \times 0.87 \times 44}{12} = 3.19 \times Gas$$

Where, CO_2 emission and gas combusted are in weight, a constant 0.87 is due to 87% carbon and 13% hydrogen in gasoline by weight, the 44 is molecular weight of CO_2, and the 12 is molecular weight of carbon. One gallon of gasoline weights about 6.3 lb. Therefore, one-gallon gasoline can produce approximately 20 lb (9.07 kg) CO_2 when burned (DOE, 2016).

Another factor, the CO_2 generated from the material production, should be considered as well. Life-cycle assessment (LCA) is a good tool to evaluate the CO_2 emission for material selection. If an LCA study includes the material production, the analysis results can be different from those only focusing on vehicle fuel combustion. According to the World Steel Association, steel production generates much less GHG than other materials for vehicle bodies, refer to Table 1 (ArcelorMittal, 2014). From this perspective, taking the vehicle mass reduction rate (say up to 50% if using aluminum) into count and the CO_2 from material production, steels can be competitive for the purpose of CO_2 emission reduction.

Table 1. Greenhouse gas generated from material production

Material	GHG (kg CO_2e/kg)
Steels	2.0–2.5
Aluminum Alloys	11.2–12.6
Magnesium Alloys	18–45
Carbon Fiber	21–23

MATERIALS FOR LIGHTWEIGHT BIW

BIW is a body frame structure of a vehicle without paint. Traditionally, BIW was composed of mild sheet steels. Using new materials for BIW is a main approach to reduce the vehicle weight. There are three types of materials for BIW: new steels, non-ferrous metals, and plastics.

New High-Strength Steels

Steels have been used for vehicle bodies since day one. Conventional mild steels had dominated vehicle BIW material usages until about 15 years ago. New steels, often called Ultra High-strength Steels (UHSS) or Advanced High-Strength Steels (AHSS), have been playing an increasingly important role in vehicle weight reduction because of thinner sheet steels with higher strengths. About ten years ago, UHSS unitization was about 15% in BIW, but now is more than 50% for steel BIW. For example, the Jeep Cherokee's BIW has an approximately 65% high-strength steel content for the 2019 model (FCA, 2019).

The various types of steels have distinctive physical properties. Steels are normally recognized by their tensile strength and elongation. If the tensile strength is less than 270 MPa, then the steel is considered low strength. Some professionals state that the steels with 700 MPa (other say 780 MPa) are UHSS. With advances in steel development, the strength of some latest steels is higher than 1000 MPa. The steels that are typically used in BIW include mild steels, Bake Hardenable (BH), High-strength Low Alloy (HSLA), Isotropic (IS), Interstitial-free (IF), and Carbon-Manganese (CMn), etc. The UHSS often used for BIW are Dual Phase (DP), Transformation Induced Plasticity (TRIP), Complex Phase (CP), and Martensitic (Mart). Figure 2 illustrates a general comparison of these steels.

However, UHSS have a few unfavorable characteristics, such as a low elongation for part stamping processes. DP and Martensitic steels have similar forming characteristics with conventional mild steels. However, new boron-treated steels need a hot stamping process; meaning the sheet is heated first and then formed and cooled. New DP and CP steels are in development. For new developed UHSS, the hot

Figure 2. Property comparison of steels

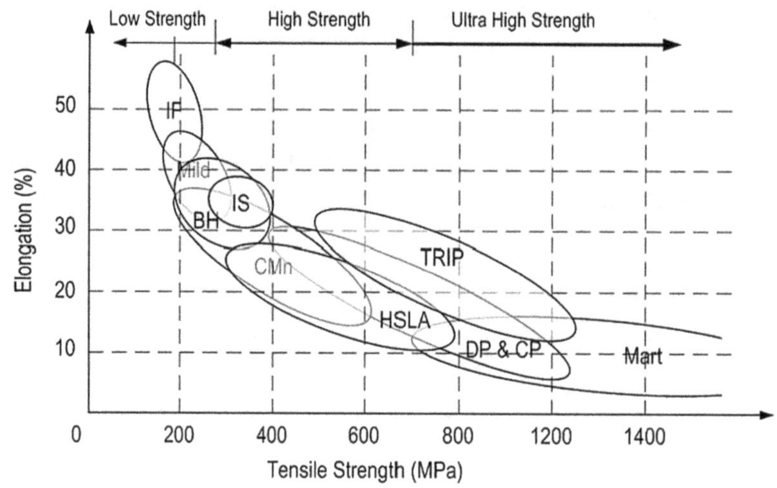

stamping can have some challenges to resolve, such as production efficiency, piercing method, shape constraints, anti-corrosion property, ductility after forming, and guaranteed strength (Funakawa, 2019).

Aluminum Alloys

Aluminum alloys (Al) may be the most promising non-ferrous metal for light-weighting because the specific gravity of aluminum alloys is 2.7 g/cm^3, about 1/3 of that of steels. Consequently, using aluminum alloys can significantly reduce vehicle weight. For instance, the weight of Chevrolet Malibu Maxx's lift-gate was reduced from 39 lb (17.8 kg) with steel to 20 lb (9.1 kg) with an aluminum alloy. Ford redesigns its bestselling pickup F-150 for its 2015 model, switching from the original steel body to all aluminum, which reduced vehicle weight by over 700 lb (318 kg). As another example, 2013 Range Rover body is reduced by 42% by switching to aluminum alloys, or 463 lb (210 kg) lighter than the earlier model of 1098 lb (498 kg).

Many industry experts predicted that using aluminum alloys for BIW parts would keep growing. One report predicted that the aluminum volume share of body and closure components would increase to 26.6% in 2025 from 9.3% in 2016 (Docker, 2014). Another report stated that 85% hoods, 46% doors, and 18% whole BIW parts would be aluminum alloys by 2025 (Boney, 2014).

In fact, aluminum alloys are a big family, with distinctive material properties and manufacturability. The 1xxx series are considered as pure aluminum. The main aluminum alloy grades used for vehicle body parts are the 5xxx, 6xxx, and 7xxx series. Table 2 lists their basic characteristics.

For example, the 5xxx series of aluminum alloys are often used for structural applications because of their good formability, strength, and weldability. BIW parts with 5xxx series alloys are produced mainly as sheet and plate and occasionally as extrusions. Similarly, the 6xxx alloys, e.g. 6111, are for closure panels, because of their surface quality, strength of finished parts, and formability. The 6xxx series alloys can be extruded economically.

The burden of using aluminum alloys is the cost, which is about three times more than steels in terms of weight for raw materials. The actual cost, due to the large variation of aluminum and steel grades, can

Table 2. Overview of some aluminum alloys for BIW

Grade	Primary Alloying Elements	Characteristics
1xxx	Pure aluminum (99%)	Corrosion resistant, high electrical and thermal conductivity, good workability, very low strength
5xxx	+ Magnesium	Good corrosion resistance, weldability, high-strength
6xxx	+ Magnesium + Silicon	Good weldability, machinability, formability, corrosion resistance, and medium strength
7xxx	+ Zinc + Magnesium and/or Copper	Very high strength, various welding performance

be much higher than that of the raw materials. In addition, vehicle structural requirements and manufacturing cost challenge the usage of aluminum alloys. The cost of assembly manufacturing can be about 20% higher using aluminum parts than using steel counterparts. For example, a passenger car door can be designed and built with an aluminum alloy. Compared with the equivalent steel door, the weight of the aluminum door is reduced from 22.72 to 10.43 kg. At the annual production volume of 100,000 vehicles, the cost of an aluminum door is a $20.53 soar (Seksaria, 2015), including material and manufacturing.

The design of aluminum parts and the corresponding manufacturing processes can be considerably different from those of steels. For instance, aluminum parts have to be thicker than the steel counter parts. That is the reason that the weight reduction by using aluminum alloys is about 40-50% from using steels, not proportional to the difference of the specific gravities.

Magnesium Alloys

Magnesium (Mg) alloys, as a non-ferrous metal, are a rising star for automotive applications as even 35% lighter than aluminum alloys. Table 3 lists the main material properties of Mg, Al, and steel for a comparison purpose. Magnesium has a lower ultimate tensile strength, fatigue strength, and creep strength compared with aluminum. Magnesium alloys with other elements, such as aluminum, manganese, and zinc, can have better mechanical properties.

Table 3. Physical property comparison

Material Property	Mg	Al	Iron
Specific gravity at 20°C (g/cm^3)	~ 1.74	~ 2.7	~ 7.9
Coef. of Thermal Expansion (10^{-6}/°C)	~ 27	~ 23	~ 12
Electrical Resistivity (10^{-8} Ω•m)	~ 4.5	~ 2.7	~ 9.7
Thermal Conductivity (W/m•°C)	~ 95	~ 121	~ 16
Modulus of Elasticity (GPa)	~ 43	~ 74	~ 195
Melting Point (°C)	~ 650	~ 660	~ 1,536

Existing Mg applications include crankcases, oil pans, gearbox housings, and wheels, as Mg has good applications in thin-wall die castings. Some BIW parts, such as pillars, hoods, deck lids, and inner doorframes, are good candidates. For example, the lift-gate inner of the Lincoln MKT is an Mg structure with a weight reduction of 22 pounds (Ashley, 2010). BMW used magnesium to cast the instrument panel, which was 4 pounds lighter than its steel predecessor was (Huetter, 2016). A recent study showed that there are technical solutions for the stiffness design, crash performance, an assembly process, a coating system, and mitigation strategies for galvanic corrosion (Weiler, 2019).

On the other hand, the high cost, roughly six times of steels in terms of weight, and some manufacturing technologies still under development are the constraints for Mg applications. A review showed that the auto industry target is to use an average of 159 kg of Mg components per vehicle by 2020 (Shuldiner, 2010). The world Mg production is about 1000 thousand tons in 2018; 80% made in China (Bray, 2019).

In addition to the high cost, under development assembly processes, and limited supply, other concerns of using Mg include its corrosion resistance that does not suit conventional phosphate and E-coat processes in vehicle paint operations. Costly hot forming for wrought parts, lack of design experience, and intensive energy for material production are also often cited. Therefore, all these factors should be evaluated before adopting magnesium alloys for vehicle parts.

Carbon Fiber Reinforced Polymers

Plastics and composites include thermosets, thermoplastics, glass fiber and carbon fiber reinforced polymers or plastics (CFRP). One of the main reasons of using plastics is their lightweight.

Both carbon fiber and glass fiber, similar in many ways, can be used for vehicle structural parts. Relatively, glass fiber has a lower price, about one third of carbon fiber, but lower stiffness-to-weight ratio. CFRP has been long used in race cars and aircrafts. For automotive, Lexus LFA uses the material for its main passenger cell, transmission tunnel, floor module, roof, hood, C-pillars and real load floor. Glass fiber filled plastics is also used to substitute structural components in vehicle bodies. The specific tensile strength, or tensile strength-to-weight ratio, of fiber-reinforced plastics are significantly higher than steels. Using the glass fiber filled polymers, for example, BMW M3 Front Bumper Beam achieves 40% weight reduction. In the same way, Audi A8 (D4) uses the material for several components and gets 30% weight reduced (Brennan, 2011).

The assembly manufacturing challenges of the material applications are the high cost of fiber, processing, slow cycle time, and new joining technologies. At mass production volumes, a hood, for example, can cost about $1,000, compared with $10 for steel and $30 for aluminum (Brooke, 2012). The cost of CFRP continuously decreases as expected but seems still prohibitive for mass production vehicles.

The conventional molding technology for the plastic components of vehicles, depending on part designs and materials, can take over one hour. A newly developed process, called resin-transfer molding (RTM), can reduce the cycle time to less than 10 min. For mass production vehicles at about one minute of cycle time, thus, multiple parallel equipment and processes for the CFRP parts are required. Due to the material characters and unique design, a study found that the positioning of the joints could be made contrary to conventional composite joining guidelines (Mårtensson, 2015). The unique designs can be challenging for manufacturing process and tooling.

Regarding the tooling costs, CFRP is favorable because a composite-molded part can replace a multiple-piece assembly of sheet metal parts. The investment on the equipment, in terms of cost per part, can be lower than that on stamping dies for low-volume vehicles. For a high volume production, however, the composite materials may lose their edge on the tooling cost perspective. That is one of the reasons that the composite materials, except Sheet Molded Compound (SMC) for body panels, may be economically justifiable for high-end, low-volume cars for now.

Considerations for Material Selection

Lightweight is an important goal but not the sole factor for material selection. Material selection is a substantial business decision based on various technical requirements, financial analysis, and resources. Considering all the factors can lead the weight reduction to the best overall results. For example, Nissan LEAF changed the materials of doors to steel in 2017 model from aluminum in 2011 model and gained 19 kg (Erriquez, 2017), considering safety and other factors. Table 4 lists general factors for material selection. An optimal selection result may be obtained based on a thorough and balanced study on these requirements and factors.

Table 4. Factors in materials selection

General Consideration	• Corrosion resistance • Weight/density • Economics • Recyclability	**Stamping Process**	• Formability/drawing quality • Elasticity • Hardenability • Spring-back severity
Structural Performance	• Stiffness/strength • Fatigue resistance • Tear strength • Wear resistance	**Overall Manufacturing**	• Processing time • Tooling costs • Dimension tolerance
Surface Parts	• Dent-resistance • Impact resistance • Paintable • Texture	**Other Factors**	• Part consolidation ability • Performance of noise, vibration, and harshness • Temperature resistance

As discussed, the obstacles to use new, lightweight materials are often their costs, which include the total cost of materials and manufacturing processes. Table 5 lists the potential weight reduction (for equal stiffness as steel designs) and relative cost per part. This table illustrates an overall picture that an advance on weight reduction is normally with cost penalty. For example, an aluminum vehicle body can cost $1500 more than a steel one. Please note that the unit cost, or the cost per vehicle, varies with production volume.

Table 5. Overall cost-effectiveness of material applications

Lightweight Material	Material Replaced	Mass Reduction (%)	Relative Cost
Ultra-High-strength Steel	Mild Steel	10-15	1.2-1.5
Aluminum alloy	Steel, Cast Iron	40–50	1.3–2
Magnesium alloy	Steel, Cast Iron	50–70	2–2.5
Magnesium alloy	Aluminum alloy	20–30	1.3–1.6
Fiber-reinforced polymers	Steels	60-80	6-8
Titanium alloy	Steels	30–50	5–10+

Furthermore, depending on the types of vehicles, the weight reduction may have different priorities. Because of the cost constraint, entry-level cars are unfavorable candidates for using light and expensive materials. That explains that aluminum alloys have been primarily utilized on luxury cars because their values to customers can be justifiable.

Therefore, the material utilization trend depends on the lightweight demand, material cost, and manufacturing feasibility. It is predicted that aluminum alloys and UHSS would continue their quick grow, magnesium alloys and CFRP might have a moderate grow, and conventional mild and regular high-strength steels would keep losing their shares (Modi, 2016).

Direction of Using Multiple Materials

Because of various requirements and considerations on the vehicle design, the materials selected may be and should be different for parts to be optimal applications. For example, aluminum alloys are light but have only about 1/3 strength and 70% stiffness of steels. It can be a good idea to use aluminum alloys for non-load bearing parts, such as hood and lift-gate. In fact, aluminum hoods have been common for many steel-dominated vehicle bodies. The next frontier to move to aluminum may be for doors.

For the load bearing parts of BIW, using aluminum alloys is feasible but steels may be a better choice to take the advantages of high strengths. For the same reason, the selection of expensive UHSS should be carefully evaluated and compared with the regular strength steels. UHSS should be used for the components subject to heavy bending and torsion requirements, such as front rails of underbody, etc. for significant weight reduction.

Therefore, using different materials, instead of single type of materials, for vehicle bodies can be a better way to produce light-weighted, good performance, and affordable vehicles.

Current industry practice shows several cases using multiple or hybrid materials than that with 100% aluminum alloys on vehicle BIW. For example, the 2015 Cadillac CT6 uses 64% aluminum and 13 different types of metals and alloys in its structure and body, which saves 198 lb (89.8 kg) compared with if it had been built from steel (Holmes, 2015). Another example of hybrid material applications is Audi R8 body that uses aluminum and CFRP (Birch, 2015). It is interesting that original Acura NSX was an all-aluminum car. The 2017 Acura NSX is changed to multi-material space frame composed primarily of aluminum, with the strategic use of steel and carbon fiber in select areas (Honda, 2016).

A comprehensive study was on a door assembly consisting of five components: outer, inner, bean, reinforcements, and hinge brackets. The study tried a dozen combinations of different grades of steels, aluminum, CFRP, etc. for the five components (Stevens, 2016). The weight reduction can be up to 65% from the baseline material design of 2011 Chevrolet Silverado, depending on the material selection and combination.

A prediction made by Toyota showed that their new models would be based on UHSS BIW frame. For closure panels, such as hood, doors, and deck lid, they would use aluminum as outers with CFRP being inners. The roof panel could be made of either aluminum or CFRP. Aluminum die-casting parts would be used as well (Shiga, 2014). Toyota prediction was based on new joining technologies. With similar prediction, another report also indicated the variations of material usages between models, automakers, and market regions (Shankar, 2016).

Another example is Ford. Ford switched its 2015 F-series pickups to an all-aluminum body. Five years later, Ford changed again and adopted a "mixed materials" strategy on the 2020 Explorer SUV.

Using multiple materials for vehicle BIW is challenging for assembly and joining process in manufacturing, as some joining technologies may not be suitable for joining some combinations of different materials. The joining technologies for the BIW materials are discussed in the next section.

ASSEMBLY JOINING REVIEW

Feasibility of Assembly Processes

Selecting assembly joining technologies for light-weighted BIW parts is material and design driven. The manufacturing processes, including part fabrication and assembly, of vehicle BIW structure can be different. For instance, aluminum body parts need unique designs, rather than similar to their steel counter parts, to meet structural requirements. Many of the structural aluminum parts are designed in the form of extrusion and casts. Even for the same stamping processes, stamping dies for aluminum parts may be more expensive than those for steel parts may.

When a UHSS is considered, the challenge is often on the assembly joining processes. The initial investment, production readiness, and reliability of the new joining processes, in addition to the cost of material itself, can be high and even prohibitive for new material introduction. Therefore, selecting a workable and economic joining technology often is a key for new material applications for lightweighting BIW.

The capacities of the existing manufacturing systems and joining processes are an important factor in material selection. On this point, new steels have a significant benefit over non-ferrous materials, as the steels do not impose major changes to the existing processes, equipment, supply, and recycling infrastructures. For the same reason, introducing aluminum alloys are easier than composites. Vehicle body closures, such as doors, hoods, and trunk lids, are built in separate subassembly areas and then attached to vehicle BIW. Therefore, the manufacturing processes of closures are less disruptive to the main vehicle assembly processes. These closures are good candidates to switch to non-ferrous or even plastic materials. In addition, the manufacturing similarity between steel and aluminum can make using aluminum with a higher confidence than using composites. If there were a major issue, it would be possible to use steel to back up aluminum, but more difficult to back up composites.

Joining technologies for BIW parts can be grouped based on the temperature when apply. If it is above the melting point of a material, the joining technology is called fusion welding, such as resistance spot welding (RSW) and laser beam welding (LBW). At the room temperature, several mechanical technologies, e.g., self-pierce riveting (SPR), clinching, and drill screws, can be used. Adhesive bonding is another important joining process at a slightly higher than the room temperature. At an elevated temperature, but not at a melting point, friction stir welding (FSW) is a rising star, particularly for joining parts made of aluminum alloys. Table 6 lists the feasible joining technologies used in production environments with different types of materials. The technologies are also illustrated in Figure 3.

Using different materials and with various design configurations, the multiple joining technologies are often used in the same BIW. For example, the Corvette aluminum body uses 15.4 m arc welding, 439 RSW spots, 482 flow drill screws, and 11.3 m LBW. Cadillac ATS, designed with many parts using new steels and several aluminum parts, has 6181 RSW, 8.3 m arc welding, 5.6 m laser brazing, 90 SPR, and 109 adhesive bonding beads (Regiec, 2013). Table 7 shows a comparison study (Holmes, 2015) of steel and aluminum for vehicle bodies with different BIW architectures, material contents, and joining processes.

Table 6. Feasible joining technologies for common materials

Joint of	Materials	RSW	LBW	FSW	SPR	Clinching	Bonding
Same materials	Steel + Steel	√	√				√
	Aluminum + Aluminum	√	√	√	√	√	√
	CFRP + CFRP				√		√
Dissimilar materials	Aluminum + Steel	√		√		√	√
	Steel + CFRP				√		√
	Aluminum + CFRP				√		√

Figure 3. Commonly used joining technologies for lightweight materials

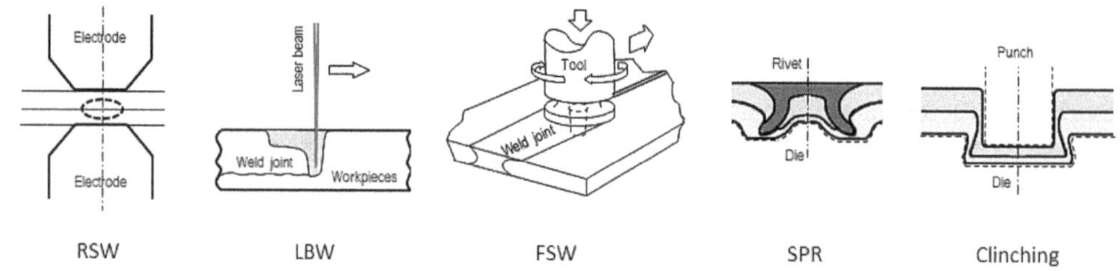

Table 7. A comparison of material and manufacturing processes

Material	Steel	Al	Al (1)	Al (2)	Al (3)
Architecture	Unibody		Unibody + Spaceframe		
Mass reduction	Baseline	45%	43%	42%	42%
Stamping parts	100%	100%	89%	79%	71%
Casting parts	0%	0%	2%	8%	13%
Extrusion parts	0%	0%	9%	13%	16%
Joining (RSW)	3028	0	0	0	0
Joining (Projection welding)	100	100	95	86	77
Joining (SPR)	0	1510	1430	920	741
Joining (arc welding, 20mm/stitch)	0	243	217	322	370
Joining (Plasma-Laser welding, m)	0	54.6	54.6	61.2	68.8
Joining (Bonding, m)	23.4	23.4	23.4	23.4	26.7

Considerations in Fusion Welding

Main fusion welding technologies for BIW are RSW, LBW, as well as arc welding. Almost all types of fusion welding can be used for metals, including aluminum alloys.

Since its accidental invention in 1877 (Stanley, 1950), RSW has become a reliable and economic joining process, predominately used for BIW assemblies made of steels. RSW applies the principle of electricity flowing through work pieces, generating Joule heat to melt metal. Then, a solid weld nugget is formed after cooling.

Introducing UHSS challenges RSW as the welds may fail in a unique mode called interfacial fracture (IF). The advantage of using UHSS may be lessened as the IF degrades the weld properties in terms of static peak load and energy absorption. In addition, the RSW process of UHSS is less robust than that of conventional steels. RSW for UHSS also needs longer welding cycle time.

RSW for aluminum can be more difficult than that for steels because of material properties of aluminum alloys. For example, the electrical and thermal conductivities of aluminum alloys are approximately three times of those of steels; the melting temperature of aluminum alloys (around 660 °C) is much lower than that of steels' (about 1540 °C). Accordingly, RSW for aluminum alloys requires 2.5 to 3 times of the electrical current, but roughly one-third of the weld time for welding steels. The high demand on welding current means larger, about three times, welding gun transformers. Additionally, industrial practices show electrodes deteriorate much more quickly than welding steels and in a different form from steel welding, which needs more frequent dressing on electrodes.

The median frequency direct current (MFDC) power source of RSW is now widely used. Compared with the conventional AC, the MFDC has faster response time of feedback control at the millisecond level. As less sensitive to the surface condition of aluminum alloys, the MFDC welding is also suitable for aluminum sheet parts.

There are study reports on the joining of the steels to specific aluminum alloys using RSW, such as (Qiu, 2010). After fusion, $FeAl_3$ and Fe_2Al_5 are formed at the interface between steels to aluminum alloys. General Motors announced in 2016 that it developed a new RSW technology for joining the dissimilar materials. The aluminum-to-steel spot welding system makes use of a uniquely designed welding tip patented by GM. GM began using the RSW process in production of the Cadillac CT6 (Winner, 2016).

Overall, RSW is still the main application for the joining process for the steel dominated BIW of mass production vehicles. With new and different materials introduced and used in BIW structures, RSW applications would be reduced to the level of about 50% in 15-20 years. While in the same time, the applications of laser beam welding, adhesive bonding, and mechanical fastening will growth significantly (Tang, 2017).

Laser beam welding (LBW) for the vehicle body assembly was first introduced in the mid-1980s. As an effective joining process, LBW has become increasingly used in vehicle body assembly manufacturing. Table 8 lists the common industrial laser types and characteristics (EAA, 2015). However, with new materials and emerging later technology, LBW has been still a research topic for UHSS. For example, new laser is for dissimilar DP steel sheets (Tunçel, 2019) and CO_2 remote laser for the welding of steel to aluminum (Kotadia, 2019).

Table 8. Common industrial laser types and characteristics

Laser	Wavelength	Beam Quality	Electrical Efficiency
CO_2	10.6 µm	High/Medium	Low (10%)
Fiber	1.075	High	High (25 – 35%)
Nd:YAG	1.064	Medium	Low / Medium (5 – 25%)
Disk	1.030	High	High (25 – 35%)
Diode	0.9 – 1.07	Low	Highest (45%)

LBW is very capable of welding deep and narrow joints, which is important for BIW assemblies with thick metal, extrusions, and castings. Its single-sided accessibility is another advantage, essential to weld the hydro-formed parts and the parts in closed sections. LBW can be applied to various materials, not only different types of steels but also aluminum, magnesium, and some dissimilar materials. Moreover, LBW needs a very small or even no part flange, which means the savings of materials and vehicle weight. Remote laser welding with CO_2 laser uses scan mirrors for positioning the beam precisely at the desired weld locations and quick decoupling in between welds. The joining efficiency of remote laser welding can be several times higher than RSW. The high efficiency of laser welding, resulting in fewer robots and smaller floor space, can offset most of the initial high cost of laser welding equipment.

It was optimistically predicted that LBW would replace RSW. However, the initial capital investment required for LBW seems to be a major obstacle. In addition, to meet the requirements of better part fit and more accurately positioning part, a specially designed fixture is required. Thus, LBW requires an additional tooling investment at extra cost. When competing against RSW, the LBW's initial investment is not favorable to the new technology. A research study reported that the initial investment for the major portions of a body shop could be about doubled if LBW replacing RSW. At a production volume 200,000 units/year for five years, LBW would result in \$21.79 more per vehicle (Klinger, 2012).

For aluminum welding, CO_2 laser 10.6 μm wavelength has a very low absorption or a high reflectivity of aluminum, meaning high laser high power density required. To increase the laser energy absorption, coating the welding surface is a solution, but increases the process complexity and cost. The 0.8-μm wavelength of direct diode laser coincides with the absorption peak, which maximizes heating efficiency and reduces the power required. However, the initial investment on the diode laser is higher than that on CO_2 and Nd:YAG lasers.

Arc welding is still used in many vehicle assembly plants. For example, the Audi A8 tubular space frame is primarily welded by gas metal arc welding (GMAW). An arc welding process uses an electric arc formed between a consumable wire electrode and the parts. The arc heats the parts, causing them to melt, and joins them together. Among several types, GMAW is often used in the automotive manufacturing, such as for body frames, engine cradles, and a few vehicle body applications.

Compared with LBW, the advantages of arc welding are a low cost and simple process. However, there is a main concern on the uncertainty of the assembly dimensional quality due to the heat introduced during welding process. The heat can cause dimensional distortion of the parts. Compensating the heat-introduced dimensional distortion of vehicle BIW subassemblies is often a trial and error based on experience.

Friction Stir Welding for Aluminum

Invented in 1991, friction stir welding (FSW) has been implemented in the aircraft industry. FSW is based on friction heat and mechanical movement. During the process, heat is created by frictional movement, rotating in 500-2000 rpm and traveling at 50-500 mm/min between a rotating tool and work pieces. Joining is achieved by the plastic flow of heated material to generate a joint.

As a solid-state welding process, FSW can have a wider range of applications than the conventional fusion welding processes. One of the advantages is that FSW can be used to join aluminum, copper, magnesium, and similar non-ferrous metal combinations.

The common pattern of FSW is linear joins. In addition, FSW can also make spot welds, which is called friction stir spot welding (FSSW). FSSW for aluminum alloys has been developed. Mazda reported the first application of FSSW on its RX-8, a mass production vehicle (Mazda, 2003).

Regarding joining multiple types of materials, FSW is a good candidate, too. One of the challenges is that aluminum has a much lower melting temperature than steel. Therefore, the process parameters should be different. A study showed that tool spun at 450 r/min with a tilt angle of 1.5°, welding speeds at 45 mm/min, and an axial force at 1574 lbf (7 kN). The tensile strength of the FSW welds can be about 90% of the base aluminum alloy (Ramachandran, 2015).

Applications of Mechanical Joining

One of mechanical joining technologies used for BIW assemblies is self-piercing riveting (SPR). SPR is a piercing-forming process, refer to Figure 3. It can be used to join two or more layers of sheet materials by driving a rivet through the top layer and upsetting the bottom layer with the rivet without piercing through the bottom layer. Depending on the join design, the join strengths, in terms of shear strength and fatigue strength, are normally lower than RSW joins. The diameters of rivets commonly used are 3 mm and 5 mm.

Combined with adhesive bonding, SPR is competitive with RSW for aluminum sheet joining. For instance, Ford uses SPR and bonding for its aluminum F-150 bodies.

SPR can be used for joining mixed materials, for instance aluminum alloys with steels. In the SPR applications, the steel sheet should be placed on the top with the aluminum part on the die side. For the parts made of high-strength steels, it may be challenging to pierce a thin steel sheet and form a lock in an aluminum base sheet. One solution is to increase the speed of piercing. A study showed that a high tool speed of 5 to 10 m/s versus conventional <1 m/s could improve the feasibility of applying SPR on steel-to-aluminum joints (Neugebauer, 2011).

Clinching, like SPR, is another mechanical joining technology and drawing attention as a possible replacement of aluminum RSW. Unlike SPR, clinching does not require fasteners, which is a cost and weight saving compared with a SPR process. Comparison studies show different strength characteristics among RSW, SPR, and clinch under different working conditions and with different base materials. For example, the fatigue performance of a clinched joint was found 60% stronger than that of an RSW spot (Cai, 2005). The results are significantly affected by base materials and joint design. Figure 4 shows some examples.

Figure 4. Examples of mechanical joining strength comparison

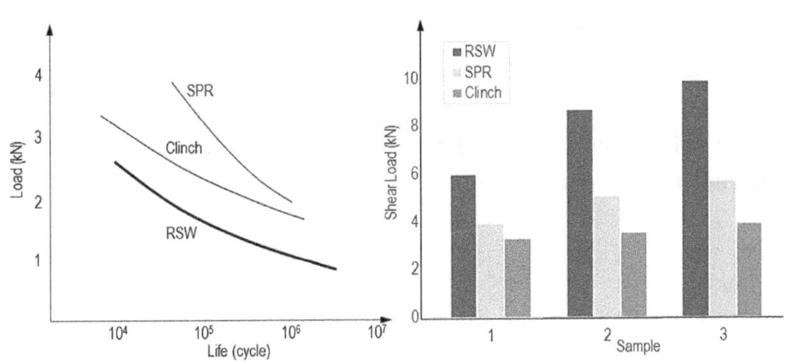

Advantages of Adhesive Bonding

Adhesive bonding has been increasingly used in vehicle body assembly for improved BIW structural stiffness, crash resistance, noise, vibration, and harshness (NVH), water leak prevention, and corrosion resistance. The total length of adhesive bonding can be long. For better structural and NVH performance, luxury passage vehicles use more adhesive than mass production cars. In 2016 luxury vehicles, the average length of adhesive beads is 590 ft (180 m) (Benevento, 2016).

Compared with the processes of welding and mechanical joining discussed, the bonding process is unique because of its homogeneous stress distribution between two joined materials without local stress concentrations. Moreover, for non-weldable materials, such as dissimilar materials and plastics, adhesive bonding shows its unique advantage. For example, Audi and Lamborghini utilize carbon fiber and bonding of structural parts with coated aluminum (Gehm, 2016).

For applications, the adhesives need cure time to develop its bonding strength. Most types of adhesives used to bond automotive panels require at least 30 minutes reaching a handling strength at room temperature. Therefore, the bonded joints normally come with RSW spots for weldable materials or mechanical joints for non-weldable materials to hold the sheet panels together until the adhesive cures. Such hybrid bonding is also advantageous as it improves the structural stiffness and crash performance of vehicles.

CONCLUSION

Lightweight of vehicles is a goal, driven by government regulations, environment concerns, and customer expectations. One of main approaches is to reduce the weight of BIW by way of using new materials. Many factors affect the material selections. In general, the key factors are the material cost and assembly joining feasibility, in addition to the functionality and performance requirements of vehicle designs. Based on the comprehensive assessment and discussion in this chapter, the following points can be made.

- Certain levels of lightweight can be achieved by using new materials, such as high-strength steels, aluminum alloys, plastics, and their combinations.
- The most promising light materials are aluminum alloys. Other light materials with increasing usages are CFRP and magnesium alloys.
- New ultra high-strength steels are competitive to aluminum alloys for lightweight vehicle BIW.
- Multiple or mixed material applications can be the better solutions and the direction, to use the "right" materials optimally for various applications, for lightweight mass production vehicle models.
- Main concerns of using light materials are the material cost and joining technologies. Some assembly joining technologies are available and under additional development for specific materials and applications.
- Specific applications of joining technologies are material and design dependent. A comprehensive analysis, considering design requirements, costs, manufacturing feasibility, etc., is needed for all specific cases.

REFERENCES

ArcelorMittal. (2014). Driving Advanced Automotive Steel solutions. *ArcelorMittal North America.*

Ashley, S. (2010). Shedding Pounds on a Magnesium Diet. *Automotive Engineering International*, 34–36.

Benevento, M. (2016). *Adhesives and Sealants for Automotive Weight Reduction, Safety, and Performance, Industrial Market Insight.* https://www.adhesives.org/resources/knowledge-center/aggregate-single/adhesives-and-sealants-for-automotive-weight-reduction-safety-and-performance

Birch, S. (2015). New-generation Audi R8: The Race Driver's View. *SAE International*. Available from: http://articles.sae.org/14469

Boney, T., & Richman, D. (2014). *The Aluminum Association: 2015 Ducker Worldwide Automaker Survey.* Available from http://www.aluminumintransportation.org/research-resources/PDF/Webinars/2014/2015-ducker-worldwide-survey-of-automakers

Bray, L. (2019). Mineral Commodity Summaries - Magnesium Metal 2019. *U.S. Geological Survey.* https://www.usgs.gov/centers/nmic/magnesium-statistics-and-information

Brennan, J. M. (2011). *Intelligent Simulation Technology Delivering Weight Efficient Vehicles.* CAR Management Briefing Seminars 2011, Traverse City, MI.

Brooke, L. (2012, November). Creating the 54.5 mpg Car, SAE. *Automotive Engineering International*, 32–36.

Cai, W., Wang, P. C., & Yang, W. (2005). Assembly Dimensional Prediction for Self-Piercing Riveted Aluminum Panels. *International Journal of Machine Tools & Manufacture*, *45*(6), 695–704. doi:10.1016/j.ijmachtools.2004.09.023

Ducker Worldwide LLC. (2014). *2015 North American Light Vehicle Aluminum Content Study, June, 2014*. Author.

Erriquez, M. (2017). What did we learn from a teardown and benchmarking of ten EV models? *McKinsey Center for Future Mobility*. https://www.mckinsey.com/industries/automotive-and-assembly/our-insights/trends-in-electric-vehicle-design

European Aluminium Association (EAA). (2015). EAA Aluminium Automotive Manual – Joining. Brussels: Author.

Fiat Chrysler Automobiles (FCA). (2019). *2019 Jeep® Cherokee: Authentic, Premium Jeep Design and New Advanced Global Turbo Engine Enhance Evolution of Renowned Mid-size SUV*. https://media.fcanorthamerica.com/newsrelease.do?id=18756&mid=426

Funakawa, Y., & Nagataki, Y. (2019). *High Strength Steel Sheets for Weight Reduction of Automobiles.* JFE Technical Report, No. 24. https://www.jfe-steel.co.jp/en/research//report/024/pdf/024-02.pdf

Gehm, R. (2016). Composite Bonding Adhesive. *SAE International*. Available from: http://articles.sae.org/14862/

Holmes, L. (2015). Cadillac CT6 Aluminum Construction Saves 198 Pounds Compared to Steel. *Automobile Magazine*. Available from: http://www.automobilemag.com

Honda North America. (2014). *The Second-Generation NSX: 2017 Model Media Information*. https://hondanews.com/releases/2017-acura-nsx-press-kit-overview

Huetter, J. (2016). *New BMW 5 Series loses 137 lbs.; has aluminum, HS steel, magnesium — but no carbon fiber*. https://www.repairerdrivennews.com/2016/10/14/new-bmw-5-series-loses-137-lbs-has-aluminum-hs-steel-magnesium-but-no-carbon-fiber/

Klinger, J. F. (2012). *Automotive Body Structure Assembly - Mass & Cost Saving Potential of Laser Welding Compared to Spot Welding* (Master's thesis). KTH Royal Institute of Technology.

Kotadi, H. R. (2019). Challenges and Opportunities in Remote Laser Welding of Steel to Aluminium. *MATEC Web of Conferences*, *269*, 02012. doi:10.1051/matecconf/201926902012

Mårtensson, P., Zenkert, D., & Åkermo, M. (2015). Cost and Weight Efficient Partitioning of Composite Automotive Structures. *Polymer Composites*, *38*(10), 2174–2181. doi:10.1002/pc.23795

Mazda News Release. (2003). *Mazda Develops World's First Aluminum Joining Technology Using Friction Heat*. Available from: https://www.mazda.com/publicity/release/200302/0227e.html

Modi, S. (2016). Material Qualification in the Automotive Industry. Center for Automotive Research.

Neugebauer, R. (2011). *Mechanical Joining With Self Piercing Solid-Rivets at Elevated Tool Velocities*. International Conference on Material Forming 2011, Belfast, UK.

Peterson, G. (2013). An Analysis of Impact Performance and Cost Considerations for a Low Mass Multi-Material Body Structure. NHTSA Mass/Size/Safety Workshop, Washington, DC.

Qiu, R., Shi, H., Zhang, K., Tu, Y., Iwamoto, C., & Satonaka, S. (2010). Interfacial Characterization of Joint Between Mild Steel and Aluminum Alloy Welded by Resistance Spot Welding. *Materials Characterization*, *61*(7), 684–688. doi:10.1016/j.matchar.2010.03.015

Ramachandran, K. K. (2015). Friction Stir Welding of Aluminum Alloy AA5052 and HSLA Steel. *Welding Journal*, *94*(9), 219s–300s.

Redelbach, M. (2012). Impact of lightweight design on energy consumption and cost effectiveness of alternative powertrain concepts. *European Electric Vehicle Congress*, Brussels, Belgium.

Regiec, M. (2013). Vehicle lightweight: Manufacturing Impact. *2013 CAR Management Briefing Seminar*.

Seksaria, D. (2015). A Cost Effective Solution for Automotive Doors - Alcoa's Thin Door Technology. *International Automotive Body Congress, Proceedings of Automotive Materials: Interior/Exterior & Safety Systems*, *3*, 38-53.

Shankar V., (2016). *Lightweight Options and Forecast of Material Types: Global snapshot Of Lightweight Solutions*. Frost & Sullivan.

Shiga, T. (2014). Challenge for Light Weight. *2014 CAR Management Briefing Seminar*.

Shuldiner, H. (2013). *Lower Cost Key to Magnesium's lightweight Uses*. Available from: http://wardsauto.com/vehicles-amp-technology/lower-cost-key-magnesium-s-lightweight-uses

Smock, D. (2010, November 16). More Composites for Cars? *Design News*, 132–135.

Stanley, W. A. (1950). *Resistance Welding*. New York: McGraw-Hill Book Company.

Stevens, M. (2016). *Mixed Materials Solutions: Alternative Materials for Door Assemblies*. https://www.cargroup.org/wp-content/uploads/2017/02/Mixed-Materials-Solutions-Alternative-Materials-for-Door-Assemblies.pdf

Tang, H. (2017). *Joining Processes for Body Assembly, Automotive Vehicle Assembly Processes and Operations Management*. SAE International. https://www.sae.org/publications/books/content/r-456/ doi:10.4271/R-456

Tunçel, O. (2019). Pulsed Nd:YAG laser welding of similar and dissimilar DP steel sheets. Alternative Energy Sources, Materials & Technologies (AESMT'19), 1.

US Department of Energy (DOE). (2016). *Physical and Chemical Properties of Gasoline, Alternative Fuels Data Center (AFDC)*. Available from: https://www.fueleconomy.gov/feg/contentIncludes/co2_inc.htm

US Department of Energy (DOE) Vehicle Technologies Office (VTO). (2015). *Lightweight Materials for Cars and Trucks*. Available from: https://energy.gov/eere/vehicles/vehicle-technologies-office-lightweight-materials-cars-and-trucks

US Department of Transportation (DOT). (2018). *U.S. DOT and EPA Propose Fuel Economy Standards for MY 2021-2026 Vehicles*. https://www.transportation.gov/briefing-room/dot4818

US Environmental Protection Agency (EPA), (2019). *The 2018 EPA Automotive Trends Reports: Greenhouse Gas Emissions, Fuel Economy, and Technology since 1975*. EPA-420-R-19-002.

Weiler, J. P. (2019). A review of magnesium die-castings for closure applications. *Journal of Magnesium and Alloys*, *7*(2), 297–304. doi:10.1016/j.jma.2019.02.005

Winner, D. (2016). Lightweight Innovations: Unsung Heroes of the Weight-loss Battle. Wards Auto, 1-18.

KEY TERMS AND DEFINITIONS

Body in White (BIW): BIW is a body frame structure of a vehicle without paint. Traditionally, a body structure is built with steel sheet metal parts. The "white" refers to the metallic appearance or color of steel surface even the color is not actually white. A BIW often does not include closure panels, such as doors, fenders, hood, and deck-lid.

Corporate Average Fuel Economy (CAFE): CAFE standards are the regulations of the National Highway Traffic Safety Administration (NHTSA), the US Department of Transportation. CAFE standards regulate how far our vehicles must travel on a gallon of fuel. In 2019, NHTSA and EPA establish the One National Program for fuel economy regulation Safer Affordable Fuel-Efficient (SAFE).

Greenhouse Gas (GHG): GHG mainly refers to CO_2 and includes other gases. CO_2 contributes to the greenhouse effect. Once emitted, GHG rises into the air, stays in the atmosphere for a period, and traps energy from the sun as heat. Thus, excessive GHG causes earth to warm up.

Ultra-High-Strength Steels (UHSS): UHSS are the steels that have very high strength, such as over 700 or 780 MPa. They are also called Advanced High-Strength Steel (AHSS).

Vehicle Body Assembly Processes: Body assembly processes are primarily joining processes. In the processes, the parts are positioned properly first, and then joined using various types of joining processes. For a traditional vehicle body made of steel sheet metal parts, the processes are commonly welding and adhesive bonding. Adopting new materials and body design can significantly change the processes.

Vehicle Mass Production: For vehicle assembly, mass production may be called high-volume production as well. Two main characteristics of mass production are standardized products and processes, and a high level of automation. For vehicle BIW assembly, the automation level can be higher than 90% for mass production. The annual production output of a vehicle assembly plant, running two shifts a day, five days a week, is normally 150,000 – 300,000 units a year.

Chapter 9
Selection of Suitable Control Parameters for Proper High-Resolution Deposition Performance of E-Jet Microfabrication Process:
A Comparative Analysis

Raju Das
Department of Mechanical Engineering, National Institute of Technology, Durgapur, India

Shibendu Shekhar Roy
Department of Mechanical Engineering, National Institute of Technology, Durgapur, India

ABSTRACT

E-jet is a high-resolution microfabrication technology. Its operation depends on several process control parameters like applied voltage, flow rate of the material, stand-off height, type of ink material. High-resolution deposition along with high ejection frequency is the performance parameter of E-jet. In the present study, an attempt is made to design the process parameters in such a way that it could simultaneously satisfy both the performance characteristics of the E-jet process. A Taguchi robust design-based utility concept is applied to optimize the multi-response E-jet process through a case study. The optimization result is compared with another multi-objective optimization method called Grey relational grade analysis. The comparative analysis showed that both the methodologies identified the same process parameter combination for improved E-jet performance. Analysis of variance found applied voltage is the most influential control factor in the investigated region. The anticipated improvement in performance shows the successful implementation of the aforesaid methodology.

DOI: 10.4018/978-1-7998-4939-1.ch009

INTRODUCTION

Electrohydrodynamic (EHD) inkjet or E-jet printing is a high-resolution, direct write, non-contact patterning technology for micro fabrication. High registration accuracy, vacuum free operation along with the ability to handle flexible stretchable substrate makes this new fabrication methodology a suitable candidate for flexible electronics application (Onses et al., 2015). The capability of dispensing a wide range of materials both liquid as well as colloidal solutions of solid nanoparticles (Kim et al., 2011), makes this emerging technology diverse. It finds its application in printed electronics (Kamyshny et al., 2014), biotechnology (Jayasinghe et al., 2005), human wearable flexible sensors (Nothnagle et al., 2015), display devices (Park et al., 2013), combinatorial chemistry (Kim et al., 2010) etc. Being a simple process with no complicated pre or post processing operation, it could be a suitable alternative to lithography technology. Layered manufacturing or additive printing is also an added feature of this technology (Qin et al., 2017). In conventional inkjet, there are mainly two kinds of actuation mechanisms used for material ejection namely thermal actuation and piezo-based actuation. But limited resolution operation is one of the inherent shortcoming with these technologies. It has been reported that the maximum achieved resolution in the range of 10-20 μm (Derby et al., 2010). To get higher resolution deposition small size nozzles are required, as the printing dimension is about two times of the nozzle dimension (Back et al., 2012). But incorporating smaller size nozzle creates nozzle clogging problem with high viscous ink material. Moreover, with small size nozzles, the ejection pressure becomes very high. The fabrication of these small size nozzles itself is a costly affair. The aforementioned limitations can be successfully mitigated by E-jet printing. The functional material gets ejected at the tip of the Taylor cone (Taylor, 1964) instead of the nozzle tip. And as the tip of cone dimension is smaller than the nozzle dimension, the high-resolution operation can be achieved with conventional nozzle configuration. Both continuous as well discrete features can be produced with this direct write technology.

In EHD inkjet printing a potential difference is applied between the nozzle and the substrate. The applied electrical field causes the charges in the ink to move towards the tip of the nozzle. Due to mutual columbic repulsion in the pendant liquid meniscus, it produces an electric stress which deforms the liquid meniscus into a conical form towards the substrate. This deformed shape is known as Taylor's cone. When the applied electrostatic force overcomes the restraining surface tension force of the liquid material, a droplet or continuous jet of functional material is produced (Yudistira et al., 2011; Nguyen & Byun, 2009). The main components of EHD printing system consist of: a pressure chamber to supply

Figure 1. Schematic Diagram of EHD Jet Printing

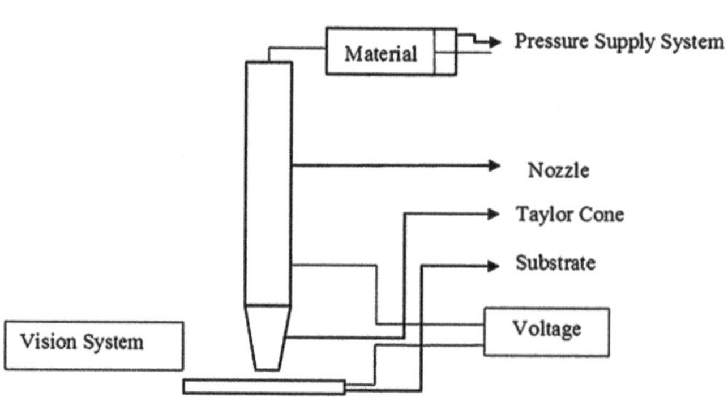

the ink at the tip of the nozzle, a voltage source to apply the potential difference, a conducting nozzle to dispense the material, a substrate on which printing is done, a CNC stage for mounting the substrate and a vision system to observe the droplet or jet formation process. Depending on the geometry of the product, CNC stage moves. The following schematic diagram shows different components of EHD inkjet system.

Chen et al. (2006) successfully deposited droplets whose dimensions had much smaller than the nozzle dimension in drop on demand paradigm by E-jet. The achieved resolution has been in the range of 2 μm from 50 μm nozzles. It eliminates the need of a very narrow size nozzles to get high-resolution operation. Park et al. (2007) successfully printed droplets of functional materials of sub-micron range. They also successfully demonstrated the capability of EHD printing for potential electronic applications. Investigators (Chen et al., 2006; Choi et al., 2008) developed the scaling laws associated with this technology. These laws provide a better understanding of the process. The manipulation of jet diameter or droplet size has traditionally been done with voltage modulation. Li (2006) studied the effect of applied voltage on liquid meniscus deformation. Three different kinds of ejection modes have been observed namely dripping mode, pulsating jet, and stable cone-jet. Dripping mode results in the drop on demand mode of operation, while stable cone-jet is responsible for continuous mode of ejection. He also characterizes the transition of different ejection modes with respect to applied voltage. Generally, pulsed DC voltage offers better control over continuous DC voltage mode of operation. The effect of different parameters like duty cycle, pulsing frequency, pulse voltage amplitude on printing performance has been reported by researchers (Kim et al., 2008; Lee et al., 2012). Mishra et al. (2010) presented a high-speed printing regime in drop-on-demand mode. They reported printing speed i.e. deposition rate of droplets as high as 1000 drops per second. They also used pulsed DC voltage for ejection. Kim et al. (2009) used hybrid ejection methodology for getting higher throughput of the process. They mentioned the usage of a piezo-based actuator to get the material at the tip of the nozzle then applied electric filed as usual to get the ejection of material. Park et al. (2014) developed a model to predict the droplet dimension. It helps to control the process in a better way. Reports on physical properties of the ink material on printing performance have also been found in the literature. Yu et al. (2016) studied the effect of viscoelasticity of ink material. They have developed separate models each for elasticity and viscosity of the ink material while keeping other properties constant like surface tension, conductivity etc. They propose an operating map of stable Taylor cone generation region in terms of applied voltage and the flow rate of the material while changing the viscosity and elasticity of the material one by one. An et al. (2014) tried to establish a relationship between the deposition rate and ink properties like viscosity, surface tension, and conductivity. They proposed a dimensionless coefficient which is a function of ink properties to predict the droplet ejection rate. The objective was to optimize the ink properties for the betterment of the ejection rate.

Different kinds of materials have been processed by E-jet for various applications. Awais et al. (2013) employed EHD to fabricate resistive switch. They had used Ag/ZrO2/Ag material to produce a sandwich kind of structure on a polyimide substrate. Another research group (Park et al., 2014) used Ag nanoparticles as a functional material for flexible electronics application. They had printed Ag nanoparticles as a transparent electrode on a flexible substrate. The produced electrodes can be employed in many optoelectronic applications like such as touch panels, organic light emitting diode (OLED), organic solar cells etc. As EHD is a non-contact fabrication method, researchers successfully replace brittle Indium Tin oxide (ITO) by Ag nanoparticles as a transparent electrode material for flexible electronic applications. Graphene-based humidity sensors for health and environmental applications have also been developed by EHD printing (Ali et al., 2016). They also printed on a flexible substrate for wearable electronics application. Work has also been done to produce microelectronic gas sensors for detection of toxic gases such as NO2, CO, H2S with the help of E-jet (Kang et al., 2017). Nano-fibers of several types of metal oxides such as SnO2, NiO etc. have been used as an ink material for electrode printing. The reported achieved resolution is about less than 50 μm by this method. Das et al. (2018) conducted an initial case study to tune the process environment of E-jet for better performance.

From the literature, it is found that most of the investigations have been devoted towards the successful implementation of electrohydrodynamic principle for micro droplet generation. Voltage modulation studies on the printing performance have been explored. Characterization of this new emerging micro-fabrication technique related to ink-material physical properties like viscosity, elasticity, surface tension has also been reported. Successful implementation of this new direct write technique in the area of micro-manufacturing and material processing has been demonstrated quite remarkably.

Parameter design can be viewed as an effective way to improve the process performance. With the help of parametric optimization study, useful information regarding the controllable factors and their relative importance on the process response can be gathered. This information could lead to a better-controlled operation of the process. Proper selection of process parameters in the investigated region could result in improved performance of the process. Present work takes up this issue and tries to design the controllable parameters of E-jet process in such a way that it could lead to a favorable working environment. In E-jet micro-fabrication process, creating micron size droplets along with high deposition rate are the objectives. So it is a multi-objective response problem. The operating condition should be such that it can generate high- resolution droplets of functional materials with a reasonable printing frequency. Both the objectives can be achieved by a multi-response optimization study. Traditionally Taguchi method finds its application in the single response variable scenario. But most of the production process involves multi-response situation. A particular combination of the controllable variables may work for one response but it may not be useful to other responses. From a process engineer's perspective, a single combination of the process parameters should be selected, which can be used to optimize the entire process as a whole. In this work, utility concept is integrated with the Taguchi approach to overcome this limitation. Utility concept is employed to convert multi-response optimization scenario into a single response optimization problem, by calculating overall utility value. These overall utility values serve as a single objective function which ultimately optimized by Taguchi signal to noise ratio (S/N) analysis. In the present study, ink type, applied pressure, nozzle to substrate stand-off height and the applied voltage selected as process control parameters of E-jet micro-fabrication process. Deposition frequency and deposited droplet diameter identified as a process performance characteristics. The case study data, is also analyzed with another multi-objective optimization method called Grey relational grade analysis (GRA). The results of both the optimization techniques (i.e., Taguchi coupled utility analysis and GRA) are compared. The objective of such comparative analysis is to find out which method is more suitable for E-jet optimization. With the variation in the optimization technique does the solution differs or not? The aim of the research is to propose the level combination of the chosen process control parameters in such a way, that it could simultaneously optimize both the responses with a reasonable trade-off. Analysis of variance (ANOVA) a statistical technique to measure the source of variation in the parametric domain, is employed to find out the most important parameters which affect the process.

The proposed hybridization of the Taguchi methodology is described in section 2. The collection of experimental data and its analysis is presented in section 3. And lastly, some concluding remarks drawn from the study are indicated in the last section.

ADOPTED METHODOLOGY

Utility Concept

Utility is the measure of how useful a product or process is with respect to the consumer's expectation. Usually, a product or a process involves several quality characteristics or attributes. The overall performance of the process or product is composed of the performance of each response variables or attributes. To make a proper judgement on the overall performance, all the individual response variables should be integrated. Such integrated index is known as the overall utility of the product or process. So it transforms the assessment of several performance characteristics into a single performance index termed as overall utility value. Hence the proper selection of process control parameters which leads to favorable overall utility value would aslo satisfy the several performance characteristics or attributes simultaneously.

If X_i represents the performance of a response variable or an attribute i and there are n attributes to evaluate the overall performance, the overall utility value is given by (Chakraborty et al., 2017; Badkar et al., 2011) as follows:

$$U(X_1, X_2, ..., X_n) = \sum_{i=1}^{n} W_i U_i(X_i) \tag{1}$$

where $U_i(X_i)$ is the individual utility of the ith attribute or response variable and W_i represents priority weights of the individual responses. The summation of all individual weight values is equal to 1.

Procedural Steps of Conducting Utility Analysis

- Evaluate the constant value for calculation of preference scale
- Construct the preference scale for individual response variable or attribute
- Calculate the overall utility value

To construct the preference scale following equation has been used:

$$P_i = A \log \frac{X_i}{X_i'} \tag{2}$$

where X_i is the actual value of the attribute or response variable i, X_i' is the minimum or just acceptable value of response i and A is a constant. This preference scale is used to calculate the utility values. Two arbitrary values (preference number) 0 and 9 are assigned to just acceptable and most acceptable value of attribute or response variable respectively (Kumar et al. 2000). The value of the constant A can be determined as follows:

$$A = \frac{9}{\log \frac{X^*}{X_i'}} \tag{3}$$

The value of A is chosen such that at $X_i = X^*$, $P_i = 9$ where X^* is the optimal or best value of the response or attribute i.

The next step is to assign priority weights to the individual response and to calculate the performance index of the methodology i.e. overall utility value. The weight values should be such that it satisfies the following condition:

$$\sum_{i=1}^{n} W_i = 1 \tag{4}$$

In this study, equal weight value i.e. 0.50 is assigned to both the responses. The reason behind such choice is that both the responses i.e. deposited droplet diameter and the deposition frequency are equally important for printing performance. To evaluate overall utility value following relation is used:

$$U = \sum_{i=1}^{n} W_i P_i \tag{5}$$

Taguchi Analysis

Taguchi robust design approach for parameter design has been successfully implemented over the years (Patowari et al., 2011; Bagci et al., 2006) for the improvement of product quality and process optimization. Taguchi methodology is a powerful tool for designing a high-quality system. It involves experimentation with the help of specially designed orthogonal arrays (OA). The numbers of experiments with OA's are usually less than the full factorial investigations thus reducing the experimental complexity and resources associated with it. The computations for the analysis are also very simple; this is another reason for its overwhelming popularity among the process engineers. This method was developed by a Japanese engineer and statistician Genichi Taguchi. It involves of calculation of quality loss function and then it's transformation in logarithmic scale. The transformed quality measure is known as the signal to noise ratio (S/N) ratio (Phadke, 1995). It represents the relationship between response and the deviation of the response from the desired condition. Signal indicates the mean of the response variable and noise represents the standard deviation of the response. According to the methodology, the optimal parameter level combination is the one which gives the maximum signal to noise ratio. It is quite understandable also; more the value of signal to noise (S/N) ratio lesser will be the effect of undesired factors (noise terms). There are three kinds of signal to noise (S/N) ratios available namely- larger the better, smaller the better and nominal is the best. The selection of a specific signal to noise (S/N) ratio is problem dependent. In this work as the objective is to maximize the overall utility value, larger the better signal to noise (S/N) ratio is employed. The following formula is used for calculation of larger the better criteria signal to noise (S/N) ratio (Phadke, 1995):

$$S/N\ ratio = -10\log\left[\frac{1}{n}\sum_{i=1}^{n}\frac{1}{y_i^2}\right] \tag{6}$$

Where y_i represents the value of the response variable or attribute i, and n represents the no. of replications.

As the conventional Taguchi approach cannot solve the multi-response optimization problem, utility concept is integrated with Taguchi approach in the present work. The multiple attributes or response variables are converted into a single representative performance index i.e. overall utility value. Thus the multi-response problem is converted into a single response problem. These overall utility values serve as an objective function which is finally optimized by Taguchi method.

Grey Relational Grade Analysis (GRA)

GRA was first established by Dr. Deng in 1982, and since then it finds its employability for performing prediction, relation analysis, and decision-making in many areas. In GRA (Deng 1989) grey means insufficient information and white means full information is available about the relationship between variables. It is essentially a method to study the complicated interrelationship between variables. This method can be employed to study the multi-performance characteristics evaluation (Lin et al. 2002; Ho et al. 2003; Fung 2003). The decision parameter of this analysis is defined by grey relation grade (GRG). GRG converts the multiple objectives into a single parameter of interest. Thus optimization of GRG value leads to the optimization of multi-performance characteristics of the given problem. Several researchers (Tosun 2006; Lin et al. 2009; Malik et al. 2018; Eskandari et al. 2018; Nair et al. 2018, Anand et al. 2018) employed this methodology in an attempt to get better system performance through optimization route. The sequential steps to be carried out while conducting GRA is given below.

Data Normalization

The first step in the GRA is to normalize or scale the response variable in the range of 0 and 1. Normalization is done to convert the all the response variables values into comparable sequence values. These normalized values of response variables are then used to calculate the grey relation coefficients (GRC) values. The following equations are used for normalization purpose in this study:

$$X_i\left(k\right) = \frac{x_i\left(k\right) - \min x_i\left(k\right)}{\max x_i\left(k\right) - \min x_i\left(k\right)} \tag{7}$$

$$X_i\left(k\right) = \frac{\max x_i\left(k\right) - x_i\left(k\right)}{\max x_i\left(k\right) - \min x_i\left(k\right)} \tag{8}$$

where $X_i(k)$ is the normalized response variable value, $x_i(k)$ is the actual response variable value, and min $x_i(k)$, max $x_i(k)$ represents the minimum and maximum value of the $x_i(k)$ respectively. The number of observations is given by *i* (*i*= 1, 2, 3...), and *k* (*k*=1, 2, 3…) represents number of response variables. Equation (7) is used for the droplet frequency normalization purpose. Equation (7) is generally employed when a larger value of the performance criterion is required. To normalize the droplet diameter values equation (8) is used, where smaller response value is desirable.

Grey Relation Coefficient (GRC) and Grey Relation Grade (GRG) Computation

After pre-processing the output data, GRC values of each response variables are calculated with the help of the following equation:

$$\xi_i(k) = \frac{\Delta_{min} + \xi * \Delta_{max}}{\Delta_i(k) + \xi * \Delta_{max}} \tag{9}$$

where $\xi_i(k)$ is the GRC value, $\Delta_i(k)$ represents the deviation sequence, which is the absolute difference between the reference sequence and comparability sequence given by X_i*(k) and X_i (k) respectively. The following equation is used to calculate deviation sequence value:

$$\Delta_i(k) = \left| X^*_i(k) - X_i(k) \right| \tag{10}$$

Δ_{max} and Δ_{min} are maximum and the minimum values of $\Delta_i(k)$ respectively. ξ is distinguishing factor or identification factor, which is generally taken as 0.5.

After calculating the GRC values for each observation, the next step is to find out the performance characteristics MPI (multi-performance index) value of the methodology which is the GRG value. In the GRA, the overall process performance is based on GRG value, which converts the multi-objective problem into single response problem, i.e., GRG value. The optimal process parameters combination corresponds to the maximum GRG value, which ultimately optimizes the actual multi-response problem. The GRA determines the absolute difference between the reference sequence and comparability sequence. A higher value of the GRG means a stronger relationship between the sequences. It means that the comparability sequence is much closer to the reference sequence which is the ideal one. So for a given parameter combination if GRG value is the highest, it means that corresponding process parameters are close to the optimal one. The following equation is used to calculate GRG value:

$$\gamma_i = \sum_{k=1}^{n} w_k \xi_i(k), \ \sum_{k=1}^{n} w_k = 1 \tag{11}$$

where γ_i is the GRG and w_k denotes the weight of the individual response variables. Since, both the response variable is equally important for better E-jet performance, an equal weight of 0.50 each is assigned to droplet diameter and droplet deposition frequency also.

DATA ANALYSIS AND DISCUSSION

Graf (2011) conducted an experimental study to investigate E-jet printing performance. Deposition frequency of droplets and the average diameter of ejected droplets were taken as the response variables of the study. During the investigation, the main effects of four control factors or process parameters were analyzed. The chosen controllable factors were as follows: ink material (A), applied pressure

(B), nozzle to substrate gap or stand-off height (C) and applied voltage (D). Two levels of each control factor were studied. The experiments were conducted according to the design of experiments (DOE) technique, instead of one factor at a time method. The chosen experimental layout was L16 orthogonal array (OA). Norland optical adhesive (NOA) was used as an ink material for analysis. It is an excellent optical bond for various substrates like plastic, glass. The levels of the each controllable variable along with the experimental layout are given in Table no. 1 & 2 respectively.

Table 1. Process parameters and parametric space

Process Parameters	Designation	Levels	
		Level 1	Level 2
Ink-type	A	NOA 74	NOA 89
Applied pressure (psi)	B	0.25	0.50
Stand-off height(μm)	C	30	50
Applied Voltage (V)	D	low	high

Table 2. Experimental layout

Experiment No.	A	B	C	D
	Ink-type	Applied Pressure (psi)	Stand-off Height(μm)	Applied Voltage (V)
1	1	1	1	1
2	1	1	1	2
3	1	1	2	1
4	1	1	2	2
5	1	2	1	1
6	1	2	1	2
7	1	2	2	1
8	1	2	2	2
9	2	1	1	1
10	2	1	1	2
11	2	1	2	1
12	2	1	2	2
13	2	2	1	1
14	2	2	1	2
15	2	2	2	1
16	2	2	2	2

Table 3 represents the experimentally obtained values of the chosen response variables or attributes. The same experimental data are analyzed here with the help of proposed methodology. The experimentally obtained values of deposition frequency and droplet diameter were used to calculate the preference scale.

Table 3. Experimental values of attributes or response variables (Graph 2011)

Experiment No.	Deposition Frequency (Hz)	Droplet Diameter(μm)
1	83.38	1.79
2	285.14	1.47
3	39.77	2.19
4	253.13	1.33
5	50.88	1.63
6	383.64	2.07
7	74.08	1.32
8	326.69	2.16
9	1754.16	2.45
10	7584.82	1.95
11	748.14	2.18
12	5129.83	2.70
13	2915.22	2.80
14	6407.23	2.55
15	1307.91	2.88
16	5558.34	2.85

The effect of different control factors on individual performance characteristic or response variable is shown in Table 4. The level averages are computed and tabulated. The results showed that parameter combination of *A2B2C1D2* yields higher printing deposition rate while the level combination of *A1B1C1D2* results lower deposited droplet diameter. At the favorable operating condition, the values of deposition rate and droplet diameter were predicted using additive model (Kumar et al., 2000; Phadke

Table 4. Main effects of process control variables

Control Parameters	Designation	Deposition Frequency (Hz)		Droplet Diameter (μm)	
		Level 1	Level 2	Level 1	Level 2
Ink-type	A	187.08	*3925.70*	*1.74*	2.54
Applied Pressure	B	1984.79	*2127.99*	*2.00*	2.28
Stand-off height	C	*2433.05*	1679.73	*2.08*	2.20
Applied voltage	D	871.69	*3241.10*	2.15	*2.13*

Table 5. Favourable combination levels of process control parameters

Response Variables	Favourable Settings of Process Parameters	Predicted Desirable Value of Response Variables
Deposition frequency (Hz)	*A2 B1 C1 D2*	*5558.67*
Droplet Diameter(μm)	*A1 B1 C1 D2*	*1.54*

et al., 1995). The predicted values of performance characteristics or response variables along with their individual favorable parameter combination are presented in Table 5.

Before calculating the preference scale for each response variable, the value of constant A has to be determined. The following calculations have been done for calculating the constant value and for constructing preference scale for each response variable or attribute:

- Deposition frequency (Hz)

X^*= favourable value of deposition frequency = 5558.67 Hz (Table 5)

X_i' = minimum or just acceptable value of deposition frequency = 39.77 Hz (Table 3)

After putting these values in the equation (3) the value of constant A for frequency becomes 4.194991. Using this value in equation (2) the preference scale for frequency becomes:

$$P_1 = 4.194991 \log \frac{X_i}{X_i'} \tag{12}$$

- Droplet diameter (μm)

X^*= favourable value of ejected droplet diameter = 1.54 μm (Table 5)

X_i' = minimum or just acceptable value of droplet diameter = 2.88 μm (Table 3)

Similar steps have been followed while calculating the value of A for droplet diameter using equation (2). For droplet diameter maximum value is taken (Table 3) because it represents the minimum or just acceptable limit for diameter. The objective is to get a diameter as low as possible. The preference scale equation for droplet diameter becomes:

$$P_2 = -33.1468 \log \frac{X_i}{X_i'} \tag{13}$$

After calculating the individual preference scale for each attribute the next step is to calculate the overall utility values. These preference scale values represent the individual utility values of response variables. In this work equal weight of 0.50 is assigned to each response variable. The equation for calculating overall utility value is:

$$U = 0.50 \times P_1 + 0.50 \times P_2 \tag{14}$$

The calculated values of preference scale for each response variable along with overall utility values are displayed in Table 6.

After calculating the overall utility values, the signal to noise (S/N) ratio values have been computed by using equation (6) and presented in Table 7. The objective is to maximize the overall utility value of the process. This is the reason behind the selection of "larger the better" criteria type signal to noise (S/N) ratio for analysis purpose (Routara et al., 2010). According to the methodology, the optimal fac-

Table 6. Preference scale and overall utility index

Experiment No.	Preference Scale		Overall Utility (U)
	Deposition Frequency (Hz)	Droplet Diameter (μm)	
1	1.34	6.84	4.09
2	3.58	9.68	6.63
3	0	3.94	1.97
4	3.37	11.12	7.24
5	0.44	8.19	4.32
6	4.12	4.75	4.44
7	1.13	11.23	6.18
8	3.83	4.14	3.98
9	6.89	2.32	4.61
10	9.56	5.61	7.58
11	5.34	4.00	4.67
12	8.85	0.92	4.89
13	7.82	0.40	4.11
14	9.25	1.75	5.50
15	6.36	0	3.18
16	8.99	0.15	4.57

tor level combination is the one which gives the maximum signal to noise (S/N) ratio value. The results of Taguchi analysis on the S/N ratio values are given in Table 8. The mean of the signal to noise (S/N) ratio for each level of process parameters were computed and tabulated in Table 8. From Table 8 it can be seen that the optimal level combination of process control parameters is *A2B1C1D2* (in coded form). It means that second type of ink material, low level of applied pressure, smaller nozzle to substrate stand-off height and high value of applied voltage result in most desirable operating condition for E-jet printing process. The aforesaid level combination satisfies both the attributes namely droplet deposition rate or frequency and ejected droplet diameter simultaneously with a reasonable trade-off. The order of significance of process parameters based on signal to noise (S/N) ratio values are presented in the rank column of Table 8. It can be seen that applied voltage is the most influential process parameter followed by the nozzle stand-off height. This finding is further validated by a statistical analysis called ANOVA. The anticipated improvement in the S/N ratio value under the proposed operating condition is reported in Table 9.

From Table 9 it is clear that under optimal parameter design the signal to noise ratio value is increased by .40. This shows the successful implementation of the aforesaid methodology aiming to improve the performance of the E-jet micro-fabrication process.

Since the experimental layout used in this study (Table 2) with four process control factors and sixteen experiments is a full factorial design, no other confirmation experiments were required (Llo et al., 2012). The experimental domain search the entire parametric space. There is no other design point left for examination. Furthermore, in the present work, the proposed optimal operating setting is already

Table 7. Signal to noise (S/N) ratio analysis based on utility values

Experiment No.	Signal to Noise Ratio
1	12.25
2	16.43
3	5.89
4	17.20
5	12.71
6	12.95
7	15.82
8	12.01
9	13.28
10	17.60
11	13.40
12	13.78
13	12.28
14	14.81
15	10.05
16	13.20

Table 8. Response table of signal to noise (S/N) ratio values

Factor	Level 1	Level 2	Difference	Rank
Ink-type (A)	13.16	*13.55*	0.39	4
Applied pressure (B)	*13.73*	12.98	0.74	3
Stand-off height (C)	*14.04*	12.67	1.362	2
Applied voltage (D)	11.96	*14.75*	2.79	1

Table 9. Estimated Improvement in signal to noise (S/N) ratio values

	Initial Process Parameters	Optimal Process Parameters
Level combination	*A1B1C2D2*	*A2B1C1D2*
Signal to Noise ratio	17.20	17.60
Anticipated improvement	0.40	

present in the experimental layout i.e. experiment no. 10 (Table 2). Therefore there is no need to conduct any confirmation experiment.

A statistical analysis named analysis of variance (ANOVA) (Kumar et al., 2013) is conducted to identify influential control factors in multi-response optimization study for deposition frequency and ejected droplet diameter of E-jet printing. The computed values of ANOVA analysis are presented in Table 10. It is a variance study, which searches for the source of variation present in the data. The control

Table 10. Analysis of variance (ANOVA) on signal to noise (S/N) ratio values

Source	D.F.	S.S.	M.S.	F Ratio
Ink material (A)	1	0.62	0.62	0.08
Applied pressure (B)	1	2.24	2.24	0.30
Stand-off height (C)	1	7.49	7.49	1.02
Applied voltage (D)	1	31.14	31.14	4.25
Error	11	80.57	7.32	
Total	15	122.07		

D.F. = degrees of freedom, S.S. = sum of squares, M.S. = mean square, F ratio= variance ratio

Table 11. Normalized response variables

Experiment No.	Deposition Frequency	Droplet Diameter
1	0.01	0.70
2	0.03	0.90
3	0.00	0.44
4	0.03	0.99
5	0.00	0.80
6	0.05	0.52
7	0.00	1.00
8	0.04	0.46
9	0.23	0.28
10	1.00	0.60
11	0.09	0.45
12	0.67	0.12
13	0.38	0.05
14	0.84	0.21
15	0.17	0.00
16	0.73	0.02

factor which is responsible for the generating more variation in the response is the most critical control factor among the others. In the analysis, F ratio measures the variation in the data. For a given factor more the value of F ratio more will be the effect of that factor in the outcome of the process. From the result of the analysis, it is clear that applied voltage is the most influential control factor which affects the printing performance of E-jet micro-fabrication process followed by the nozzle to substrate stand-off height. The fabrication process is more sensitive to the applied voltage and the stand-off height when compared to the ink type and applied pressure.

For conducting GRA, first raw response variable values are normalized. The desired droplet diameter should be as small as possible therefore equation (8) is used for data pre-processing. Moreover, as for droplet deposition frequency equation (7) is employed for normalization purpose in GRA.

The normalization results are shown in Table 11. The normalized values are used to calculate the GRC values. For computation purpose of GRC Eq. (9) is employed. Moreover, to calculate the deviation sequence which is required for the calculation of GRC values, Eq. (10) is used. Table 12 and Table 13 shows the deviation sequence values and GRC values for all observations. An equal weight of 0.50 each is assigned to both the response variables. These weight values of the performance characteristics reflect their relative importance in the outcome of the GRA, i.e., GRG value.

Table 12. Deviation sequence

Experiment No.	Deviation Sequence (Δ)	
	Deposition Frequency	Droplet Diameter
1	0.30	0.99
2	0.10	0.97
3	0.56	1.00
4	0.01	0.97
5	0.20	1.00
6	0.48	0.95
7	0.00	1.00
8	0.54	0.96
9	0.72	0.77
10	0.40	0.00
11	0.55	0.91
12	0.88	0.33
13	0.95	0.62
14	0.79	0.16
15	1.00	0.83
16	0.98	0.27

Table 14 indicates the decision criterion value of the analysis, i.e., GRG value. The last column of Table 14 indicates the ranking of different operating conditions (or, alternatives) of E-jet. The ranking is based on GRG values. Maximum GRG value is desirable, this why first rank is assigned to an observation which has the maximum GRG value.

In GRA, the optimal process setting is the one which gives the maximum GRG value. From Table 14 it can be seen that parameter combination in experiment no. 10 yields maximum GRG value. Therefore, the proposed optimum parameter setting for multiple quality characteristics is *A2B1C1D2* (in coded form), namely ink type: NOA 89, an applied pressure: 0.25 psi, the gap between the nozzle and the substrate: 30 μm and high level of applied voltage.

Table 13. Computed Grey relation coefficient (GRC) values

Experiment No.	Grey Relation Coefficients (GRC)	
	Deposition Frequency	Droplet Diameter
1	0.62	0.33
2	0.84	0.34
3	0.47	0.33
4	0.99	0.34
5	0.72	0.33
6	0.51	0.34
7	1.00	0.33
8	0.48	0.34
9	0.41	0.39
10	0.55	1.00
11	0.48	0.36
12	0.36	0.61
13	0.35	0.45
14	0.39	0.76
15	0.33	0.38
16	0.34	0.65

Table 14. Evaluated Grey relation grades (GRG) & ranking of alternatives

Experiment No.	Grey Relation Grades (GRG)	Rank
1	0.479	9
2	0.590	4
3	0.403	13
4	0.664	3
5	0.525	6
6	0.427	10
7	0.667	2
8	0.412	12
9	0.401	14
10	0.777	1
11	0.416	11
12	0.483	8
13	0.396	15
14	0.575	5
15	0.354	16
16	0.494	7

Comparative Study of Two Optimization Approaches

In the literature, several multi-objective optimization methods are available for determining optimal process parameters combination. Suitability of each of them is problem specific. Different algorithm may generate an alternate solution, and therefore a comparative study is required to examine whether the performance has improved or not by adopting different techniques. Hence, in the current case study, two optimization methodology is first implemented and then compared for finding out which method gives the improved performance of E-jet. The results of both optimization methodologies are presented in the tabular form in Table 15. The results show very good agreement between Taguchi coupled utility analysis and GRA method.

Table 15. Comparative study

Optimization Method	Optimal Process Parameters
Taguchi coupled Utility analysis	*A2B1C1D2*
Grey relational grade analysis	*A2B1C1D2*

CONCLUSION

In the present study, Taguchi based utility approach is used to design the parametric combination of high-resolution E-jet micro-fabrication process with multiple performance characteristics. Deposition frequency and ejected droplet diameter were selected as the performance characteristics or response variables. Hybrid Taguchi approach is adopted for optimization purpose, as simple Taguchi analysis is unable to handle the multi-response scenario. Multi-performance characteristics converted into a single performance index called overall utility value. The analysis is done on the overall utility value and optimal process parameter combination is selected based on the signal to noise (S/N) ratio value. According to the methodology, the favorable process environment is the one, which corresponds to the highest signal to noise (S/N) ratio value. The optimum parameter level combination is *A2B1C1D2*. The anticipated improvement in the signal to noise (S/N) ratio is 0.40. Thus parameters are successfully designed in the investigated region to achieve the desired objective. The Taguchi based analysis result is also compared with GRA. Both the analysis showed good agreement to each other. This further validates the suitability of the adopted methodology for solving the multi-objective optimization problem of E-jet. The influential control variables are identified with the help of a statistical technique called Analysis of variance (ANOVA). Both signal to noise (S/N) ratio and ANOVA draw similar conclusions. The applied voltage is found to be the most influential control parameter followed by the nozzle to substrate gap or stand-off height. The proposed methodology is simple, useful, easy to implement and provide an efficient way to handle the multi-response problem. The presented case study can be used as a template for better control of the process. The effects of other control factors like surface wettability, nozzle coating etc. could be studied in the future. The adopted design recipe can be applied for off-line quality control and process improvement to other processing and manufacturing operation.

ACKNOWLEDGMENT

The authors acknowledge the continuous support from the department of Mechanical Engineering, NIT Durgapur and would like thanks to the reviewers for their valuable comments to make the work a good contribution.

REFERENCES

Ali, S., Hassan, A., Hassan, G., Bae, J., & Lee, C. H. (2016). All-printed humidity sensor based on graphene/methyl-red composite with high sensitivity. *Carbon*, *105*, 23–32. doi:10.1016/j.carbon.2016.04.013

An, S., Lee, M. W., Kim, N. Y., Lee, C., Al-Deyab, S. S., James, S. C., & Yoon, S. S. (2014). Effect of viscosity, electrical conductivity, and surface tension on direct-current-pulsed drop-on-demand electrohydrodynamic printing frequency. *Applied Physics Letters*, *105*(21), 214102–214105. doi:10.1063/1.4902241

Anand, G., Alagumurthi, N., Elansezhian, R., Palanikumar, K., & Venkateshwaran, N. (2018). Investigation of drilling parameters on hybrid polymer composites using grey relational analysis, regression, fuzzy logic, and ANN models. *Journal of the Brazilian Society of Mechanical Sciences and Engineering*, *40*(4), 214. doi:10.100740430-018-1137-1

Awais, M. N., Kim, H. C., Doh, Y. H., & Choi, K. H. (2013). ZrO_2 flexible printed resistive (memristive) switch through electrohydrodynamic printing process. *Thin Solid Films*, *536*, 308–312. doi:10.1016/j.tsf.2013.04.003

Back, S. Y., Song, C. H., Yu, S., Lee, H. J., Kim, B. S., Yang, N. Y., ... Ahn, H. (2012). Drop-on-Demand Printing of Carbon Black Ink by Electrohydrodynamic Jet Printing. *Journal of Nanoscience and Nanotechnology*, *12*(1), 446–450. doi:10.1166/jnn.2012.5370 PMID:22524000

Badkar, D. S., Pandey, K. S., & Buvanashekaran, G. (2011). Parameter optimization of laser transformation hardening by using Taguchi method and utility concept. *International Journal of Advanced Manufacturing Technology*, *52*(9-12), 1067–1077. doi:10.100700170-010-2787-z

Bagci, E., & Aykut, S. (2006). A study of Taguchi optimization method for identifying optimum surface roughness in CNC face milling of cobalt-based alloy (stellite 6). *International Journal of Advanced Manufacturing Technology*, *29*(9-10), 940–947. doi:10.100700170-005-2616-y

Chakraborty, S., Kar, S., Ghosh, S. K., & Dey, V. (2017). Parametric optimization of electric discharge coating on Aluminium-6351 alloy with green compact silicon carbide and copper tool: A Taguchi coupled utility concept approach. *Surfaces and Interfaces*, *7*, 47–57. doi:10.1016/j.surfin.2017.02.001

Chen, C., Saville, D. A., & Aksay, I. A. (2006). Electrohydrodynamic "drop-and-place" particle deployment. *Applied Physics Letters*, *88*(25), 154104–3. doi:10.1063/1.2191733

Chen, C. H., Saville, D. A., & Aksay, I. A. (2006). Scaling laws for pulsed electrohydrodynamic drop formation. *Applied Physics Letters*, *89*(2), 124103–3. doi:10.1063/1.2356891

Choi, H. K., Park, J.-U., Park, O. O., Ferreira, P. M., Georgiadis, J. G., & Rogers, J. A. (2008). Scaling laws for jet pulsations associated with high-resolution electrohydrodynamic printing. *Applied Physics Letters*, *92*(12), 123109–3. doi:10.1063/1.2903700

Das, R. (2018). Optimization of E-Jet Based Micro-manufacturing Process Using Desirability Function Analysis. In S. Bhattacharyya, S. Sen, M. Dutta, P. Biswas, & H. Chattopadhyay (Eds.), Industry Interactive Innovations in Science, Engineering and Technology, Lecture Notes in Networks and Systems (Vol. 11, pp. 477-484). Singapore: Springer. doi:10.1007/978-981-10-3953-9_46

Deng, J.L. (1989). Introduction to Grey System Theory. *J Grey Syst, 1*(1), 1-24.

Derby, B., & ... . (2010). Inkjet Printing of Functional and Structural Materials: Fluid Property Requirements, Feature Stability, and Resolution. *Annual Review of Materials Research*, *40*(1), 395–414. doi:10.1146/annurev-matsci-070909-104502

Eskandari, B., Davoodi, B., & Ghorbani, H. (2018). Multi-objective optimization of parameters in turning of N-155 iron-nickel-base superalloy using gray relational analysis. *Journal of the Brazilian Society of Mechanical Sciences and Engineering*, *40*(4), 233. doi:10.100740430-018-1156-y

Fung, C. P. (2003). Manufacturing process optimization for wear property of fiber-reinforced polybutylene terephthalate composites with grey relational analysis. *Wear*, *254*(3-4), 298–306. doi:10.1016/S0043-1648(03)00013-9

Graf, P. G. (2011). *A 2nd generation electrohydrodynamic jet (e-jet) printing system, parametric studies of e-jet nozzles and integrated electrode e-jet deposition* (M.S. Thesis). University of Illinois at Urbana-Champaign. https://www.ideals.illinois.edu/handle/2142/24288

Ho, C. Y., & Lin, Z. C. (2003). Analysis and Application of Grey Relation and ANOVA in Chemical-Mechanical Polishing Process Parameters. *International Journal of Advanced Manufacturing Technology*, *21*(1), 10–14. doi:10.1007001700300001

Ilo, S., Just, C., & Xhiku, F. (2012). Optimisation of multiple quality characteristics of hardfacing using grey-based Taguchi method. *Materials & Design*, *33*, 459–468. doi:10.1016/j.matdes.2011.04.050

Jayasinghe, S. N., Qureshi, A. N., & Eagles, P. A. M. (2005). Electrohydrodynamic Jet Processing: An Advanced Electric-Field-Driven Jetting Phenomenon for Processing Living Cells. *Small*, *2*(2), 216–219. doi:10.1002mll.200500291 PMID:17193023

Kamyshny, A., & Magdassi, S. (2014). Conductive Nanomaterials for Printed Electronics. *Small*, *10*(17), 3515–3535. doi:10.1002mll.201303000 PMID:25340186

Kang, K., Yang, D., Park, J., Kim, S., Cho, I., Yang, H.-H., ... Park, I. (2017). Micropatterning of metal oxide nanofibers by electrohydrodynamic (EHD) printing towards highly integrated and multiplexed gas sensor applications. *Sensors and Actuators. B, Chemical*, *250*, 574–583. doi:10.1016/j.snb.2017.04.194

Kim, D. S., Khan, A., Rahman, K., Khan, S., Kim, H. C., & Choi, K. H. (2016). Drop-on-Demand Direct Printing of Colloidal Copper Nanoparticles by Electrohydrodynamic Atomization. *Materials and Manufacturing Processes*, *26*(9), 1196–1201. doi:10.1080/10426914.2011.551956

Kim, J., Oh, H., & Kim, S. S. (2008). Electrohydrodynamic drop-on-demand patterning in pulsed cone-jet mode at various frequencies. *Journal of Aerosol Science*, *39*(9), 819–825. doi:10.1016/j.jaerosci.2008.05.001

Kim, K., Lee, B. U., Hwang, G. B., Lee, J. H., & Kim, S. (2010). Drop-on-Demand Patterning of Bacterial Cells Using Pulsed Jet Electrospraying. *Analytical Chemistry*, *82*(5), 2109–2112. doi:10.1021/ac9027966 PMID:20143835

Kim, Y. J. (2009). On-demand electrohydrodynamic jetting with meniscus control by a piezoelectric actuator for ultra-fine patterns. *J. Micromech. Microeng.*, *19*(10). doi:10.1088/0960-1317/19/10/107001

Kumar, P., Barua, P. B., & Gaindhar, J. L. (2000). Quality optimization (multi-characteristics) through Taguchi's technique and utility concept. *Quality and Reliability Engineering International*, *16*(6), 475–485. doi:10.1002/1099-1638(200011/12)16:6<475::AID-QRE342>3.0.CO;2-0

Kumar, S., Meenu, & Satsangi, P. S. (2013). Multiple-response optimization of turning machining by the taguchi method and the utility concept using uni-directional glass fiber-reinforced plastic composite and carbide (k10) cutting tool. *Journal of Mechanical Science and Technology*, *27*(9), 2829–2837. doi:10.100712206-013-0731-5

Lee, M. W., Kang, D. K., Kim, N. Y., Kim, H. Y., James, S. C., & Yoon, S. S. (2012). A study of ejection modes for pulsed-DC electrohydrodynamic inkjet printing. *Journal of Aerosol Science*, *46*, 1–6. doi:10.1016/j.jaerosci.2011.11.002

Li, J. L. (2006). On the meniscus deformation when the pulsed voltage is applied. *Journal of Electrostatics*, *64*(1), 44–52. doi:10.1016/j.elstat.2005.04.005

Lin, C. L., Lin, J. L., & Ko, T. C. (2002). Optimisation of the EDM Process Based on the Orthogonal Array with Fuzzy Logic and Grey Relational Analysis Method. *International Journal of Advanced Manufacturing Technology*, *19*(4), 271–277. doi:10.1007001700200034

Lin, Y. C., & Lee, H. S. (2009). Optimization of machining parameters using magnetic-force-assisted EDM based on gray relational analysis. *International Journal of Advanced Manufacturing Technology*, *42*(11-12), 1052–1064. doi:10.100700170-008-1662-7

Malik, A., & Manna, A. (2018). Multi-response optimization of laser-assisted jet electrochemical machining parameters based on gray relational analysis. *Journal of the Brazilian Society of Mechanical Sciences and Engineering*, *40*(3), 148. doi:10.100740430-018-1069-9

Mishra, S. (2010). High-speed and drop-on-demand printing with a pulsed electrohydrodynamic jet. *J. Micromech. Microeng.*, *20*(9). doi:10.1088/0960-1317/20/9/095026

Nair, A., & Kumanan, S. (2018). Optimization of size and form characteristics using multi-objective grey analysis in abrasive water jet drilling of Inconel 617. *Journal of the Brazilian Society of Mechanical Sciences and Engineering*, *40*(3), 121. doi:10.100740430-018-1042-7

Nguyen, V. D., & Byun, D. (2009). Mechanism of electrohydrodynamic printing based on ac voltage without a nozzle electrode. *Applied Physics Letters*, *94*(17), 173509–3. doi:10.1063/1.3126957

Nothnagle, C., Baptist, J. R., Sanford, J., Lee, W. H., Popa, D. O., & Wijesundara, M. B. J. (2015). EHD printing of PEDOT: PSS inks for fabricating pressure and strain sensor arrays on flexible substrates. Next-Generation Robotics II; and Machine Intelligence and Bio-inspired Computation: Theory and Applications IX. *Proceedings of the Society for Photo-Instrumentation Engineers*, *9494*, 949403–949408. doi:10.1117/12.2177415

Onses, M. S., Sutanto, E., Ferreira, P. M., Alleyne, A. G., & Rogers, J. A. (2015). Mechanisms, Capabilities, and Applications of High-Resolution Electrohydrodynamic Jet Printing. *Small*, *11*(9), 4237–4266. doi:10.1002mll.201500593 PMID:26122917

Park, H. G., Byun, S.-U., Jeong, H.-C., Lee, J.-W., & Seo, D.-S. (2013). Photoreactive Spacer Prepared Using Electrohydrodynamic Printing for Application in a Liquid Crystal Device. *ECS Solid State Lett.*, *2*(12), R52–R54. doi:10.1149/2.010312ssl

Park, J., Kim, B., Kim, S.-Y., & Hwang, J. (2014). Prediction of drop-on-demand (DOD) pattern size in pulse voltage-applied electrohydrodynamic (EHD) jet printing of Ag colloid ink. *Applied Physics. A, Materials Science & Processing*, *117*(4), 2225–2234. doi:10.100700339-014-8650-6

Park, J. U., Hardy, M., Kang, S. J., Barton, K., Adair, K., Mukhopadhyay, D., ... Rogers, J. A. (2007). High-resolution electrohydrodynamic jet printing. *Nature Materials*, *6*(10), 782–789. doi:10.1038/nmat1974 PMID:17676047

Park, J. (2014). Fabrication of a flexible Ag-grid transparent electrode using ac based electrohydrodynamic Jet printing. *J. Phys. D: Appl. Phys.*, *47*(40). doi:10.1088/0022-3727/47/40/405102

Patowari, P. K., Saha, P., & Mishra, P. K. (2011). Taguchi analysis of surface modification technique using W-Cu powder metallurgy sintered tools in EDM and characterization of the deposited layer. *International Journal of Advanced Manufacturing Technology*, *54*(5-8), 593–604. doi:10.100700170-010-2966-y

Phadke, M. S. (1995). *Quality Engineering Using Robust Design*. Prentice Hall.

Qin, H., Dong, J., & Lee, Y.-S. (2017). Fabrication and electrical characterization of multi-layer capacitive touch sensors on flexible substrates by additive e-jet printing. *Journal of Manufacturing Processes*, *28*(3), 479–485. doi:10.1016/j.jmapro.2017.04.015

Routara, B. C., Mohanty, S. D., Datta, S., Bandyopadhyay, A., & Mahapatra, S. S. (2010). Optimization in CNC end milling of UNS C34000 medimum leaded brass with multiple surface roughnesses characteristics. *Sadhana*, *35*(5), 319–629. doi:10.100712046-010-0041-x

Taylor, G. (1964). Disintegration of Water Drops in an Electric Field. *Proceedings of the Royal Society of London. Series A*, *280*(1382), 383–397. doi:10.1098/rspa.1964.0151

Tosun, N. (2006). Determination of optimum parameters for multi-performance characteristics in drilling by using grey relational analysis. *International Journal of Advanced Manufacturing Technology*, *28*(5-6), 450–455. doi:10.100700170-004-2386-y

Yu, M., Ahn, K. H., & Lee, S. J. (2016). Design optimization of ink in electrohydrodynamic jet printing: Effect of viscoelasticity on the formation of Taylor cone jet. *Materials & Design*, *89*, 109–115. doi:10.1016/j.matdes.2015.09.141

Yudistira, H. T., Nguyen, V. D., Tran, S. B. Q., Kang, T. S., Park, J. K., & Byun, D. (2011). Retreat behavior of a charged droplet for electrohydrodynamic inkjet printing. *Applied Physics Letters*, *98*(8), 083501–083503. doi:10.1063/1.3555346

Chapter 10
Triple Bottom Line-Focused Optimization of Oblique Turning Processes Based on Hybrid Modeling:
A Study Case on AISI 1045 Steel Turning

Ramón Quiza
University of Matanzas, Cuba

Iván La Fé-Perdomo
Pontifical Catholic University of Chile, Chile & University of Matanzas, Cuba

Marcelino Rivas
University of Matanzas, Cuba

Veena Ramtahalsing
Fusecon NV, Suriname

ABSTRACT

This chapter proposes a hybrid approach for modelling and optimizing the oblique turning processes. Analytical modelling and statistical regressions are combined for predicting the values of the most important parameters involved in the oblique cutting process. The predictions of the model were validated by using experimental data, showing coincidence for a 95% confidence level. Then, an a posteriori multi-objective optimization is carried out by using a genetic algorithm. Three conflicting objectives, which represent the three pillars of the sustainability as defined in the triple bottom line, are simultaneously considered: the carbon dioxide emissions, the cost, and the cutting time. The outcome of the optimization process is a set of non-nominate solutions, which are optimal in the wide sense that no other solution in the search space can improve one objective without worsening the other one. Finally, the decision-maker chooses the most convenient solution depending on the actual workshop conditions.

DOI: 10.4018/978-1-7998-4939-1.ch010

INTRODUCTION

Optimization is a very important issue in mechanical industry. Specially, in machining processes, where different aspects must be considered, selecting the most proper cutting conditions plays a key role for obtaining efficient and competitive products.

From the first published cutting processes optimization work (Taylor 1907), to the most current literature (Chethan et al. 2019, Radha Krishnan, & Ramesh 2020, La Fé et al. 2020) a big amount of research has been expended on this topic. Three main concerns arise from the reported studies: the optimization objectives, the models relating decision variables with targets and constraints, and optimization techniques. With regard to the first issue, most of the considered goals have been focused on economic aspects, such as cost and time. Sometimes, quality criteria, such as obtained surface roughness have been also included. As sustainability has become an important matter in contemporary industry, objectives focused on this point of view, have also appeared in the specialized literature. Specially, the co-called Triple Bottom Line (Peralta et al. 2016), which defines the three pillars of sustainability (i.e., environmental, economic and social aspects) has been considered as a useful tool for these analysis (La Fé et al. 2020, Hegab, Darras, & Kishawy 2018).

On the other hand, models portrait an important function into de optimization, as they guaranty the accuracy of the obtained solutions. The high complexity of the phenomena involved on the cutting processes makes it very hard to obtain reliable accurate models for describing the behavior of the involved variables. Most of the optimization approaches use empirical models which cannot be applied outside the data intervals used for fitting these models. Moreover, obtaining experimental data from cutting processes is expensive and time consuming. Other approaches have been proposed for modelling the cutting processes, such as finite element method (Bartarya, & Choudhury 2011, Abouridouane et al. 2016, Weng et al. 2017) and mechanistic approaches (Vinogradov 2014, Zhang, & Guo 2015, Abouridouan et al. 2016, Baohai, et al. 2016, Bai et al. 2017, D'Acunto et al. 2017, Fu et al. 2018, Gao et al. 2018, Weng et al. 2018). Both offer reasonable accuracy with few experimental data, but requires a lot of time for computing the outcomes (specially, the finite element method).

Finally, due to the complex nature of these models, the objectives functions and constraints do not fulfil the conditions of continuity, differentiability and unimodality, required for most of the optimization tools (Quiza et al. 2012). Therefore, heuristics techniques have been applied for solving these problems. They include genetic algorithms (Ganesan, & Mohankumar 2013, Batish et al. 2014, Kübler et al. 2015), simulated annealing (Wang et al. 2006, Baseri 2011), particle swarm optimization (Marko et al. 2014) and ant colony optimization (Vijayakumar et al. 2003), but all of them rely on the accuracy and reliability of the underlying models.

Furthermore, the use of a single objective model is not enough for depicting the complex nature of the cutting processes (Kovačević et al. 2014, Saha, & Majumder 2018). Neither the a priori multi-objective optimization, which combines the different goals into a single one, can solve this problem (Quiza et al. 2014).

The present paper proposes a framework for optimizing the sustainability of oblique turning process, by combining a hybrid modelling with a multi-objective heuristic-based optimization approach. This method not only allows to improve the three aspects of the sustainability defined in the Triple Bottom Line, but also avoids the use of large and expensive amount of experimental work and allows carrying out the decision-making with accuracy and flexibility.

The chapter is structured in five sections. After this introduction, the general framework of the proposed approach is outlined. The third section presents the mechanistic model of the oblique cutting process. Section four presents a study case exemplifying the proposed methodology, including an experimental validation. Finally, conclusions are exposed at the fifth section.

FRAMEWORK DESCRITION

The proposed hybrid approach (see Fig. 1) is aimed to optimize, simultaneously, the unit carbon dioxide emission, ε_0, which constitute the capital environmental impact; the unit cost, Z_0, which summarizes the economic impact; and the unit cutting time (i.e., the productivity), τ_0, which is the most significant social impact (centered in the satisfaction and motivation of workers). In these objectives, the term unit refers to removing of a volumetric unit of workpiece material.

Figure 1. General optimization framework

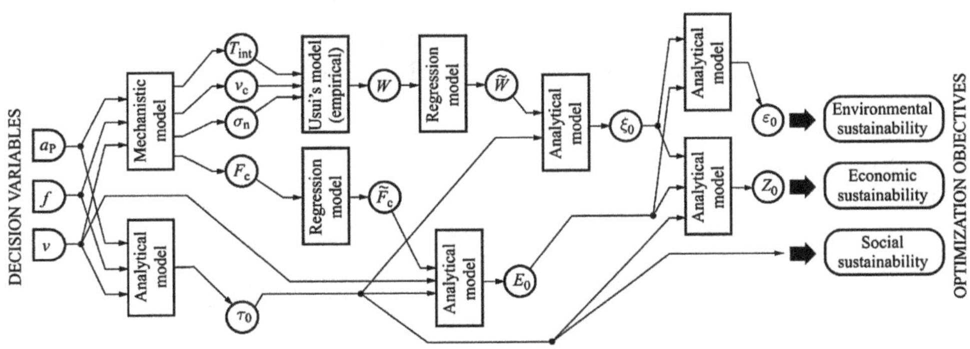

The chosen decision variables are the cutting parameters: cutting speed, V; feed, f; and depth of cut, a_p. A mechanistic model (which is detailed in the next section) (La Fé et al. 2018), allows to compute the values of tool-chip interface temperature, T_{int}; chip velocity, V_c; normal pressure, σ_n; and cutting force, F_c. The three first variables, through the use of the empirical Usui´s model, give the wear rate, W. Whit the values of both, the cutting force and the wear rate, regression models are fitted, in order to predict these variables as functions of the cutting parameters:

$$\tilde{F}_c = \phi_1(a_p, f, V)[\mathrm{N}]; \tag{1}$$

$$\tilde{W} = \phi_2(a_p, f, V)[\mathrm{mm/min}]. \tag{2}$$

On the other hand, an analytical model relates the unit cutting time, τ_0, with the cutting parameters:

$$\tau_0 = \frac{1000}{a_p f V}[\text{min}/\text{dm}^3]. \tag{3}$$

By combining the predicted tool wear rate, $\tilde{W}$, and the unit cutting time, another analytical relationship gives the unit tool waste, ξ, which is the ratio between the unit cutting time and the tool life, T:

$$\xi_0 = \frac{\tau_0}{T}[1/\text{dm}^3]; \tag{4}$$

being, the tool life given by:

$$T = \frac{\Omega^{\max} n_E}{\tilde{W}}[\text{min}]; \tag{5}$$

where Ω^{max} is the maximum value of the tool wear and n_E, the number of usable cutting edges.

The unit wasted energy, E_0, can be also computed by using the corresponding analytical model, from the predicted cutting force, $\tilde{F}_c$, the cutting speed, V, and the unit cutting time:

$$E_0 = \frac{F_c V \tau_0}{3.6 \times 10^6}[\text{kW} \cdot \text{h}/\text{dm}^3]. \tag{6}$$

The unit carbon dioxide emissions can be analytically computed by adding the emissions from wasted electric power and from manufacturing of wasted cutting tool:

$$\varepsilon_0 = e_E E_0 + e_T m_T \xi_0 [\text{kg}/\text{dm}^3]; \tag{7}$$

where e_E and e_T are the emission factors for the energy and tool, respectively, and m_T is the tool mass. The unit total cost, Z_0, is computed by including the labor, energy and tool costs:

$$Z_0 = \frac{\zeta_L \tau_0}{60} + \zeta_E E_0 + \zeta_T \xi_0 [\$/\text{dm}^3]; \tag{8}$$

where ζ_L, ζ_E and ζ_T are the specific costs of labor, energy and tool, respectively.

The optimization is carry out from a multi-objective a posteriori approach, where the Pareto front is obtained. This Pareto front contains the non-dominated solutions, which are optimal in the wide sense that no other solution in the search space can improve any of the considered objective without worsening another one. From the Pareto front, the most convenient solution can be selected, depending on the how the three sustainability pillars are valuated in the specific workshop condition,

MECHANISTIC MODELLING OF OBLIQUE CUTTING PROCESS

Geometry of the Oblique Cutting

The first step in the model formulation is to establish a set of geometric planes for transforming vectors (mainly, forces and velocities) from one coordinate system to another one (see Fig. 2).

Figure 2. Significant planes in the oblique cutting

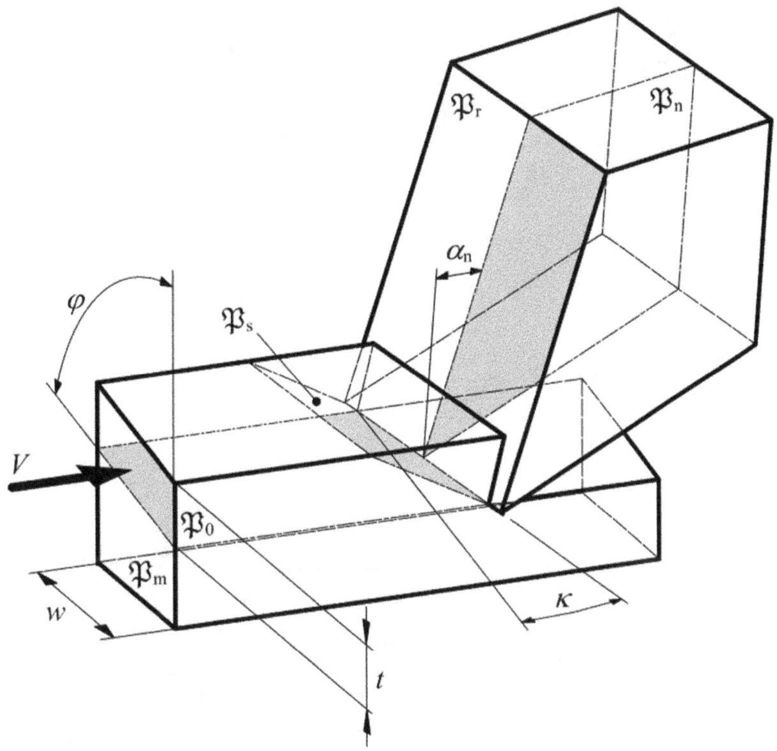

The first one is the machine plane, $\mathfrak{P}_m$, corresponding to the coordinate system of the machine tool (see Fig. 3), which is defined by the unit vectors $\boldsymbol{R}_m \equiv \{x_m, y_m, z_m\}$. $\mathbf{x}_m$ is in the direction of the spindle axis, $\mathbf{y}_m$ is directed vertically upward, while $\mathbf{z}_m$ is perpendicular to the previous ones.

The actual cutting width, *w'*, and un-deformed chip thickness, *t'*, can be computed from the technological cutting width (which is equal to the depth of cut, $w = a_p$) and the un-deformed chip thickness (which is equal to the feed per revolution, $t = f$):

$$w' = w \, / \, \sin\varphi, t' = t\sin\varphi; \tag{9}$$

Figure 3. Machine plane definition

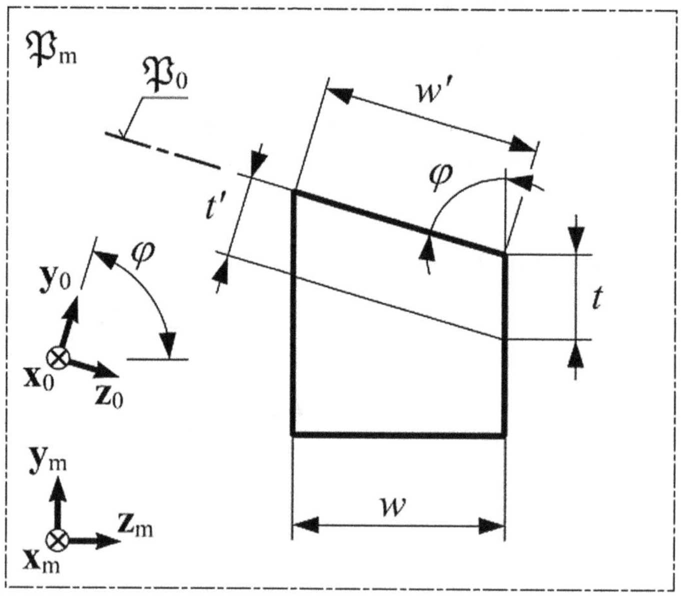

Figure 4. Reference plane definition

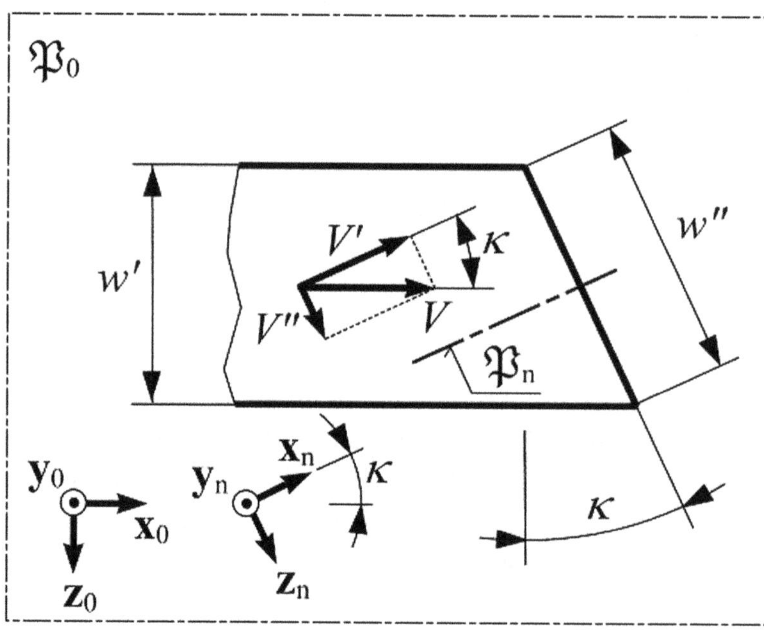

Closely related with the machine plane, is the reference plane, $\mathfrak{P}_0$, defined by the vectors $\boldsymbol{R}_0 \equiv \{x_0, y_0, z_0\}$ which can be obtained by rotating the system $\mathbf{R}_m$, about the $\mathbf{x}_m$-direction, an angle $\pi/2 - \varphi$, where φ if the tool edge position angle. The transformation from the system $\mathbf{R}_m$ to $\mathbf{R}_0$ can be done by using the matrix:

$$\boldsymbol{T}_{m0} = \begin{bmatrix} 1 & 0 & 0 \\ 0 & \cos(\frac{\pi}{2} - \varphi) & \sin(\frac{\pi}{2} - \varphi) \\ 0 & -\sin(\frac{\pi}{2} - \varphi) & \cos(\frac{\pi}{2} - \varphi) \end{bmatrix} = \begin{bmatrix} 1 & 0 & 0 \\ 0 & \sin\varphi & \cos\varphi \\ 0 & -\cos\varphi & \sin\varphi \end{bmatrix}. \tag{10}$$

On the reference plane (see Fig. 4), the cutting velocity, V, can be divided into two components; one of them, V', perpendicular and the other one, V'', parallel to the cutting edge, which is inclined an angle, κ, with actual cutting width ($\mathbf{z}_0$-direction). Consequently, the cutting velocity components can be computed by the equations:

$$V' = V\cos\kappa, V'' = V\sin\kappa; \tag{11}$$

Also, the contact cutting width, w'', is given by:

$$w'' = \frac{w'}{\cos\kappa} = \frac{w}{\cos\kappa\sin\varphi}; \tag{12}$$

From the reference plane, the normal plane, $\mathfrak{P}_n$, can be defined (see Fig. 5).This plane is perpendicular to the cutting edge and is given by the unit vectors $\boldsymbol{R}_n \equiv \{x_n, y_n, z_n\}$, which corresponds to a rotation of $\mathbf{R}_m$, an angle κ about the $\mathbf{y}_0$ direction. The transformation matrix from $\mathfrak{P}_0$ to $\mathfrak{P}_n$ can be written as:

$$\boldsymbol{T}_{0n} = \begin{bmatrix} \cos\kappa & 0 & \sin\kappa \\ 0 & 1 & 0 \\ -\sin\kappa & 0 & \cos\kappa \end{bmatrix}. \tag{13}$$

Equations 2 and 5 can be combined for obtaining the total transformation from the machine plane, $\mathfrak{P}_m$, to the normal plane, $\mathfrak{P}_n$:

$$\boldsymbol{T}_{mn} = \boldsymbol{T}_{m0}\boldsymbol{T}_{0n} = \begin{bmatrix} 1 & 0 & 0 \\ 0 & \sin\varphi & \cos\varphi \\ 0 & -\cos\varphi & \sin\varphi \end{bmatrix} \begin{bmatrix} \cos\kappa & 0 & \sin\kappa \\ 0 & 1 & 0 \\ -\sin\kappa & 0 & \cos\kappa \end{bmatrix};$$

$$T_{mn} = \begin{bmatrix} \cos\kappa & 0 & \sin\kappa \\ -\cos\varphi\sin\kappa & \sin\varphi & \cos\varphi\cos\kappa \\ -\sin\varphi\sin\kappa & -\cos\varphi & \sin\varphi\cos\kappa \end{bmatrix}. \tag{14}$$

On the normal plane (see Fig. 5), the orthogonal cutting process takes place. The shearing process occurs at the shearing plane, $\mathfrak{P}_s$, which forms the angle ϕ_n (normal shear angle) with the reference plane, $\mathfrak{P}_0$, and the rake face plane, $\mathfrak{P}_r$, forms and angle α_n (normal rake angle) with the vertical.

Figure 5. Normal plane definition

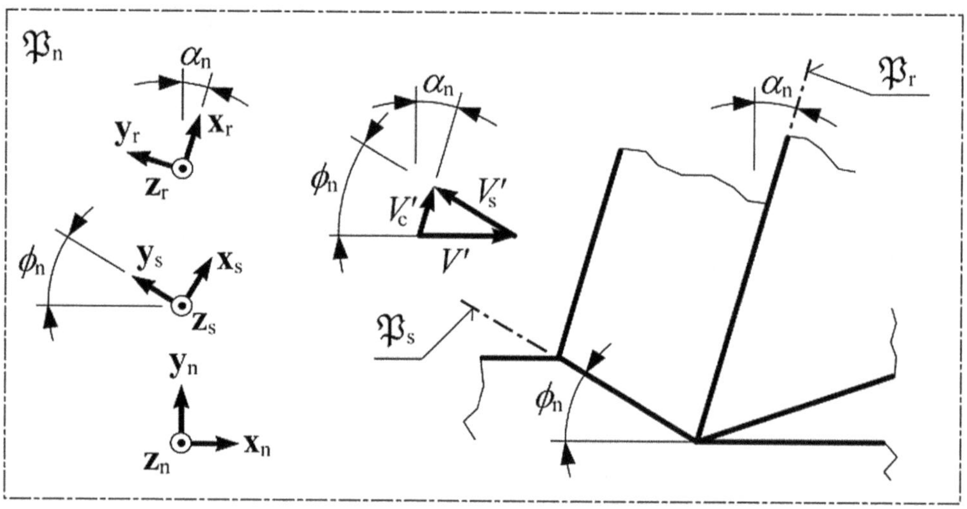

Thus, on the normal plane, the equilibrium of the cutting velocity, V', the shear velocity, V_s', and the chip velocity, V_c', is given by the equation 7:

$$\frac{V'}{\cos(\phi_n - \alpha_n)} = \frac{V_c'}{\sin\phi_n} = \frac{V_s'}{\cos\alpha_n}. \tag{15}$$

The orthogonal cutting process at the normal plane is complemented by an edgewise chip sliding and an edgewise shearing process. The chip sliding velocity, V_c (see Fig. 6), is formed not only by the normal chip velocity, V_c', but also by the edgewise chip velocity, V_c''.

Figure 6. Rake face plane definition

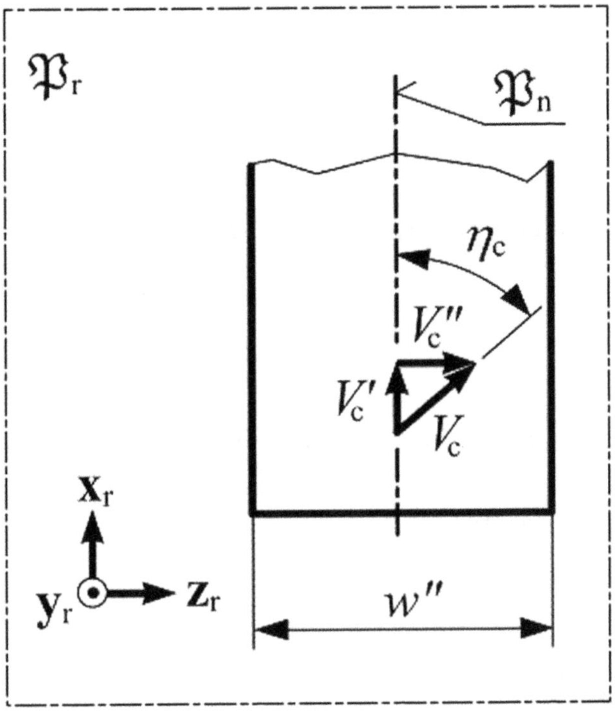

Figure 7. Shear plane definition

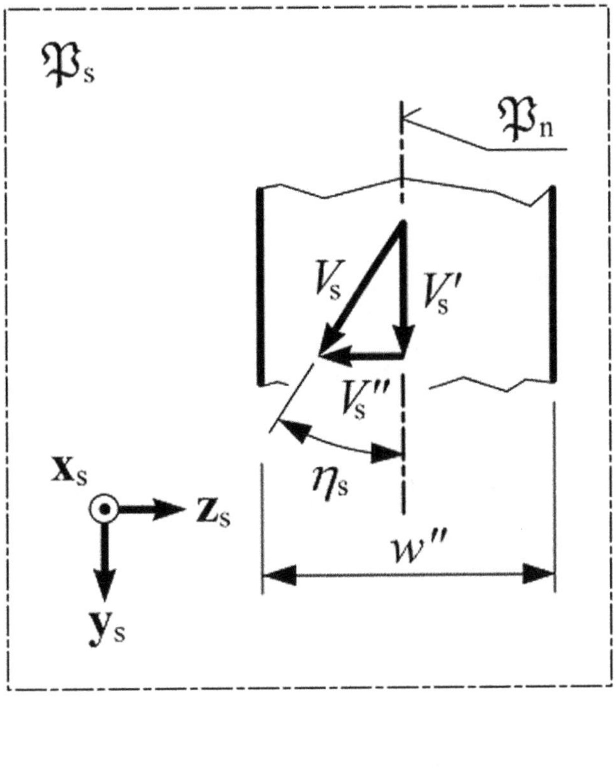

The angle between the normal plane and the chip velocity, V_c, is called the chip angle, η_c, and can be determined by:

$$\eta_c = \tan^{-1}\left[\frac{\tan\kappa\cos\alpha_n}{\tan(\phi_n + \lambda_n)} + \sin\alpha_n \tan\kappa\right]. \quad (16)$$

On the other hand, the shearing velocity, V_s, can be decomposed into a normal shear velocity, V_s', and a edgewise shear velocity, V_s'' (see Fig. 7), which are related by the expressions:

$$V_s' = V_s\cos\eta_s, V_s'' = V_s\sin\eta_s; \quad (17)$$

where η_s is the shear angle, which can be determined by the implicit equation (Moufki et al. 2004):

$$\eta_s = \tan^{-1}\left[\frac{\tan\eta_c\sin\phi_n - \tan\kappa\cos(\phi_n - \alpha_n)}{\cos\alpha_n}\right]. \quad (18)$$

Orthogonal Cutting Model

An important part of the proposed oblique model is the orthogonal process that carries out on the normal plane. It is based on the approach developed by (Lalwani et al. 2009). The first part is the computation of the parameters at the shear plane (i.e., the strain, ε_{AB}, strain rate, $\dot{\varepsilon}_{AB}$, temperature, T_{AB}, and flow shear stress, k_{AB}. The strain is computed by:

$$\varepsilon_{AB} = \frac{\gamma_{AB}}{\sqrt{3}} = \frac{1}{2\sqrt{3}}\frac{\cos\alpha_n}{\sin\phi_n\cos(\phi_n - \alpha_n)}; \quad (19)$$

while the strain rate can be computed by:

$$\dot{\varepsilon}_{AB} = \frac{C_0 V\cos\kappa\sin\phi_n\cos\alpha_n}{\sqrt{3}t\sin\varphi\cos(\phi_n - \alpha_n)}; \quad (20)$$

The flow shear stress, k_{AB}, is computed by merging the Johnson-Cook model and the von Misses yield criterion:

$$k_{AB} = \frac{\sigma_{AB}}{\sqrt{3}} = \frac{1}{\sqrt{3}}\left(A + B\varepsilon_{AB}^n\right)\left(1 + \nu\ln\frac{\dot{\varepsilon}_{AB}}{\dot{\varepsilon}_0}\right)\left[1 - \left(\frac{T_{AB} - T_0}{T_M - T_0}\right)^m\right]. \quad (21)$$

The temperature increment at the shear zone, T_{SZ}, is computed from equation:

$$T_{SZ} = \frac{(1-\beta)k_{AB}\cos\alpha_n}{\rho(T_{AB})S(T_{AB})\cos(\phi_n-\alpha_n)\sin\phi_n}; \tag{22}$$

being the mass density, ρ, the thermal conductivity, K, and the specific heat, S, functions of the temperature, T_{AB}, in the most general case:

Temperature, T_{AB}, coefficient β; and dimensionless number R_T, can be computed as:

$$T_{AB} = T_0 + \eta T_{SZ}; \tag{23}$$

$$\beta = \begin{cases} 0.5 - 0.35\log\left(R_T \tan\phi_n\right), & 0.04 \leq R_T \tan\phi_n \leq 10 \\ 0.3 - 0.15\log\left(R_T \tan\phi_n\right), & R_T \tan\phi_n > 10 \end{cases} \tag{24}$$

$$R_T = \frac{\rho S V' t'}{K} = \frac{\rho(T_{AB})S(T_{AB})Vt\cos\kappa\sin\varphi}{K(T_{AB})}. \tag{25}$$

The equivalent strain hardening exponent, n_{eq}, can be directly obtained from equation:

$$n_{eq} \simeq \frac{nB\varepsilon_{AB}^n}{A+B\varepsilon_{AB}^n}; \tag{26}$$

while the force angle, θ_n, and the friction angle, λ_n, are obtained from equations:

$$\theta_n = \tan^{-1}\left(1+\frac{\pi}{2}-2\phi_n - C_0 n_{eq}\right); \tag{27}$$

$$\lambda_n = \theta_n + \alpha_n - \phi_n. \tag{28}$$

The normal stress at the tool edge, σ_n, is computed from:

$$\sigma_n = k_{AB}\left(1+\frac{\pi}{2}-2\alpha_n - 2C_0 n_{eq}\right). \tag{29}$$

The second part of the orthogonal cutting is the calculation of all the parameter at the tool-chip interface. The strain at the tool-chip interface, ε_{int}, can be obtained from:

$$\varepsilon_{int} = 2\varepsilon_{AB} + \frac{1}{2\sqrt{3}\delta}\frac{(\sin\theta_n + \frac{1}{3}C_0 n_{eq}\cos\theta_n)}{\cos\lambda_n \cos(\phi_n - \alpha_n)}. \tag{30}$$

The strain rate at the tool-chip interface, $\dot{\varepsilon}_{int}$, is computed by:

$$\dot{\varepsilon}_{int} = \frac{V\sin^2\phi_n \cos\kappa}{\sqrt{3}\delta t\cos^2(\phi_n - \alpha_n)\sin\varphi}. \tag{31}$$

The shear flow stress, at the tool-chip interface, k_{int}, can be computed from:

$$k_{int} = \frac{\sigma_{chip}}{\sqrt{3}} = \frac{1}{\sqrt{3}}\left(A + B\varepsilon_{int}^n\right)\left(1 + \nu\ln\frac{\dot{\varepsilon}_{int}}{\dot{\varepsilon}_0}\right)\left[1 - \left(\frac{T_{int} - T_0}{T_M - T_0}\right)^m\right]. \tag{32}$$

The average temperature rice in the chip, ΔT_c, can be obtained from expression:

$$\Delta T_c = \frac{k_{AB}\sin\lambda_n}{\rho(T_c)S(T_c)\cos(\phi_n - \alpha_n)\cos\theta_n}; \tag{33}$$

The maximum temperature at the tool-chip interface, ΔT_M can be calculated from:

$$\Delta T_M = \Delta T_c 10^{0.06 - 0.195\delta\sqrt{\frac{R_T\cos(\phi_n - \alpha_n)\cos\lambda_n}{\sin\theta_n + \frac{1}{3}\cos\theta_n C_0 n_{eq}}} + 0.5\lg\left(\frac{R_T\cos(\phi_n - \alpha_n)\cos\lambda_n}{\sin\theta_n + \frac{1}{3}\cos\theta_n C_0 n_{eq}}\right)}; \tag{34}$$

while R_T is computed from equation:

$$R_T = \frac{\rho S V' t'}{K} = \frac{\rho(T_c)S(T_c)Vt\cos\kappa\sin\varphi}{K(T_c)}; \tag{35}$$

and T_{int}, from:

$$T_{int} = T_0 + \Delta T_{SZ} + \psi\Delta T_M. \tag{36}$$

The shear stress at the tool-chip interface, τ_{int}, is computed by:

$$\tau_{\text{int}} = \frac{3k_{AB}\sin(2\lambda_n)}{3\sin(2\theta_n) + 2C_0 n_{eq}\cos^2\theta_n}. \tag{37}$$

Similarly, the normal stress at the tool-chip interface, σ_N, is computed from equation:

$$\sigma_N = \frac{6k_{AB}\cos^2\lambda_n}{3\sin(2\theta_n) + 2C_0 n_{eq}\cos^2\theta_n}. \tag{38}$$

Transformation from Orthogonal to Oblique Cutting

From its orthogonal components, the oblique force components can be determined by:

$$R' = \frac{k_{AB}wt}{\cos\theta_n \cos\kappa \sin\phi_n}; \tag{39}$$

$$F_C' = R'\cos(\theta_n - \phi_n), F_T' = R'\sin(\theta_n - \phi_n); \tag{40}$$

$$P_f = R'\sin\lambda_n \tan\eta_c; \tag{41}$$

$$\eta_c = \tan^{-1}\left[\frac{\tan\kappa\cos\alpha_n}{\tan(\phi_n + \lambda_n)} + \sin\alpha_n \tan\kappa\right]; \tag{42}$$

where, R', is the resultant force on the normal plane, F_C' and F_T' are force components on the orthogonal plane, P_f is perpendicular component and η_c is the chip angle.

The final step is to compute the force components in the machine coordinate system by using the corresponding transformation matrix:

$$\begin{Bmatrix} F_C \\ F_T \\ F_R \end{Bmatrix} = \begin{bmatrix} \cos\kappa & 0 & \sin\kappa \\ -\cos\varphi\sin\kappa & \sin\varphi & \cos\varphi\cos\kappa \\ -\sin\varphi\sin\kappa & -\cos\varphi & \sin\varphi\cos\kappa \end{bmatrix}^{-1} \begin{Bmatrix} F_C' \\ F_T' \\ P_f \end{Bmatrix}. \tag{43}$$

STUDY CASE

Process Conditions Definition

In order to show and validate the proposed hybrid modeling and optimization approach, a case of study was carried out. The considered process was a cylindrical turning of an AISI 1045 (HB 197) steel work-piece. The relationship between the stress, σ, and the strain, ε, strain rate, $\dot{\varepsilon}$, and temperature, T, for this material can be described is described by the corresponding Johnson-Cook model (Jasper, & Dautzenberg, 2002):

$$\sigma = (553.1 + 600.8\varepsilon^{0234})(1 + 0.034 \ln \dot{\varepsilon})\left(1 - \frac{T - 293}{1733 - 293}\right); \tag{44}$$

and the physical properties (heat capacity, S, and thermal conductivity, K) can be expressed by the equations:

$$S = 420 + 0.504T[\mathrm{J/(kg{\bullet}K)}]; \tag{45a}$$

$$K = 52.61 - 0.0281T[W/(\mathrm{m{\bullet}K})]; \tag{45b}$$

while the mass density was taken as a constant, $\rho = 7\ 860\ \mathrm{kg/m^3}$.

The cutting tool was an insert TNMG 22 04 08-PM, with a tool holder PTNGR 2525M 22. The tool material was carbide ISO P25, coated with TiN+Al2O3+TiC.

The allowable intervals for the cutting parameters, by using this cutting tool, are:

$$0.5\mathrm{mm} \le a_p \le 2.0\mathrm{mm}; \tag{46a}$$

$$0.15\mathrm{mm/rev} \le f \le 0.35\mathrm{mm/rev}; \tag{46b}$$

$$200\mathrm{m/min} \le V \le 300\mathrm{m/min}. \tag{46c}$$

Model Execution

In order to cover the entire cutting parameters interval, a three level full factorial experimental design was selected. Therefore, 27 runs were carried out. Table 2 (Appendix 1) shows the values of some of the most important parameters for the simulations.

Experimental Validation

In order to validate the model outcomes, experimental data were taken from (Quiza 2004). In these experiments, the used machine-tool was a lathe C11MT. The accuracy and stiffness requisites were verified by the standard ISO 1708:1989. Spindle speed was measured by using a digital tachometer ONO Sokki HT-413, with an accuracy of ±0.5 min^{-1}. The feed rate was measured with a displacement transducer Kyowa DT-100, with an accuracy of ±0.01 mm.

Cutting and feed forces were measured through two strain-gages Wheatstone bridges. Al the signals were magnified by using a digital amplifier DPM-602B and then introduced in a personal computer through a data capture card, at a rate of 820 measurements per second.

A three-level full factorial design was used for carrying out the experimental study. The cutting force was selected for validating the model as it is one of the most representative parameters in any cutting process. Three replicates were measured at each experimental point and the mean value, $\mu(F_C)$, and standard deviation, $\sigma(F_C)$, were also computed (see Table 3, Appendix 1).

Comparisons between predicted and measured values for the cutting forces are represented in Figure 8. The experimental values are shown by its confidence intervals (computed for 95% confidence level). As can be noted, all the simulated values, but point 13, are inside the confidence interval of the respective experimental values.

Figure 8. Comparison between experimental and modelled values of the cutting force

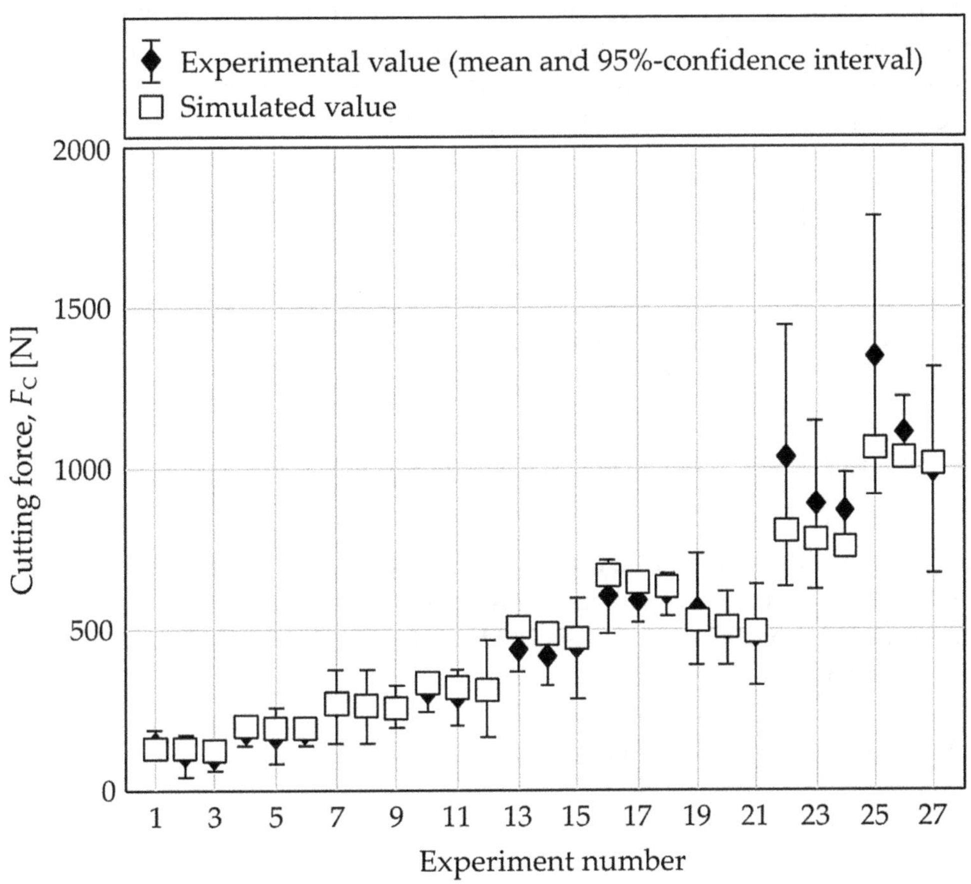

Consequently, it can be concluded that the predictions of the proposed analytical model agree with the experimental data, for the given confidence level, and they can be used in the optimization process.

Regression Models

For establishing the mathematical relationship between the studied variable (cutting force) with the experimental factors (cutting speed, feed and depth of cut), the first choice are the statistical regressions because they are simple and have a solid mathematical foundation. By applying the multiple regression technique, the following logarithmic model was obtained for the cutting force:

$$F_c = \frac{2800 a_p^{1.002} f^{0.8435}}{V^{0.1453}}; \tag{47}$$

Since the value of probability associated with the ANOVA analysis is less than 0.01, we can say that there is a statistically significant relationship between the variables, with a level of 99.9% reliability. The value of R^2 ensures that the adjusted model explains more than 99.9% of the variability in F_C. For its part, the analysis of the sensitivity of the coefficients shows that all terms of the expression have influence statistically significant with a confidence level of 99%.

From the physical point of view, the obtained model corresponds to the theoretical issues (Tschätsch 2008). The most influent parameter is the depth of cut, followed by the feed. Both parameters have a positive influence. On the contrary, the cutting speed has a lower influence and it is negative.

In order to estimate the tool life, the wear rate, *w*, in the rake surface is computed from values of contact pressure, sliding velocity and temperature, which are evaluated for different points on the tool-chip interface, by using Usui's expression (Quiza et al. 2012);

$$W = A' \sigma_n v_c \exp\left(-\frac{B'}{T_{\text{int}}}\right); \tag{48}$$

where, for the plain carbon steel versus uncoated tools:

$$\left.\begin{array}{ll} A' = 7.80 \times 10^{-9} m^2 / MN, B' = 5302 K^{-1} & : T_{\text{int}} < 1150K \\ A' = 1.198 \times 10^{-9} m^2 / MN, B' = 2195 K^{-1} & : T_{\text{int}} \geq 1150K \end{array}\right\}; \tag{49}$$

From the corresponding computed values of wear rate (see Table 4, Appendix 1), a multiple regression was applied. The depth of cut was removing from the model as it was not significant with a 90% of confidence level (the probability value associated to the *t*-Student test for its coefficient was higher than 0.9). The final model is:

$$W = -0.06844 + 0.1170f + 3.633 \times 10^{-4} V. \tag{50}$$

Since the value of probability associated with the ANOVA table is less than 0.01, we can say that there a statistically significant relationship between the variables, with a level of 99.7% reliability. The value of R^2 ensures that the adjusted model explains more than 95.2% of the variability in w. For its part, the analysis of the sensitivity of the coefficients shows that all terms of the expression have influence statistically significant with a confidence level of 99%.

From the physical point of view, the removal of the depth of cut from the model is full of sense, as this parameter has not influence on the parameters determining the wear rate (temperature, contact pressure and sliding velocity). Also, the coefficients of speed and feed are both positive, indicating a direct effect of these variables on the wear rate, which completely agree not only with the theoretical background (Bonny, De Baets et al. 2010) but also with previous experimental work (Arsecularatne et al. 2006).

Optimization

For executing the optimization process, the non-dominated sorting genetic algorithm II (NSGA-II) (Deb et al. 2002) was selected. It is very well reputed and widely used optimization heuristic (Emmerich, & Deutz 2018). The optimization was carried out with a population of 5000 solutions and a generation number of 100. Table 1 shows a selection of the obtained non-dominated solutions.

Table 1. Optimal cutting parameters

a_p [mm]	f [mm/rev]	V [m/min]	ε_0 [kg/dm^3	E_0 [kW·h/dm^3]	τ_0 [min/dm^3]
2.0	0.35	200	0.452	5.19	7.14
2.0	0.35	212	0.451	5.31	6.74
2.0	0.35	225	0.450	5.43	6.35
2.0	0.35	238	0.449	5.53	6.00
2.0	0.35	250	0.448	5.61	5.71
2.0	0.35	262	0.447	5.69	5.45
2.0	0.35	275	0.446	5.77	5.19
2.0	0.35	288	0.445	5.83	4.96
2.0	0.35	300	0.444	5.89	4.76

The obtained Pareto front (objective spaces), with the corresponding Pareto set (decision variables space), are shown in Fig. 9. Analysis these outcomes, it can be noted a relationship between the environmental and social sustainability indicators, because there is a coincidence between their respective minimum values (point A). On the contrary, the minimum economic sustainability value (point B) corresponds to the maximum environmental and social performance.

In the Pareto set, the most remarkable fact is that all the non-dominated solutions corresponds to the higher values of feed and depth of cut, and a variation in all the interval of allowed cutting speeds can be seen. The lower cutting speeds correspond to better environmental and social sustainability while the higher speeds offer a better economic sustainability. The main fact behind this behavior is the sudden increase in both the energy and tool waste when higher cutting speeds are used.

Figure 9. Obtained Pareto front and Pareto set

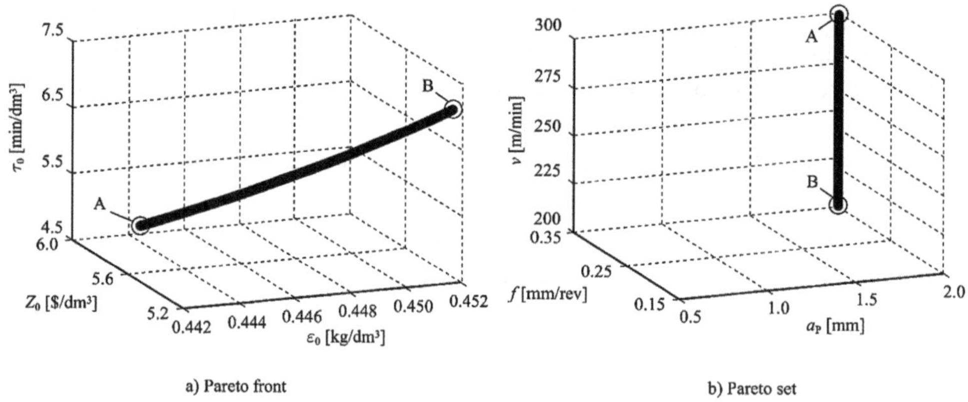

For deciding which cutting parameters will be finally chosen, the difference between higher and lower level of the objectives should be also analyzed. Difference between higher and lower values in carbon dioxide emissions are less than 2%, while the corresponding differences for cost and productivity are 12% and 33%, respectively. Therefore, environmental impact can be neglected.

Point B, then, should be chosen if economy must be ranked over productivity. On the contrary, point A is the most suitable choice for maximizing productivity, which can take place on several situations. Finally, other solutions are trade-off parameters which can be used for balancing both goals.

CONCLUSION

The proposed hybrid approach, based on a combination of analytical-empirical modelling and genetic algorithm-based multi-objective optimization, can be used for determining the optimal cutting parameters in oblique turning processes.

The used mechanistic model provided accurate outcomes (validated with experimental results), which can be obtained with few empirical data. The regression models, fitted from the analytical simulations, can summarize the mechanistic model outcomes into single mathematical expressions that can be quickly evaluated and, consequently, can be used in a heuristic-based optimization where a high number of evaluations are required.

The proposed a posteriori multi-objective optimization scheme included two important goals of the oblique cutting process: unit time (representing productivity) and tool wear rate (representing tool waste). It was carried out through the micro genetic algorithm, showing fast performance and feasible outcomes.

The obtained Pareto front included a wide spectrum of solutions, which can be used under different workshop conditions. This approach allows selecting the cutting parameters in a more flexible way as there is no need of e new optimization execution if the working conditions change.

ACKNOWLEDGMENT

This research was partially supported by the Program of Basic Science, from the Cuban Ministry of Science, Technology and Environment [grant number P223LH001-068]. Authors are also highly grateful to the CONICYT PFCHA/DOCTORADO BECAS CHILE/2019-21190520 for all the financial support.

REFERENCES

Abouridouane, M., Klocke, F., & Döbbeler, B. (2016). Analytical temperature prediction for cutting steel. *CIRP Annals - Manufacturing Technology*, *65*, 77-80. doi:10.1016/j.cirp.2016.04.039

Arsecularatne, J. A., Zhang, L. C., & Montross, C. (2006). Wear and tool life of tungsten carbide, PCBN and PCD cutting tools. *International Journal of Machine Tools & Manufacture*, *46*(6), 482–491. doi:10.1016/j.ijmachtools.2005.07.015

Bai, W., Sun, R., Roy, A., & Silberschmidt, V. V. (2017). Improved analytical prediction of chip formation in orthogonal cutting of titanium alloy Ti6Al4V. *International Journal of Mechanical Sciences*, *133*, 357–367. doi:10.1016/j.ijmecsci.2017.08.054

Baohai, W., Di, C., Xiaodong, H., Dinghua, Z., & Kai, T. (2016). Cutting tool temperature prediction method using analytical model for end milling. *Chinese Journal of Aeronautics*, *29*(6), 1788–1794. doi:10.1016/j.cja.2016.03.011

Bartarya, G., & Choudhury, S. K. (2011). State of the art in hard turning. *International Journal of Machine Tools & Manufacture*, *53*(1), 1–14. doi:10.1016/j.ijmachtools.2011.08.019

Baseri, H. (2011). Simulated annealing based optimization of dressing conditions for increasing the grinding performance. *International Journal of Advanced Manufacturing Technology*, *59*(5-8), 531–538. doi:10.100700170-011-3518-9

Batish, A., Bhattacharya, A., Kaur, M., & Cheema, M. S. (2014). Hard turning: Parametric optimization using genetic algorithm for rough/finish machining and study of surface morphology. *Journal of Mechanical Science and Technology*, *28*(5), 1629–1640. doi:10.100712206-014-0308-y

Bonny, K., De Baets, P., Perez, Y., Vleugels, J., & Lauwers, B. (2010). Friction and wear characteristics of WC–Co cemented carbides in dry reciprocating sliding contact. *Wear*, *268*(11–12), 1504–1517. doi:10.1016/j.wear.2010.02.029

Chethan, Y. D., Ravindra, H. V., & Krishnegowda, Y. T. (2019). Optimization of machining parameters in turning Nimonic-75 using machine vision and acoustic emission signals by Taguchi technique. *Measurement*, *144*, 144–154. doi:10.1016/j.measurement.2019.05.035

D'Acunto, A., Le Coz, G., Moufki, A., & Dudzinski, D. (2017). Effect of cutting edge geometry on chip flow direction – analytical modelling and experimental validation. *16th CIRP Conference on Modelling of Machining Operations*, *58*, 353-357. 10.1016/j.procir.2017.03.327

Deb, K., Pratap, A., Agarwal, S., & Meyarivan, T. (2002). A fast and elitist multiobjective genetic algorithm: NSGA-II. *IEEE Transactions on Evolutionary Computation*, *6*(2), 182–197. doi:10.1109/4235.996017

Emmerich, M. T. M., & Deutz, A. H. (2018). A tutorial on multiobjective optimization: Fundamentals and evolutionary methods. *Natural Computing*, *17*(3), 585–609. doi:10.100711047-018-9685-y PMID:30174562

Fu, Z., Chen, X., Mao, J., & Xiong, T. (2018). An analytical force mode applied to three-dimensional turning based on a predictive machining theory. *International Journal of Mechanical Sciences*, *136*, 94–105. doi:10.1016/j.ijmecsci.2017.12.021

Ganesan, H., & Mohankumar, G. (2013). Optimization of machining techniques in CNC turning centre using genetic algorithm. *Arabian Journal for Science and Engineering*, *38*(6), 1529–1538. doi:10.100713369-013-0539-8

Gao, Y., Sun, R., & Leopold, J. (2018). An analytical force model for modulation-assisted turning. *Journal of Manufacturing Processes*, *31*, 712–730. doi:10.1016/j.jmapro.2017.12.024

Hegab, H. A., Darras, B., & Kishawy, H. A. (2018). Towards sustainability assessment of machining processes. *Journal of Cleaner Production*, *170*, 694–703. doi:10.1016/j.jclepro.2017.09.197

Jasper, S. P. F. C., & Dautzenberg, J. H. (2002). Material behaviour in conditions similar to metal cutting: Flow stress in the primary shear zone. *Journal of Materials Processing Technology*, *122*(2-3), 322–330. doi:10.1016/S0924-0136(01)01228-6

Kovačević, M., Madić, M., Radovanović, M., & Rančić, D. (2014). Software prototype for solving multi-objective machining optimization problems: Application in non-conventional machining processes. *Expert Systems with Applications*, *41*(13), 5657–5668. doi:10.1016/j.eswa.2014.03.037

Kübler, F., Böhner, J., & Steinhilper, R. (2015). Resource efficiency optimization of manufacturing processes using evolutionary computation: A turning case. *The 22nd CIRP conference on Life Cycle Engineering*, *29*, 822-827. doi: 10.1016/j.procir.2015.02.208

La Fé, I., Quiza, R., Rivas, M., & Ramtahalsing, V. (2018). Hybrid modelling and optimization of the oblique cutting of AISI 1045 steel. *International Journal of Manufacturing, Materials, and Mechanical Engineering*, *8*(4). doi:10.4018/IJMMME.2018100101

La Fé Perdomo, I., Quiza, R., Haeseldonckx, D., & Rivas, M. (2020). (in press). Sustainability-focused multi-objective optimization of a turning process. *International Journal of Precision Engineering and Manufacturing-Green Technology*. doi:10.100740684-019-00122-4

Lalwani, D. I., Mehta, N. K., & Jain, P. K. (2009). Extension of Oxley's predictive machining theory for Johnson and Cook flow stress model. *Journal of Materials Processing Technology*, *209*(12-13), 5305–5312. doi:10.1016/j.jmatprotec.2009.03.020

Marko, H., Simon, K., Tomaz, I., Matej, P., Joze, B., & Miran, B. (2014). Turning Parameters Optimization using Particle Swarm Optimization. *Procedia Engineering*, *69*, 670–677. doi:10.1016/j.proeng.2014.03.041

Moufki, A., Devillez, A., Dudzinski, D., & Molinari, A. (2004). Thermomechanical modelling of oblique cutting and experimental validation. *International Journal of Machine Tools & Manufacture*, *44*(9), 971–989. doi:10.1016/j.ijmachtools.2004.01.018

Peralta, M. E., Marcos, M., & Aguayo, F. (2016). A review of sustainable machining engineering: Optimization process through Triple Bottom Line. *Journal of Manufacturing Science and Engineering*, *138*(10), 100801. doi:10.1115/1.4034277

Quiza, R. (2004). *Optimización multiobjetivo del proceso de torneado* (Ph.D. thesis), Universidad de Matanzas, Matanzas, Cuba.

Quiza, R., Beruvides, G., & Davim, J. P. (2014). Modeling and optimization of mechanical systems and processes. In J. P. Davim (Ed.), *Modern mechanical engineering* (pp. 169–198). Springer Berlin Heidelberg. doi:10.1007/978-3-642-45176-8_8

Quiza, R., López-Armas, O., & Davim, J. P. (2012). *Hybrid modeling and optimization of manufacturing: Combining artificial intelligence and finite element method*. Heidelberg, Germany: Springer. doi:10.1007/978-3-642-28085-6

Radha Krishnan, B., & Ramesh, M. (2019). Optimization of machining process parameters in CNC turning process of IS2062 E250 Steel using coated carbide cutting tool. *Materials Today: Proceedings*. doi:10.1016/j.matpr.2019.05.460

Saha, A., & Majumder, H. (2018). Performance analysis and optimization in turning of ASTM A36 through process capability. index. *Journal of King Saud University-Engineering Sciences*, *30*(4), 377–383. doi:10.1016/j.jksues.2016.11.002

Taylor, F. W. (1907). On the art of cutting metals. *Transactions of the ASME*, *28*, 310-350.

Tschätsch, H. (2008). *Applied machining technology* (8th ed.). Heidelberg, Germany: Springer.

Vijayakumar, K., Prabhaharan, G., Asokan, P., & Saravanan, R. (2003). Optimization of multi-pass turning operations using ant colony system. *International Journal of Machine Tools & Manufacture*, *43*(15), 1633–1639. doi:10.1016/S0890-6955(03)00081-6

Vinogradov, A. A. (2014). On the mechanism of chip formation in oblique cutting of metals. *Journal of Superhard Materials*, *36*(6), 428–434. doi:10.3103/S1063457614060094

Wang, Z. G., Wong, Y. S., Rahman, M., & Sun, J. (2006). Multi-objective optimization of high-speed milling with parallel genetic simulated annealing. *International Journal of Advanced Manufacturing Technology*, *31*(3-4), 209–218. doi:10.100700170-005-0191-x

Weng, J., Zhuang, K., Chen, D., Guo, S., & Ding, H. (2017). An analytical force prediction model for turning operation by round insert considering edge effect. *International Journal of Mechanical Sciences*, *128-129*, 168–180. doi:10.1016/j.ijmecsci.2017.04.018

Weng, J., Zhuang, K., Zhu, D., Guo, S., & Ding, H. (2018). An analytical model for the prediction of force distribution of round -insert considering edge effect and size effect. *International Journal of Mechanical Sciences*, *138-139*, 86–98. doi:10.1016/j.ijmecsci.2018.01.024

Zhang, G., & Guo, C. (2015). Modeling of Cutting Force Distribution on Tool Edge in Turning Process. *Procedia Manufacturing*, *1*, 454–465. doi:10.1016/j.promfg.2015.09.001

APPENDIX 1

Nomenclature

A: yield stress (Johnson-Cook model) [MPa]
A': coefficient (Usui wear model) [m^2/MN]
a_p: depth of cut [mm]
B: hardening coefficient (Johnson-Cook model) [MPa]
B': temperature factor (Usui wear model) [K^{-1}]
C_0: ratio of the shear plane length to the PSZ thickness
E_0: unit wasted energy [kW·h/dm^3]
e_E: energy emission factor [kg/(kW·h)]
e_T: energy emission factor [kg/(kW·h)]
F_C: main component of the cutting force [N]
F'_C : main component of the cutting force on the normal plane [N]
F_R: radial component of the cutting force [N]
F_T: thrust component of the cutting force [N]
F'_T : thrust component of the cutting force on the normal plane [N]
f: feed [mm/rev]
K: thermal conductivity [W/(m·K)]
k_{AB}: flow shear stress at the shear plane [MPa]
k_{int}: flow shear stress at the tool-chip interface [MPa]
m: temperature exponent (Johnson-Cook model)
m_T: tool mass [kg]
n: hardening exponent (Johnson-Cook model)
n_E: number of cutting edges
n_{eq}: equivalent strain hardening exponent
P_f: perpendicular component of the force [N]
R': resultant force on the normal plane [N]
R_T: dimensionless thermal number
S: specific heat [J/(kg·K)]
T: tool life [min]
T_{AB}: temperature at the shear plane [K]
T_{int}: tool-chip interface temperature [K]
T_M: melting temperature [K]
T_{SZ}: temperature increment at the shear zone [K]
T_0: reference temperature at the Johnson/Cook model [K]
$\mathbf{T}_{mn}$: transformation matrix from machine to normal coordinate systems
$\mathbf{T}_{m0}$: transformation matrix from machine to reference coordinate systems
$\mathbf{T}_{0n}$: transformation matrix from reference to normal coordinate systems
t: un-deformed chip thickness [mm]
t': actual un-deformed chip thickness [mm]
V: cutting speed [m/min]

V': perpendicular component of the cutting speed [m/min]
V'': parallel component of the cutting speed [m/min]
V_c: chip sliding velocity [mm/s]
V_c' : normal chip sliding velocity [mm/s]
V_c'' : edgewise chip sliding velocity [mm/s]
V_s' : normal shear velocity [mm/s]
V_s: shear velocity [mm/s]
V_s'' : edgewise chip shear [mm/s]
W: wear rate [mm/min]
w: technological width of cut [mm]
w': actual width of cut [mm]
$\mathbf{x}_m$, $\mathbf{y}_m$, $\mathbf{z}_m$: unit vectors in the machine coordinate system
$\mathbf{x}_n$, $\mathbf{y}_n$, $\mathbf{z}_n$: unit vectors in the normal coordinate system
$\mathbf{x}_r$, $\mathbf{y}_r$, $\mathbf{z}_r$: unit vectors in the rake plane coordinate system
$\mathbf{x}_s$, $\mathbf{y}_s$, $\mathbf{z}_s$: unit vectors in the shear coordinate system
$\mathbf{x}_0$, $\mathbf{y}_0$, $\mathbf{z}_0$: unit vectors in the reference coordinate system
Z_0: unit cost [$/dm³]
$\mathfrak{P}_m$: machine plane
$\mathfrak{P}_n$: normal plane
$\mathfrak{P}_r$: rake face plane
$\mathfrak{P}_s$: shearing plane
$\mathfrak{P}_0$: reference plane
α: rake angle [°]
α_n: normal rake angle [°]
β: heat partition coefficient
ΔT_c: average temperature rice in the chip [K]
ΔT_M: maximum temperature at the tool-chip interface [K]
ΔT_{SZ}: maximum temperature at primary shear zone [K]
δ: ratio of the thickness of the secondary shear zone to the chip thickness
ε_0: unit carbon dioxide emission [kg/dm³]
ε_{AB}: strain at the shear plane
ε_{int}: strain at the tool-chip interface
$\dot{\varepsilon}_{AB}$: strain rate at the shear plane [s^{-1}]
$\dot{\varepsilon}_{int}$: strain rate at the tool-chip interface [s^{-1}]
$\dot{\varepsilon}_0$: reference strain rate at the Johnson-Cook model [s^{-1}]
ζ_E: specific electric energy cost [$/(kW·h)]
ζ_L: specific labor cost [$/h]
ζ_T: specific tool cost [$/unit]
η_c: chip angle [°]
η_s: shear angle [°]
θ_n: normal force angle [°]
κ: inclination angle [°]

λ_n: normal friction angle [°]
ν: strain rate coefficient (Oxley and Johnson-Cook models)
ξ_0: unit tool waste [mm/min]
ρ: mass density [kg/m^3]
σ_{int}: normal stress at the tool-chip interface [MPa]
σ_n: normal pressure [MPa]
τ_{int}: shear stress at the tool-chip interface [MPa]
τ_0: unit cutting time [min/dm^3]
φ: tool edge position angle [°]
ϕ_n: normal shear angle [°]
ψ: factor in the tool-chip interface temperature
Ω^{max}: maximum tool wear [mm]

APPENDIX 2

Simulated and Experimental Data

Table 2. Mechanistic model outcomes

Decision Variables			Response Variables						
a_P [mm]	f [mm/rev]	V [m/min]	ϕ_n [°]	$\mu = \tan(\lambda_n)$	F_C [N]	F_T [N]	V_c [m/min]	T_{int} [K]	σ_n [MPa]
0.50	0.15	200	29.1	0.510	131	45	104	1138	849
0.50	0.15	250	30.3	0.472	126	39	136	1183	877
0.50	0.15	300	31.6	0.440	122	35	172	1219	906
0.50	0.25	200	31.7	0.430	200	55	115	1236	899
0.50	0.25	250	32.9	0.396	194	47	149	1280	924
0.50	0.25	300	33.7	0.370	189	41	185	1315	945
0.50	0.35	200	33.3	0.382	266	61	122	1299	929
0.50	0.35	250	34.4	0.351	259	52	158	1341	953
0.50	0.35	300	35.2	0.326	253	45	195	1374	974
1.25	0.15	200	29.1	0.510	328	113	104	1138	849
1.25	0.15	250	30.3	0.472	316	99	136	1183	877
1.25	0.15	300	31.3	0.443	307	88	170	1219	900
1.25	0.25	200	31.7	0.430	501	137	115	1231	899
1.25	0.25	250	32.9	0.396	485	118	149	1280	924
1.25	0.25	300	33.7	0.370	473	103	185	1315	945
1.25	0.35	200	33.3	0.382	666	153	122	1299	929
1.25	0.35	250	34.4	0.351	647	130	158	1341	953
1.25	0.35	300	35.2	0.326	633	112	195	1374	972
2.00	0.15	200	29.1	0.510	526	181	104	1138	849
2.00	0.15	250	30.3	0.472	507	158	136	1183	877
2.00	0.15	300	31.6	0.440	490	138	172	1219	906
2.00	0.25	200	31.7	0.430	802	219	115	1236	899
2.00	0.25	250	32.9	0.396	776	188	150	1280	924
2.00	0.25	300	33.7	0.370	758	165	185	1315	945
2.00	0.35	200	33.3	0.382	1065	244	122	1299	929
2.00	0.35	250	34.4	0.351	1036	207	158	1341	953
2.00	0.35	300	35.2	0.326	1013	180	195	1374	972

Table 3. Experimental values of cutting force

a_P [mm]	f [mm/rev]	V [m/min]	F_c [N]			$\mu(F_c)$ [N]	$\sigma(F_c)$ [N]
			No. 1	No. 2	No. 3		
0.50	0.15	200	130	150	140	140	10.0
0.50	0.15	250	120	110	90	107	15.3
0.50	0.15	300	100	90	110	100	10.0
0.50	0.25	200	180	190	170	180	10.0
0.50	0.25	250	150	160	190	167	20.8
0.50	0.25	300	190	170	180	180	10.0
0.50	0.35	250	240	290	250	260	26.5
0.50	0.35	300	240	260	270	257	15.3
1.25	0.15	200	290	300	320	303	15.3
1.25	0.15	250	310	270	280	287	20.8
1.25	0.15	300	350	310	280	313	35.1
1.25	0.25	200	450	430	420	433	15.3
1.25	0.25	250	440	400	410	417	20.8
1.25	0.25	300	430	480	410	440	36.1
1.25	0.35	200	630	590	580	600	26.5
1.25	0.35	250	590	600	570	587	15.3
1.25	0.35	300	620	610	590	607	15.3
2.00	0.15	200	600	520	570	563	40.4
2.00	0.15	250	520	470	510	500	26.5
2.00	0.15	300	440	510	490	480	36.1
2.00	0.25	200	1130	1040	940	1037	95.0
2.00	0.25	250	820	900	940	887	61.1
2.00	0.25	300	850	860	900	870	26.5
2.00	0.35	200	1370	1440	1240	1350	101.5
2.00	0.35	250	1090	1110	1140	1113	25.2
2.00	0.35	300	920	990	1070	993	75.1

Table 4. Computed wear rates

a_p [mm]	f [mm/rev]	V [m/min]	V_c [m/min]	T_{int} [K]	σ_n [MPa]	w [mm/min]
0.50	0.15	200	212	1138	849	0.0134
0.50	0.15	250	268	1183	877	0.0440
0.50	0.15	300	323	1219	906	0.0577
0.50	0.25	200	216	1236	899	0.0395
0.50	0.25	250	273	1280	924	0.0545
0.50	0.25	300	330	1315	945	0.0704
0.50	0.35	200	219	1299	929	0.0450
0.50	0.35	250	276	1341	953	0.0615
0.50	0.35	300	334	1374	974	0.0788
1.25	0.15	200	212	1138	849	0.0134
1.25	0.15	250	268	1183	877	0.0440
1.25	0.15	300	323	1219	900	0.0577
1.25	0.25	200	216	1231	899	0.0395
1.25	0.25	250	273	1280	924	0.0545
1.25	0.25	300	330	1315	945	0.0704
1.25	0.35	200	219	1299	929	0.0450
1.25	0.35	250	276	1341	953	0.0615
1.25	0.35	300	334	1374	972	0.0788
2.00	0.15	200	212	1138	849	0.0134
2.00	0.15	250	268	1183	877	0.0440
2.00	0.15	300	323	1219	906	0.0577
2.00	0.25	200	216	1236	899	0.0395
2.00	0.25	250	273	1280	924	0.0545
2.00	0.25	300	330	1315	945	0.0704
2.00	0.35	200	219	1299	929	0.0450
2.00	0.35	250	276	1341	953	0.0615
2.00	0.35	300	334	1374	972	0.0788

Chapter 11
Optimization of Seat Frame With Dissimilar Materials for Lightweight Materials

Euy Sik Jeon
Department of Mechanical Engineering, Kongju National University, South Korea

Yeong Jo Ju
Department of Mechanical Engineering, Kongju National University, South Korea

ABSTRACT

High-strength and lightweight methods for vehicle parts include methods such as optimization and application of lightweight materials by reflecting load or material characteristics. Safety regulations have been established in accordance with the loads affecting the vehicle to secure the safety of the vehicle. In order to reduce the weight, high strength materials such as high strength steel (HSS) or high tensile strength steel (AHSS) have been studied. In addition, research on additional lightweight optimization is actively performed by removing parts that do not require high strength or replacing them with plastics. The process of designing a vehicle or part with different properties and considering various loads is costly and time consuming. In order to secure safety and light weight, the authors propose an approximate model for the optimal design of the seat frame that has a direct impact on occupants among the parts of the vehicle, and reduces the development cost, time, and intuitive design through the procedure.

INTRODUCTION

Consider a Variety of Materials

As the automobile industry advances, lightweight vehicles are being developed to improve fuel efficiency while increasing the strength of the vehicle's body and the internal components, in addition to enhancing passenger safety (Jeong, Oh, & Cheon, 2016). Among the numerous important components of a vehicle, the seat not only supports the human body, but also has a direct impact on the passenger due to

DOI: 10.4018/978-1-7998-4939-1.ch011

external shocks. Therefore, the seat frame should have sufficient strength (Cho et al., 2013). Through shape and strength design of the seat frame, we are actively developing a new seat that conforms to the automobile test regulations. A study on the optimization of the seat frame made of lightweight material that can replace the typical steel seat frame is necessary.

The research on the weight reduction of automobiles is divided into the method of increasing the strength and reducing the weight by changing the material applied to the existing automobile, the method of reducing the weight of the automobile body structure, and the method of reducing the weight by applying the new construction method in the manufacturing process (Jang, Yeom, & Lim, 2018). Structural weight reduction can be achieved by improving the design of the vehicle body through optimization of components and modularization, and can be made lighter by utilizing new construction methods for welding and joining in manufacturing. In the case of material weight reduction, a method of applying to a component by applying a lightweight material having excellent physical properties mainly on a composite material such as plastic, aluminum, and magnesium has been developed. In addition to the method of lightening the structure of the car, among the lightening methods of the car, the part about the lightening of the car seat is also an important lightening method.

A lightweight seat frame not only has a reduced mass, but it also has the strength that is appropriate to the material. Thus, the aforementioned factors have stimulated the investigation of lightweight seat frames. Compared the frame stiffness with thickness using high-strength steel (HSS), which has a strength greater than that of typical steel (Kim, Cho, Kim, & Cho, 2014). Confirmed that high-strength plastic materials could be applied to the seat frame to make it lightweight, while achieving a strength comparable to steel (Kim, Lee, Yang, & Kang, 2016). Performed optimization experiments for the seat cushion structure with HSS using design-of-experiment (D.O.E), based on the thickness of the HSS (Jung, Lee, Kwon, & Han 2010). Conducted a lightweight design for bulk trailers using topology optimization to reduce fuel costs and effectively respond to environmental regulations (Kim et al. 2017). Conducted an optimal design using topology optimization for the review of structures that can be lightweight while ensuring the safety of railway vehicles and parts as transportation machines (Han & Jung, 2011; Hwang et al., 2003).

Consider a Variety of Loads

As environmental regulations for automotive exhaust gas are intensified, there is growing interest in high-strength and lightweight design methods for automotive components. In general, such high-strength and lightweight design requirements can be met using high-strength and lightweight materials. As safety can be lowered by material or structural changes, however, it must be ensured after applying the combined tensile, compressive, shear, twisting, bending, and impact loads in accordance with the ASTM, ISO, EEC, ECE, and FMVSS regulations.

An example of a structure under combined loads is a seat frame.

The load applied to the top of a seatback frame generates a moment that leans the seat backward; the combined loads considered in this test are tensile, compressive, and bending loads. Regulations for assessing safety against excessive tensile, compressive, and bending loads are FMVSS 207 (Salim et al., 2013; Lim, 2017; Kim, & Lee, 1999; Lee et al., 2005) and FMVSS 202a (Jang & Min, 2005; InterRegs Ltd, 2012). Tests for tensile, compressive, and bending loads are classified as static tests, and they assess safety using the maximum displacement determined by the regulations.

The load applied to the seat belt fastening points of a seat cushion twists the entire seat and generates plastic deformation. The combined loads considered in this test are twisting and bending loads. The regulations for assessing safety against excessive twisting and bending loads are FVMSS 210 and ECE R14. Tests for twisting and bending loads are classified as quasi-static tests. They assess safety through the maximum displacement determined by component regulations and fractures (InterRegs Ltd, 2012; Klaus Hessenberger, 2016; Vikas Partwardhan, 2013).

In the crash and dynamic safety test of a seat frame, both an anthropometric dummy, such as Hybrid III and a seat belt are considered. In this test, the fracture of all seat components is induced and severe acceleration is generated for the main parts of the dummy. A regulation for assessing crash safety is the European Economic Community (EEC) Directive 2005/40/EC. Tests for crash and impact loads are classified as dynamic tests and they assess safety through the maximum displacement determined by the regulation and fracture of components (Lee, 2004; Heo et al., 2011; Lee, 2014).

Combined loads are applied to the seats, body of the white (BIW) frame, front and rear bumpers, as well as the fuel tanks of automobiles. They can also be applied to airplanes, trains, smartphones, and building members with similar components.

Propose a method to satisfy the strength and light weight by changing the topology and material of the seat frame with the characteristics of the existing steel. Static analysis and topology optimization identified unnecessary parts and investigated the application of materials using existing and lightweight materials.

Static analysis and topology optimization of the sheet frame selected the appropriate material that can be applied as a dissimilar material and confirmed the contribution of the sheet parts. In addition, D.O.E. can be used to identify the weight loss ratio of the dissimilar material and its application rate. First, deformation of the seat frame was confirmed by static analysis using people according to safety criteria and using static test specifications. Subsequently, a method of performing topological optimization of the frame and applying three different materials to a typical frame based on the density ratio was proposed, and the contribution was confirmed for each part of the frame. As a result of the analysis of the frame using three dissimilar materials, it was confirmed that the deformation was reduced compared to the frame using the general steel, and the application ratio of the dissimilar material for light weight optimization was derived. Lightweight optimizations using D.O.E and algorithms have been altered depending on the frame thickness and application rate of dissimilar materials, resulting in reduced weight and improved strength compared to ordinary steel.

MAIN FOCUS OF THE CHAPTER

Required Material Properties for Each Type of Load

Technical advances in various areas that make human lives convenient are closely related to the development of suitable materials as an understanding of a certain material often leads to innovative technical advances. For example, it would have been difficult to make automobiles without the development of inexpensive steel materials or comparable alternative materials. The properties of a material refer to the type and degree of a specific response to an external stimulus. In general, such properties are not related to the geometry and size of the material. The important properties of a solid material belong to the following six categories: mechanical, electrical, thermal, magnetic, optical, and deteriorative properties. Mechanical properties represent deformation responses to the applied force, such as elastic modulus and strength.

In general, the material property that represents the tensile load is the elastic modulus. The elastic modulus can be derived through tests such as ASTM E8-E8M, ASTM E646-16, and ASTM D638-14. Poisson's ratio is a material property that can be derived by measuring the change in the axial direction to which a load is applied and the anisotropy in the cross-sectional direction in the process of deriving the elastic modulus.

A material property that can be additionally obtained in the tensile test is the strain-stress (S-S) curve. The S-S curve is a data set that can represent the elastic region expressed by the elastic modulus, the yield point at which plastic deformation begins, the plastic region that exceeded the elastic region, and the ultimate stress of the material. The S-S curve can also represent the compressive property. Cases in which both tensile and compressive properties are considered at the same time can be mainly found in the test cases of hyperelastic materials.

The weight of a structure is significantly affected by the density of the material. Changing the density can reduce the weight, but it is rare because the strength of the material is proportional to its density.

For the bending load, three-point bending tests such as the FMVSS 214 quasi-static test method, are used (Jeon et al., 2010; Lee, Lee, & Jang, 2015; Yoo et al., 2006; Song et al., 2002; Yoon et al., 2014). The bending test results can also be expressed using the S-S curve.

In addition, the resistance to the shear load can be expressed using the shear modulus, and the crash resistance can be expressed using the S-S curve or maximum energy that can be absorbed by the material.

Geometry Optimization Method

Geometry optimization to meet the optimal strength performance under limited weight reduction conditions has been researched as the most important issue in the optimization research field. This method is considered for deriving the optimal strength underweight constraints and is generally divided into material changes and geometry changes (Yang, & Jeon, 2016; Jeon, & Jeon, 2017; Oh, & Jeon, 2017; Na, & Jeon, 2017; Song, & Jeon, 2017, Yang, & Jeon 2018).

Studies on the optimal design that use lightweight materials have a larger weight reduction effect than those that use geometry changes. Material changes, however, involve changes in safety, machinability, and economic efficiency. Studies on the optimal design that use geometry changes, on the other hand, can easily predict safety, machinability, and economic efficiency despite a low weight reduction effect. In addition, they are highly useful when the selection of materials is limited. They are largely divided into studies for optimizing only the cross-sectional geometry and those for optimizing the entire geometry.

Studies on the cross section are targeted at the conceptual geometry with the interior filled or that use the mid surface, which is the open cross section of a thin plate. The interior-filled conceptual geometry uses the topology optimization method, and the mid surface of a thin plate uses the size optimization method that incorporates the shape factor (Yang, & Jeon, 2017; Yang, Kim, & Jeon, 2018; Yang, Jeon, & Park, 2018; Shin, Song, & Jeon, 2018; Na, & Jeon, 2019).

The topology optimization method can change both the material and geometry of the conceptual geometry; however, it can significantly lower machinability and economic efficiency due to the geometry difficulties in its application; therefore, many studies use the open cross section. As the open cross section considers a thin plate, it can easily predict machinability and economic efficiency despite the difficulties in significantly improving strength.

As the engineering values of products with open cross-sectional geometry increase, more research on the optimization of open cross-sectional geometry is attempted in various fields.

Presented a general study on the optimization of open cross-sectional geometry. The BIW model was assumed to be a wireframe model and the cross section of the wireframe was used as the research target by setting it as an open cross section. The node positions were set as parameters and GA was applied in the optimization process. As GA requires many iterations, in the optimization process, improved GA is utilized for the optimization of open cross-sectional geometry (Qin Huan, 2018).

The design of experiments (DOE), however, can be used in the process of repeated experiments. If the response surface generated by the DOE is applied to GA, the number of repeated experiments can be significantly reduced.

Method to Apply Dissimilar Materials

Carbon fiber Reinforced Plastic (CFRP) is one of the materials used for automobile weight reduction in the field of transport machinery (Lim, 2017). However, because the manufacturing method and strength of the existing automobile parts are different, the study on the material properties and the shape design and strength evaluation considering the manufacturing method should be done at the same time (Martinsen, hu, & Carlson, 2015).

The seat frame, which directly affects the occupant of the automobile parts, is mainly made of a conventional steel material, and a study for applying a lightweight and high strength material to reduce the weight of the seat frame was performed (Shin, & Jeon, 2017; Ju, & Jeon, 2019). For the seat frame with Mg instead of the existing steel material, the analysis method of the seat frame was proposed through the static strength analysis and the endurance analysis using the finite element model, and the applicability of the Mg material was confirmed (Shin, 2016). Multi-materials such as Mg and Al were applied to analyze the economic feasibility, static strength analysis, and seat frame evaluation test using the finite element model (Jung, Seo, & Kim, 2016). The finite element model for the seat frame designed with Al was used to study the optimal analysis and the improvement direction for the analysis results (Park, Kook, & Cho, 2016), Static strength analysis and durability analysis were performed using finite element model for the seat frame made of CFRP. (Song, 2015)

Optimization Method That Considers Both Thickness and Material

To meet various environmental pollution regulations, research on transport machinery aims to improve fuel economy. For all automotive components, weight reduction has been performed by changing geometry, materials, and thickness. The weight reduction of components that are directly related to the safety of users, however, involves the deterioration of safety performance. Meeting both weight reduction and safety performance is a long-standing research subject of academia and industry.

A representative case for high-strength weight reduction is the automobile seat frame. The seat frame must meet various static, quasi-static, and dynamic safety standards and must also consider weight reduction to improve fuel economy.

In previous studies on the weight reduction of seat frames, topology or size optimization methods have been performed. Studies on the optimal design of structures using topology optimization deal with basic geometry in the concept design stage. They apply high-strength lightweight materials to the entire structure and determine the layout of the conceptual geometry. Studies on the optimal design of structures using size optimization deal with commercial models of limited geometry. They are mostly utilized to adjust strength and weight by changing the shape factor and thickness. In studies on size optimization, real numbers that cannot be machined can also be derived as design parameters. To address this problem, efficient algorithms that replace a real number with an adjacent commercial value by calculating differences between commercial values and the real number have been researched.

Studies that consider discrete integers as design parameters within the design range from the beginning can also be considered. Methods for the optimal design using discrete parameters have been mostly researched in the field of composite materials. For composite materials, the thickness per cell and the lamination direction are mainly set as discrete parameters. This theory can also be considered in structural optimization. Discrete material and thickness optimization (DMTO), in which optimization is performed by arranging materials and thicknesses in order of strength and setting discrete IDs as parameters, can also be considered.

SOLUTIONS AND RECOMMENDATIONS

Topology Example That Considers Shape Applied Dissimilar Materials

Finite Element Analysis

Applied the finite element analysis method through static analysis as a basic simulation to examine the application of dissimilar materials. The seat cushion frame model was a finite element model of the Honda Accord vehicle, one of the vehicles that officially uses finite element analysis. The materials of the cushion frame are structural steel and plastic. The finite elements of the cushion frame model consisted of a frame and rail components with shell elements. In the static analysis method, a standard 980 N force was applied while adhering to safety standards per person, so that the deformation does not exceed 5 mm. The boundary conditions of the finite element analysis were rigidly bonded to the cushion frame parts, and the lower rail mounts of the cushion frame were fixed as shown in Figure 1. The upper frame was subjected to a load of 980 N, and finite element analysis was performed by applying the property values as shown in Table 1.

As a result of the finite element analysis, it was confirmed that the deformation of the front frame of the seat cushion for the structural steel material was 3.74 mm and did not exceed 5 mm. As shown in Figure 2(b), a stress of 484.3 MPa occurred locally in the cushion front frame, and it was confirmed that the yield strength of the material was exceeded.

Figure 2 shows that the front frame of the cushion frame deforms significantly, and the topology optimization is applied to reinforce the weak part of this frame through material modification. Topology optimization techniques are the most commonly used basic-shape-decision method based on the density method (SIMP) along with finite elements analysis.

Topology Optimization

Topology optimization was performed to apply dissimilar materials of the seat cushion frame. As shown in Figure 3., the plastic and structural steel materials comprising the seat-cushion frame are constrained to not exceed the yield stress as shown in Table 2 This is facilitated by applying the easy-to-apply den-

Figure 1. Finite elements model of Accord seat cushion

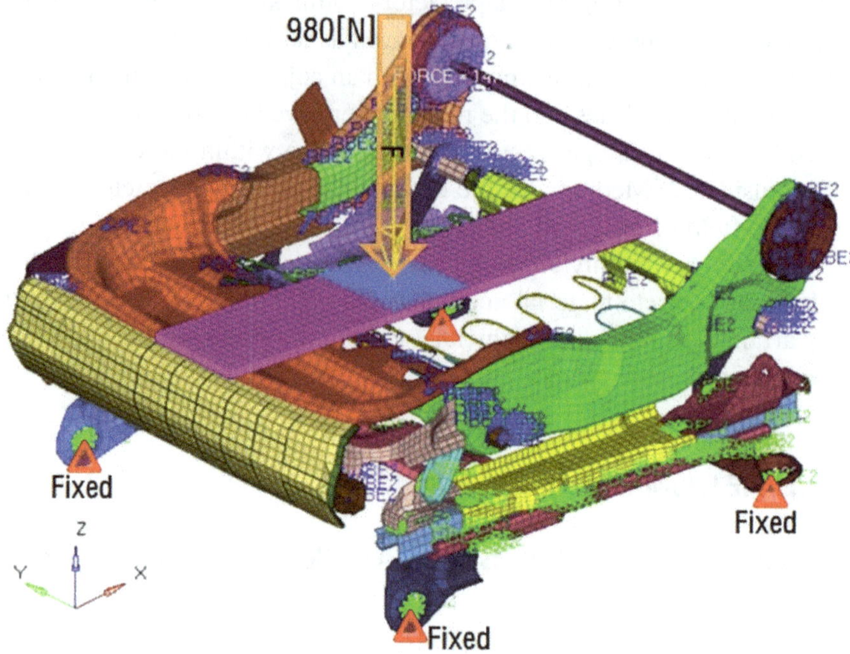

Table 1. Property of cushion frame model

Material	Yield Stress [MPa]	Young's Modulus [MPa]	Poisson's Ratio	Density [kg/mm³]	Description
ABS	45	2800	0.3	1.0×10^{-6}	Plastic
Structural Steel	250	210,000	0.3	7.8×10^{-6}	Steel 1
	285	210,000	0.3	7.8×10^{-6}	Steel 2
	350	210,000	0.3	7.8×10^{-6}	Steel 3

Figure 2. Finite elements analysis results of cushion frame

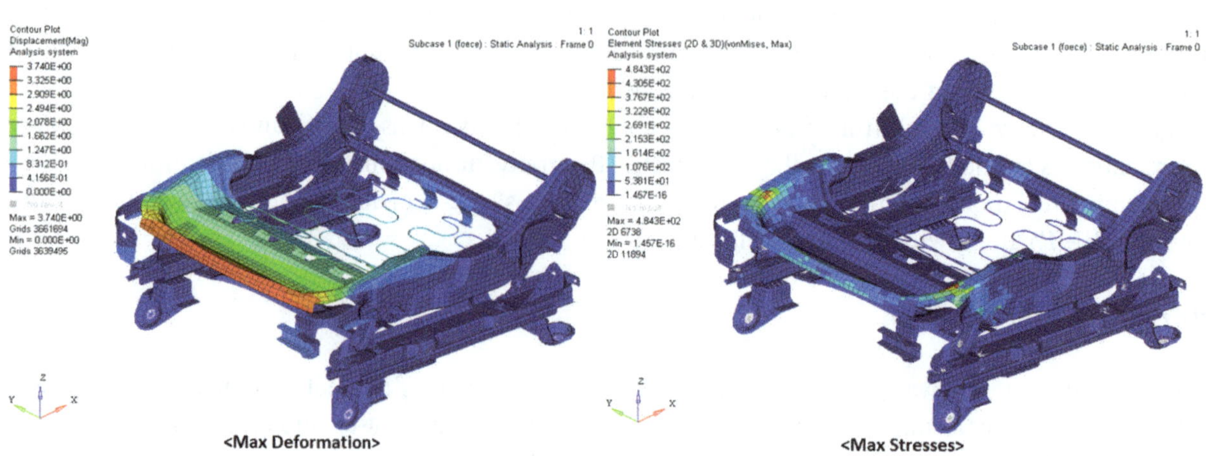

sity method (Kim et al., 2012) to topology optimization. The constraint is to determine the density ratio depending on the load condition by setting the deformation of the cushion frame as an objective function.

The deformation and stresses of the cushion frame were 2.563 mm and 397.5 MPa, respectively. The deformation and the stress of the cushion frame relative to the static analysis were 30% and 18%. It is seen that the topology optimization results converge after a total of 33 iterations, and as shown in Figure 4., the density ratio is confirmed to be concentrated on the front frame. More than 70% of the cushion frame was not affected by the density ratio of the cushion frame.

The density ratio indicator of the cushion frame's finite elements according to topology optimization can be represented by an indicator function. Within the range of each exponential function, the elements that may not be affected by the load relative to the number of elements in the cushion frame is expected to be 78.37%. When an indicator function of 0.0 to 0.2 is applied to dissimilar materials, it is expected that materials with lower strength than typical steel can be acquired. From the finite element analysis results, it was confirmed that the front frame of the seat cushion frame exceeded the yield strength of the material. And we want to apply the dissimilar material to satisfy the strength and light weight of typical material of cushion frame.

Figure 3. Finite elements analysis results of cushion frame

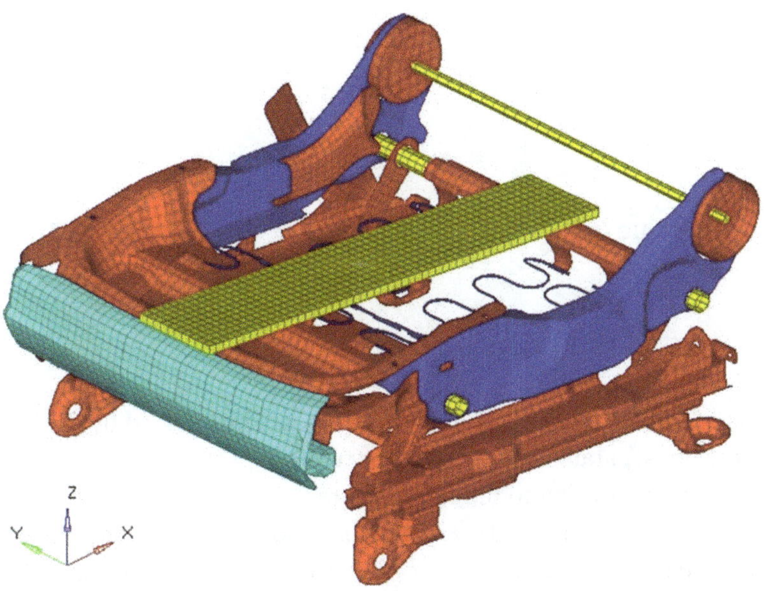

Table 2. Condition of properties

Symbol	Value	Material
$\sigma_{plastic}$	Under 45 [MPa]	Plastic
$\sigma_{steel\ 1}$	Under 250 [MPa]	Structural Steel 1
$\sigma_{steel\ 2}$	Under 285 [MPa]	Structural Steel 2
$\sigma_{steel\ 3}$	Under 350 [MPa]	Structural Steel 3

Figure 4. Analysis results according to topology optimization

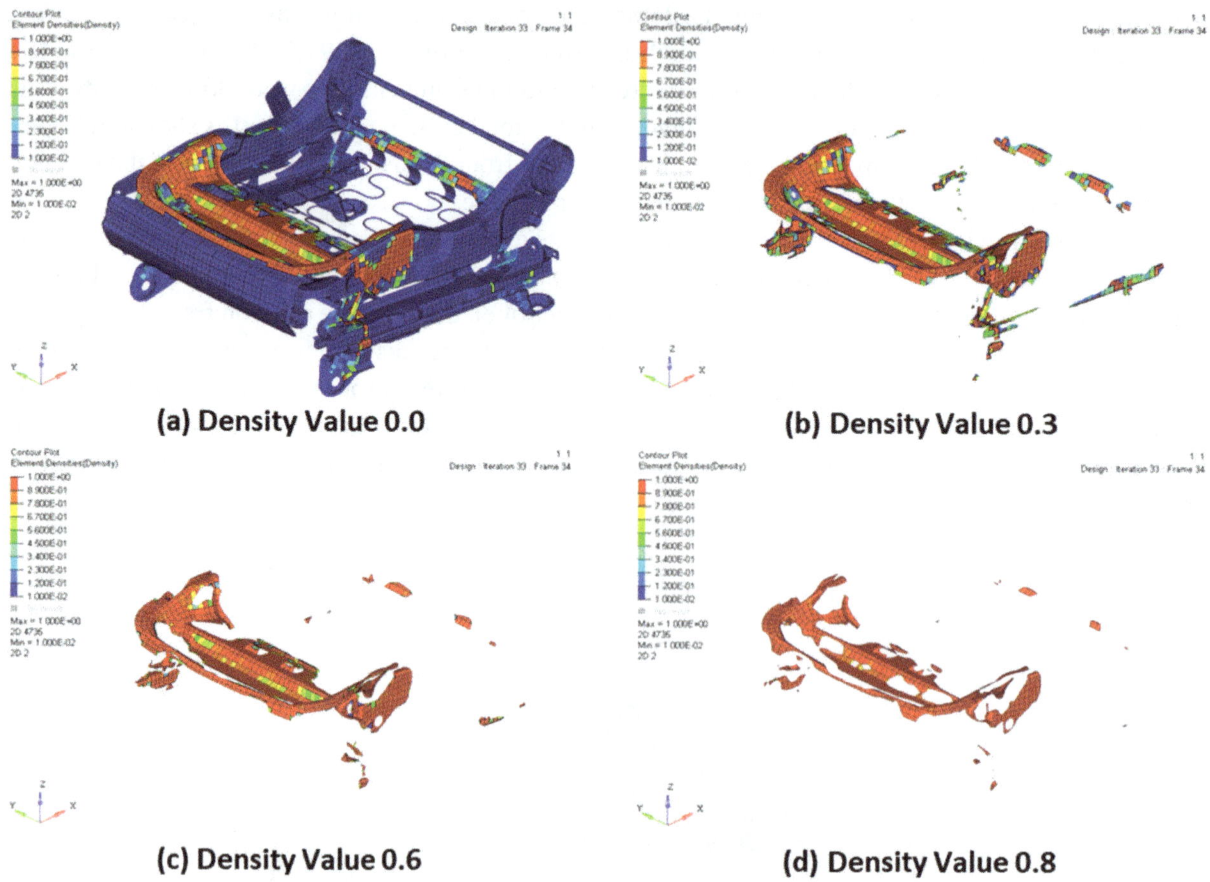

Finite Elements Analysis Using Dissimilar Materials

In the topology results of the seat cushion frame, the elements influenced by the load and the elements influenced by the load were separated by an indicator function. In order to check the strength and light weight of the cushion frame according to the ratio of three materials to the ratio of the indication function, three materials were set at different ratios and a case study was conducted. The method of selecting the material is to select the unnecessary parts according to the density ratio through the topology and to select the materials that can cope by comparing the stress values according to the static analysis results. After the number of cases in which dissimilar materials can be applied to the indication function is set, CFRP (Plastic) and HSS (steel material) are applied as shown in Table 3, in addition to the existing seat-cushion frame material. Owing to the dissimilar materials based on the ratio of the indication function, plastic material is applied to the cushion front frame, side frame, side frame excluding mount, sliding rail and bracket, lift link, front cover, and seatback joint considering the real model. Also, the design of the seat cushion frame was not modified but only the material of the main part in the frame was changed. The material of the cushion front frame, the side frame, and the rear shaft frame, which are the parts that can change the material among the parts of the seat cushion frame, are changed.

Figure 5. Density ratio of cushion side frame

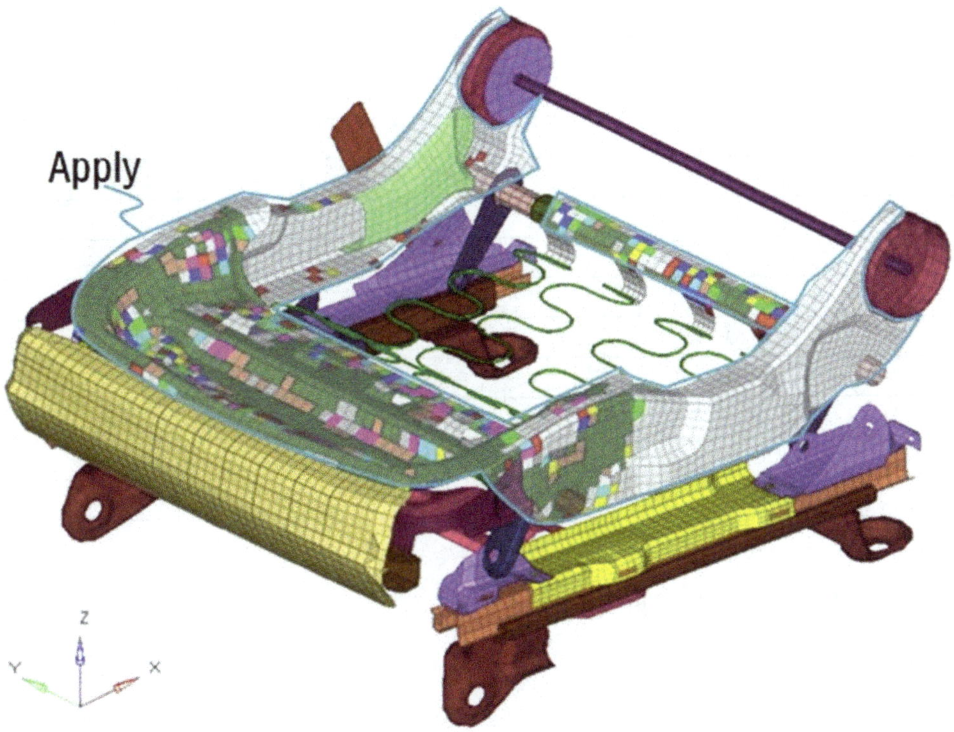

Table 3. Typical property and additional property

Material	Yield Stress [MPa]	Young's Modulus [MPa]	Poisson's Ratio	Density [kg/mm3]	Description
CFRP	-	60,000	0.08	1.5×10^{-6}	Plastic
SPFC 780	699	342,200	0.3	7.8×10^{-6}	HSS

As shown in Figure 5., when the dissimilar materials were applied only to the front frame, side frame, and side frame of the cushion, the ratio of the indicator function 0.0 to 0.1 was reduced to 50%. The experimental table to perform the static analysis was set according to the indicator function ratio and interpreted as 12 cases (see Table 4).

As a result of analyzing each case of the indicator function, two application ratios of dissimilar materials were selected for lightweight optimization as shown in Figure 6. First, Case 2 shows that stresses have a minimum deviation from the average stress and the deformation has a small deviation from the average deformation. Second, in Case 8, the deformation is a minimum and the stresses are less than the average stress, hence the reasons for selecting the two application ratios.

Deformation was reduced by 20% in Case 2 and by 37% in Case 8 compared to the cushion frame with existing structural steel.

Table 4. Condition of properties

Cases	CFRP Ratio [%]	Steel Ratio [%]	HSS Ratio [%]
Case 1	49.32	17.19	33.49
Case 2	49.32	14.67	36.01
Case 3	49.32	17.19	33.49
Case 4	49.32	8.58	42.10
Case 5	53.53	15.25	31.22
Case 6	53.53	12.97	33.49
Case 7	53.53	10.46	36.01
Case 8	53.53	7.64	38.83
Case 9	49.32	50.68	0
Case 10	49.32	0	50.68
Case 11	53.33	46.47	0
Case 12	53.33	0	46.47

Figure 6. Analysis results of Case2 and Case8

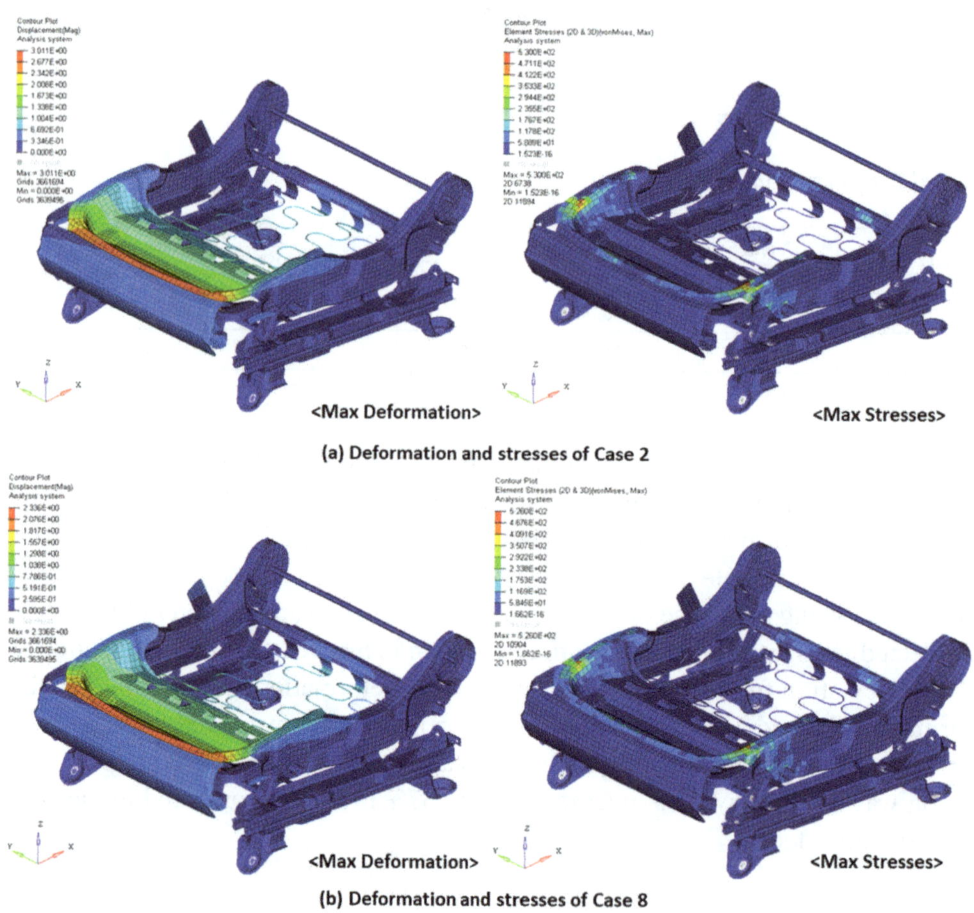

Finite Element Analysis Using D.O.E

Lightweight optimization was performed by changing the thickness of the cushion front frame, side frame, and rear shaft frame depending on the ratio of two dissimilar materials of the cushion frame. As shown in Table 5, the thickness of each frame changed by ± 30% from the existing thickness considering interference between frames. The thickness of the frame was changed by applying the D.O.E and deformation and stress values were confirmed by analyzing the thickness of each frame applied with the three materials.

Table 5. Condition of thickness changes

Symbol	Value	Material	Part
$t_{plastic\_0.7}$	$t_{plastic\_0.7} \pm 30[\%]$	Plastic	Front
$t_{steel\_0.7}$	$t_{steel\_0.7} \pm 30[\%]$	Steel	Front
$t_{HSS\_0.7}$	$t_{HSS\_0.7} \pm 30[\%]$	HSS	Front
$t_{plastic\_1.5}$	$t_{plastic\_1.5} \pm 30[\%]$	Plastic	Side
$t_{steel\_1.5}$	$t_{steel\_1.5} \pm 30[\%]$	Steel	Side
$t_{HSS\_1.5}$	$t_{HSS\_1.5} \pm 30[\%]$	HSS	Side

The D.O.E uses the Latin-Hypercube, which has the advantage of achieving results with a small number of experiments when there is a large number of factors, or when designing alternatives in the design of static structures (Kim, Hong, & Jang, 2011). The total number of interpretations was 28 times for each of the two dissimilar materials, 56 total. The thickness of the dissimilar materials derived from the experimental design method is shown in Table 1, and it can be seen that the thickness is set to within 30% of the error range according to the reference thickness of the dissimilar materials (see also Table 6).

Case 2 analysis by D.O.E showed a minimum deformation of 2.125 mm at experiment point 15 and a minimum stress of 238.51 MPa at experiment point 7. The thickness of the front cushion frame and the side frame of each material were confirmed as shown in Table 7 at the experiment point 7 where the minimum stresses were derived and at the experiment point 15 where the minimum deformation was derived.

Analysis of Case 8 by D.O.E shows a minimum deformation 1.796 mm at experiment point 15 and a minimum stress of 351.5 MPa at experiment point 14. The thickness of the front cushion frame and the side frame of each material were confirmed as shown in Table 7 at the experiment point 14 where the minimum stress was derived, and the experiment point 15 where the minimum deformation was derived.

The results for Case 2 and Case 8 are shown in Figure 7. The thickness values of the plastic, steel, and HSS indicating the minimum deformation are shown in Table 8. It was confirmed that the strength of the seat cushion frame was improved and was lightweight within ± 10% of the typical frame thickness, by varying the thickness of the two dissimilar materials and applying D.O.E. First, the stress increased by about 4% when the thickness was changed than the cushion frame applied to the dissimilar materials with the reference thickness, while the deformation was reduced by approximately 30%. This confirms that the strength of the seat cushion frame was improved. Second, it was confirmed that the weight of the cushion frame applied with dissimilar materials and thickness modification was reduced by about 10%

Table 6. The thickness of the dissimilar materials by design-of-experiment

Number	$t_{plastic\_0.7}$	$t_{steel\_0.7}$	$t_{HSS\_0.7}$	$t_{plastic\_1.5}$	$t_{steel\_1.5}$	$t_{HSS\_1.5}$
No. 1	0.496	0.640	0.837	1.408	1.532	1.542
No. 2	0.669	0.554	0.657	1.684	1.929	1.816
No. 3	0.550	0.718	0.760	1.945	1.610	1.693
No. 4	0.781	0.690	0.580	1.593	1.673	1.283
No. 5	0.760	0.776	0.498	1.540	1.535	1.861
No. 6	0.593	0.886	0.796	1.802	1.434	1.310
No. 7	0.875	0.843	0.825	1.072	1.579	1.136
No. 8	0.819	0.657	0.684	1.146	1.122	1.256
No. 9	0.734	0.615	0.723	1.312	1.371	1.432
No. 10	0.616	0.640	0.561	1.522	1.716	1.082
No. 11	0.568	0.548	0.608	1.851	1.329	1.890
No. 12	0.713	0.593	0.736	1.485	1.495	1.190
No. 13	0.516	0.905	0.519	1.916	1.078	1.594
No. 14	0.523	0.868	0.881	1.152	1.249	1.703
No. 15	0.595	0.856	0.860	1.114	1.218	1.918
No. 16	0.723	0.748	0.646	1.286	1.280	1.105
No. 17	0.627	0.681	0.524	1.710	1.371	1.453
No. 18	0.840	0.826	0.538	1.249	1.742	1.641
No. 19	0.770	0.603	0.616	1.738	1.646	1.602
No. 20	0.555	0.714	0.705	1.379	1.834	1.228
No. 21	0.796	0.500	0.631	1.189	1.819	1.737
No. 22	0.858	0.807	0.868	1.339	1.458	1.481
No. 23	0.692	0.511	0.808	1.224	1.892	1.779
No. 24	0.643	0.580	0.695	1.448	1.767	1.376
No. 25	0.896	0.737	0.900	1.767	1.863	1.369
No. 26	0.676	0.526	0.593	1.868	1.191	1.153
No. 27	0.825	0.765	0.760	1.641	1.096	1.513
No. 28	0.882	0.804	0.782	1.599	1.151	1.836

Table 7. Analysis results according to thickness change

Response	Symbol			
	Case 2, No. 7	Case 2, No. 15	Case 8, No. 14	Case 8, No. 15
Max Deformation[mm]	3.031	2.125	2.064	1.796
Max Stresses[MPa]	238.6	399.2	351.5	365.4
Mass[kg]	9.383	9.523	9.254	9.434

Figure 7. Analysis results according to thickness change of Case2 and Case8

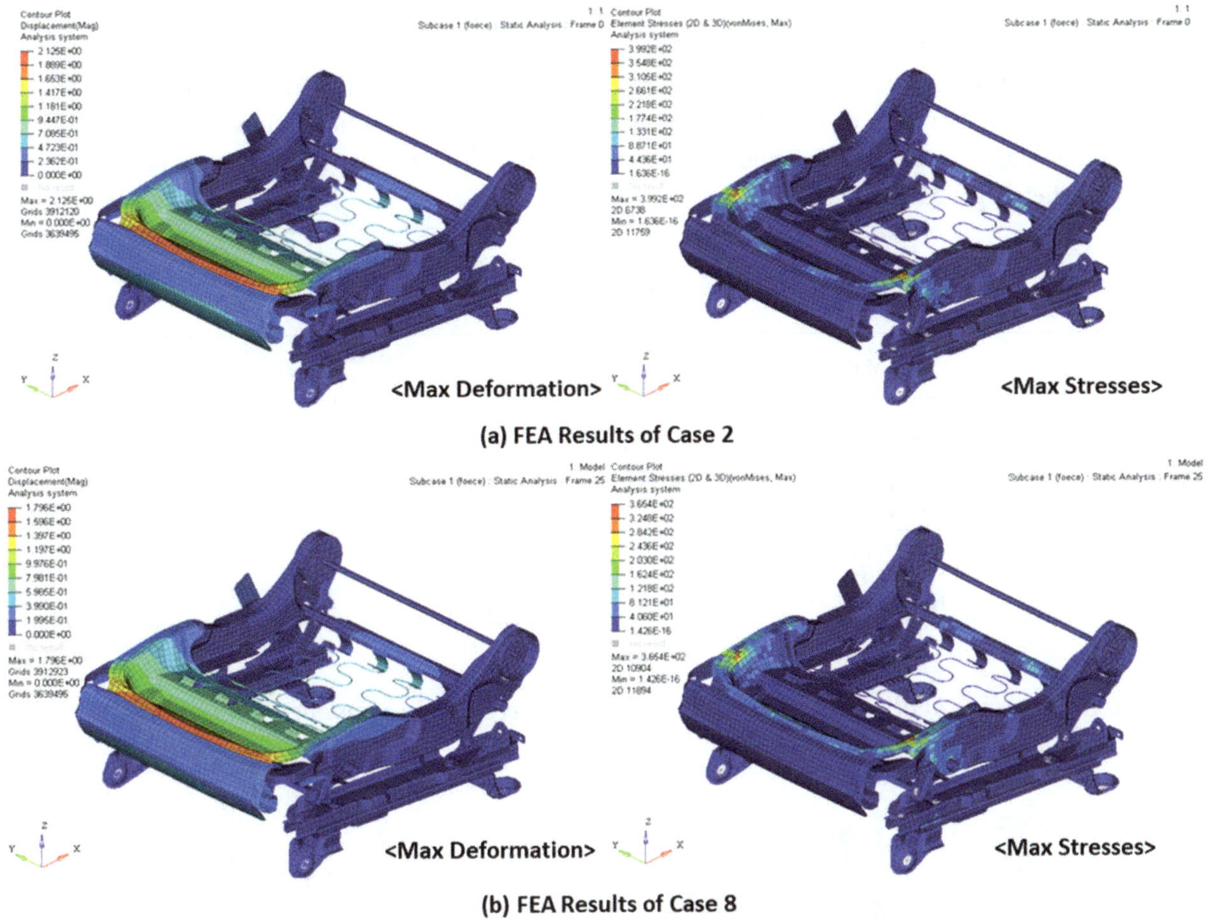

Table 8. Analysis results according to thickness change for Case 2 and Case 8

Response	Symbol			
	Case 2, No. 7	Case 2, No. 15	Case 8, No. 14	Case 8, No. 15
Max Deformation[mm]	3.031	2.125	2.064	1.796
Max Stresses[MPa]	238.6	399.2	351.5	365.4
Mass[kg]	9.234 £ Mass £ 9.596		9.074 £ Mass £ 9.494	
$t_{plastic\_0.7}$[mm]	0.875	0.590	0.523	0.590
$t_{steel\_0.7}$[mm]	0.843	0.856	0.868	0.856
$t_{HSS\_0.7}$[mm]	0.852	0.860	0.881	0.860
$t_{plastic\_1.5}$[mm]	1.072	1.114	1.152	1.114
$t_{steel\_1.5}$[mm]	1.579	1.218	1.249	1.218
$t_{HSS\_1.5}$[mm]	1.136	1.918	1.703	1.918

compared to the seat cushion frame using typical steel. It is possible to design a lightweight because it is larger than the difference of stresses between deformation and weight even when the thickness of the seat is different from that of the seat cushion frame using the existing material and reference thickness. As a result, we confirmed that lightweight design is possible by changing the frame thickness according to the application ratio of the dissimilar materials shown in Table 9. Figure 8 shows the results of the finite element analysis and the case study of the seat cushion frame using the existing material before applying the heterogeneous material.

Table 9. Result of initial value and optimum value

Type	Material	Deformation [mm]	Stresses [MPa]	Weight [kg]
Typical	Structural Steel	3.74	484.3	10.304
Dissimilar Materials for Case 2 using D.O.E	Plastic 49.32% Steel 14.67% HSS 36.01%	3.543 (average)	503.59 (average)	9.398 (average)
Dissimilar Materials for Case 8 using D.O.E	Plastic 53.53% Steel 7.64% HSS 38.83%	2.501 (average)	513.75 (average)	9.312 (average)

Figure 8. Comparison of displacement, stress and weight according to optimization results

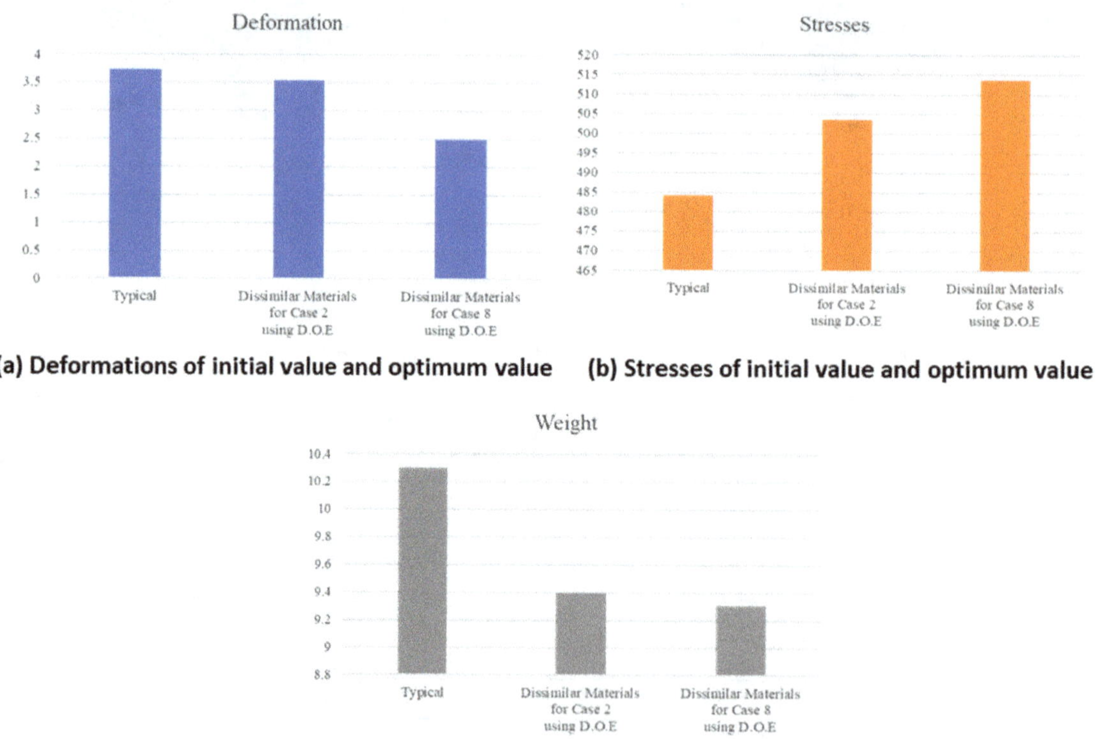

(a) Deformations of initial value and optimum value **(b) Stresses of initial value and optimum value**

(c) Weights of initial value and optimum value

DMTO Example That Considers Both Thickness and Material

Finite Element Analysis

This example presents the lightweight optimal design method for the seat cushion frame by applying a method that sets the material and thickness as discrete parameters and selecting them at the same time (DMTO). The Honda Accord 2017 model provided by the NHTSA was used as the seat cushion frame model. Three material levels (1: Steel, 2: HSS, and 3: AHSS) and four thickness levels (1: 0.5t, 2: 1.0t, 3: 1.5t, and 4: 2.0t) were set.

The model of the sinking seat consists of 44 components, using Hyperworks to create model geometry and finite element models as shown in Figure 9. A finite element model was constructed with 2D elements that are favorable for shortening the analysis time. The size of the element was composed of 2mm and 4mm, the shape was composed of Hexa Elements Type. The load was applied in accordance with the FMVSS207 test regulations, and according to the safety standards the strain should not exceed 50mm. Material property applied structural steel as shown in Figure 10.

Figure 9. Finite elements model (F.E.M) of sinking seat frame

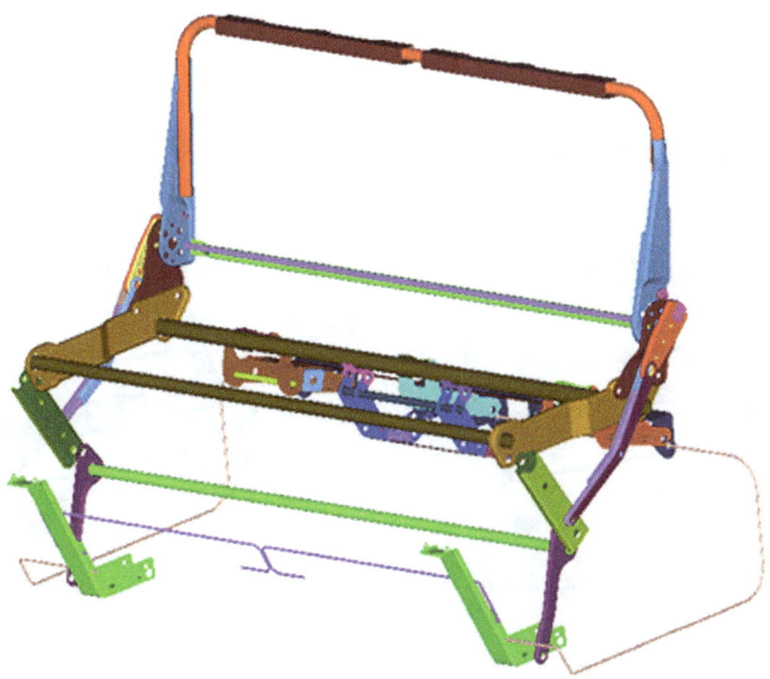

In the basic analysis stage, 12 basic analyses were conducted to consider the three material levels and four thickness levels. When the finite element analysis resulting in the existing thickness and physical properties when the forward moment force is applied, it was confirmed that the deformation amount does not exceed 50 [mm] to 48.8 [mm]. The yield stress of 243.5 [MPa] in the cushion frame side frame

Figure 10. Boundary conditions of F.E.M

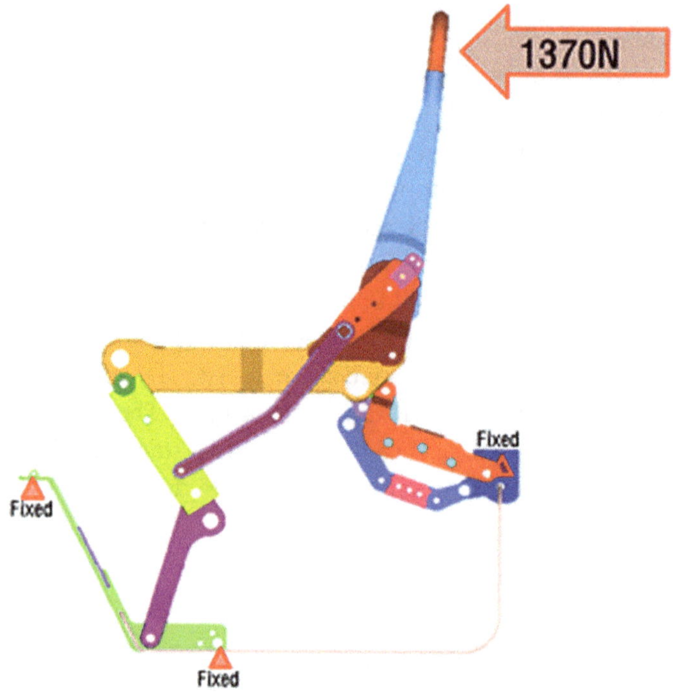

Figure 11. Results of static analysis using FMVSS

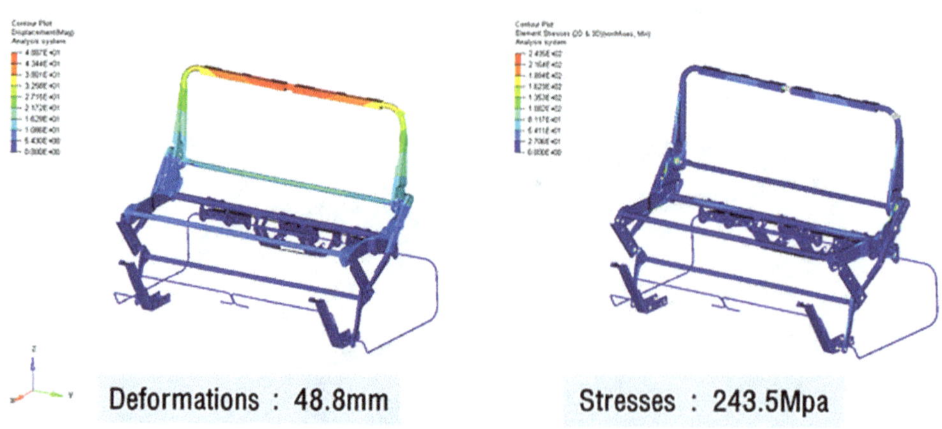

was confirmed that it does not exceed the yield strength of the material. Results of static analysis using FMVSS as shown in Figure 11.

Applying a 30% increased load to the initial analysis conditions and compared to the strain on the basis that the displacement does not exceed the safety standard to 66mm was confirmed to vary in intensity. In this process, it is possible to adjust the error of the finite element model, analysis accuracy, and analysis time.

DMTO Optimization

Optimization was performed twice. In all optimization stages, three material levels and four thickness levels were set as parameters in the same manner as the initial conditions. The stress was concentrated in the side frame as a result of the finite element analysis, and in addition, the main component of the laminated shape for light weight was selected as shown in Figure 12. The total weight of the sinking seat is 22.26kg, and the three heaviest parts are selected and selected 3.537kg, accounting for 15.7% of the total. Part weight and ratio of sinking seat frame as shown in Table 10. Thickness and young's modulus of seat frame part as shown in Table 11.

Figure 12. Part select of sinking seat frame using weight comparison

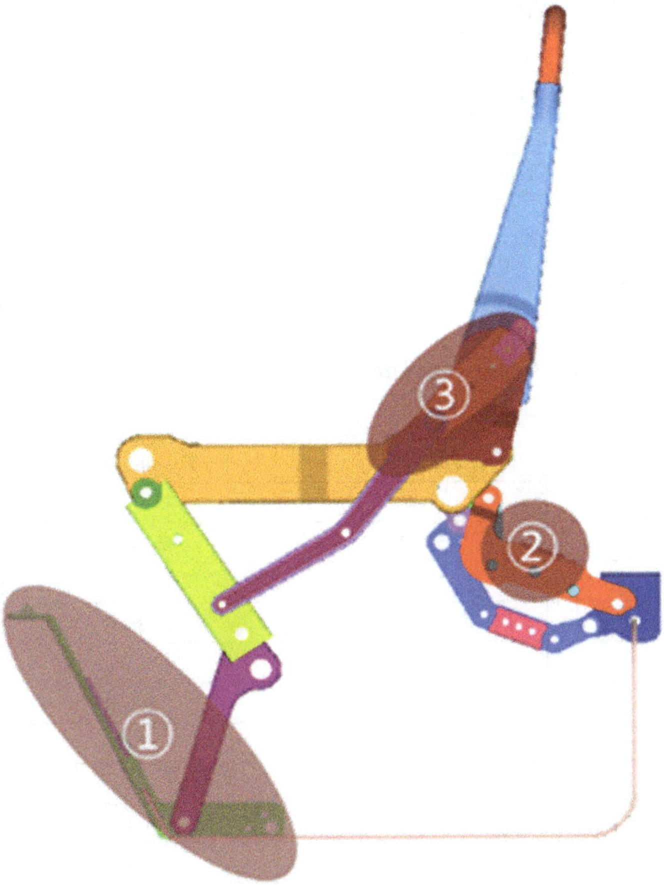

Parameter and Boundary were set. Mass and Displacement, and the observed subjects were set to mass minimize and displacement minimize. The limit condition exceeds the Displacements to less than 0mm and was set to less than 50mm. Using the reaction surface method (GRSM) of the experimental planning method was to create an orthogonal arrangement table to perform optimization. D.O.E. table according to part thickness as shown in Table 12.

Table 10. Part weight and ratio of sinking seat frame

Part	Mass [kg]	Ratio [%]
All part	22.26	100
Base Frame Front (1)	1.807	4.8
Rear Leg Plate (2)	1.386	6.2
Cushion Frame (3)	1.064	4.7

Table 11. Thickness and young's modulus of seat frame part

Parameter	Unit	Lower Bound	Upper Bound
Base Frame Front (1)	mm	1.4	2.6
Rear Leg Plate (2)	mm	2.8	5.2
Cushion Frame (3)	mm	1.4	2.6
Young's Modulus	MPa	210,000	314,400

Table 12. D.O.E. table according to part thickness

Number	Base Frame Front [mm]	Rear Leg Plate [mm]	Cushion Frame [mm]	Young's Modulus [MPa]
1	2	4	2	210,000.0
2	2.36960	5.14240	2.4464	312,748.4
3	1.40192	3.94048	1.56128	305,714.0
4	2.32544	2.91136	2.33696	303,685.2
5	2.55776	4.10944	1.45184	296,736.3
6	1.44416	5.03104	1.50944	209,018.1
7	1.4	3.71738	1646183	298,038.8
8	1.4	2.80001	1.4	303,443.5
9	1.4	2.8	1.4	299,407.5
10	1.4	2.8	1.4	308,404.4
11	1.4	2.8	1.4	303,971.8
12	1.4	2.8	1.4	306,829.0
:	:	:	:	:
49	1.4	2.8	1.4	304808.6
50	1.4	2.8	1.4	307323.1

Table 13. Mass and displacements according to part thickness

Number	Base Frame Front [mm]	Rear Leg Plate [mm]	Cushion Frame [mm]	Young's Modulus [MPa]	Mass [kg]	Displacements [mm]
1	2	4	2	210,000.0	22.3	67.1
2	2.36960	5.14240	2.4464	312,748.4	23.1	43.7
3	1.40192	3.94048	1.56128	305,714.0	21.7	48.0
4	2.32544	2.91136	2.33696	303,685.2	22.3	45.3
5	2.55776	4.10944	1.45184	296,736.3	22.4	50.0
6	1.44416	5.03104	1.50944	209,018.1	22.1	67.4
7	1.4	3.71738	1646183	298,038.8	21.6	49.9
8	1.4	2.80001	1.4	303,443.5	21.3	49.3
9	1.4	2.8	1.4	299,407.5	21.3	50.0
10	1.4	2.8	1.4	308,404.4	21.3	48.2
11	1.4	2.8	1.4	303,971.8	21.3	49.2
12	1.4	2.8	1.4	306,829.0	21.3	48.8
:	:	:	:	:	:	:
49	1.4	2.8	1.4	304808.6	21.3	49.1
50	1.4	2.8	1.4	307323.1	21.3	48.7

In the first stage, components that significantly affect the results were selected through sensitivity analysis. The first stage is utilized to reduce the number of parameters. Each parameter was tested through the application of the Latin Hypercube DOE. The use of this DOE can significantly reduce the number of repeated analyses, more so than the direct use of GA. The analysis results derived through the DOE

Figure 13. Factor analysis according to D.O.E. table

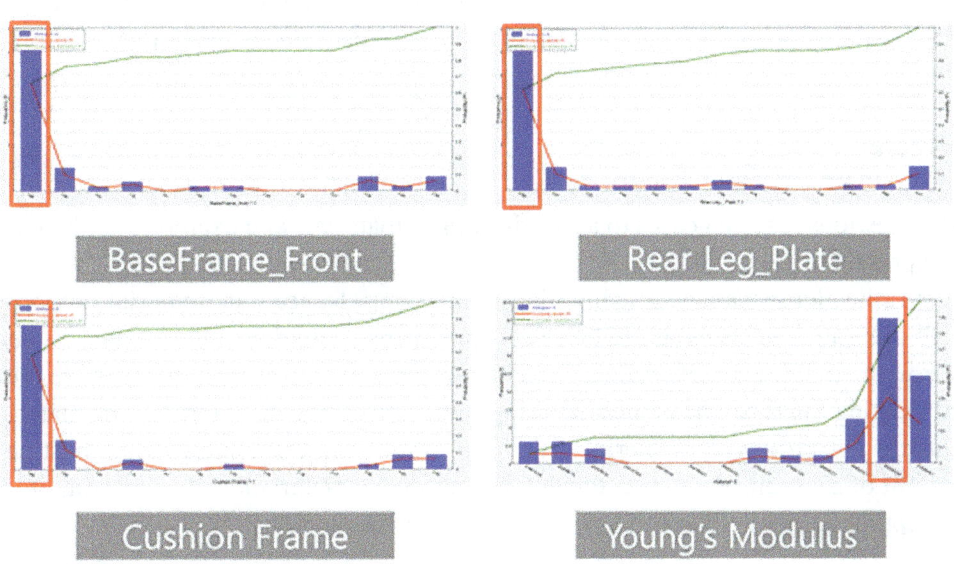

were utilized to generate the response surface. Sensitivity analysis can be conducted on the generated response surface and parameters that have a large impact on the maximum displacement and weight can be reduced to less than 50%. Mass and displacements according to part thickness as shown in Table 13. And Factor analysis according to D.O.E. table as shown in Figure 13.

In the second stage, optimization was performed using the reduced parameters. As there were fewer parameters to be considered, the D-Optimal DOE was used. The analysis results derived through DOE were utilized to generate the response surface. GA was applied to the generated response surface to search for the optimum point that meets the lightweight high-strength condition. If there is a large difference between the predicted value of the response surface and the analysis result of the optimum point, the parameter range is reduced based on the optimum point; then, optimization is performed again. This process can generally be repeated 2–30 times.

The use of DMTO makes it possible to reduce the weight by 5–20%.

FUTURE RESEARCH DIRECTIONS

By applying the dissimilar material application method for weight reduction, it is possible to perform the optimal design according to the dynamic characteristics, nonlinear dynamic response, thermal characteristics, and electrical characteristics as well as the optimization of the shape of the seat frame, the thickness and the material. As the era of electric vehicles nears, thermal or electrical characteristics are very important factors, and dissimilar materials must be applied to improve fuel efficiency, stability and light weight of vehicles. Therefore, it is possible to save development cost and time by applying to the main parts of the vehicle requiring safety and light weight rather than the optimization method limited to the seat frame.

Among the materials used as dissimilar materials, if applied to engineering plastic materials and applied to the entire car instead of the seat frame, the weight is about 40% lighter and excellent moldability can be utilized. In addition, it is expected that various designs can be applied, and engineering plastics have an opportunity to reduce manufacturing costs through the integral molding and modularization of parts. However, according to the application of dissimilar materials, it is necessary to develop the joining process and the welding point derivation technology for joining parts or unit parts.

CONCLUSION

In this chapter, previous research cases to apply dissimilar materials and examples of the latest research methods were presented.

To conduct finite element analysis using dissimilar materials, accurate material properties must be considered. To this end, it is necessary to fabricate specimens using the applied materials and test them in accordance with ASTM and ISO standards. The behavior of a material under tensile, compressive, bending, and twisting loads can be expressed using the strain-stress (S-S) curve. The material property that represents the anisotropy of a material is Poisson's ratio, and the density value must be derived to obtain the weight of a structure. To investigate the fracture characteristics of a material, it is necessary to derive the rupture point of either the S-S curve or the shear modulus.

For high-strength weight reduction, material and thickness can be set as parameters. Previous studies had the problem of deriving optimal values that are difficult to commercialize. For example, real numbers that are difficult to commercialize were derived as the optimization results. In recent studies, such optimization results have been rounded off and replaced with integer values. Therefore, the examples in this chapter present methods for effectively reducing experimental points by managing all parameters as integers (ID).

First, the dissimilar material application example, which used topology optimization, considered the optimization of the compliance and material that belong to the ID of an element. Previous studies on topology optimization had to consider compliance and material separately.

The deformation of the cushion frame was confirmed by static analysis of the seat cushion frame of the typical steel using the standard per safety standard. The topology optimization of the seat cushion frame with typical steel showed a reduction of 30% for deformation and 18% for stress compared to the typical steel frame, and a method that can apply three dissimilar materials based on the density ratio was proposed. It was confirmed that the strength of the cushion frame using three dissimilar materials according to the density ratio was reduced by about 20% to 37% compared to the frame using typical steel. Moreover, an application ratio of dissimilar materials for lightweight optimization was derived. As a result of lightweight optimization by changing the thickness of the cushion's upper frame by applying the D.O.E according to the application ratio of the dissimilar materials, a reduction of approximately 10% weight was obtained compared to a typical steel cushion frame, and the deformation decreased by 30%. Both of these factors improve strength. In this paper, it is considered that the optimal design that satisfies the strength and the lighter weight of the structure or parts of the automobile can be performed by applying the different material of the seat cushion frame through the topology.

Thus, they required a large number of resources for optimization. The use of this example can reduce the resources required for optimization because compliance and material are considered concurrently.

In the discrete material and thickness optimization (DMTO) application example, which used size optimization, the optimization of the thickness and material that belong to the ID of a component by setting them as parameters, was considered. Previous studies encountered the problem of deriving thickness values that could not be commercialized. They also generated many experimental points to derive optimal values. In this example, however, the optimal values can be directly applied to the mass production model because the thickness value is managed as the ID; furthermore, experimental points can be effectively reduced.

The use of this example makes it possible to apply dissimilar materials to one component and to determine the thicknesses and materials of several components at the same time.

The use of these examples in all cases, however, is not a suitable design approach. The required material properties may vary depending on the test method, and the type of design of experiments (DOE) may also vary depending on the number of parameters. In addition, the interpolation method of the response surface may vary depending on the number of experimental points. To consider such variability in the research method, it is recommended that papers on optimization theories be referred to.

ACKNOWLEDGMENT

This research was financially supported by the Ministry of Trade, Industry and Energy (MOTIE), Korea, under the "Regional Specialized Industry Development Program" supervised by the Korea Institute for Advancement of Technology (KIAT-No. R0006048).

REFERENCES

Cho, H. S., Kim, K. S., Choi, D. S., Kim, Y. C., Park, S. H., Oh, B. S., ... Kook, J. H. (2013). Structure analysis on automotive seat recliner housing with high tension steel plate. *Journal of the Korea Academia-Industrial Cooperation Society, 14*(8), 3644–3649. doi:10.5762/KAIS.2013.14.8.3644

Han, S. W., & Jung, H. S. (2011). Weight reducing of aluminum extrusion profiles of a railway-car body based on topology and size optimization. *Transactions of the Korean Society of Mechanical Engineers-A, 25*(2), 213–221. doi:10.3795/KSME-A.2011.35.2.213

Heo, U., Kim, S. K., Song, M. J., Yang, I. Y., & Im, K. H. (2011). Effectiveness Evaluation for Seat Parts by Seat Belt Anchorage Strength Analysis. *IJPEM, 12*(6), 1031–1034. doi:10.100712541-011-0137-1

Hessenberger. (n.d.). Strength Analysis of Seat Belt Anchorage According to ECE R14 and FMVSS. *4th Europeon LS-DYNA Users Conferrnce: Crash / Automotive Applications,* (2), 15-20.

Hwang, Y. J., Kang, S. K., Kim, J. B., Lee, S. S., Choi, C. G., & Son, J. H. (2003). Topology optimal design for lightweight shape of the vehicle mechanical component. *Journal of the Korean Society for Precision Engineering, 20*(7), 177–184.

InterRegs Ltd. (2012). Federal Motor Vehicle Safety Standard: FMVSS 202 Head Restraints. 77(4), 1-5.

Jang, I. S., & Min, B. J. (2006). A Study on the Weight Optimization for the Passenger Car Seat Frame Part. *Transaction of the Korean Society of Automotive Engineers, 14*(5), 155–163.

Jang, J. Y., Yeom, E. S., & Lim, D. B. (Eds.). (2018). *Moving the center of the light weight of automobiles, lightening the material.* Samsung KPMG Issue Monitor.

Jeon, D., & Jeon, E. S. (2017). Lightweight Optimization of Headrest Modules Including Flat-Plate Sliding Brackets. *International Journal of Applied Engineering Research, 12*(8), 1719–1724.

Jeon, S. J., Kim, M. H., Lee, G. H., & Lee, M. Y. (2010). Experimental Study on Side Impact Characteristics for Automotives Door module. *Journal of the Korean Society for Technology of Plasticity, 10*(10), 391–394.

Jeong, C. H., Oh, H. S., & Cheon, S. S. (2016). Strength characterization of automotive composite seat frame by FE analysis based on FMVSS 207. *Journal of Korean Society Precision Engineering, 16*(12), 158–159.

Ju, Y. J., & Jeon, E. S. (2017). Lightweight Optimization of Joint Type Seatback including Plastic Materials. *International Journal of Applied Engineering Research, 12*(15), 5173–5180.

Ju, Y. J., & Jeon, E. S. (2019). Topology Optimization of Seat Cushion Frame With Dissimilar Materials for Lightweight Strength Optimization. *International Journal of Manufacturing, Materials, and Mechanical Engineering*, *9*(1), 1–12. doi:10.4018/IJMMME.2019010101

Jung, H. J., Lee, D. S., Kwon, Y. D., & Han, D. H. (2010). A study of optimum design and analysis with D.O.E for automotive seat frame. *Transaction of the Korean Society of Automotive Engineerss*, *10*(11), 2598–2603.

Jung, S. H., Seo, B. T., & Kim, T. H. (2016). A Study of safe and lightweight car seat by applying multi material design. *Proceedings of the Korea Auto-Vehicle Safety Association autumn conference except collection*, 4-5.

Kim, H. S., Lee, Y. S., Yang, S. M., & Kang, H. Y. (2016). Structural analysis on variable characteristics of automotive seat frame by FEA. *International Journal of Precision Engineering and Manufacturing-Green Technology*, *3*(1), 75–79. doi:10.100740684-016-0010-x

Kim, H. Y., & Lee, S. K. (1999). Prediction of Impact Characteristics for Automobile Seat. *Transaction of the Korean Society of Automotive Engineers*, *7*(3), 196–204.

Kim, J. S., Roh, H. S., Min, Y. J., Hui, S. H., Shin, S. J., & Jang, I. G. (2017). Lightweight design of a bulk trailer using topology optimization. *Transaction of the Korean Society of Automotive Engineers*, *25*(5), 548–554. doi:10.7467/KSAE.2017.25.5.548

Kim, K. S., Cho, H. S., Kim, Y. C., & Cho, J. U. (2014). Structural study of automotive seat frame with high tension steel plate using analysis and experiment. *Journal of the Korea Academia-Industrial Co-operation Society*, *15*(1), 27–31. doi:10.5762/KAIS.2014.15.1.27

Kim, S. S., Kim, K. S., Choi, D. S., Park, S. H., Kim, S. H., & Cho, J. U. (2012). Structural strength analysis at cushion frame and back frame of automotive seat. *Journal of the Korea Academia-Industrial Cooperation Society*, *13*(11), 4956–4962. doi:10.5762/KAIS.2012.13.11.4956

Kim, Y. C., Hong, J. K., & Jang, G. W. (2011). Lightweight crane design by using topology and shape optimization. *Transactions of the Korean Society of Mechanical Engineers-A*, *35*(7), 821–826. doi:10.3795/KSME-A.2011.35.7.821

Lee, D. J., Park, C. S., Lee, K. T., Kim, S. B., & Kim, H. Y. (2005). Evaluation of Static/Dynamic Structural Strength for Automotive Round Recliner. *Transaction of the Korean Society of Automotive Engineers*, *13*(1), 140–146.

Lee, I. C., Lee, T. K., & Jang, D. H. (2015). Development Process of Side Impact Beam for Automotive Light- Weighting Door using Sheet Type. *Transactions of Materials Processing*, *24*(2), 130–137. doi:10.5228/KSTP.2015.24.2.130

Lee, K. S. (2014). *A study on improvement of the correlation between frontal sled FEA and test of vehicle seatbelt system*. Department of Mechanical Engineering, Graduate School of Engineering, Ajou University.

Lee, K. T. (2004). *CAE for Development of Round Recliner in Automotive seat System. Department of Mechanical Engineering, Graduate School Kangwon National University.*

Lim, J. Y. (2017). *Lightweight Design of the Automotive Seat Back Frame using FEM*. Department of Mechanical Engineering, Graduate School of Chonbuk National University.

Martinsen, K., Hu, S.J., & Carlson, B.E. (2015). Joining of dissimilar materials. *CIRP Annals - Manufacturing Technology, 64*(2), 679-699. doi:10.1016/j.cirp.2015.05.006

Na, S. J., & Jeon, E. S. (2017). Lightweight Corrugated Plate for Cargo box by using DOE and Topology Optimization. *International Journal of Applied Engineering Research*, *12*(22), 12564–12569.

Na, S. J., Kim, S. Y., & Jeon, E. S. (2019). Size optimization design of threewheeled motorcycle frame with cargo box. *International Journal of Mechanical Engineering and Technology*, *10*(5), 345–356.

Oh, H. T., & Jeon, E. S. (2017). Lightweight Optimal Design of Tailgate Lift Link under Eccentric Loading. *International Journal of Applied Engineering Research*, *12*(17), 6627–6632.

Park, J. W., Kook, J. H., & Cho, J. U. (2016). A study on the Durability of Automotive Seat Frame with Aluminium. *Journal of Korean Society of Mechanical Technology*, *18*(4), 505–555.

Partwardhan. (n.d.). Simulation and Validation of FMVS 207/210 Using LS-DYNA. *7th International LS-DYNA Users Conferrnce: Simulation Technology*, (2), 9-13.

Salim, D. M., Sanjay, D. Y., Shinde, D., & Deshpande, G. (2013). Importance of Federal Motor Vehicle Safety Standards 207,210 in Occupant Safety - A Case Study. *Procedia Engineering*, *64*, 1099–1108. doi:10.1016/j.proeng.2013.09.188

Shin, D. S., & Jeon, E. S. (2017). The s-s curve approximation for GFRP tensile specimen by using inverse method and optimization. *Journal of Engineering and Applied Sciences (Asian Research Publishing Network)*, *12*(20), 5702–5707.

Shin, D. S., Song, Y. J., & Jeon, E. S. (2018). Lightweight design of seat cushion extension modules using the properties of plastic and HCA-SIMP. *International Journal of Mechanical Engineering and Technology*, *9*(5), 624–632.

Shin, K. H. (2016). Development of Mg seat FEM using Accelerated Durability test. Department of Mechanical Engineering, Graduate School of Kongju National University.

Song, H. D. (2015). Light-Weight Design and Analysis in car Seat Frame. Department of Carbon Fusion Engineering, Graduate School Jeonju University.

Song, S. I., Cha, I. R., Lee, K. H., & Park, G. J. (2002). Optimization of the Automotive Side Door Impact Beam Considering Static Requirement. *Transaction of the Korean Society of Automotive Engineers*, *10*(3), 176–184.

Song, Y. J., & Jeon, E. S. (2017). Optimal Design of Lightweight Seat Extension Equipment Using Topology Optimization and Design of Experiment. *International Journal of Applied Engineering Research*, *12*(9), 1855–1859.

Unknown. (2010). *TEST METHOD 210: Seat Belt Anchorages*. Transport Canada.

Yang, Y. S., & Jeon, E. S. (2016). Tailgate Hinge Stiffness Design using Topology Optimization. *International Journal of Applied Engineering Research*, *11*(19), 9776–9781.

Yang, Y. S., & Jeon, E. S. (2017). Structural Design of Automatic Seatback Table. *International Journal of Applied Engineering Research*, *12*(19), 8901–8906.

Yang, Y. S., & Jeon, E. S. (2018). Design Parameters and Dynamic Analysis of Unfolded Headrests for Vehicles. *International Journal of Mechanical & Mechatronics Engineering*, *18*(3), 1–5.

Yang, Y. S., Jeon, E. S., & Park, D. H. (2018). Optimization of Design Parameters of Spiral Spring for Active Headrest Deployment. *MATEC Web of Conferences IC4M & ICDES 2018, 167*, 1-8. doi:10.1051/matecconf/201816702017

Yang, Y. S., Kim, Y. S., & Jeon, E. S. (2018). Design for headrest including guides slot to reduce the neck force in rear end impact. *International Journal of Automotive Technology*, *19*(6), 1041–1048. doi:10.100712239-018-0101-5

Yoo, D. Y., Kim, J. Y., Cho, K. S., Beom, H. K., & Kwon, O. S. (2006). A Study of Numerical Analysis for Improvement of US NCAP Side Impact Performance. *Journal of the Korean Society of Automotive Engineers*, *6*(6), 1088–1093.

Yoon, T. H., Kim, H. Y., Heo, C., Shin, S. G., & Kwon, J. W. (2014). FE Dynamic Simulation Application for Performance Evaluation of Car Side Impact Test by One-Body Door Impact Beam Development. *Journal of the Korean Society of Mechanical Engineers*, *14*(11), 926–931.

Chapter 12
Simulation of a Hydropneumatic Suspension for Agricultural Working Vehicles

Nicola Bosso
Politecnico di Torino, Italy

Nicolò Zampieri
Politecnico di Torino, Italy

Aurelio Somà
Politecnico di Torino, Italy

Francesco Mocera
Politecnico di Torino, Italy

Emanuele Conte
Politecnico di Torino, Italy

ABSTRACT

The chapter shows the study and simulation of a hydropneumatic suspension to be adopted for a telescopic handler vehicle. The hydropneumatic suspension system with independent wheels and with quadrilateral architecture has been studied to improve comfort and productivity of the existing vehicle, which has a suspended rigid axle on the front and a rigid axle on the rear, limiting the comfort and the grip. After the choice of the architecture and the kind of suspension, the chapter shows the design of the suspension kinematics. The optimization of the characteristic angles of the suspension has been performed using Adams/Car and Adams/Insight. The kinematic model optimized is subsequently reproduced in Adams/View to simulate the dynamics of the complete vehicle. Simulation results are used to evaluate vehicle performance in terms of comfort and stability according to the methods proposed by the standards.

DOI: 10.4018/978-1-7998-4939-1.ch012

INTRODUCTION

In the last years, working vehicles manufacturer are investing money and efforts to make their machines compliant with the current quality standards, especially when dealing with safety and comfort (Zampieri et al., 2013; Kittusamy & Buchholz, 2004; Kuijt-Evers et al., 2003) of the operator during working operations. Several studies in the literature show that long working sessions (lifting, moving, pulling, etc..) can affect the operator comfort inside the cab due to the exposure to prolonged vibrations (Akinnuli et al., 2018; Caffaro et al., 2016; Zampieri et al., 2013). According to the European Agency for Safety and Health at Work (European Agency for Safety and Health at Work, 2008) todays mobile machinery operators can be exposed to prolonged vibrations with accelerations in the range from 0.2 up to 2 m/s^2. The frequency content of the vibration can affect the human body in different ways: low frequency vibrations, below 10 Hz usually lead to human motion sickness while frequencies up to 100 Hz are associated to a general discomfort for the operator (Guignard, 1971). In the June of 1989, the European Directive 89/391/EEC (European Council Directive, 1989) introduced for the first time the need for special attention on safety and health of workers, especially against mechanical vibrations levels to which they can be exposed during their daily activities. Then, several other directives stated the Whole Body Vibrations (WBV) levels to which worker can be exposed without risks (European Parliament and European Council, 2002). Generally speaking, vibration acceleration should not exceed 3.2 m/s^2 for short operations (30 minutes) and 0.8 m/s^2 long operations (> 1 hour).

On a Working Vehicle, vibrations can derive from different sources: a bad filtered Diesel engine, tools attached and propelled by the vehicle and the road asperities. One of the key elements which affect comfort and safety of a vehicle is the type of suspension adopted to introduce a degree of freedom between the body of the vehicle and the wheels directly in contact with the road surface. It is not unusual to find working vehicles with no real suspension mechanism (Nicolini et al, 2018; Mocera & Nicolini, 2017; Peachey et al., 1989; Lines et al., 1989). The ground asperities can be mitigated by the stiffness and damping provided by the deformation of the tyres (Siefkes, 1989; Rasmussen & Cortese, 1968) but are practically entirely transmitted to the main body if the machine is equipped with tracks (Zhang et al., 2018; Cai et al. 2012). The lack of a suspension mechanism provides robustness and cost effectiveness for the simplicity of the solution, but at the cost of a lower comfort for the operator and limited dynamic performance of the vehicle. To provide a minimum level of filtering from vehicle vibrations, the most common solution adopted is the suspension of the cab by mean of rubber elements placed at the interface connection with the chassis (Zehsaz et al., 2011; Hansson, 1995; Hilton & Moran, 1975). Considering the most common commercial solution, this approach can well filter vibrations in the high frequency range but does not provide good comfort levels when it comes to low frequencies. A possible solution to improve comfort consists in the adoption of active or semi-active suspensions for the cabin (Deprez et al., 2005) which however does not improve the overall dynamic behaviour of the vehicle. In this work, a semi-active suspension system between the tyres and the chassis will be investigated considering a telescopic handler as case study. The suspension architecture will be discussed to understand the benefits and the overall impact of the solution at vehicle level in terms of performance.

Generally speaking, vehicle suspension system can be grouped in four relevant classes:

- Passive suspension (Grott, 2010);
- Adaptive suspension (Cebon, 2000);

- Semi-active suspension (Reimpell et al., 2001; Gries, 1966; Gorabal et al., 2011; Ferraresi et al., 1997; Giliomee & Els, 1998);
- Active suspension (Genta 2008; Williams, 1997)

Passive suspensions are the most used for their simplicity and their low cost. The characteristics in terms frequency response of a passive suspensions cannot be modified during working operations. They are designed to provide the best compromise between comfort and good handling of the vehicle. Figure 1 shows the typical model of a passive suspension. The main body or suspended mass is connected to the tyre by mean of a spring and a damper As described in (Genta, 2009) common values for the stiffness of spring k and k_t can differ for an order of magnitude justifying the use of simplified models with just 1DOF to have a first approximation of the vertical movement of the suspended mass M. The damping effect is usually given using a shock absorber where the kinetic energy is dissipated into heat via viscous friction.

Figure 1. Passive suspension model

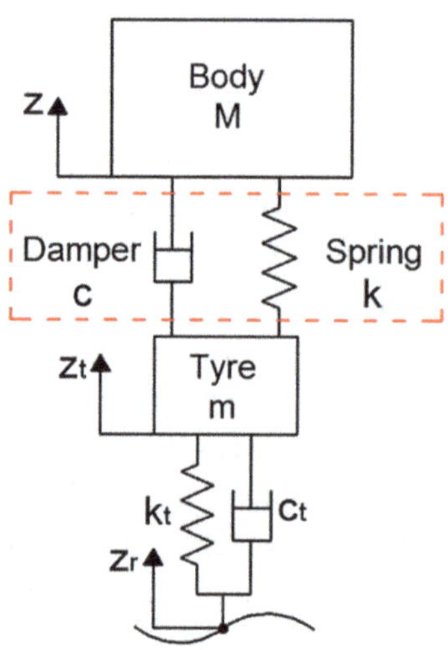

Adaptive and semi-active suspensions are systems where damping characteristic of the shock absorber can be changed during working operations. Usually an adaptive suspension has a slower response and is designed to propose different fixed driving modes. They are controlled in open loop by the driver which set the desired suspension characteristics for the type of work to be approached. Semi active (Figure 2a) suspensions quickly change their damping characteristics with a good range of damping values. They represent the best compromise between performance and costs. They are usually controlled in closed loop to automatically adapt to different road conditions maintaining the comfort level chosen by the driver. A well-known type of semi-active suspension is the Sky-Hook (Figure 2b). To be precise, Sky-Hook refers to the name of the control algorithm of the semi-active suspension which consider the sprung mass connected to a fixed frame by mean of a damping element. Controlling the variable section of the orifices of the shock absorber unit or using magnetorheological fluids which viscosity changes according the external magnetic field applied, a control unit can change the damping according to the relative

Figure 2. Semi-active suspension model: a) general layout; b) Skyhook control strategy

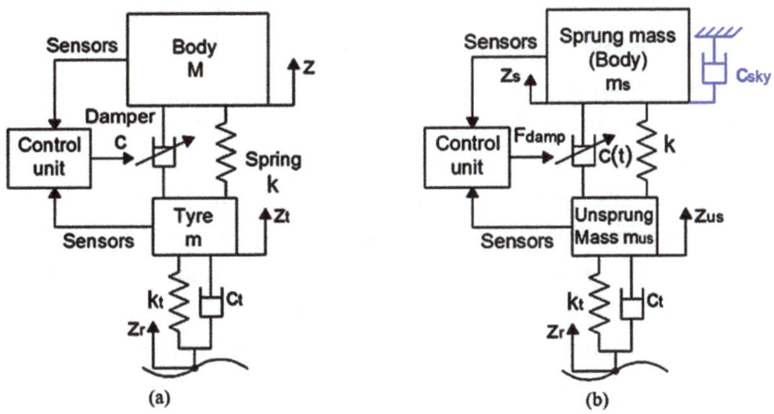

position between the sprung and unsprung mass. Although controllable, semi-active suspension still are dissipative systems, so the range of possible configurations is limited.

Finally, active suspensions represent the most effective solution when it comes to controllability of the vertical movement and handling of a vehicle. Using electric or hydraulic actuators these solutions can apply energy to the system increasing the level of controllability and obtaining better comfort, grip, reduction of pitch and roll of the suspended mass during traction, braking or when approaching curves. This high level of controllability comes at a higher economic cost which limits consistently the number of possible applications.

The Sky-Hook approach is often a common control strategy between semi-active and active suspension. However, since such architecture cannot be realized in practice, the control system calculates the force exerted by the shock absorber element, proportional to the absolute speed of the vehicle body with respect to an inertial reference frame and the relative speed of the unsprung mass with respect to the vehicle body. This technique allows for a reduction of displacement transmissibility with respect to frequency, if compared to conventional damping system. It is important to note that the absolute speed of the sprung mass can be opposite to the one of the un-sprung mass. In this case, to apply the skyhook theory, it is necessary to add energy in the suspension system because the actuation force versus is opposite with respect to the damping force versus. This is possible only with active system. For these reasons the skyhook theory is usually applied only in part by means of semi-active systems that allow a variable damping. In this case, it is possible to manage all situations requiring an actuating force that have the same versus of relative speed between sprung mass and unsprung mass. In the other case the actuation force should have an opposite versus and cannot be reproduced. For semi-active system, being this configuration physically impossible, the damping is adjusted to provide the minimum force in this case, minimizing the difference between the theoretical skyhook actuation force and the real actuation force.

Semi-active solutions represent the best compromise between performance, cost and comfort representing a suitable option for industrial applications where the Total Cost of Ownership (TCO) is a crucial characteristic for profitable machine. Among the different semi-active solutions available in the literature, the hydropneumatic suspension can achieve a good level of comfort thanks to the adjustability of its stiffness and damping characteristics. As shown in Figure 3 (Savaresi et al., 2005; Bauer 2011;

Citroen 1977), the system is filled with incompressible fluid (oil) and compressible fluid (nitrogen). The former transfers the load to the elastic element of the suspension, while the latter is the elastic element (spring). The diaphragm (7) separates accumulator in two hemispheres (2) and (8), containing nitrogen and oil respectively. At the base of the accumulator there is a resistance valve that produces the damping effect. By adding and subtracting oil from the cylinder (10) it is possible to modify the height of the suspension, (Murrenhoff & Wallentowitz, 1998; Ellmann & Loser, 2002).

Hydropneumatic and pneumatic semi-active suspension systems are well known for their flexibility and progressivity in the regulation of the stiffness characteristic. In fact, for a traditional suspension this characteristic is linear. If stiffness grows more than linear, the mechanical limit switch is more difficult to reach in compression or extension extremes. Hydropneumatic suspensions have a quick response to the driver settings, especially when it comes to the height level adjustment. The main reason is the use of an incompressible fluid which can increase in pressure quickly and with a relatively low amount of energy if compared with pneumatic systems. Hydropneumatic suspension systems are more efficient and require less amount of space if compared to other controllable systems. However, the current cost of this solution is higher than pure pneumatic and traditional systems.

When it comes to the design of a suspension system for working vehicles, several benefits could be achieved by using a hydropneumatics suspension system. In this work, a telescopic handler is considered as case study. Generally, telescopic handlers are designed with rigid axles and no suspension systems other than elastic bushing elements between the cab and the vehicle chassis. Thus, high frequency vibrations can be mitigated but both low frequency vibrations and vehicle dynamic performance are strictly related to the stiffness and damping characteristics of the tyres. The use of a hydropneumatics suspension on such type of vehicle would allow to:

- Increase the overall stiffness of the suspension when the vehicle is driving on the road for better longitudinal and lateral dynamic behaviour (handling)
- Adjust the height level to overcome rocks or mounds of soil in agricultural fields
- Introduce a certain level of controllability without a high amount of space requirements
- Achieve a good level of robustness and reliability of the suspension system

Traditional passive suspensions systems require a certain amount of space for the installation of the mechanical spring. Pneumatic suspensions require more on-board space and have a lower reliability due to the inevitable pneumatic losses in the systems. In this work, performance of a telescopic handler with hydropneumatic suspensions with double wishbone architecture are explored. The choice of the double wishbone architecture with rigid bushing elements is mainly related to the need of a stiff and robust design solution able to give a certain flexibility on the vertical travel of the wheel. Comfort and stability can be greatly improved adjusting the height level of the vehicle body. This work describes the kinematic model of the suspension, which has been realized using the Adams Car software and used to optimize the suspension characteristics. The optimized suspension has been then introduced in the full vehicle model, developed in Adams View, to analyse vehicle performance in terms of stability and comfort. In order to estimate the improvement of the proposed solution, the full model of the vehicle, adopting the hydro-pneumatic suspension, has been compared with the model of the traditional architecture with rigid axles.

Figure 3. Section of a hydropneumatic suspension (Citroën): (1) filling screw; (2) the upper hemisphere; (3) input high-pressure fluid; (4) piston; (5) overflow; (6) shield rod; (7) diaphragm; (8) lower hemisphere; (9) the damper (10) cylinder; (11) pad thrust; (12) the sealing system (13) protective cover dust.

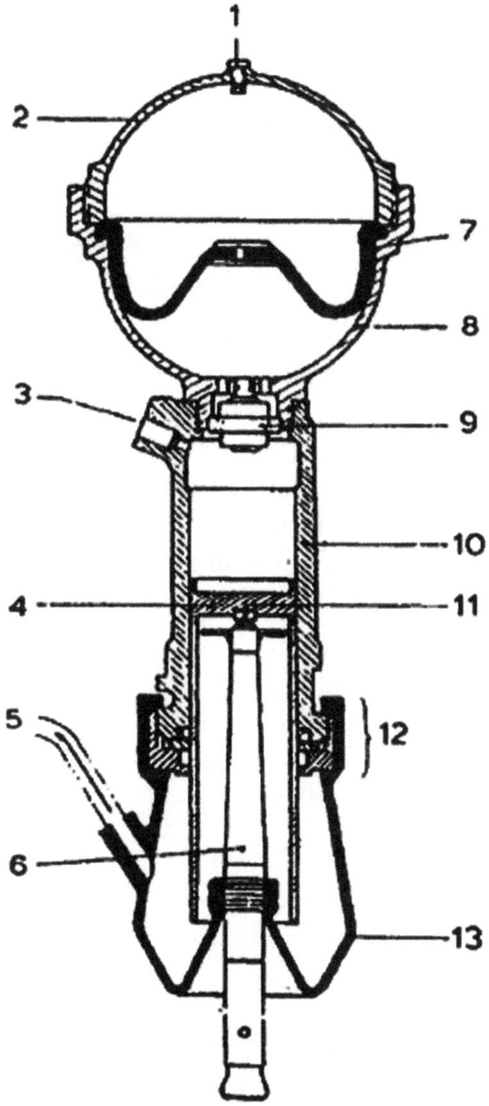

KINEMATICS DESIGN OF THE SUSPENSION

The kinematic model of the wishbone suspension has been developed by means of the multibody software MSC Adams/Car. Figure 4 shows the kinematics of the suspension. It consists of five main components:

1. Upper control arm
2. Lower control arm
3. Upright
4. Outer tie rod
5. Inner tie rod

Figure 4. Scheme of the front axle suspension modelled in MSC Adams / Car

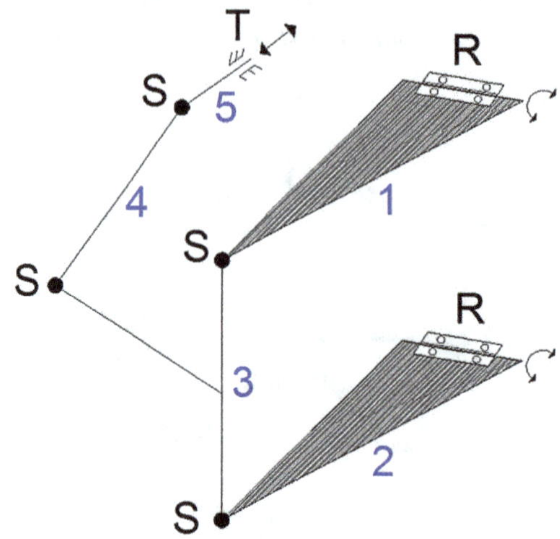

The bodies 1 and 2 shown in Figure 4 are linked to the vehicle chassis by means of two revolute joints R. All the main bodies of the suspensions are connected through several spherical joints S. Finally, body 5 is connected to the vehicle steering system with a translational joint T. The kinematic system so obtained has 3 degrees of freedom as stated by the Kutzbach formula shown in Equation 1.

$$n = 6m - 5R - 5T - 4C - 3S \tag{1}$$

Where:

m= number of bodies;
R= number of revolute joints;
T= number of translational joints;
C= number of cylinder joints;
S= number of spherical joint.

The system shown in Figure 4, has 3 degrees of freedom but only two of them are responsible of the "effective state" of mechanism. In other words, the axial degree of freedom of link 4 is not responsible for the variation of the suspension characteristic angles. Considering the model of Figure 4, the upper (1) and lower (2) control arms are linked to chassis with revolute joints, whose free rotations are limited by elastic bushings to reduce vibrations. The spring-damper element of the suspension is modelled as a point to point force between the lower control arm and the chassis. The non-linear characteristics of stiffness and damping of this element are shown in Figure 5 and Figure 6. The stiffness characteristic used is linear in the working range, while it shows a hardening behaviour to simulate the bumpstop action. The damping characteristic is linear with saturation. These characteristics are introduced by means of expressions and input functions. Table 1 shows the final coordinates of the suspension hardpoints, that are shown in the scheme of Figure 7 a and b. These values have been obtained from the optimization of the suspension that is described in the following paragraph. Given this configuration of hardpoints, the static configuration of the suspension is fixed, especially in terms of the Toe, Camber, Caster angles, roll centre height and overall operating range of the spring and shock absorber.

Figure 5. Stiffness characteristics of the spring

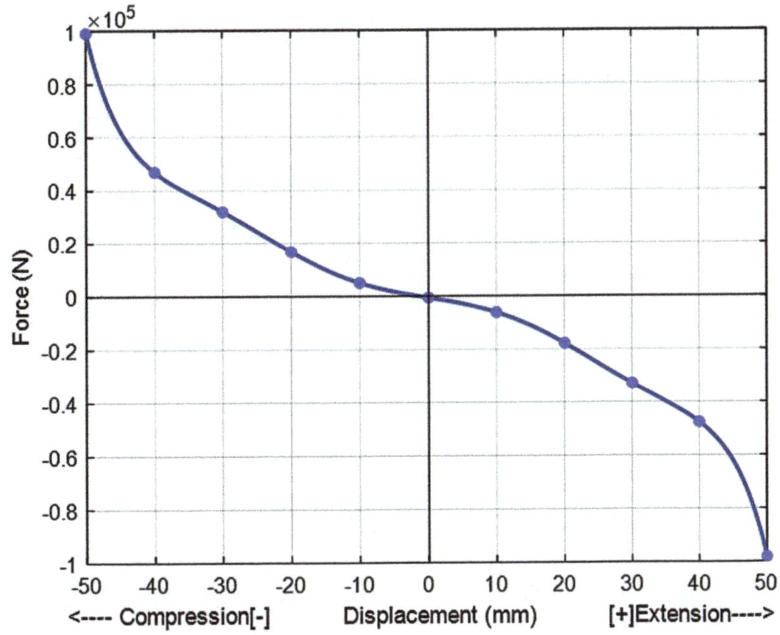

Figure 6. Damping characteristics of the flow resistor

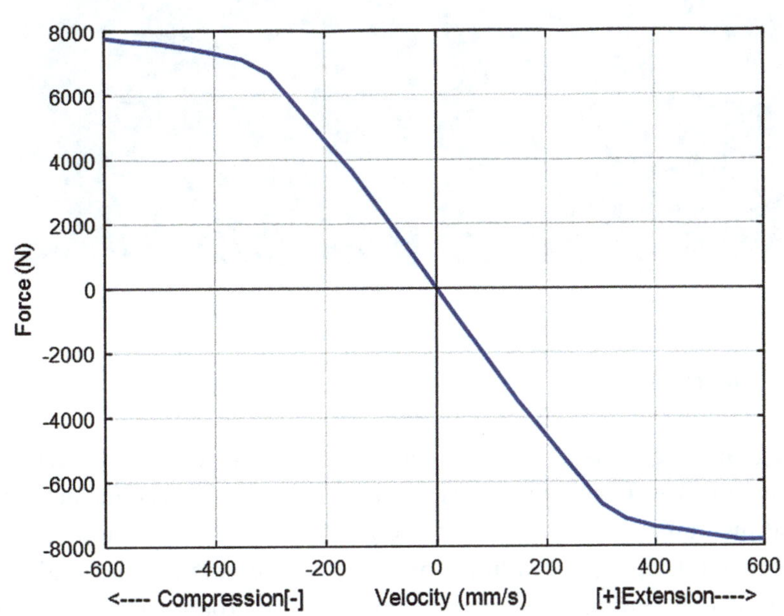

Table 1. Coordinates of the suspension hardpoints

Hardpoint	x [mm]	y [mm]	z [mm]
hpr_drive_shaft_inner	0.0	183.0	212.0
hpr_Ica_front	-187.0	217	24.0
hpr_Ica_outer	0.0	740.0	-72.0
hpr_Ica_rear	187.0	217	24.0
hpr_Iwr_strut_mount	-93.0	578.0	-46.0
hpr_tierod_inner	200.0	400	82.5
hpr_tiroed_outer	218.0	654.0	82.5
hpr_top_mount	-93.0	319.0	332.0
hpr_uca_front	-31.0	319.0	290.0
hpr_uca_outer	0.0	677.0	284.0
hpr_uca_rear	140.0	319.0	290.0
hpr_wheel_center	0.0	921.5	62.5

Figure 7. Model of the suspension in MSC Adams / Car: a) front axle view; b) hardpoint legend

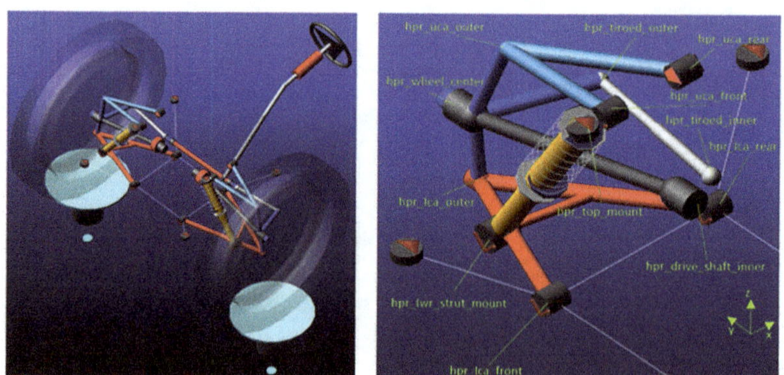

(a) (b)

Kinematic Analysis Results

The most important geometrical characteristics of a suspension, which highly affect the overall vehicle dynamic behaviour are the Toe, Camber and Caster angle. The toe angle is the angle between the wheel equatorial plane and the vehicle longitudinal axis, considering the plane view. This angle influences the ability of the vehicle to keep a straight trajectory. The camber angle is the angle between the wheel equatorial plane and the longitudinal plane perpendicular to the ground. This angle influences the performances of the vehicle when approaching a curve. Finally the caster angle is the angle formed by the steering axis of the suspension (also known as king pin, in this case defined connecting the hpr_uca_outer and hpr_Ica_outer hardpoints) with respect to the vertical axis in the side view of the vehicle. This angle influences the force that the driver has to apply to the steer to turn the wheels. In addition, it influences the ability of the wheel to return to the starting position (straight) after rotation.

Figure 8. Toe, Camber, Caster angle variations with respect to the vertical shaking of the wheel

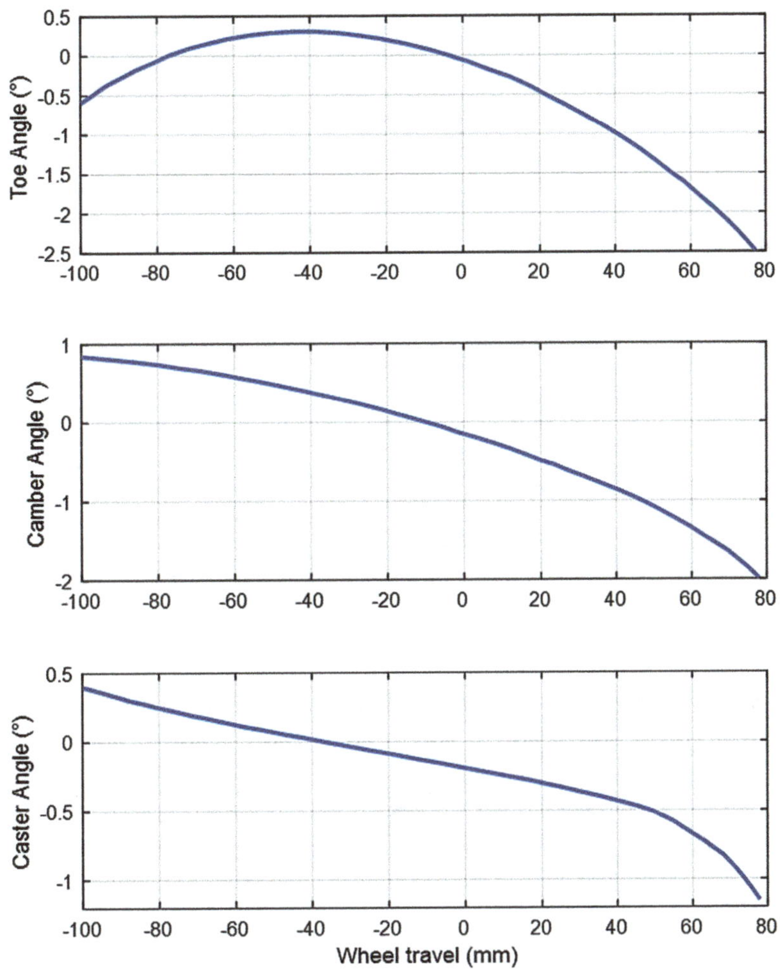

The numerical model of the suspension, shown in Figure 7, is used to perform a parallel wheel travel simulation, to explore the kinematic characteristics of the suspension. In this phase, a first optimization of the characteristic angles has been performed. Figure 8 show the main trends of the suspension angles versus the vertical displacement of the wheel, obtained with a test stand numerical simulation performed with Adams/Car.

Kinematic Optimization of the Suspension

Using the Adams/Insight software the effect of each hardpoint on the characteristic angles were studied evaluating the changes in the control variables (mainly Toe and Camber angles) when changing the hardpoints position. The attention focused in the optimization of the of the camber angle (Figure 9) characteristic with respect to the wheel vertical displacement when changing the position of the hardpoints hpr_uca_front and hpr_uca_rear (see Table 1 and Figure 7b). However, the improvement of the camber angle shown in Figure 9 led to a deterioration of the toe angle, as shown in the behaviour of the

Figure 9. Effect of variation of hpr_uca_front and hpr_uca_rear hardpoints on the camber angle curve

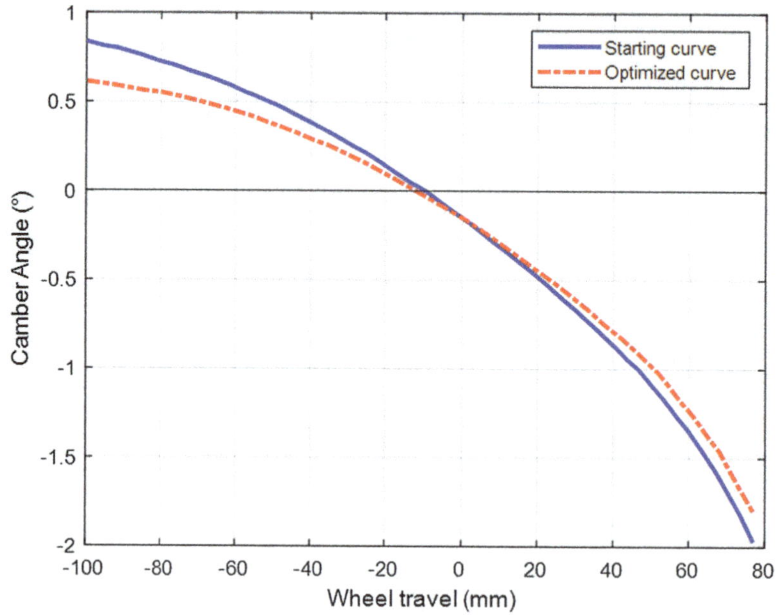

Figure 10. Effect of camber angle optimization on the toe angle and toe angle curve optimization

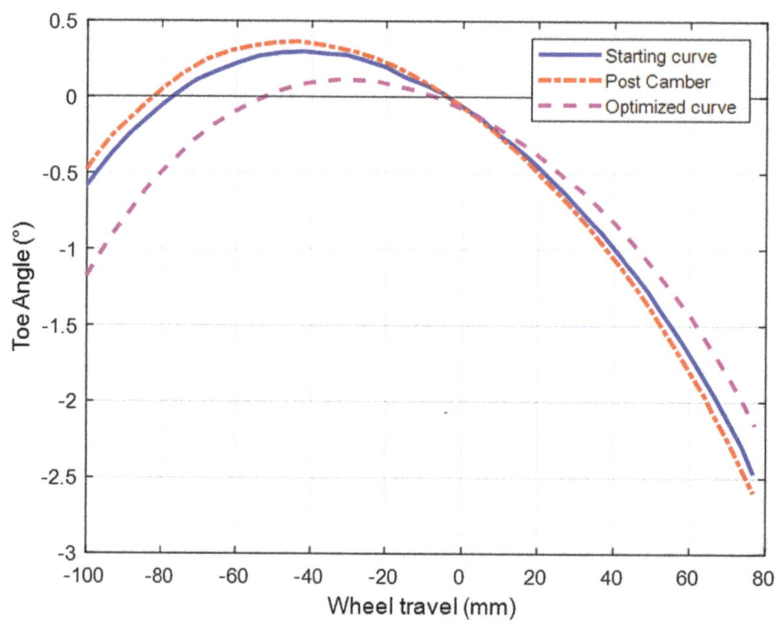

toe angle depicted by the dot-dashed red curve in Figure 10. To solve this problem the hardpoint position of the steering system (hpr_tierod_inner hardpoint) has been further optimized, as shown magenta dashed curve in Figure 10.

In conclusion, the main angles have been optimized in order to obtain a good compromise of suspension performance (handling) and uneven tire wear usually related to high camber angles (in absolute values). Similar considerations can be made about the toe angle. In this case, excessive toe angles values can cause wear on the inner/outer part of the tyre surface. The optimized position of the suspension hardpoints is shown in Table 1.

FULL VEHICLE DYNAMIC SIMULATION

The characteristics of the suspension designed in the previous section were included in a full vehicle multibody model of a telescopic handler to study the effects on the overall dynamic behaviour of the machine. The geometry of the suspension shown in Figure 7, including the same steering system, and the overall vehicle were modelled in the multibody code Adams View. The suspensions on all the four wheels, on the front and rear axle, were modelled with the same approach and characteristics. For the purpose of this analysis, no steering behaviour was considered for the rear axle. The assumption was also justified by the fact that this functionality is usually involved when approaching very narrow curves at low speed. Figure 11 shows a view on the full multibody vehicle model (a) where the inertia characteristics of each vehicle subsystem was taken into account and a representation of a simplified vehicle model seen from a lateral view (b). The simplified model allows to identify the main elements of interest that will affect the longitudinal, lateral and vertical behaviour of the multibody model:

- a spring-damper model of the wheel
- a spring-damper model of the suspension (hydropneumatic)
- an elastic element connecting the frame and cab (silent block)
- a translational joint that allows the rotation of the telescopic arm.

The main inertial masses of the involved are:

- the wheels mass m
- the chassis mass M_{ch}

Figure 11. Vehicle model: full vehicle model representation in Adams/View; b) Simplified schematic representation of the lateral view

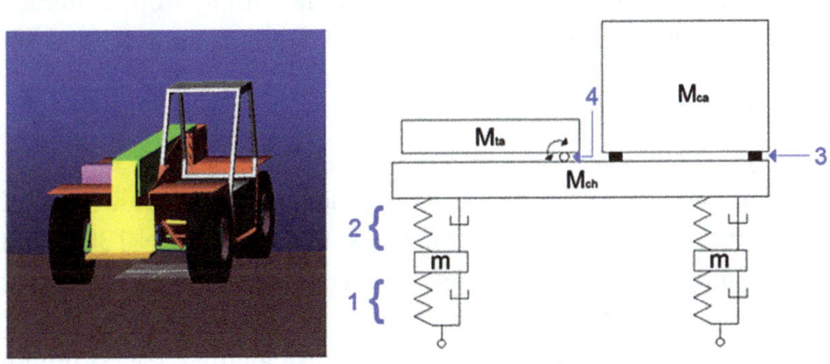

(a) (b)

- the cabin mass M_{ca}
- the telescopic arm mass M_{ta}

The stiffness of the tires was modelled considering the results of experimental tests done on a real wheel of the considered vehicle and introduced as an input function in the numerical model. With reference to Figure 11, the damping and the stiffness characteristics of the bushings (3) have been obtained experimentally and introduced into the model as input function (Somà et al., 2006). The complete model of the vehicle is used to evaluate the vehicle lateral behaviour in critical conditions and to determine the comfort level in the cab according to the considered standard. Table 2 shows the main parameters of the complete vehicle model.

Table 2. Main parameters of the complete vehicle model

Parameter	Value
Wheelbase	2800 mm
Track	2250 mm
Type of tire	405/70-24''
Empty mass	8350 Kg

Step Steer Simulation

To study the vehicle roll behaviour, a step steer simulation was performed. The vehicle, at tare condition, runs at constant speed (50Km/h) and then it forces a step steer manoeuvre. In this way, it is possible to understand the vehicle behaviour when it is subjected to high lateral accelerations. The steering remains straight until the time t=20s after that the step steer manoeuvre occurs. The engine provides torque to the four-wheel drives until the time t= 22 s. Figure 12 shows the vehicle trajectory during simulation, while Figure 13 show the vehicle behaviour in terms of lateral acceleration, roll angle and yaw speed, respectively. The results of roll angle are compatible with the inertia and other characteristics of the vehicle. Figure 14 shows the comparison of load transfer between the model with and without suspension during the same step steer manoeuvre. Considering the results, it is possible to conclude that the vehicle model without suspensions has a higher load transfer. In particular, the wheels on the external side are completely unloaded and detach from the ground. The vehicle with hydropneumatic suspension system is instead more stable and the wheel unload on the external side it is about of the 50%.

Measure of Comfort

Another interesting comparison between the traditional vehicle suspension solution and the hydropneumatic one is related to the evaluation of the comfort level. The standard UNI EN 13059-2004 was considered for the calculation methodology of the comfort level. According to the mentioned standard, a comfort index can be defined as the combination of the triaxial acceleration in terms for RMS values (Somà et al, 2006). The RMS of acceleration in the generic direction is calculated according to Equation 2.

Figure 12. Trajectory of the center of mass of the vehicle on the plane x-y during the step-steer manoeuvre

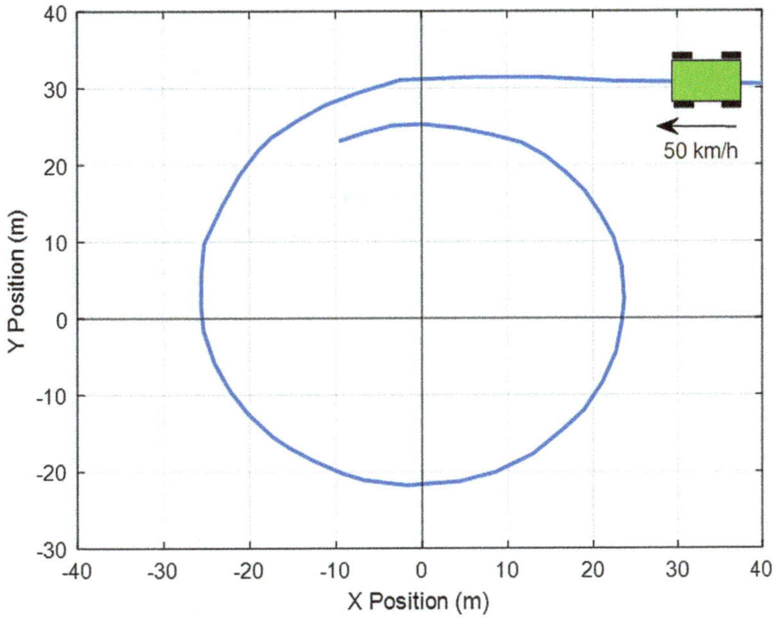

Figure 13. Lateral acceleration, Roll angle and Yaw speed during the step-steer manoeuvre

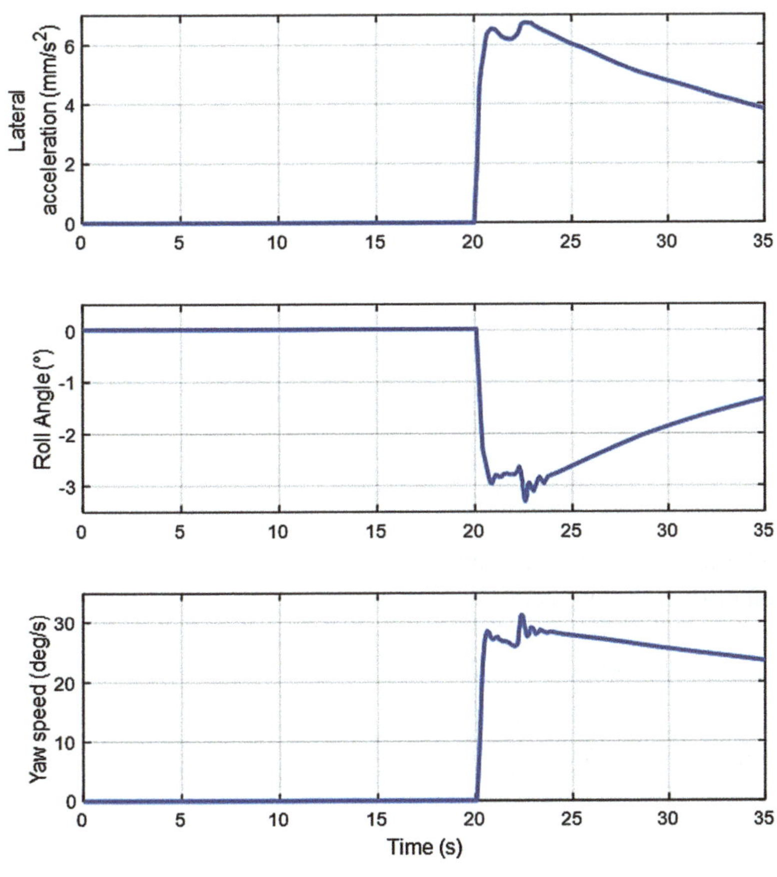

Figure 14. Comparison of load transfer between the vehicle with and without suspensions

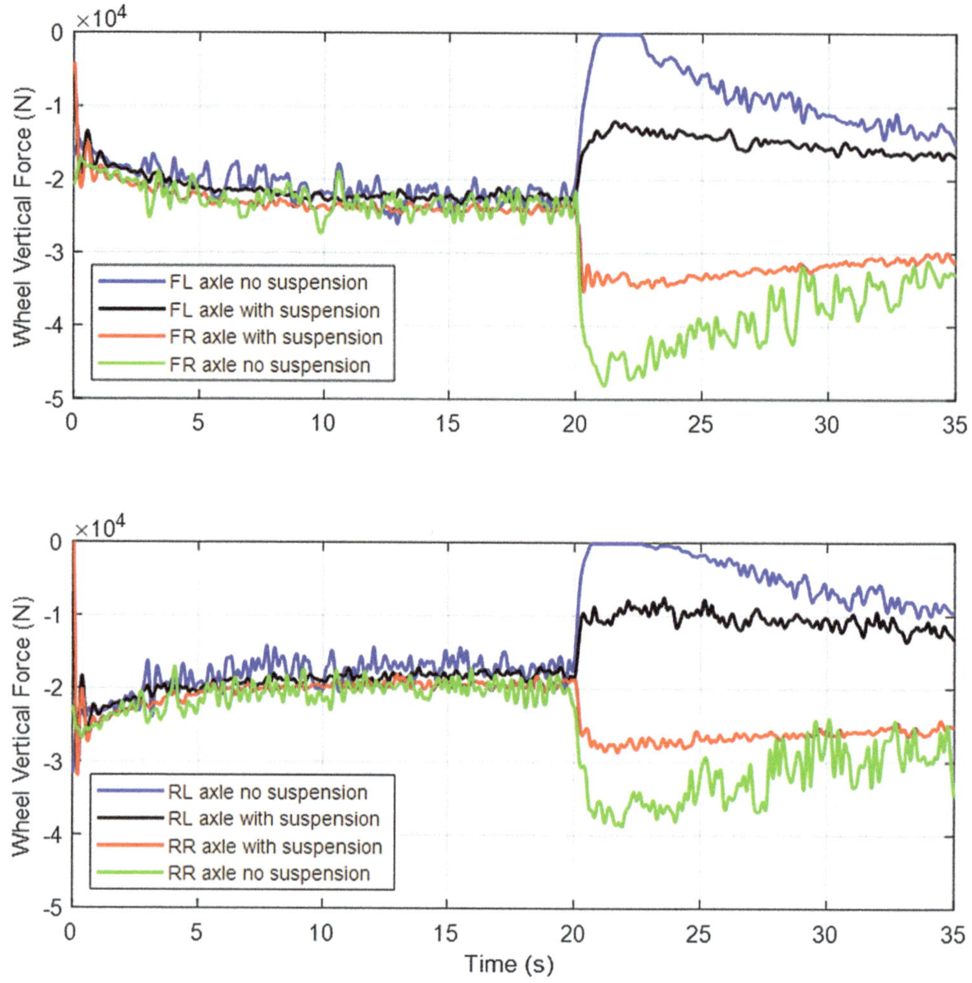

$$a_{RMS} = \left[\frac{1}{T} \int_0^T a_w^2 \left(t\right) dt \right]^{1/2} \tag{2}$$

Where:

- $a_w(t)$ is the acceleration value in the generic direction w
- T is the duration of the measurement

The comfort index is evaluated as

$$a_v = \left(k_x^{\ 2} a_{RMS,x}^{\ \ 2} + k_y^{\ 2} a_{RMS,y}^{\ \ 2} + k_z^{\ 2} a_{RMS,z}^{\ \ 2} \right)^{1/2} \tag{3}$$

With

- k_x, k_y and k_z are the coefficients that "weight" the acceleration, respectively in the x, y and z directions (Zampieri et al., 2013).
- $a_{RMS,x}$, $a_{RMS,y}$ and $a_{RMS,z}$ are the RMS accelerations in the x, y and z directions.

The standard does not directly consider the comfort evaluation, since it examines the effects of exposure of the human body only to vibrations. This standard in fact does not consider the effect of other factors that can influence the comfort perceived by the passenger. For the purpose of this analysis, accelerations were measured on marker placed at a position inside the cabin equivalent to the position of the driver seat.

The accelerations required to evaluate the comfort index must be measured during a specific test consisting in the approach of two obstacles of 30 mm height and 150 mm width placed at a distance of 10 meters between each other. The overall setup of the test is shown in Figure 15 which shows the characteristics of the test track as well as the implementation in the multibody environment. The obstacles were approached with the vehicle at tare condition, with no additional weight on the telescopic arm. The test was performed at three different speeds: 5 km/h, 15 km/h and 25 km/h. To calculate the comfort index, the coefficient k_x=1.4, k_y=1.4 and k_z=1 were used. Figure 16 shows the comparison between the comfort index evaluated on the model with the hydropneumatics suspension and the model with the traditional rigid axles. The benefits in terms of comfort are clearly stated and increase with the work-

Figure 15. UNI EN 13059-2004 obstacles layout and geometry

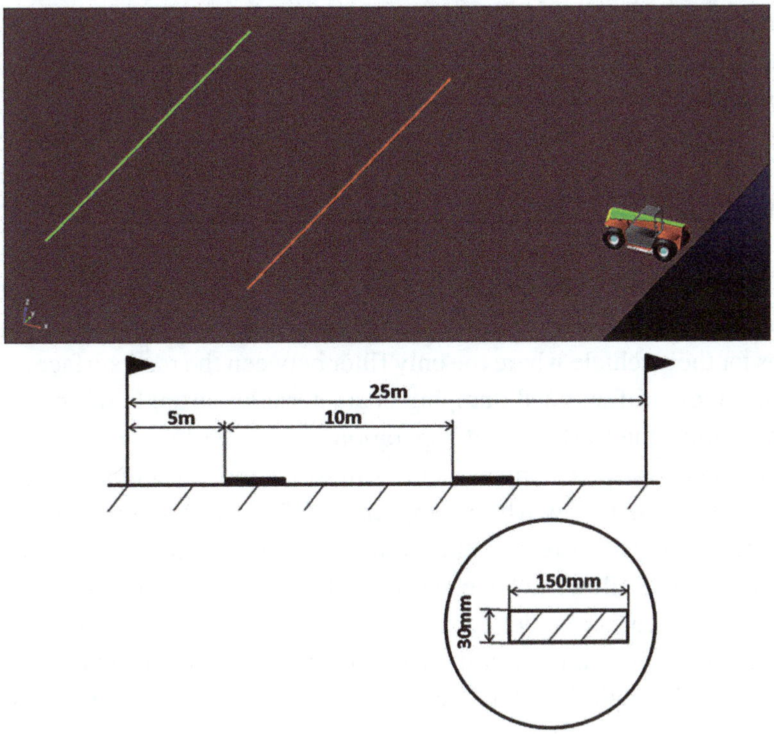

Figure 16. Comparison of the index of comfort of the cabin between model with hydropneumatic suspension and rigid suspension

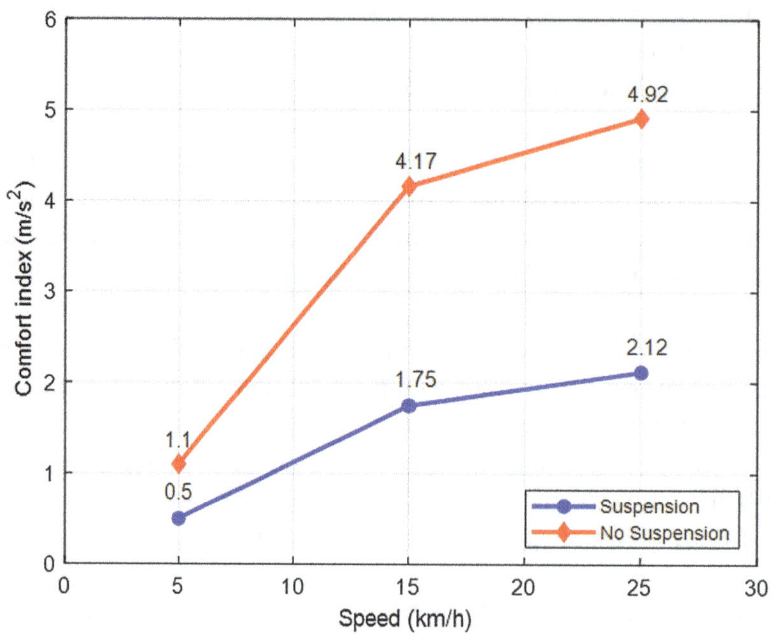

ing speed of the vehicle. Moreover, the numerical model considered just one static configuration of the hydropneumatic suspensions. As discussed in previous sections, this type of semi-active suspension can be controlled in closed loop in terms of damping. This means that the behaviour, especially at high speed could be improved more than the showed results. The control strategy for this type of suspension will be addressed in future works because of the relevant impact that such solution can have both on the vehicle performance and comfort level.

DISCUSSION AND CONCLUSION

Industrial working vehicles are subjected today to a continuous important evolution process which aims to improve safety, usability/productivity and comfort of the operator. Manufacturers usually adopt rigid suspension solutions for their vehicle where the only filter between the road surface and the vehicle chassis are the tyres. Thus, their stiffness and damping characteristics strongly affect the overall behaviour of the vehicle. Such solution can lead to unsafe operations during certain type of working conditions on agricultural soil and asphalt roads. The manufacturer must balance different mission profiles, thus the overall resulting vehicle performance will be a compromise. The introduction of suspensions in industrial vehicle can have a great impact. Moreover, the availability of semi-active solution which best trades between cost and performance allows to have controllable solutions in very small form factors. In this paper, performance of a telescopic handler equipped with a hydropneumatic suspension is shown. This solution is currently used on some agricultural tractor and mainly on the front axle, leaving the rear axle suspension to the tyres. Telescopic handlers have usually suspended rigid axles both on the front and on the rear one. This design solution is robust and reliable but does not allow to improve traction of the wheel nor vehicle dynamic behaviour. The proposed hydropneumatics suspension solution was designed and optimized in terms of hard points to improve as much as possible the toe and camber angles providing the best compromise between handling and a uniform wear of the tyre. The geometry of the double wishbone architecture was then included on a full vehicle model developed in a multibody environment. The solution with hydropneumatic suspensions and the model with rigid axles were compared in terms of dynamic behaviour and comfort level. The suspension provided better weight distribution on the four wheels during a step steer manoeuvre which translates in higher safety levels when approaching turns at high speed. The rigid axles model was at its limits during this manoeuvre having one of the four wheels detached from the ground for a certain duration of the simulation. The solution with hydropneumatics suspensions better managed the turn with all the wheels always attached to the ground. The two numerical models were then used to estimate the differences in terms of comfort levels inside the cab. Accelerations measured according to the test prescribed by the UNI EN 13059-2004 standard were combined to evaluate a comfort index starting from the RMS value on the three directions. The comparison between the comfort index evaluated on the models with and without the suspension systems reached improvements around 50%, averaging the results obtained at different speeds. Thus, the proposed suspension system introduces a real benefit also for the comfort of the driver especially in terms of vibrations to which it is subjected during daily activities. In case of long working sessions this translates in a real health benefit for the driver. Future works will also approach possible control strategies which this type of semi-active suspensions allow.

REFERENCES

Akinnuli, B.O., Dahunsi, O.A., Ayodeji S.P. & Bodunde, O.P. (2018) Whole-body vibration exposure on earthmoving equipment operators in construction industries, *Cogent Engineering*, *5*(1).

Bauer, W. (2011). *Hydropneumatic suspension system*. Springer. doi:10.1007/978-3-642-15147-7

Caffaro, F., Cremasco, M. M., Preti, C., & Cavallo, E. (2016). Ergonomic analysis of the effects of a telehandler's active suspended cab on whole body vibration level and operator comfort. *International Journal of Industrial Ergonomics*, *53*, 19–26. doi:10.1016/j.ergon.2015.10.009

Cai, Z. Z., Li, H. M., & An, G. (2012). Vibration Simulation of Tracked Vehicle Suspension and Damper Consumption Analysis. *Applied Mechanics and Materials*, *184-185*, 748–751. doi:10.4028/www.scientific.net/AMM.184-185.748

Cebon, D. (2000). *Handbook of Vehicle-Road Interaction*. Swets & Zeitlinger.

Citroen. (1977). Hydraulic course notes, Man. 008440. *Historic Archive Citroen*.

Deprez, K., Moshou, D., Anthonis, J., De Baerdemaeker, J., & Ramon, H. (2005). Improvement of vibrational comfort on agricultural vehicles by passive and semi-active cabin suspensions. *Computers and Electronics in Agriculture*, *49*(3), 431–440. doi:10.1016/j.compag.2005.08.009

Ellmann, S., & Löser, F. (2002). *Elektromechanisches Leveling System*. Forum Thyssen Krupp.

Eulenbach, D. (2003). *Nivomat – Stand und Entwicklungstrends hydropneumatischer Niveauregelsysteme. Tagung: Kraftfahrzeugstoßdämpfer und ihr Einfluss auf das Fahrwerk*. Komponenten und Systeme.

European Agency for Safety and Health at Work. (2008). *Workplace exposure to vibration in Europe: an expert review*. Author.

European Council Directive. (1989). *On the Introduction of measures to encourage improvements in the safety and health of workers*. Author.

European Parliament and European Council. (2002). *On the minimum health and safety requirements regarding the exposure of workers to the risks arising from physical agents (vibration)*. Author.

Ferraresi, C., Quaglia, G., & Sorli, M. (1997). Force control laws for semi-active vehicular suspensions. *European Journal of Mechanical and Environmental Engineering*, *42*(3), 145–151.

Genta, G. (2009). *Vibration Dynamics and Control*. Springer. doi:10.1007/978-0-387-79580-5

Genta, G., & Morello, L. (2008). *The automotive chassis*. Springer.

Gies, P. E. (1996). *Development of a hydropneumatic suspension system for an earthmoving machine*. SAE Technical Papers.

Giliomee, C. L., & Els, P. S. (1998). Semi-active hydropneumatic spring and damper system. *Journal of Terramechanics*, *35*(2), 109–117. doi:10.1016/S0022-4898(98)00016-0

Gorabal, S. V., Surbet, S. N., & Appukuttan, K. K. (2011). Design and evaluation of hydro-pneumatic friction damper suspension system. *International Journal of Manufacturing, Materials, and Mechanical Engineering*, *1*(4), 65–74. doi:10.4018/ijmmme.2011100105

Grott, M. (2010). *Design of Suspension Systems and Control Algorithms for Heavy Duty Vehicles* (PhD thesis). University of Trento.

Guignard, J. C. (1971). Human sensitivity to vibration. *Journal of Sound and Vibration*, *15*(1), 11–16. doi:10.1016/0022-460X(71)90354-3

Hansson, P. A. (1995). Optimization of agricultural tractor cab suspension using the evolution method. *Computers and Electronics in Agriculture*, *12*(1), 35–49. doi:10.1016/0168-1699(94)00036-P

Hilton, D. J., & Moran, P. (1975). Experiments in reducing tractor ride vibration with a cab suspension. *Journal of Agricultural Engineering Research*, *20*(4), 433. doi:10.1016/0021-8634(75)90079-7

Kittusamy, N. K., & Buchholz, B. (2004). Whole-body vibration and postural stress among operators of construction equipment: A literature review. *Journal of Safety Research*, *35*(3), 255–261. doi:10.1016/j.jsr.2004.03.014 PMID:15288559

Kuijt-Evers, L. F. M., Krause, F., & Vink, P. (2003). Aspects to improve cabin comfort of wheel loaders and excavators according to operators. *Applied Ergonomics*, *34*(3), 265–271. doi:10.1016/S0003-6870(03)00032-2 PMID:12737927

Lines, J. A., Whyte, A. T., & Stayner, A. M. (1989). *Agricultural vehicle suspensions - suspensions for tractor cabs*. Report Presented to the third International Symposium of the International section of the ISSA for research on prevention of Occupational Risks.

Matthews, J., & Talamo, J. D. C. (1965). Ride comfort for tractor operators. *Journal of Agricultural Engineering Research*, *10*(2), 93–108. doi:10.1016/0021-8634(65)90057-0

Mocera, F., & Nicolini, A. (2017). Multibody simulation of a small size farming tracked vehicle. *Procedia Structural Integrity*, *8*, 118–125. doi:10.1016/j.prostr.2017.12.013

Murrenhoff, H., & Wallentowitz, H. (1998). *Fluidtechnik für mobile Anwendungen*. Reihe Fluidtechnik.

Nicolini, A., Mocera, F., & Somà, A. (2018). Multibody simulation of a tracked vehicle with deformable ground contact model. *Proceedings of the Institution of Mechanical Engineers, Part K: Journal of Multi-Body Dynamics*, *233*(1), 152–162.

Peachey, R. O., Lines, J. A., & Stayner, R. M. (1989). *Agricultural vehicle suspensions - tractor front axle suspension.* Report for the proceedings of the 3rd International Symposium of the International Section of the ISSA for Research on Prevention of Occupational Risks.

Rasmussen, A.E. & Cortese, A.D. (1968). *Dynamic spring rate performance of rolling tyres.* S.A.E. Technical Paper, 680408.

Reimpell, J., Stoll, H., & Betzler, J. W. (2001). *The Automotive Chassis: Engineering Principles.* Butterworth-Heinemann.

Savaresi, S. M., Silani, E., & Bittanti, S. (2005). Acceleration-driven-damper (ADD): An Optimal Control Algorithm For Comfort-Oriented Semiactive Suspensions. *Journal of Dynamic Systems, Measurement, and Control*, *127*(2), 218–229. doi:10.1115/1.1898241

Siefkes, T. (1989). *Dynamische Kennwertevon AS-Reitenim PrOfstands-versuchund im Freiland (Characteristics of agricultural tyres -laboratory and field test results).* VDI/MEG Kolloquium Landtechnik.

Somà, A., Bosso, N., Macagno, L., & Viglietti, E. (2006). *Analisi sperimentale e numerica del comfort di un sollevatore telescopico.* XXXV AIAS Convegno Nazionale.

UNI EN 13059-2004, (2004). Safety in industrial trucks: test methods and experimental measurements.

Williams, R. A. (1997). Automotive active suspension Part 1: Basic principles. *Proceedings of the Institution of Mechanical Engineers. Part D, Journal of Automobile Engineering*, *211*(6), 415–426. doi:10.1243/0954407971526551

Zampieri, N., Bosso, N., & Gugliotta, A. (2013). Design of a Dynamic Testing Machine for Railway Seats. In *Proceedings of the Fourteenth International Conference on Civil, Structural and Environmental Engineering Computing*. Civil-Comp Press. 10.4203/ccp.102.36

Zampieri, N., Bosso, N., & Gugliotta, A. (2013). A Study of the Comfort Level of Railway Vehicle for Reduced Mobility Passengers. In *Proceedings of the Fourteenth International Conference on Civil, Structural and Environmental Engineering Computing*. Civil-Comp Press. 10.4203/ccp.102.32

Zehsaz, M., Sadeghi, M. H., Ettefagh, M. M., & Shams, F. (2011). Tractor cabin's passive suspension parameters optimization via experimental and numerical methods. *Journal of Terramechanics*, *48*(6), 439–450. doi:10.1016/j.jterra.2011.09.005

Zhang, J., Wang, Q., & Jin, Q. (2018). Research on method of vibration analysis of rubber tracked vehicle based on dynamic model. *Journal of Measurements in Engineering*, *6*(1), 25–34. doi:10.21595/jme.2018.19601

Chapter 13
Stress and Strain Distribution in the Upsetting Process: A Numerical Simulation

Japheth Obiko
Jomo Kenyatta University of Agriculture and Technology, Kenya

Fredrick Madaraka Mwema
https://orcid.org/0000-0001-6116-5587
Dedan Kimathi University of Technology, Kenya

ABSTRACT

Numerical simulation of metal flow behaviour was studied using DeformTM3D software. The simulation process was done on X20 steel taken from the software database at 1073-1273K temperature, 10mm/s die speed, and 67% height reduction. From the simulation results, forging load, damage, and stress/strain distributions were obtained. The results show that the forging load increased with a decrease in temperature or decreased with an increase in temperature. The maximum damage values increased as the temperature increased. The obtained maximum damage values were 0.42 (1073K), 0.43 (1173K), and 0.45 (1273K). The damage distribution was inhomogeneous in the deformed cylinder. The stress/strain distributions were inhomogeneous in the deformed cylinder. The location of the maximum strain was at the centre of the deformed cylinder while the maximum stress occurred at the die-cylinder contact surfaces. The study showed that flow stress behaviour can be predicted using finite element method. This shows the feasibility of applying the finite element analysis to analyse the forging process.

INTRODUCTION

The demand for large forged structural components in power plant and aerospace industrial applications has increased recently (Lee, Lee, Tyne, Joo, & Moon, 2011). Open-die forging process remains the most preferred metal forming process for high quality forged components (Lu, Li, Wang, Liu, & Fu, 2013; Overstam & Jarl, 2004). Industrial forging process is complex and time-consuming (Lin, Chen, &

DOI: 10.4018/978-1-7998-4939-1.ch013

Zhong, 2008). Since metals and alloys are subjected to large plastic deformations during forging, there occurs uneven metal flow patterns on the samples (George, 1988). As such, the metal flow analysis of such processes is complex (George, 1988). The change in forging temperature and the presence of friction coefficient affect the deformation behaviour (Evans & Scharning, 2001; Li, Onodera, Matsumoto, & Chiba, 2009). Since stress and strain distribution depends on the forging conditions, the deformation process becomes inhomogeneous (Lin et al., 2008). Several techniques such as physical simulation process have been applied to simulate and predict industrial forging process (Fan, Peng, Yang, Zhou, & Yan, 2017; Mehtonen, Karjalainen, & Porter, 2014; Yin, Hua, Mao, & Han, 2013). However, the optimisation of these processes remains a challenge due to other factors such as interfacial friction and heat generation during plastic deformation (Evans & Scharning, 2001). Hence, inhomogeneous metal flow behaviour is expected, hence affect the final product quality (Kajtoch, 2007; Li et al., 2009). Therefore, the production of high-quality products requires an understanding of the effects of forging parameters and microstructure evolution during the forging process.

To improve productivity and reduce production costs, finite element tools have been widely used in the industrial metal forming process (Mohapatra & Maity, 2016; Yang & Lin, 2016). Finite element softwares that are used for the simulation of most metalworking processes include, Deform™ 3D, Marc, ANSYS and Abaqus (Ketabchi, Mohammadi, & Izadi, 2012). In the forging process, Deform™ 3D has been extensively used in the industry for analysing two- and three-dimensional metal flow behaviours (Ketabchi et al., 2012). In this study, Deform™ 3D finite element analysis software has been used to study metal flow behaviour during the forging process. X20 steel cylinder having the same geometry size as those prepared for laboratory physical simulation process was used (Zhu, 2018). The cylinder was simulated under different forging temperature and at constant die speed. From the simulation results, forging load, damage distribution, and stress and strain distribution have been reported.

Finite Element Theory

Metal forming processes induces large plastic deformations on the material during the forging process (George, 1988). This leads to a cross-sectional area and metallurgical changes in the deformed material (Han, Sun, Zhang, & Chen, 2017). The deformed material experiences triaxiality state of stress resulting in a complex metal flow pattern. Hence, analytical analysis is complex. To reduce the computational process, FEM simulation analysis has been widely used to analyse most metal forming processes (Hsia & Chou, 2015). In the simulation process, the workpiece was taken to be of visco-plastic material. According to Hartley & Pillinger, (2006) deformation behaviour of rigid visco-plastic material depends on the deformation conditions. During the simulation process the following assumptions are made according to Hsia & Chou, (2015):

1. Von-mises yield criterion

The adopted deformation behaviour results after the simulation are the temperature, effective stress and effective strain. The effective stress is considered as the function of effective strain, effective strain rate and temperature and given as:

$$\bar{\sigma} = \left(\bar{\varepsilon}.\dot{\varepsilon}.T\right) \tag{1}$$

The yield criterion model can be used to determine the effective stress of during deformation of an isotropic material. The model considers all the three axial principal stress σ_1, σ_2 and σ_3 in obtaining the effective stress $\bar{\sigma}$ as:

$$\bar{\sigma} = \frac{1}{\sqrt{2}}\left[\left(\sigma_1 - \sigma_2\right)^2 + \left(\sigma_2 - \sigma_3\right)^2 + \left(\sigma_3 - \sigma_1\right)^2\right]^{\frac{1}{2}} \quad (2)$$

2. Deformation temperature

The deformation (forging) temperature was assumed constant. The change in the deformation resulting from the plastic deformation stage during the forging process has a minimal effect on the local temperature hence ignored.

3. Friction effects

During the forging of the cylindrical specimen, friction develops at the workpiece-die interface (Kajtoch, 2007). Li et al., (2009) showed that the interfacial friction varies continuously with further deformation during the forging process. A similar observation has been reported by Mwema *et al.,* (2019). The presence of friction affects the flow stress hence hinder metal flow pattern (Kajtoch, 2007). Therefore, flow stress data obtained from the forging process should be corrected for friction. Analytical equations for flow stress correction have been developed by Evans & Scharning (2001) and widely used in the literature (Luan, Sun, Li, & Zhang, 2014; Han, Sun, Zhang, & Chen, 2017; Mostafaei & Kazeminezhad, 2012). In the present study, the interfacial friction during the simulation process was taken to be constant.

4. Fracture criterion

Friction at the workpiece and dies interface increases as the deformation process during uniaxial compression test of a cylindrical specimen (Li et al., 2009). The coefficient of friction increases from zero friction to a 'sticking friction'. The metal flow behaviour changes from homogeneous (zero friction) into inhomogeneous (sticking friction) deformation (Jeysingh, Rao, & Reddy, 2008). The lateral surface side of the specimen bulges with an increase of interfacial friction. Further deformation process results in severe barrelling effect at the lateral side of the specimen (Li et al., 2009). The central region of the barrelled side experiences maximum tensile strain. When the tensile strain reaches a certain critical value fracture occurs (Fedoriková et al., 2016). Cockcroft and Latham proposed a simple ductile fracture constitutive model (CL) for fracture analysis of uniaxial compression tests as (Fedoriková et al., 2016):

$$\int_0^{\bar{\varepsilon}} \sigma_1 d\bar{\varepsilon} = C_1 \quad (3)$$

Cockcroft-Latham fracture model (Equation 3) was further modified by normalising the maximum principal tensile stress. The normalised CL criterion is denoted as nCL (Fedoriková et al., 2016) and is given as:

$$\int_0^{\bar{\varepsilon}} \frac{\sigma_1}{\bar{\sigma}} d\bar{\varepsilon} = C_2 \quad (4)$$

where, $\bar{\varepsilon}$ is the total strain at failure, σ_1 is the maximum stress, $\bar{\sigma}$ is the maximum effective stress, C_1 and C_2 are material constants. The material constant (C_1 or C_2) indicates the workability of the material (Pang, Shang, & Zhang, 2018).

Governing Equations

A rigid-viscoplastic finite element method is widely used for the simulation of the metalworking process, (Asadi, Mahdavinejad, & Tutunchilar, 2011). The flow constitutive equations for deformation analysis are based on the variational principle given in Equation 5 (Lin et al., 2008; Asadi et al., 2011).

$$\int_v \bar{\sigma}\delta\dot{\bar{\varepsilon}} dV + K\int_v \dot{\varepsilon}_v \delta\dot{\varepsilon}_v dV - \int_{Sf} f_i \delta v_i dS = 0 \quad (5)$$

From Equation 5, K is a penalty constant which shows volume change, V and S are volume and surface of the specimen, $\dot{\bar{\varepsilon}}$ is the effective strain rate, effective flow stress and f_i is the applied external force v_i is the velocity.

The heat transfer between anvil or workpiece can be solved using the energy balance Equation 6 (Kukuryk, 2012):

$$\nabla^T.\left(k\nabla T\right) + \dot{Q} - \rho c_p \dot{T} = 0 \quad (6)$$

where T is the deformation temperature in Kelvin, k is the thermal conductivity, $\dot{Q}$ is the rate of internal heat generated during plastic deformation, ρ is the specific density and c_p is the specific heat. In Equation 2, the first and last terms of the equation represent the rate of heat transfer and generation respectively. The rate at which the internal heat is generated during plastic deformation can be obtained by solving Equation 7:

$$Q\left(T\right) = \eta\bar{\sigma}\dot{\bar{\varepsilon}} \quad (7)$$

where, η parameter represent the heat generated efficiency during the forging process. This efficiency factor (assumed as 0.9) is the fraction of mechanical energy that is converted to heat energy (Kukuryk, 2018). Substituting Equation 7 into Equation 6, using weighted residual method, Kukuryk, (2018) has shown that the energy equation (Equation 6) can be written as:

$$\int_v kT_{,i}\partial T_{,i}dV + \int_v \rho c_p \frac{\partial T}{\partial t}\partial T dV - \int_v \eta\bar{\sigma}\bar{\varepsilon}\partial T dV - \int_{sq} q_n \partial T dS = 0 \quad (8)$$

The factor q_n in Equation 8 represents the heat flux perpendicular to the workpiece boundary surface. By solving Equation 8, forging temperature variation will be obtained.

During the deformation process, friction between the workpiece and the anvil occur due to the presence on the frictional shear stress (Kukuryk, 2012) and expressed as:

$$\tau = -mk\left[\frac{2}{\pi}\arctan\left(\frac{\Delta V_s}{u_0}\right)\right]t \quad (9)$$

where m is the coefficient of friction, k is the yield stress, ΔVs is the vector of the workpiece slipping velocity relative to the die and t is the unit vector acting in the direction of workpiece slipping velocity and u_0 is a constant taken to be 10^{-5} (Kukuryk, 2018).

Simulation Modelling

The upsetting simulation process was conducted using FEM Deform 3D software. The workpiece and the die geometries were modelled using the in-built primitive geometry tool in the FEM software. The conditions under which the simulations were undertaken are shown in Table 1.

Table 1. Forging simulation parameters

Simulation Parameter	Value
Deformation temperature (K)	1073-1273
Die temperature (K)	373
Die speed (mm/sec)	50
Convectional coefficient (N/s. mm. K)	0.02
The coefficient of heat transfer (N/s. mm. K)	5.0
Height reduction (%)	67
Coefficient of interfacial friction	0.3

RESULTS AND DISCUSSION

Force-Stroke and Damage Distribution

Figure 1(a) shows forging load-stroke curves obtained after forging simulation under different forging conditions. As shown, the forging load increased with a decrease in the forging temperature and vice versa. At the initial stage of forging, the forging load increased rapidly up to a certain value for all the forging temperatures investigated. This is attributed to higher generation and multiplication of dislocation density during deformation. Hence, the deformation mechanism was dominated by strain hardening. The flow stress has been reported to increase rapidly from the start of deformation until saturation stress occurs (Shokry, Gowid, & Kharmanda, 2019; Zhu, 2018). Therefore, the forging simulation results obtained in this study can be validated with experimental results from the literature. At a stroke of approximately 0.3 mm, the forging load increased gradually with an increase in stroke (up to 5 mm) due to the occurrence of softening mechanism. The slow increase of forging load with an increase in stroke can be attributed to the balance work hardening and dynamic softening (Han et al., 2017b). With an increase in deformation (> 5 mm stroke), the forging load increased steady until the stroke of 7 mm, Figure 1(a). The increase in forging load at this stage (7 mm stroke) was attributed to an increase in the interfacial friction coefficient at the die-workpiece interface (Evans & Scharning, 2001). Friction coefficient has been reported to increase up to a value of 1 at higher strains due to static friction (Li et al., 2009). Under this condition, higher forging loads are expected. It can be concluded that the forging temperature and coefficient of friction influences the metal flow behaviour. Therefore, forging is thermally activated process. Higher forging temperature accelerates dislocation motion and grain boundary movement leading to the occurrence of dynamic softening hence, lower forging loads (Han et al., 2017b).

Figure 1. Plots of: a) Forging load versus stroke and b) Damage variation curves for the simulation process

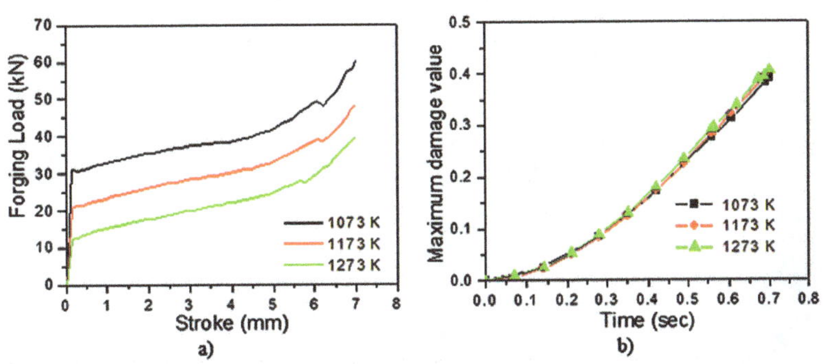

From the cumulative damage theory, the damage increases with an increase in strain during the deformation process (Quan, Luo, Mao, Liang, & Wu, 2014). Quan et al. (2014) reported that damage value increases between zero and critical value (D_c) until a fracture occurs and after rupture damage growth rate approaches zero with an increase in the strain. From the simulation process, using the normalised Cockcroft-Latham failure criterion the maximum damage versus time (sec) curves during the forging process is given in Figure 1b. The results showed that the maximum damage value increased nonlinearly with an increase in the compression time from 0 to 0.4 seconds. Thereafter, the maximum damage curved exhibited a linear curve until the final stage of deformation. At constant die speed, it was observed that the maximum damage value increased with an increase in the forging temperature and deformation. Higher damage value was observed at higher forging temperature as shown in Figure 1b. The maximum cumulative damage value has been reported to increase with an increase in the deformation temperature (Quan et al., 2014). Similar findings were observed in this study. The maximum damage occurs at the middle and lateral side of the upsetting drum while the centre region experiences the lowest damage (Pang et al., 2018). The middle lateral region of the deformed specimen also experiences higher principal tensile stress. This region is susceptible to cracking (Yang et al., 2017).

Figure 2 shows that damage distribution was inhomogeneous across the deformed samples. The maximum damage values increased with an increase in the forging temperature. These results are shown in Figure 2. However, the standard deviation was the same for the forging temperature of 1073 K (Figure 2a) and 1173 K (Figure 2b). The standard deviation for the forging temperature of 1273 K was slightly higher (0.13) than 0.12 for (1073 K and 1173 K). The inhomogeneous deformation process was further confirmed from the standard deviation values obtained. The inhomogeneous deformation process can be attributed to the variation in forging temperature and interfacial friction as reported in the literature (Obiko, Mwema, & Bodunrin, 2019; Rasti, Najafizadeh, & Meratian, 2011).

Figure 2. Effect of temperature on the damage distribution (a) 1073 K (b) 1173 K (c) 1273 K.

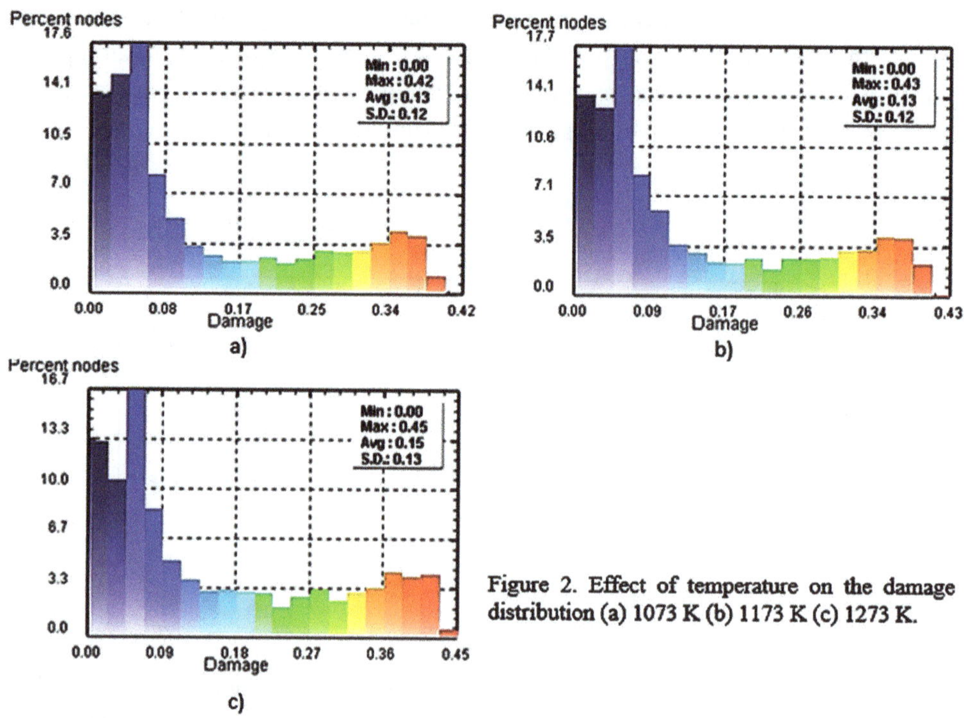

Figure 2. Effect of temperature on the damage distribution (a) 1073 K (b) 1173 K (c) 1273 K.

Effective Strain Distribution

The forging simulation time of a cylinder workpiece deformed to a height reduction of 67% was 0.74 seconds. The strain distribution curves are given in Figure 3. The deformation process was inhomogeneous across the deformed cylinder as shown in Figure 3a. From the deformed cylinder (Figure 3a), three deformation zones were observed. These zones have been categorised by Rasti et al., (2011) as the dead zone (point 1) occurs at the top and bottom die contact surfaces. This region experiences lower degree of deformation. The other regions are: the intense shear zone (point 2) where severe plastic deformation occurs and the moderate deformation zone (point 3) which is the lateral outer surface of the workpiece. The plots of effective strain distribution represented in Figure 3b-d were taken from three points (1, 2 and 3) as shown in Figure 3a. Point 1 had the least effective strain value compared to point 2 and 3. This was attributed to frictional resistance acting towards the centre of the cylinder to the metal flow during the forging process (Kajtoch, 2007). Point 2 exhibited the highest effective strain leading to severe plastic deformation during the forging process. In this region (point 2) occurrence of dynamic recrystallisation was expected, leading to grain refinement (Naderi & Roostaei, 2010; Rasti et al., 2011). Point 3 experienced the lower effective strain compared to point 2, however higher than point 1as shown in Figure 3. This region (point 3) experiences the least deformation during the forging process. However, higher principal tensile stress occurs in this region during deformation which may lead to cracking (Yang et al., 2017). The contour maps of the effective strain distribution are shown in Figure 4 for all forging temperatures simulated. Different regions of the deformed cylinder exhibited different values of the effective strain as delineated in Figure 4. The variation in the effective strain values shown in the contour maps further confirms an inhomogeneous deformation process across the cylinder during the forging process.

Figure 3. (a) Deformed sample showing data points and Effective strain distribution in the deformed cylinder at die speed of 10 mm/sec and temperature of (b) 1073 K (c) 1173 K (d) 1273 K

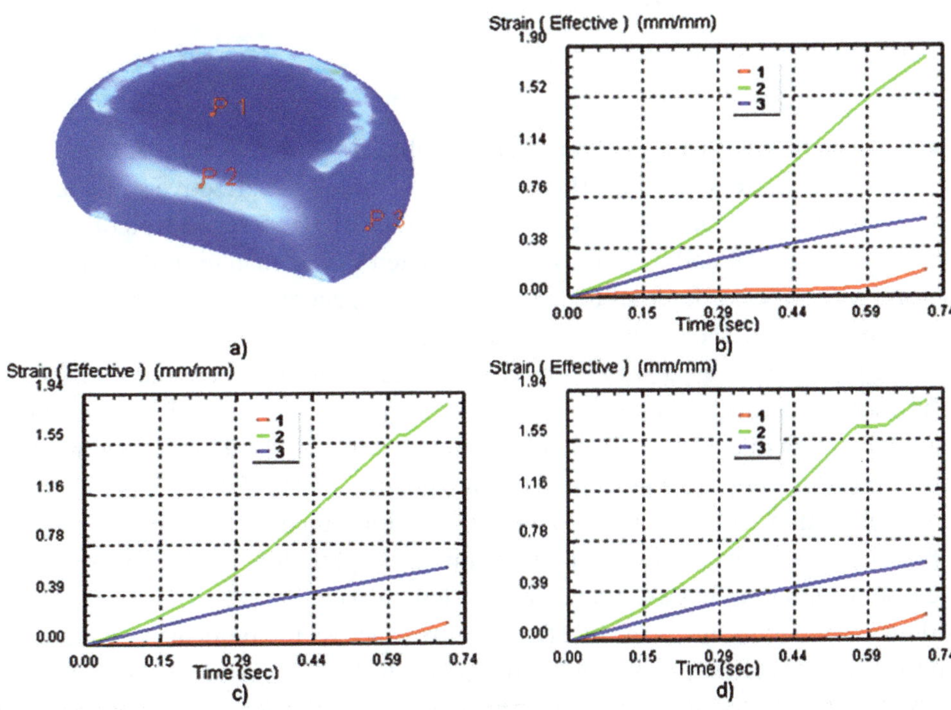

Effective Stress Distribution

The effective stress distribution curves for the whole forging simulation process are shown in Figure 5. It was observed that the effective stress distribution was inhomogeneous. Three points (1, 2 and 3) shown in Figure 3(a) were taken to track the effective stress during the whole forging process. The effective stress values decreased with an increase in the forging temperature as shown in Figure 5. The maximum effective stress occurred at the die-workpiece interface surface. This region had the lowest effective strain as shown in Figure 3. The maximum effective stress has been reported to occur at regions with lower

Figure 4. Effective strain distribution contour map of the deformed cylinder at die speed of 10 mm/sec and temperature of (a) 1073 K (b) 1173 K (c) 1273 K

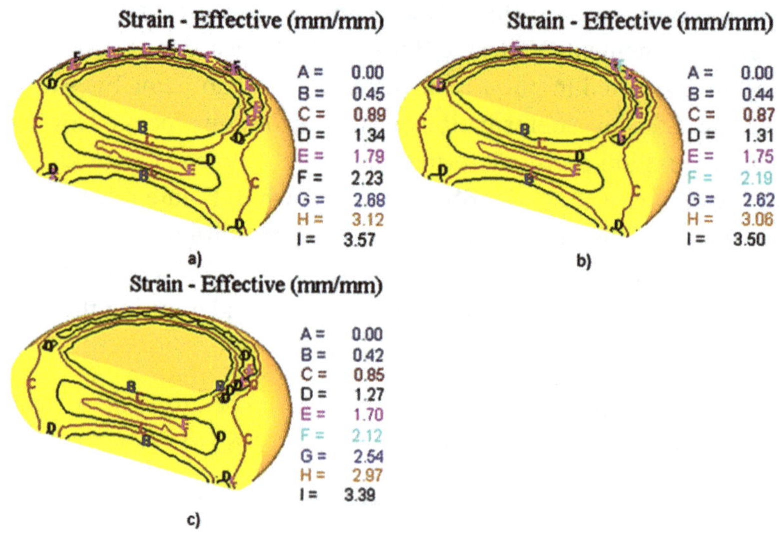

effective strain (Kingkam, Li, Zhang, & Zhao, 2017; Equbal, Talukdar, Kumar, & Ohdar, 2014). At the beginning of the forging process, in all the three points (1, 2 and 3) the effective stress increased rapidly up to the deformation time of approximately 0.15 seconds. After 0.15 seconds, a drop in the effective stress was observed at point 2 and 3. This can be attributed to the change in the forging temperature during the deformation process. Internal heat may have been generated during the plastic deformation stage leading to a reduction in the forging load hence lower effective stress (Evans & Scharning, 2001). Effective stress at point 1 increased with an increase in the deformation degree. For the three points, point 1 had the maximum effective stress in all the simulation conditions. This can be attributed to the lower forging temperature due to the chilling effect (Rasti et al., 2011). The maximum effective stress at point 1 was 786 MPa (1073 K), 684 MPa (1173 K) and 565 MPa (1273K). The other two points (2 and 3) had relatively similar effective stress values at the initial stage of deformation (deformation time < 0.15 sec). For higher deformation time, Point 3 exhibited slightly higher effective stress than point 2. This is so because point 3 does not experience metal flow restriction.

Figure 6 shows the contour maps of the effective stress across the deformed cylinder. The contour maps showed that the effective stress was inhomogeneous during the entire forging simulation process. Hence, the effective stress distribution was inhomogeneous across the deformed cylinder. The simulation analysis showed that higher effective stress occurs at point 1 and 3. In particular, point 3 experienced higher principal tensile stress which may cause forging defect such as cracking. Therefore, the optimisation of the forging process is important to study the occurrence and minimise the principal tensile stresses.

Flow Stress-Strain Curve

The effective low stress-strain curves obtained from the simulation process at constant die speed of 100 mm/sec and different forging temperature range of 1073-1273 K are shown in Figure 7. The plots of the effective flow stress-strain curves shown in Figure 7 can be described as follows:

Figure 5. Plots of Effective stress distribution after the simulation process at die speed of 10 mm/sec and temperature of (a) 1073 K (b) 1173 K (c) 1273 K

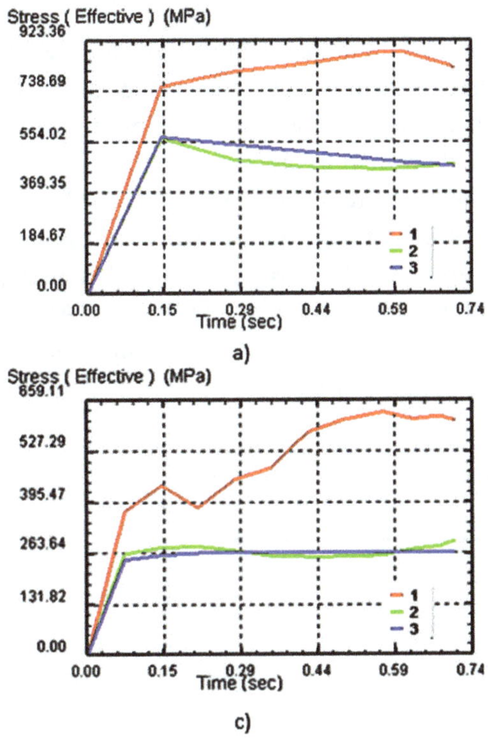

Figure 6. Effective stress distribution contour map of the deformed cylinder at die speed of 10 mm/sec and temperature of (a) 1073 K (b) 1173 K (c) 1273 K

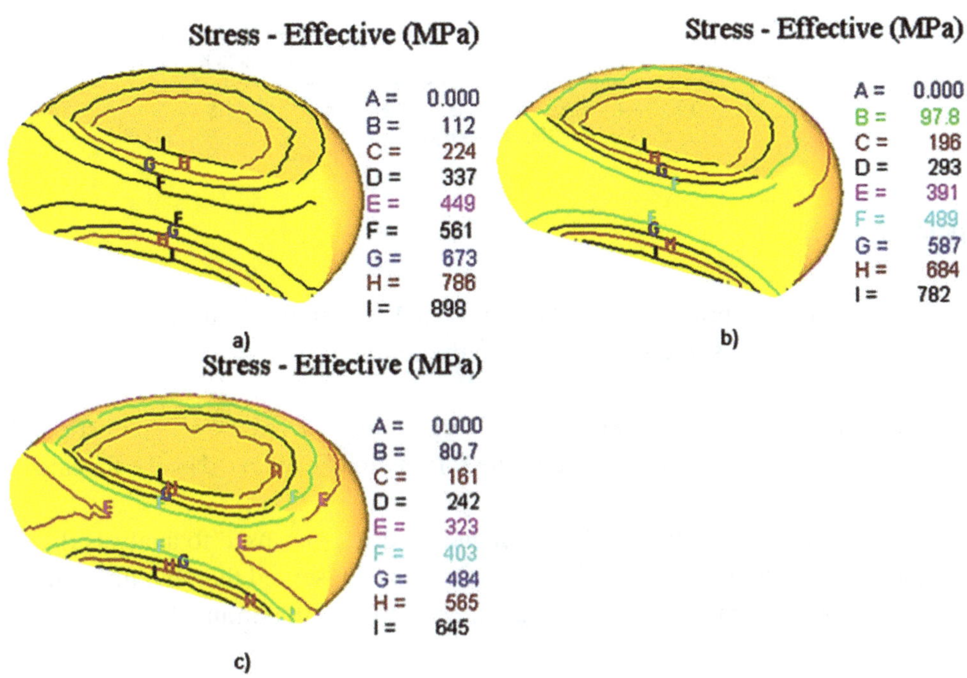

1. The effective flow stress increased with a decrease in the forging temperature at constant die speed. This indicates the dependence of the effective stress on the forging conditions (the temperature and die speed).
2. At the beginning of the forging process, the effective flow stress increased rapidly up to a peak effective stress. The higher effective stress was due to the higher resistance to the deformation process. The high generation of dislocation density leads to work hardening hence higher flow stress. Therefore, work hardening dominated the deformation process.
3. After a peak effective stress, a decrease in the effective stress was observed towards a steady-state region with an increase in the effective strain. When steady-state effective stress was attained, constant effectives flow stress was maintained up to an effective strain of 1.7. This represents a characteristic flow behaviour of a material exhibiting dynamic recrystallisation. At higher effective strain (>1.7) a slight increase in the effective stress was noticed. This can be attributed to an increase in the coefficient of friction at a higher effective strain.

Figure 7. Effective stress-strain curves at constant die speed of 10 mm/sec

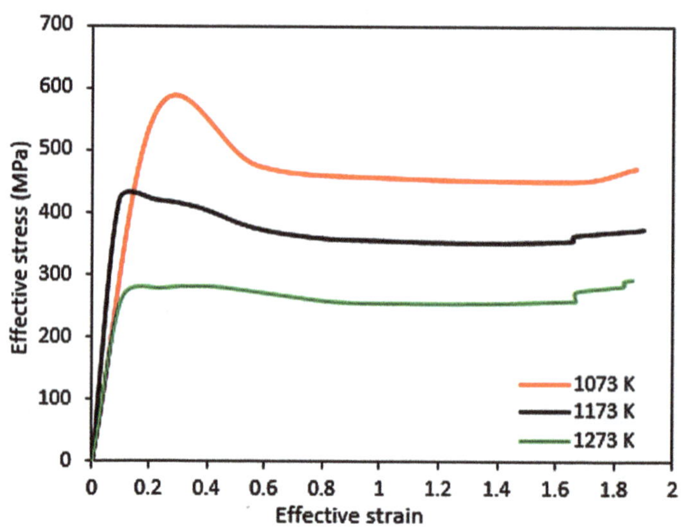

CONCLUSION

The finite element simulation process using Deform^{TM} 3D software was used to study the industrial forging process. From the results, the study observed the following:

1. The forging load-stroke curves showed that the forging load increases with a decrease in the forging temperature under a constant die speed. The forging load decreased as the forging temperature increases at a given die speed.
2. The normalised Cockcroft-Latham (nCL) damage criterion was used to analysis the damage distribution pattern during the metalworking process. It was observed that maximum damage values increased as the forging simulation temperature increased. The maximum damage values were: 0.42

(1073 K), 0.43 (1173 K) and 0.45 (1273 K). The maximum damage occurred at the outer surface of the deformed cylinder. Damage distribution across the deformed cylinder was inhomogeneous.

3. The forging process was inhomogeneous. The maximum effective strain was located at middle height of deformed cylinder. The maximum effective stress after deformation, occurred at the die-specimen interface surface. This region had the lowest effective strain.
4. The FEM simulation results have shown that the industrial production process can be optimised by using FEM Deform 3D software program. This indicates the feasibility of applying finite element analysis for the analysis of the industrial metalworking process.

REFERENCES

Asadi, P., Mahdavinejad, R. A., & Tutunchilar, S. (2011). Simulation and experimental investigation of FSP of AZ91 magnesium alloy. *Materials Science and Engineering A, 528*(21), 6469–6477. doi:10.1016/j.msea.2011.05.035

Equbal, M. I., Talukdar, P., Kumar, V., & Ohdar, R. K. (2014). Deformation Behavior of Micro-alloyed Steel by Using Thermo Mechanical Simulator and Finite Element Method. *Procedia Materials Science, 6*(Icmpc), 674–681. doi:10.1016/j.mspro.2014.07.083

Evans, R. W., & Scharning, P. J. (2001). Axisymmetric compression test and hot working properties of alloys. *Materials Science and Technology, 17*(8), 995–1004. doi:10.1179/026708301101510843

Fan, C., Peng, Y., Yang, H., Zhou, W., & Yan, H. (2017). Hot deformation behavior of Al–9.0Mg–0.5Mn–0.1Ti alloy based on processing maps. [English Edition]. *Transactions of Nonferrous Metals Society of China, 27*(2), 289–297. doi:10.1016/S1003-6326(17)60033-5

Fedoriková, A., Kvačkaj, T., Kočiško, R., Bidulský, R., Petroušek, P., Bidulská, J., & Domovcová, L. (2016). Hot compression test of 9 Cr-1 mo steel – Numerical simulation. *Acta Metallurgica Slovaca, 22*(2), 102–110. doi:10.12776/ams.v22i2.616

George, D. (1988). *Mechanical Metallurgy. McGraw-Hill Book Company.*

Han, Y., Sun, Y., Zhang, W., & Chen, H. (2017a). Hot deformation and processing window optimization of a 70MnSiCrMo carbide-free bainitic steel. *Materials (Basel), 10*(3), 318. doi:10.3390/ma10030318 PMID:28772678

Han, Y., Sun, Y., Zhang, W., & Chen, H. (2017b). Hot deformation and processing window optimization of a 70MnSiCrMo carbide-free bainitic steel. *Materials (Basel), 10*(3), 318. doi:10.3390/ma10030318 PMID:28772678

Hartley, P., & Pillinger, I. (2006). Numerical simulation of the forging process. *Computer Methods in Applied Mechanics and Engineering, 195*(48–49), 6676–6690. doi:10.1016/j.cma.2005.03.013

Hsia, S. Y., & Chou, Y. T. (2015). Fabrication improvement of cold forging hexagonal nuts by computational analysis and experiment verification. *Mathematical Problems in Engineering, 2015*, 1–11. doi:10.1155/2015/835038

Jeysingh, J., Rao, B., & Reddy, A. (2008). Development of a ductile fracture criterion in cold forming. *Materials Science Research Journal, 2*(3), 4.

Kajtoch, J. (2007). Strain in the upsetting process. *Metallurgy and Foundry Engineering, 33*(1), 1–11. doi:10.7494/mafe.2007.33.1.51

Ketabchi, M., Mohammadi, H., & Izadi, M. (2012). Finite-Element Simulation and Experimental Investigation of Isothermal Backward Extrusion of 7075 Al Alloy. *Arabian Journal for Science and Engineering, 37*(8), 2287–2296. doi:10.100713369-012-0320-4

Kingkam, W., Li, N., Zhang, H. X., & Zhao, C. Z. (2017). Hot Deformation Behavior of High Strength Low Alloy Steel by Thermo Mechanical Simulator and Finite Element Method. *IOP Conference Series. Materials Science and Engineering, 205*(1), 012001. doi:10.1088/1757-899X/205/1/012001

Kukuryk, M. (2012). Analysis of deformation and damage evolution in hot elongation forging. *Archives of Metallurgy and Materials, 57*(2), 417–424. doi:10.2478/v10172-012-0041-4

Kukuryk, M. (2018). Numerical analysis of strains and stresses in the hot cogging process. *Journal of Applied Mathematics and Computational Mechanics, 17*(3), 45–52. doi:10.17512/jamcm.2018.3.04

Lee, Y. S., Lee, S. U., Van Tyne, C. J., Joo, B. D., & Moon, Y. H. (2011). Internal void closure during the forging of large cast ingots using a simulation approach. *Journal of Materials Processing Technology, 211*(6), 1136–1145. doi:10.1016/j.jmatprotec.2011.01.017

Li, Y. P., Onodera, E., Matsumoto, H., & Chiba, A. (2009). Correcting the stress-strain curve in hot compression process to high strain level. *Metallurgical and Materials Transactions. A, Physical Metallurgy and Materials Science, 40*(4), 982–990. doi:10.100711661-009-9783-7

Lin, Y. C., Chen, M. S., & Zhong, J. (2008). Numerical simulation for stress/strain distribution and microstructural evolution in 42CrMo steel during hot upsetting process. *Computational Materials Science, 43*(4), 1117–1122. doi:10.1016/j.commatsci.2008.03.010

Lu, S. Q., Li, X., Wang, K. L., Liu, S. B., & Fu, M. W. (2013). A method for prediction of unstable deformation in hot forging process by simulation. *Transactions of Nonferrous Metals Society of China, 23*(12), 3739–3747. doi:10.1016/S1003-6326(13)62924-6

Luan, J., Sun, C., Li, X., & Zhang, Q. (2014). Constitutive model for AZ31 magnesium alloy based on isothermal compression test. *Materials Science and Technology, 30*(2), 211–219. doi:10.1179/1743284713Y.0000000341

Mehtonen, S. V., Karjalainen, L. P., & Porter, D. (2014). Modeling of the high temperature flow behavior of stabilized 12–27wt% Cr ferritic stainless steels. *Materials Science and Engineering A, 607*, 44–52. doi:10.1016/j.msea.2014.03.124

Mohapatra, S. K., & Maity, K. P. (2016). Parametric Optimization of Simulated Extrusion of Square to Square Section Through Linear Converging Die. *IOP Conference Series. Materials Science and Engineering, 115*(1), 012031. doi:10.1088/1757-899X/115/1/012031

Mostafaei, M. A., & Kazeminezhad, M. (2012). The prediction of hot flow behavior of Al-6%Mg alloy. *Materials Science and Engineering A, 535*, 216–221. doi:10.1016/j.msea.2011.12.067

Mwema, F. M., Obiko, J. O., Akinlabi, E. T., Akinlabi, S. A., & Fatoba, O. S. (2019). Effect of punch force on the upsetting deformation process using three-dimensional finite element analysis. *Journal of Physics: Conference Series*, *1378*(December), 032094. doi:10.1088/1742-6596/1378/3/032094

Naderi, M., & Roostaei, A. A. (2010). Deformation homogeneity in accumulative back extrusion processing of AZ31 magnesium alloy. *Journal of Alloys and Compounds*, *507*(1), 207–214. doi:10.1016/j.jallcom.2010.07.157

Obiko, J. O., Mwema, F. M., & Bodunrin, M. O. (2019). Finite element simulation of X20CrMoV121 steel billet forging process using the Deform 3D software. *SN Applied Sciences*, *1*(9), 1044. doi:10.100742452-019-1087-y

Overstam, H., & Jarl, M. (2004). FEM-simulation of drawing out in open die forging. *Steel Research International*, *75*(12), 812–817. doi:10.1002rin.200405847

Pang, L. J., Shang, X., & Zhang, X. (2018). Influence of Deformation Conditions on the Critical Damage Factor of AZ31 Magnesium Alloy. *Advances in Materials Science and Engineering*, *2018*, 1–8. doi:10.1155/2018/3071495

Quan, G., Luo, G., Mao, A., Liang, J., & Wu, D. (2014). Evaluation of Varying Ductile Fracture Criteria for 42CrMo Steel by Compressions at Different Temperatures and Strain Rates. *TheScientificWorldJournal*, *2014*, 1–10. doi:10.1155/2014/579328 PMID:24592175

Rasti, J., Najafizadeh, A., & Meratian, M. (2011). Correcting the stress-strain curve in hot compression test using finite element analysis and Taguchi method. *International Journal of ISSI*, *8*(1), 26–33. Retrieved from http://journal.issiran.com/article_6374_44b99e06e066fa7ee24705f05935b39d.pdf

Shokry, A., Gowid, S., Kharmanda, G., & Mahdi, E. (2019). Constitutive Models for the Prediction of the Hot Deformation Behavior of the 10%Cr Steel Alloy. *Materials (Basel)*, *12*(18), 1–18. doi:10.3390/ma12182873 PMID:31492023

Yang, C., & Lin, X. (2016). The Forming Analysis of Two-stage Extrusion for 1010 Fastener. *Journal of Mechanical Engineering and Automation*, *6*(3), 43–50. doi:10.5923/j.jmea.20160603.01

Yang, Z. N., Dai, L. Q., Chu, C. H., Zhang, F. C., Wang, L. W., & Xiao, A. P. (2017). Effect of Aluminum Alloying on the Hot Deformation Behavior of Nano-bainite Bearing Steel. *Journal of Materials Engineering and Performance*, *26*(12), 5954–5962. doi:10.100711665-017-3018-7

Yin, F., Hua, L., Mao, H., & Han, X. (2013). Constitutive modeling for flow behavior of GCr15 steel under hot compression experiments. *Materials & Design*, *43*, 393–401. doi:10.1016/j.matdes.2012.07.009

Zhu, L., He, J., & Zhang, Y. (2018). A two-stage constitutive model of X12CrMoWVNbN10-1-1 steel during elevated temperature. *Materials Research Express*, *5*(2), 026505. doi:10.1088/2053-1591/aaa911

Chapter 14
Control Optimisation of Overhead Gantry Cranes via Fuzzy Controllers

Ashwani Kharola
Tulas Institute, India

ABSTRACT

This study considers a fuzzy logic-based reasoning approach for control and optimising performance of overhead gantry crane. The objective of this study is to minimise load swing and to stabilise the crane in the least possible time. The fuzzy controllers were designed using nine Gaussian and triangular shape membership functions. The results clearly confirmed the effect of shape of memberships on performance of fuzzy controllers. Performance of overhead crane was measured in terms of settling time and overshoot ranges. The study also demonstrates the influence of varying mass of the load, mass of crane, and length of crane bar on stability of the crane. A mathematical model of the crane system has been derived to develop a simulink model of proposed system and performing simulations.

INTRODUCTION

Overhead cranes are extremely nonlinear systems extensively used in industries for carrying materials from one location to another (Solihin and Wahyudi, 2009; Sagirli et al., 2011). These machines are widely used for transportation as they provide faster transportation at a comparatively low cost with minimal safety constraints (Fedtke & Boysen, 2017). The undesirable swing due to external forces may cause unrestrained oscillations and vibrations resulting in stability and safety hazards of these machines (Kaur et al., 2014; Belunce et al., 2014). Therefore, the key objective while operating these machines is to minimise load swing in least possible travel time (Mahfouf et al., 2000; Chen et al., 2012). The overhead gantry cranes are keen source of study and research for researchers in past few decades because of major industrial applications (Santhi and Beebi, 2014). A novel adaptive hierarchical sliding mode control of three dimensional overhead crane was presented by Le et al. (2019). The proposed control has enhanced the robustness of the crane system under uncertainty conditions and was designed using

DOI: 10.4018/978-1-7998-4939-1.ch014

radial basis neural network derived from Lyapunov function. An active disturbance rejection model for the problem of load transportation and oscillation for tower crane was developed by Coral-Enriquez et al. (2019). The authors formulated a disturbance observer for the proposed system which aided in better estimation of system properties and disturbances. Singh and Ha (2019) proposed a finite control time method based on terminal sliding mode for control of underactuated gantry crane system. The authors defined hierarchical sliding surfaces comprising of two layers for underactuated system control. A model predictive control which provided faster transportation of cargo with minimal swing was proposed by Jolevski and Bego (2015). A multi-criteria optimization technique was adopted to create the solution function for the proposed control. Experiments were performed on the laboratory model of the crane system to confirmed the validity of the proposed study. Bara et al. (2009) examined an intelligent fuzzy control for designing a microcontroller of a Mechatronic system. Real-time experiments were performed on a three dimensional crane to validate the proposed technique. A fuzzy tuned Proportional-integral-derivative (PID) controller for swing control of gantry crane was proposed by Solihin et al. (2009). The parameters of PID were tuned using gain tuners for achieving robust performance. The study also discussed dynamics analysis of the system and verified it through experimental results.

An anti-swing fuzzy control of two-dimensional overhead crane with hoisting was proposed by Trabia et al. (2006). The study applied inverse dynamics technique for determining range of variables for the controllers. The simulation results confirmed that the proposed controller successfully drives the overhead cranes under diverse operating conditions. A Sliding mode anti-swing controller for control of overhead gantry cranes was developed by Lee et al. (2006). The results showed that proposed controller resulted in increased load speed with damped load swing. Mousa (2000) designed a hybrid controller based on combination of fuzzy and time-delayed feedback technique for control of rotary cranes. Computer simulations were performed for verifying the performance of proposed controller. According to Pal and Mudi (2013) an adaptive fuzzy PID controller can be successfully applied for real-time control of an overhead crane system. The output gains of the proposed controller were tuned based on current operating conditions of the process. The study further compared fuzzy control with other conventional controllers.

Simanjalam (2012) performed modeling and control of overhead gantry crane using Euler-Lagrange formulation. The study further considered a fuzzy tuned PID controller for control of the proposed system. A twin fuzzy controller for swing and position control of overhead crane was developed by Chang et al. (2005). Experiments were conducted which demonstrated the effectiveness of proposed control scheme. An adaptive sliding mode fuzzy control of a two dimensional overhead crane was proposed by Liu et al. (2004). A model of the crane system was obtained by considering it as two independent systems in X and Y directions respectively. Fuzzy logic based servo control for anti-swing and position control of overhead crane was demonstrated by Lee and Cho (2001). The results confirmed the validity of proposed approach. Wahyudi and Jalani (2006) performed comparison of an intelligent fuzzy controller with conventional PID control. The results demonstrated that fuzzy controller was more robust to parameter variation compared to PID controller.

A study on control of gantry crane using fuzzy-Proportional derivative (PD) technique was conducted by Asad et al. (2011). The authors further compared proposed control with conventional PD approach which illustrated better performance of fuzzy-PD controller. Arpaci and Ozguven (2011) developed a non-linear model of an overhead crane system. The control of proposed system has been achieved using an Adaptive neuro fuzzy inference system (ANFIS) which was tuned using results of PID controller. Smoczek and Szpytko (2008) proposed a Mechatronic approach to elaborate intelligent control of overhead crane. A fuzzy controller based on Takagi-Sugeno inference system was adopted for control proposed system.

MATHEMATICAL MODEL OF GANTRY CRANE

Gantry cranes belong to a class of overhead cranes which are widely used in industries for carrying heavy payload from location to another. These are highly nonlinear and underactuated systems which mimics dynamics of a rotary inverted pendulum system (Ranjbari et al., 2015). The system comprises of a payload of mass (m=1kg) attached to movable crane or cart of mass (M=2kg) with the help of a rigid bar of length (L=1m). When a force (u) is applied to the crane in horizontal direction (x) it swings the bar by an angle (θ) from vertical. The objective is to minimise the swing as the crane moves under action of gravity ($g=9.81m/s^2$). A schematic view of overhead crane is shown in Figure 1 (Pal and Mudi, 2013).

Figure 1. Schematic view of Gantry crane

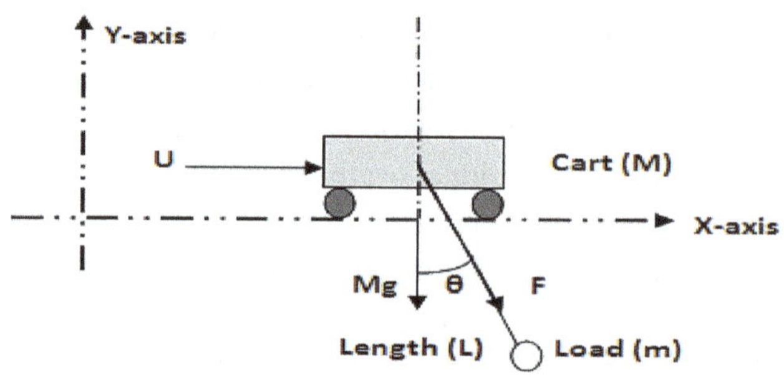

The mathematical model of system was developed using Newton's second law (Zairulazha, 2005). The gantry crane system was split into two parts i.e. cart and load for analysing dynamics of the complete system (Shcherbkov et al., 2016).

Equation of Motions for Crane

A free body diagram for crane sub-system is shown with the help of Figure 2.

Figure 2. Free body diagram of Crane

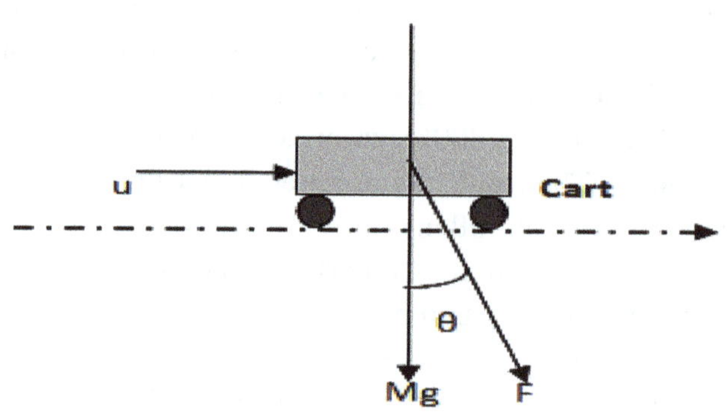

The final equations of motion for crane sub-system are given below

$$M\frac{d^2}{dt^2} = u + F\sin\theta \tag{1}$$

$$F\cos\theta + Mg = 0 \tag{2}$$

Equation of Motion for Load

A free body diagram for load sub-system is shown with the help of Figure 3.

Figure 3. Free body diagram of Load

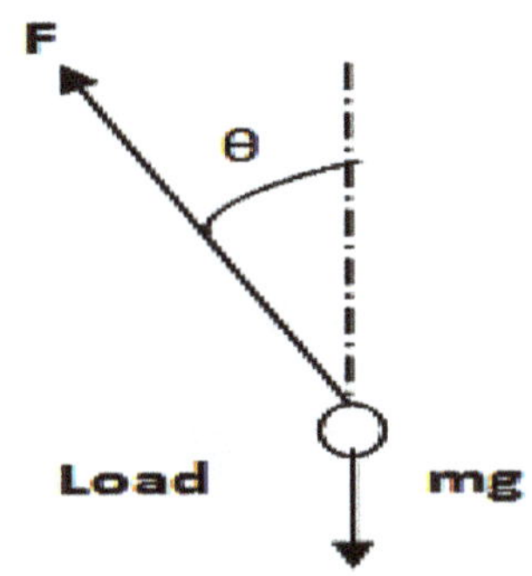

The final equations of motion for the load are given below

$$m\frac{d^2}{dt^2}\left(x + L\sin\theta\right) = -F\sin\theta \tag{3}$$

$$m\frac{d^2}{dt^2}\left(L\cos\theta\right) = -F\cos\theta + mg \tag{4}$$

Solving above equations, following values of angular acceleration of bar ($\ddot{\theta}$) and acceleration of crane ($\ddot{x}$) were obtained.

$$\ddot{\theta} = -\frac{\left[u\cos\theta + m\sin\theta\left(L\theta^2 + g\right) + Mg\sin\theta\right]}{\left(M + m\sin^2\theta\right)L} \tag{5}$$

$$\ddot{x} = \frac{\left[u + m\sin\theta\left(g\cos\theta + L\theta^2\right)\right]}{\left(M + m\sin^2\theta\right)} \tag{6}$$

The equation (5) and equation (6) were used for developing a Simulink model of overhead gantry crane system as shown in Figure 4. The simulink thus developed was further masked into a sub-system for incorporating two fuzzy controllers for controlling crane position and swing respectively as shown in Figure 5.

Figure 4. Simulink model of Overhead crane

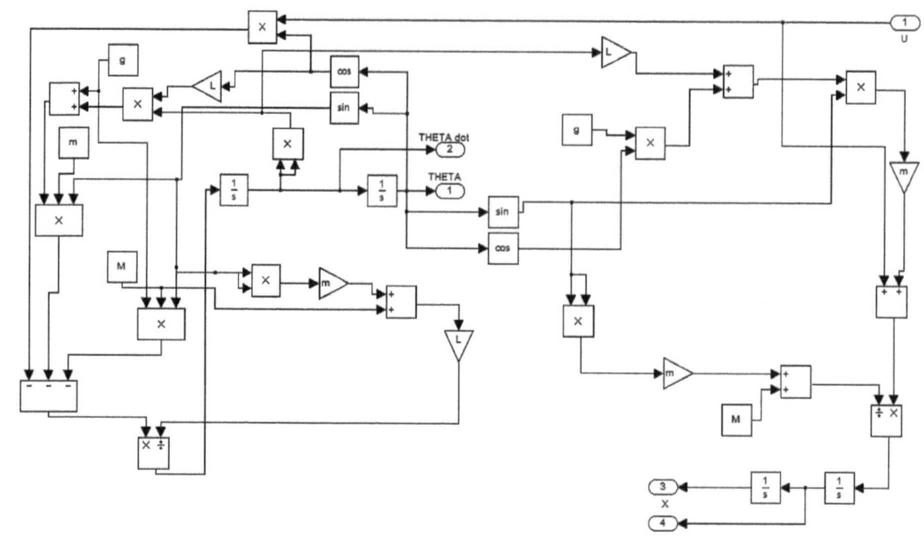

Figure 5. Sub-system model of overhead crane incorporated with fuzzy controllers

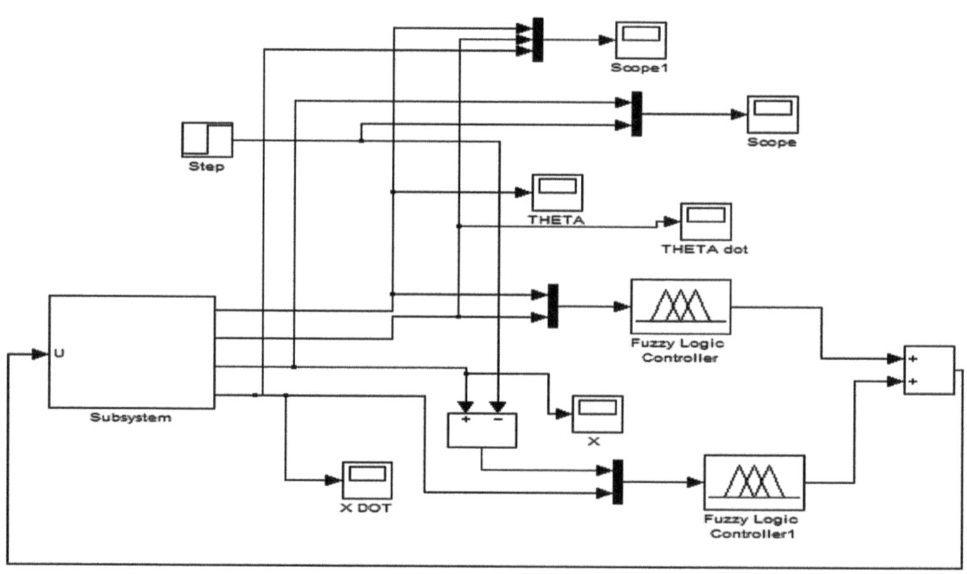

DESIGN OF FUZZY LOGIC CONTROLLERS

Fuzzy set theory was initially introduced by L.A. Zadeh in 1965. This theory is based on human logic of decision making and overcomes major drawbacks in binary logic (Chang & Chiang, 2009). It is widely used in situations when conventional binary logic fails to handle multifaceted situations (Moraga, 2005). It can be used in situations which are partially true or partially false. Fuzzy logic deals with linguistic variables which makes it user friendly and suitable for vast applications (Zadeh, 1975). In this study the fuzzy controllers for crane system has been designed using seven gaussian and triangular shape membership functions as shown in Figure 6 and Figure 7 respectively. The main objective is to illustrate effect of shape of memberships on performance of fuzzy controller. The linguistic terms which were adopted for defining memberships were negative low (NL), negative medium (NM), negative small (NS), zero (ZE), positive small (PS), positive medium (PM) and positive large (PL).

Figure 6. Gaussian shape membership functions for load angle

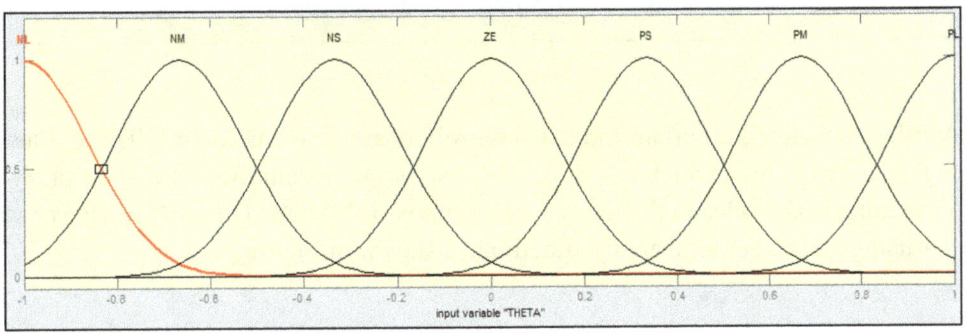

Figure 7. Triangular shape membership functions for load angle

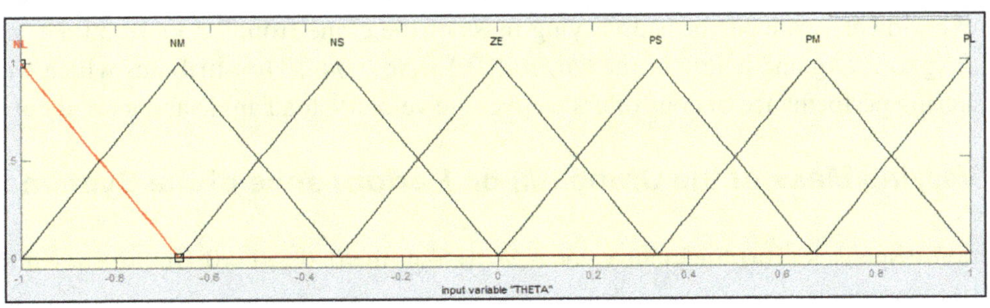

Table 1. Fuzzy control rules for crane system

Output Control Force							
Cart velocity ($\dot{x}$) or Load ang velocity ($\dot{\theta}$)	NL	NM	NS	ZE	PS	PM	PL
	NL	NL	NL	NL	NM	NS	ZE
	NL	NL	NL	NM	NS	ZE	PS
	NL	NL	NM	NS	ZE	PS	PM
	NL	NM	NS	ZE	PS	PM	PL
	NM	NS	ZE	PS	PM	PL	PL
	NS	ZE	PS	PM	PL	PL	PL
	ZE	PS	PM	PL	PL	PL	PL

Figure 8. Surface viewer for Load controller

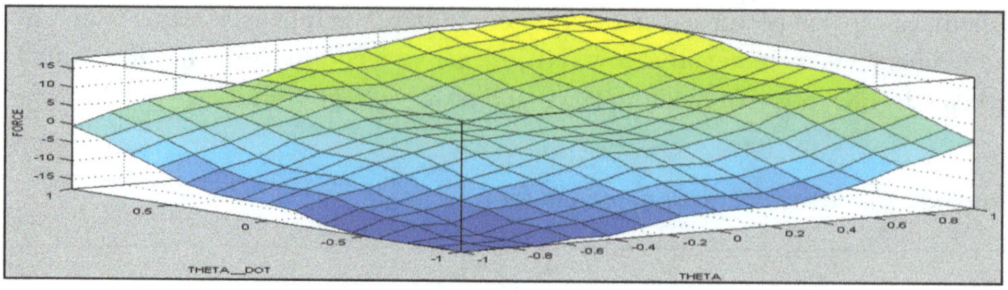

The fuzzy rules considered for crane and load controllers are given in Table 1. These rules were used for developing a surface viewer which is a three dimensional representation of if-then fuzzy rules. It is obtained after feeding fuzzy rules in the fuzzy logic toolbox of Matlab. The surface viewer obtained for load controller using gaussian membership functions is shown in Figure 8.

RESULTS AND DISCUSSION

The simulation results obtained using gaussian and triangular membership functions are demonstrated below. The experiments were performed varying mass of the crane from 0.5 kg to 3.0 kg, mass of the load from 0.2 kg to 1.2 kg and length of the bar from 0.1 m to 1.0 m. The attributes which were considered for measuring performance of controllers were crane velocity, load angle and load angular velocity.

Effect of Varying Mass of the Crane (M) on Performance of the System

The Simulation responses -1(a) for Input values: M=0.5kg, m=0.2kg, L=0.1m are shown in Figure 9 and Figure 10 respectively.

Figure 9. Simulation responses using triangular membership functions

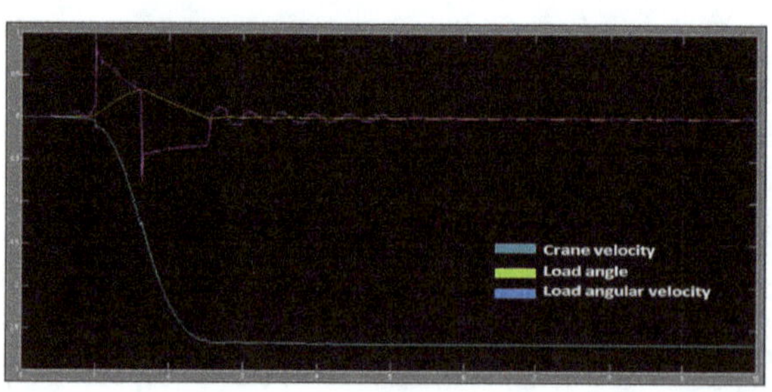

Figure 10. Simulation responses using gaussian membership functions

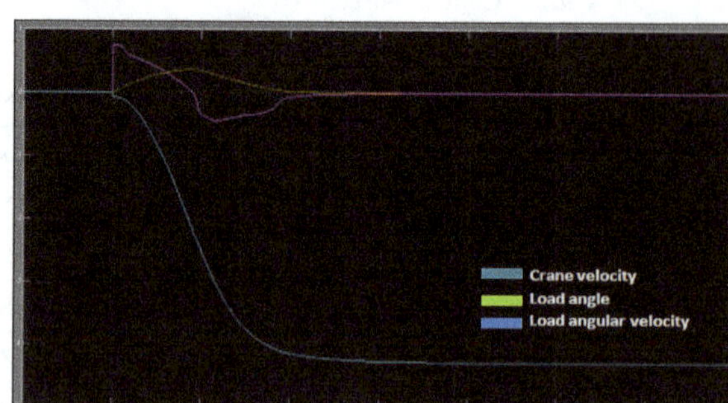

The Simulation responses -1(b) for Input values: M=1.0kg, m=0.2kg, L=0.1m are shown in Figure 11 and Figure 12 respectively

Figure 11. Simulation responses using triangular membership functions

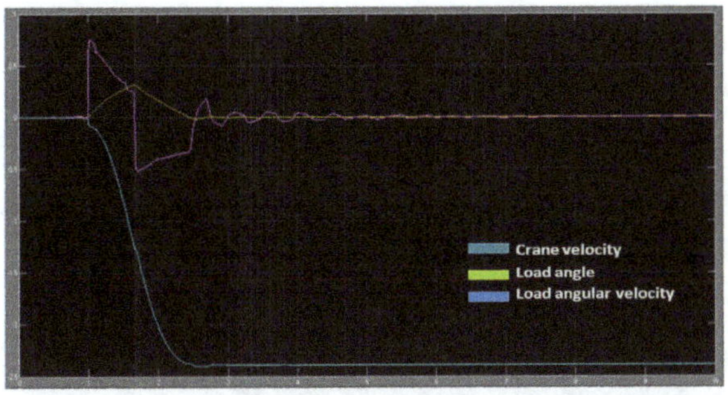

Figure 12. Simulation responses using gaussian membership functions

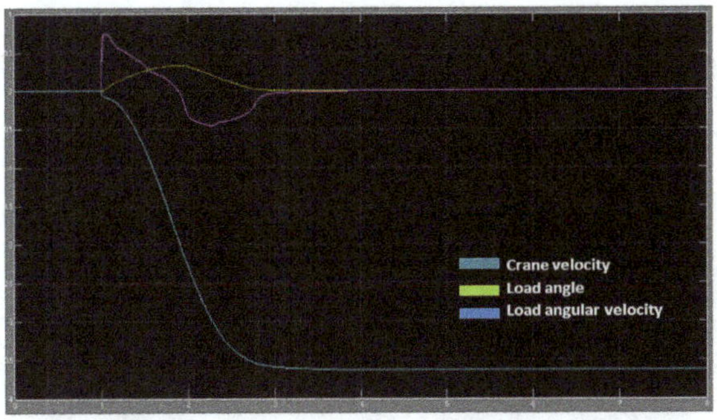

The Simulation responses -1(c) for Input values: M=2.0kg, m=0.2kg, L=0.1m are shown in Figure 13 and Figure 14 respectively

Figure 13. Simulation responses using triangular membership functions

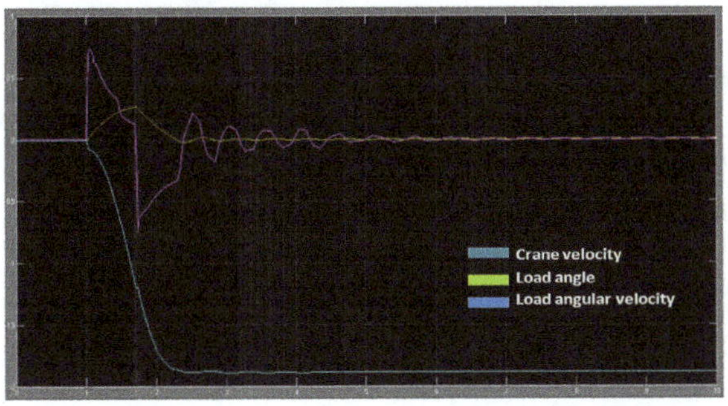

Figure 14. Simulation responses using gaussian membership functions

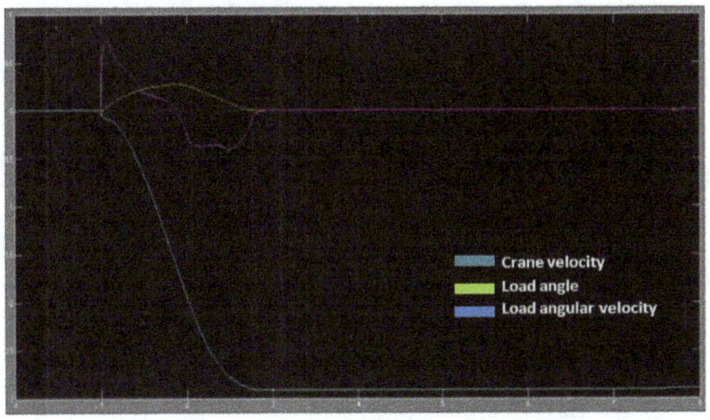

The Simulation responses -1(d) for Input values: M=2.5kg, m=0.2kg, L=0.1m are shown in Figure 15 and Figure 16 respectively.

Figure 15. Simulation responses using triangular membership functions

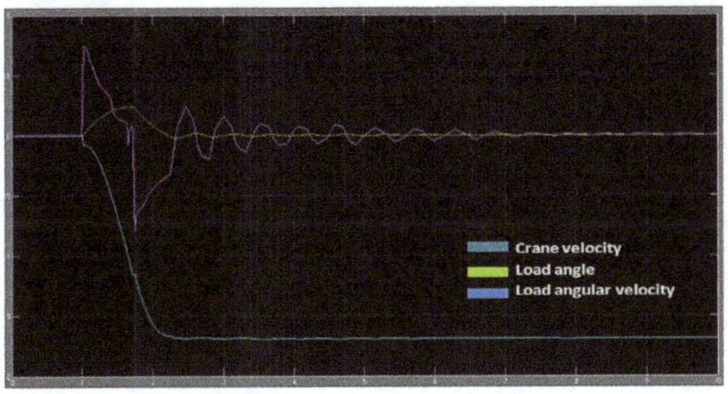

Figure 16. Simulation responses using gaussian membership functions

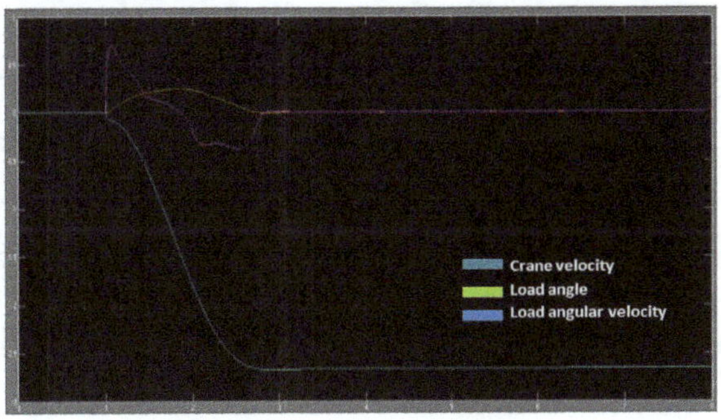

The Simulation responses -1(e) for Input values: M=3.0kg, m=0.2kg, L=0.1m are shown in Figure 17 and Figure 18 respectively

Figure 17. Simulation responses using triangular membership functions

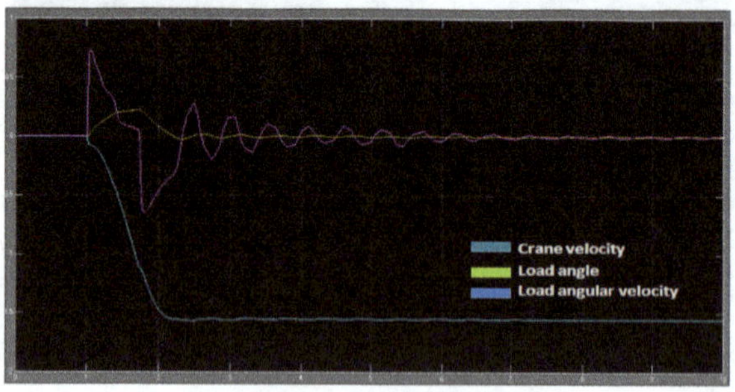

Figure 18. Simulation responses using gaussian membership functions

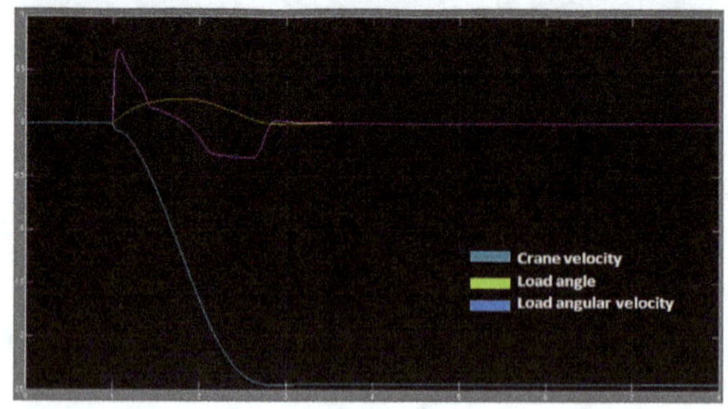

The simulation results for varying mass of the crane (M) on performance of the system are shown with the help of Table 2.

Table 2. Simulation results for varying mass of the crane (M) on performance of the system

Performance Attributes	Crane Position	Load Angle	Load Angular Velocity
Simulation Results: 1(a)			
Inputs	M=0.5Kg	M=0.2Kg	L=0.1m
Triangular Shape			
Settling time(sec)	4.0 sec	5.0 sec	6.0 sec
Max. Overshoot (degree)	-2.68°	0.34° to -0.02°	0.88° to -0.76°
Gaussian Shape			
Settling time(sec)	3.0 sec	3.0 sec	3.0 sec
Max. Overshoot (degree)	-4.25°	0.37° to -0.001	0.75° to -0.45°
Simulation Results: 1(b)			
Inputs	M=1.0Kg	M=0.2Kg	L=0.1m
Triangular Shape			
Settling time(sec)	4.5 sec	6.0 sec	7.0 sec
Max. Overshoot (degree)	-2.41°	0.32° to -0.02°	0.75° to -0.5°
Gaussian Shape			
Settling time(sec)	3.0 sec	3.0 sec	3.0 sec
Max. Overshoot (degree)	-3.70°	0.33° to -0.001	0.74° to -0.45°
Simulation Results: 1(c)			
Inputs	M=2.0Kg	M=0.2Kg	L=0.1m
Triangular Shape			

continued on following page

Table 2. Continued

Performance Attributes	Crane Position	Load Angle	Load Angular Velocity
Settling time(sec)	5.0 sec	7.0 sec	8.0 sec
Max. Overshoot (degree)	-1.88°	0.27° to -0.025°	0.72° to -0.75°
Gaussian Shape			
Settling time(sec)	3.0 sec	3.0 sec	3.0 sec
Max. Overshoot (degree)	-2.90°	0.275°	0.74° to -0.40°
Simulation Results: 1(d)			
Inputs	M=2.5Kg	M=0.2Kg	L=0.1m
Triangular Shape			
Settling time(sec)	5.5 sec	8.0 sec	9.0 sec
Max. Overshoot (degree)	-1.69°	0.24° to -0.025°	0.75° to -0.80°
Gaussian Shape			
Settling time(sec)	3.0 sec	3.0 sec	3.5 sec
Max. Overshoot (degree)	-2.70°	0.24° to 0.01°	0.70° to -0.38°
Simulation Results: 1(e)			
Inputs	M=3.0Kg	M=0.2Kg	L=0.1m
Triangular Shape			
Settling time(sec)	6.0 sec	9.0 sec	10.0 sec
Max. Overshoot (degree)	-1.57°	0.22° to -0.025°	0.72° to -0.66°
Gaussian Shape			
Settling time(sec)	3.0 sec	3.0 sec	3.5 sec
Max. Overshoot (degree)	-2.40°	0.22° to 0.01°	0.68° to -0.32°

Effect of Varying Mass of the Load (M) on Performance of the System

The Simulation responses -2(a) for Input values: M=2.0kg, m=0.2kg, L=0.1m are shown in Figure 19 and Figure 20 respectively

Figure 19. Simulation responses using triangular membership functions

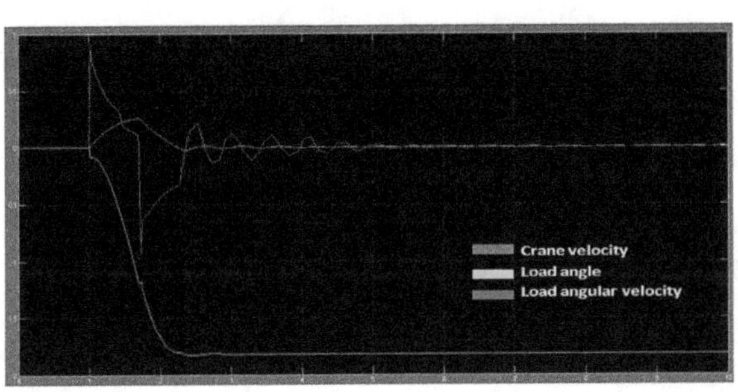

Figure 20. Simulation responses using gaussian membership functions

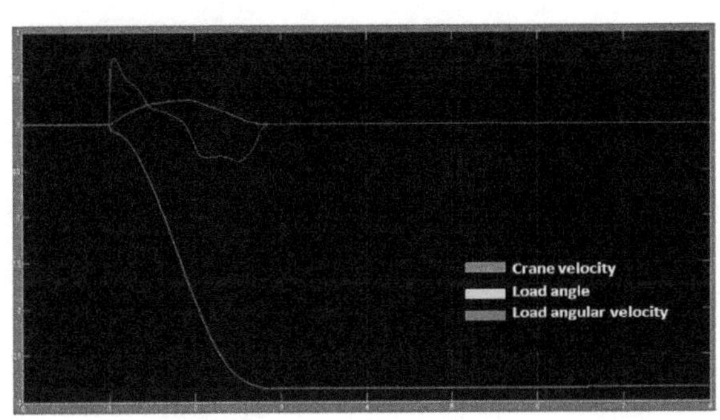

The Simulation responses -2(b) for Input values: M=2.0kg, m=0.6kg, L=0.1m are shown in Figure 21 and Figure 22 respectively

Figure 21. Simulation responses using triangular membership functions

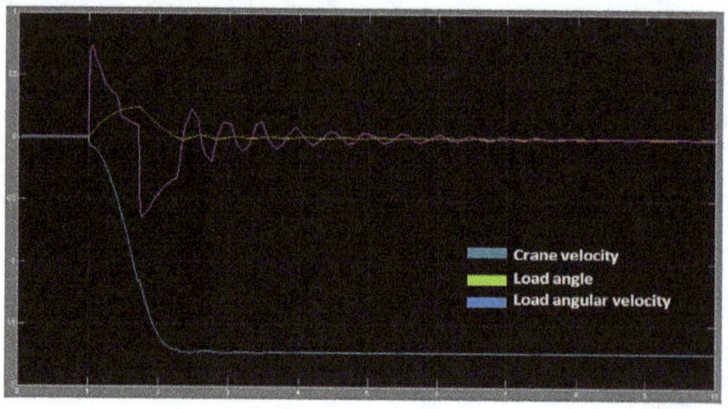

Figure 22. Simulation responses using gaussian membership functions

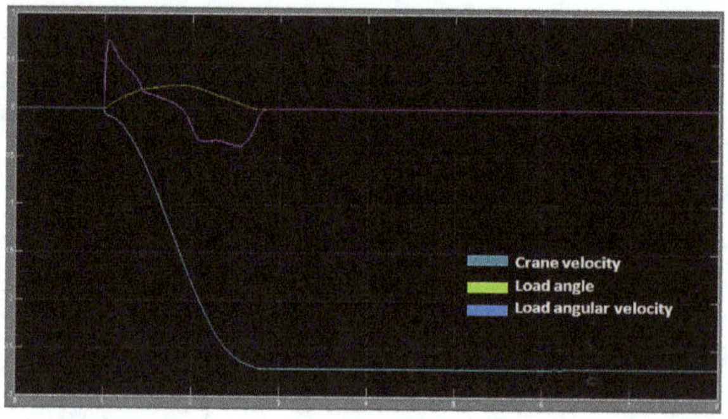

The Simulation responses -2(c) for Input values: M=2.0kg, m=0.8kg, L=0.1m are shown in Figure 23 and Figure 24 respectively

Figure 23. Simulation responses using triangular membership functions

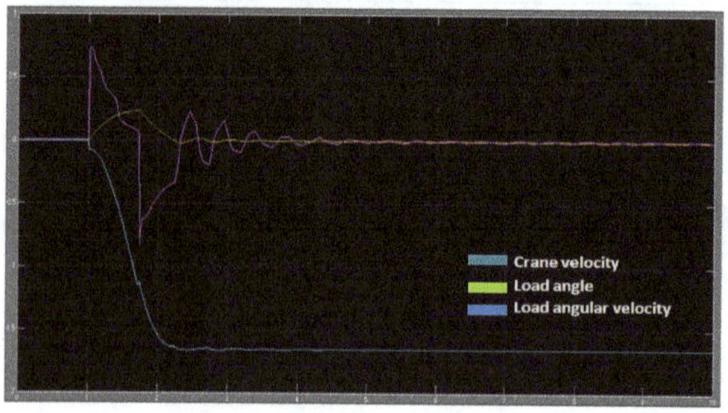

Figure 24. Simulation responses using gaussian membership functions

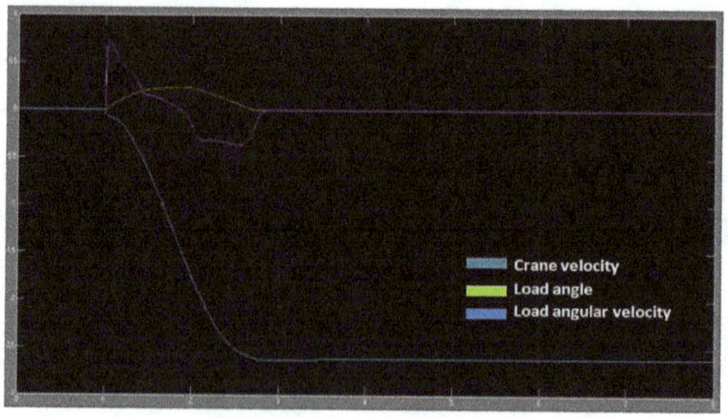

The Simulation responses -2(d) for Input values: M=2.0kg, m=1.0kg, L=0.1m are shown in Figure 25 and Figure 26 respectively

Figure 25. Simulation responses using triangular membership functions

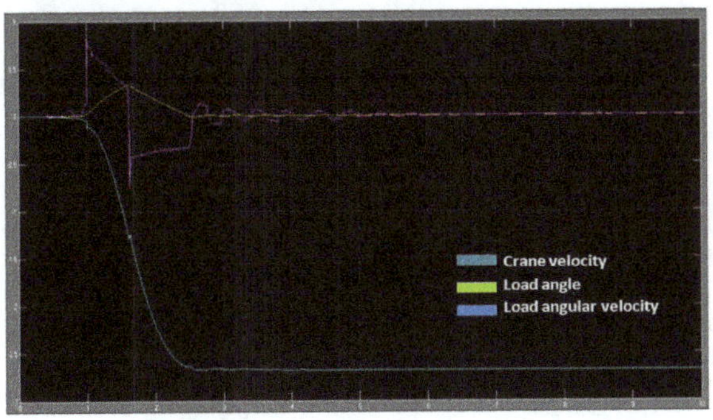

Figure 26. Simulation responses using gaussian membership functions

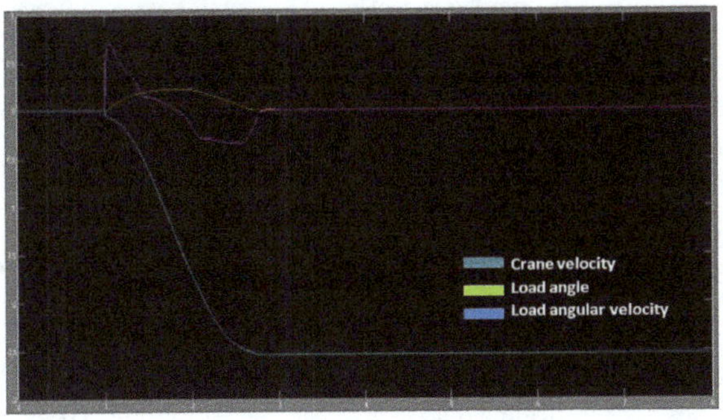

The Simulation responses -2(e) for Input values: M=2.0kg, m=1.2kg, L=0.1m are shown in Figure 27 and Figure 28 respectively

Figure 27. Simulation responses using triangular membership functions

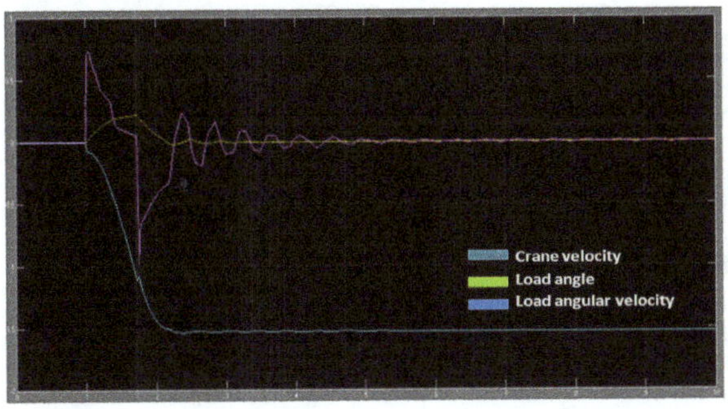

Figure 28. Simulation responses using gaussian membership functions

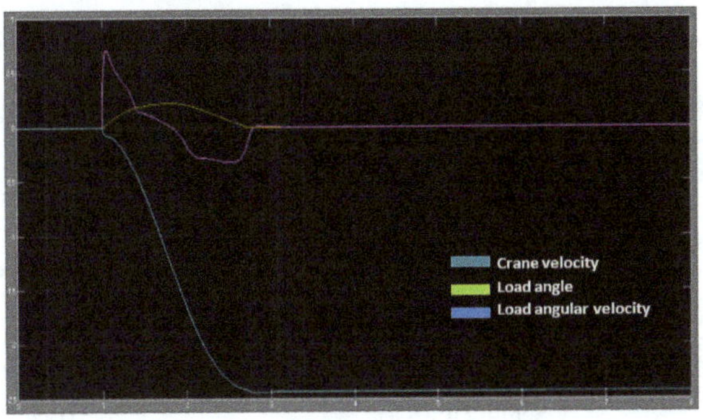

The Simulation results for varying mass of the load (m) on performance of the system are shown with the help of Table 3.

Table 3. Simulation results for varying mass of the load (m) on performance of the system

Performance Attributes	Crane Position	Load Angle	Load Angular Velocity
Simulation Results: 2(a)			
Inputs	M=2.0Kg	M=0.2Kg	L=0.1m
Triangular Shape			
Settling time(sec)	5.0 sec	4.5 sec	6.0 sec
Max. Overshoot (degree)	-1.83°	0.27° to -0.02°	0.90° to -0.90°
Gaussian Shape			
Settling time(sec)	2.8 sec	2.8 sec	2.9 sec
Max. Overshoot (degree)	-2.80°	0.27°	0.72° to -0.40°
Simulation Results: 2(b)			
Inputs	M=2.0Kg	M=0.6Kg	L=0.1m
Triangular Shape			
Settling time(sec)	5.0 sec	5.0 sec	6.0 sec
Max. Overshoot (degree)	-1.73°	0.25° to -0.02°	0.72° to -0.64°
Gaussian Shape			
Settling time(sec)	2.8 sec	2.8 sec	2.9 sec
Max. Overshoot (degree)	-2.75°	0.25°	0.70° to -0.38°
Simulation Results: 2(c)			
Inputs	M=2.0Kg	M=0.8Kg	L=0.1m
Triangular Shape			
Settling time(sec)	5.0 sec	5.0 sec	6.0 sec
Max. Overshoot (degree)	-1.66°	0.24° to -0.02°	0.72° to -0.82°
Gaussian Shape			
Settling time(sec)	2.8 sec	2.8 sec	2.9 sec
Max. Overshoot (degree)	-2.60°	0.24°	0.70° to -0.36°
Simulation Results: 2(d)			
Inputs	M=2.0Kg	M=1.0Kg	L=0.1m
Triangular Shape			
Settling time(sec)	5.0 sec	5.0 sec	6.0 sec
Max. Overshoot (degree)	-1.616°	0.22° to -0.02°	0.72° to -0.65°
Gaussian Shape			
Settling time(sec)	2.8 sec	2.8 sec	2.9 sec
Max. Overshoot (degree)	-2.55°	0.23°	0.70° to -0.35°
Simulation Results: 2(e)			
Inputs	M=3.0Kg	M=1.2Kg	L=0.1m
Triangular Shape			
Settling time(sec)	5.0 sec	5.0 sec	6.0 sec
Max. Overshoot (degree)	-1.54°	0.22° to -0.02°	0.72° to -1.0°
Gaussian Shape			
Settling time(sec)	2.8 sec	2.8 sec	2.9 sec
Max. Overshoot (degree)	-2.40°	0.22°	0.70° to -0.32°

Effect of Varying Length of the Bar (L) on Performance of the System

The Simulation responses -3(a) for Input values: M=1.0kg, m=0.2kg, L=0.2m are shown in Figure 29 and Figure 30 respectively

Figure 29. Simulation responses using triangular membership functions

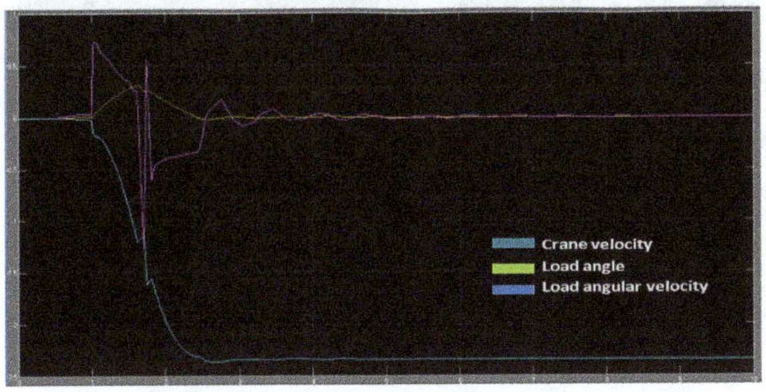

Figure 30. Simulation responses using gaussian membership functions

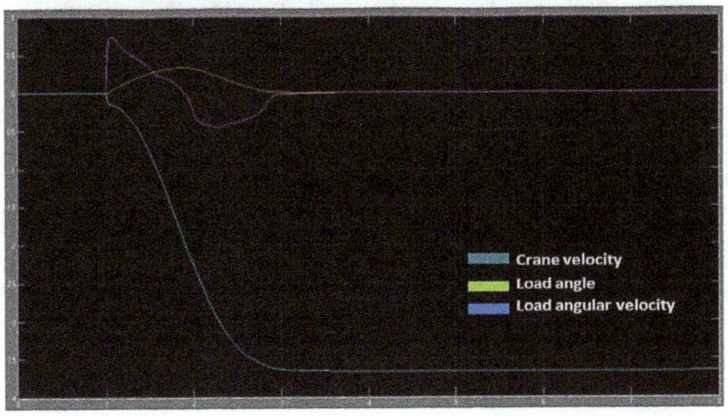

The Simulation responses -3(b) for Input values: M=1.0kg, m=0.2kg, L=0.4m are shown in Figure 31 and Figure 32 respectively

Figure 31. Simulation responses using triangular membership functions

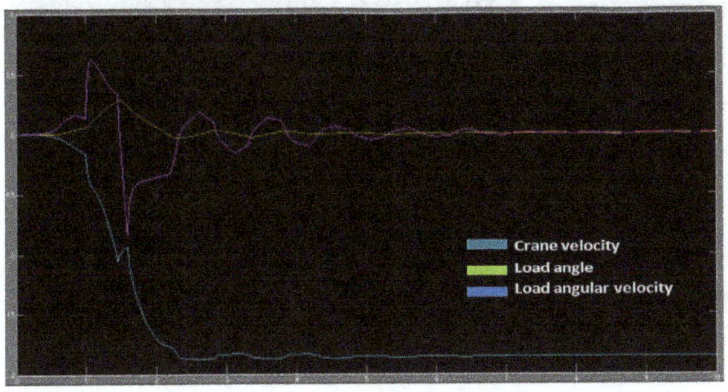

Figure 32. Simulation responses using gaussian membership functions

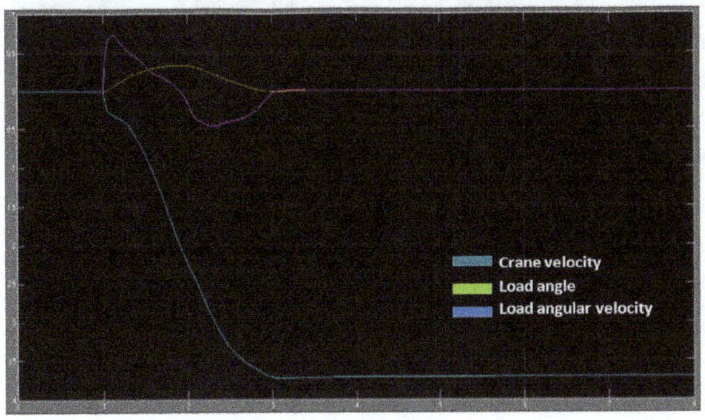

The Simulation responses -3(c) for Input values: M=1.0kg, m=0.2kg, L=0.6m are shown in Figure 33 and Figure 34 respectively

Figure 33. Simulation responses using triangular membership functions

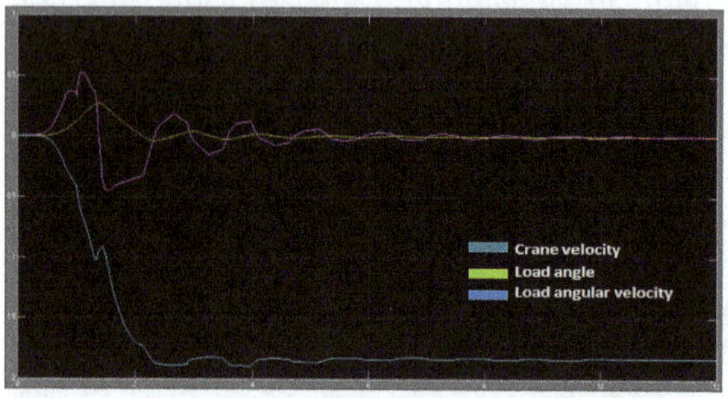

Figure 34. Simulation responses using gaussian membership functions

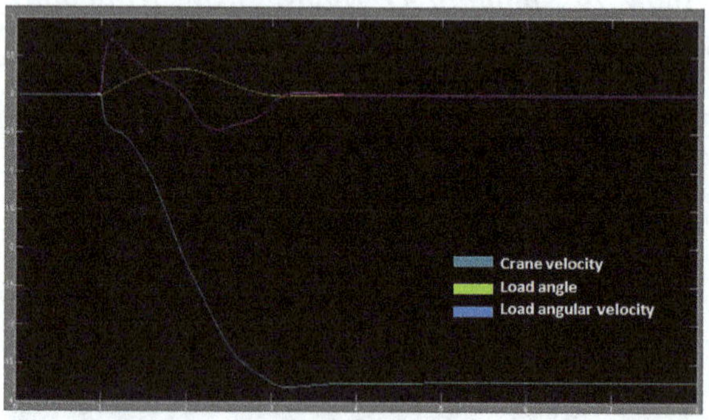

The Simulation responses -3(d) for Input values: M=1.0kg, m=0.2kg, L=0.8m are shown in Figure 35 and Figure 36 respectively

Figure 35. Simulation responses using triangular membership functions

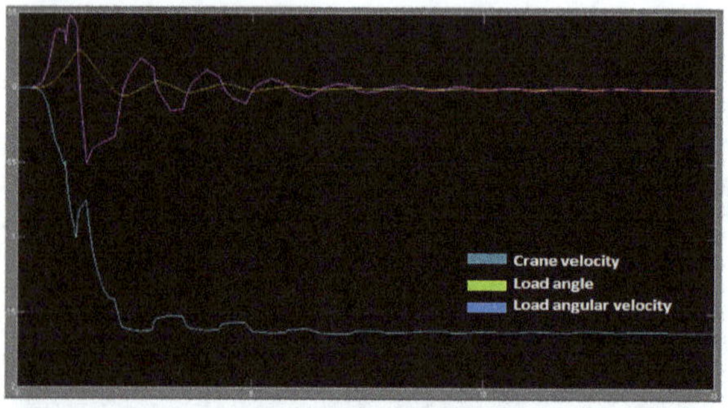

Figure 36. Simulation responses using gaussian membership functions

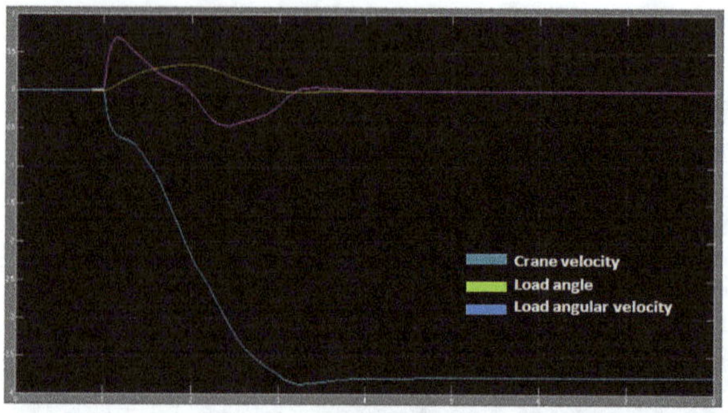

The Simulation responses-3(e) for Input values: M=1.0kg, m=0.2kg, L=1.0 are shown in Figure 37 and Figure 38 respectively

Figure 37. Simulation responses using triangular membership functions

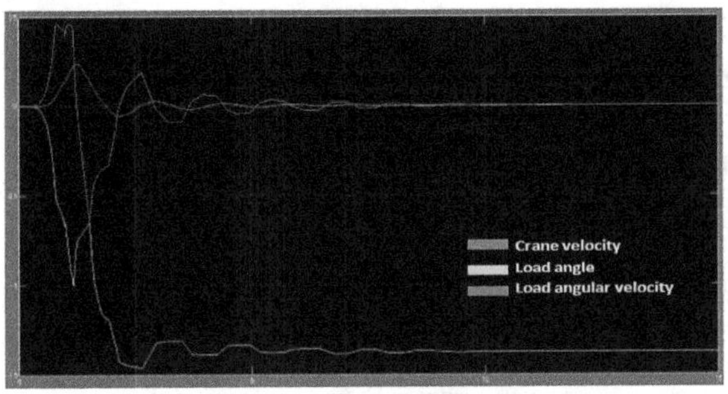

Figure 38. Simulation responses using gaussian membership functions

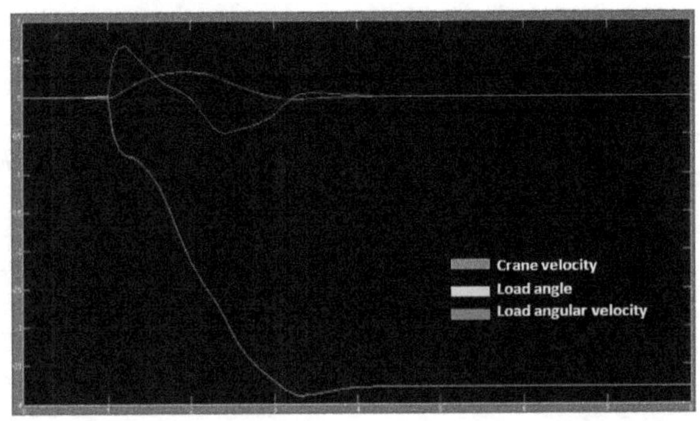

The Simulation results for varying length of the load (L) on performance of the system are shown with the help of Table 4.

Table 4. Simulation results for varying length of the load (L) on performance of the system

Performance Attributes	Crane Position	Load Angle	Load Angular Velocity
Simulation Results: 3(a)			
Inputs	M=1.0Kg	M=0.2Kg	L=0.2m
Triangular Shape			
Settling time(sec)	6.0 sec	6.0 sec	6.0 sec
Max. Overshoot (degree)	-2.37°	0.31° to -0.02°	0.70° to -1.25°
Gaussian Shape			
Settling time(sec)	3.0 sec	3.5 sec	3.5 sec
Max. Overshoot (degree)	-3.70°	0.33°	0.74° to -0.48°
Simulation Results: 3(b)			
Inputs	M=1.0Kg	M=0.2Kg	L=0.4m
Triangular Shape			
Settling time(sec)	7.0 sec	7.0 sec	7.0 sec
Max. Overshoot (degree)	-1.90°	0.28° to -0.03°	0.62° to -0.84°
Gaussian Shape			
Settling time(sec)	3.5 sec	3.8 sec	3.9 sec
Max. Overshoot (degree)	-3.70°	0.33° to 0.01°	0.72° to -0.46°
Simulation Results: 3(c)			
Inputs	M=1.0Kg	M=0.2Kg	L=0.6m

continued on following page

Table 4. Continued

Performance Attributes	Crane Position	Load Angle	Load Angular Velocity
Triangular Shape			
Settling time(sec)	8.0 sec	8.0 sec	10.0 sec
Max. Overshoot (degree)	-1.88°	0.27° to -0.04°	0.52° to -0.48°
Gaussian Shape			
Settling time(sec)	4.0 sec	4.0 sec	4.0 sec
Max. Overshoot (degree)	-3.60°	0.33° to 0.02°	0.71° to -0.46°
Simulation Results: 3(d)			
Inputs	M=1.0Kg	M=0.2Kg	L=0.8m
Triangular Shape			
Settling time(sec)	9.0 sec	9.0 sec	10.5 sec
Max. Overshoot (degree)	-1.48°	0.25° to -0.06°	0.48° to -0.50°
Gaussian Shape			
Settling time(sec)	4.0 sec	4.1 sec	4.2 sec
Max. Overshoot (degree)	-3.80°	0.33° to 0.02°	0.70° to -0.46°
Simulation Results: 3(e)			
Inputs	M=1.0Kg	M=0.2Kg	L=1.0m
Triangular Shape			
Settling time(sec)	9.5 sec	10.0 sec	11.0 sec
Max. Overshoot (degree)	-1.46°	0.24° to -0.06°	0.46° to -0.64°
Gaussian Shape			
Settling time(sec)	4.2 sec	4.2 sec	4.3 sec
Max. Overshoot (degree)	-3.90°	0.33° to 0.03°	0.68° to -0.46°

CONCLUSION

The study deals with optimsation of fuzzy logic controllers by considering more appropriate shape of fuzzy membership functions. In this study the physical attributes of crane like its mass, length and mass of load has been varied within a particular range and their impact on control of crane has been studied. The study considered a swing and position control of an overhead gantry crane system using fuzzy based reasoning technique. The mathematical equations of motion have been derived for building a simulink model of proposed system. Two different fuzzy inference controllers have been developed using gaussian and triangular shape membership functions. The study illustrates a relation between shape of membership and performance of the crane. It is observed that gaussian controllers stabilises crane system more quickly compared to triangular shape. It is also observed that better overshoot response is obtained using triangular membership's controllers. The simulation results confirmed that settling time of system increases with increase in mass and length of the crane. The mass of the load doesn't have much influence on settling time of the crane. Finally, it is inferred that mass of the crane and mass of the load are inversely proportional to the overshoot range whereas length of the bar is directly proportional to the overshoot ranges.

REFERENCES

Arpaci, H., & Ozguven, O. F. (2011). ANFIS and PID controller design and comparison for overhead cranes. *Indian Journal of Engineering and Materials Sciences*, *18*(3), 191–203.

Asad, S., Salahat, M., & Zalata, A., & Rawashdeh, A.A. (2011). Design of fuzzy PD controlled overhead crane system with anti-swinging compensation. *Journal of Engineering and Computer Innovations*, *2*(3), 51–58.

Bara, A., Dale, S., & Nagy, Z. T. (2009). Comparative real-time experimental study case for control algorithms, from implementation point of view. *Proceedings of the 8th WSEAS International Conference on System Science and Simulation in Engineering*, 302-306.

Belunce, A., Pandolfo, V., Roozbahani, H., & Handroos, H. (2015). Novel control method for Overhead cranes load stability. *Procedia Engineering*, *106*, 108–125. doi:10.1016/j.proeng.2015.06.014

Chang, C., Hsu, S., & Chiang, K. (2005). A practical fuzzy controllers scheme of overhead crane. *Journal of Control Theory and Applications*, *3*(3), 266–270. doi:10.100711768-005-0047-9

Chang, C. Y., & Chiang, T. C. (2009). Overhead cranes fuzzy control design with deadzone compensation. *Neural Computing & Applications*, *18*(7), 749–757. doi:10.100700521-009-0264-0

Chen, Z. M., Meng, W. J., & Zhang, J. G. (2012). Intellihent anti-swing control for bridge crane. *Journal of Central South University*, *19*(10), 2774–2781. doi:10.100711771-012-1341-6

Coral-Enriquez, H., Pulido-Guerrero, S. P., & Cortes-Romero, J. (2019). Robust disturbance rejection based control with extended state resonant observer for sway reduction in uncertain tower cranes. *International Journal of Automation and Computing*, *16*(6), 812–827. doi:10.100711633-019-1179-6

Fedtke, S., & Boysen, N. (2017). Gantry crane and shuttle car scheduling in modern rail-rail transshipment yards. *OR-Spektrum*, *39*(2), 473–503. doi:10.100700291-016-0461-z

Jolevski, D., & Bego, O. (2015). Model predictive control of gantry/bridge crane with anti-sway algorithm. *Journal of Mechanical Science and Technology*, *29*(2), 827–834. doi:10.100712206-015-0144-8

Kaur, A., Priyahansha, Kumari, S., & Singh, T. (2014). Position control of overhead cranes using fuzzy controllers. *International Journal of Advanced research in Electrical Electronics and Instrumentation Engineering*, *3*(5), 9341–9350.

Le, V. A., Le, H. X., Nguyen, L., & Phan, M. X. (2019). An efficient adaptive hierarchical sliding mode control strategy using neural networks for 3D overhead cranes. *International Journal of Automation and Computing*, *16*(5), 614–627. doi:10.100711633-019-1174-y

Lee, H. H., & Cho, S. K. (2001). A new fuzzy logic anti-swinging control for Industrial three dimensional overhead cranes. *Proceedings of IEEE International conference on Robotics and Automation*, 2956-2961. doi: 10.1109/ROBOT.2001.933070

Lee, H. H., Liang, Y., & Segura, D. (2006). A sliding-mode anti-swing trajectory control for overhead cranes with high-speed load hoisting. *Journal of Dynamic Systems, Measurement, and Control*, *128*(4), 842–846. doi:10.1115/1.2364010

Liu, D., Yi, J., Zhao, D., & Wang, W. (2004). Swing free transporting of two dimensional overhead crane using sliding mode fuzzy control. *Proceedings of the American control conference*, 1764-1769.

Mahfouf, M., Kee, C. H., Abbod, M. F., & Linkens, D. A. (2000). Fuzzy logic-based anti-swing control design for overhead cranes. *Neural Computing & Applications*, *9*(1), 38–43. doi:10.1007005210070033

Moraga, C. (2005). Introduction to Fuzzy logic. *Facta Universitatis*, *18*(2), 319–328.

Mousa, A. A. (2000). *Control of rotary cranes using fuzzy logic and time-delayed position feedback control* (Master thesis). Virginia Polytechnic Institute and State University, Blacksburg, VA.

Pal, A.K., & Mudi, R.K. (2013). An adaptive PD-type FLC and its real implementation to overhead crane control. *International Journal of Emerging Technologies in Computational and Applied Sciences*, *6*(2), 178-183.

Ranjbari, L., Shirdel, A.H., Aslahi-Shahri, M., Anbari, S., Ebrahimi, A., Darvishi, M., Alizadeh, M., Rahmani, R., & Seyedmahmoodian, M. (2015). Designing precision fuzzy controller for load swing of an overhead crane. *Neural Computing and Applications*, *26*(7), 1555-1560.

Sagirli, A., Azeloglu, C. O., Guclu, R., & Yazici, H. (2011). Self-tuning fuzzy logic control of crane structures against earthquake induced vibrations. *Nonlinear Dynamics*, *64*(4), 375–384. doi:10.100711071-010-9868-2

Santhi, L. R., & Beebi, L. (2014). Position control and anti-swing control of overhead crane using LQR. *International Journal of Scientific and Engineering Research*, *3*(8), 26–40.

Shcherbkov, V. S., Korytov, M. S., & Shershneva, E. O. (2016). Influence of an obstacle on load displacement by a gantry crane. *Russian Engineering Research*, *36*(3), 194–197. doi:10.3103/S1068798X16030151

Simanjalam, K. R. (2012). Position and anti-sway control for gantry crane system using fuzzy tuned PID controller. Project report submitted to University Teknologi Malaysia.

Singh, A. M., & Ha, Q. P. (2019). Fast terminal sliding control application for second-order underactuated systems. *International Journal of Control, Automation, and Systems*, *17*(8), 1884–1898. doi:10.100712555-018-0785-3

Smoczek, J., & Szpytko, J. (2008). A Mechatronics approach in Intelligent control systems of the overhead travelling cranes prototyping. *Information Technology and Control*, *37*(2), 154–158.

Solihin, M. I., Wahyudi, & Legowo, A. (2007). Fuzzy-tuned PID Anti-swing control of Automatic Gantry crane. *Journal of Vibration and Control*, *16*(1), 127–145. doi:10.1177/1077546309103421

Solihin, M. I., & Wahyudi, N. A. (2009). Sensorless anti-swing control of automatic gantry crane using dynamic recurrent neural network-based soft sensor. *International Journal of Intelligent Systems Technologies and Applications*, *6*(1), 112–127. doi:10.1504/IJISTA.2009.022691

Trabia, M. B., Renno, J., & Moustafa, K. (2006). A General anti-swing fuzzy controller of an overhead crane with hoisting. *Proceedings of IEEE International Conference on Fuzzy systems*, 627-634. 10.1109/FUZZY.2006.1681777

Wahyudi & Jalani. (2006). Robust fuzzy logic controller for an intelligent gantry crane system. *Proceedings of 1st International conference on Industrial and Information systems*, 497-502. doi: 10.1109/ICIIS.2006.365778

Zadeh, L. A. (1965). Fuzzy Sets. *Information and Control, 8*(3), 338–353. doi:10.1016/S0019-9958(65)90241-X

Zadeh, L. A. (1975). The concept of a Linguistic variable and its application to approximate reasoning-I. *Information Sciences, 8*(3), 199–249. doi:10.1016/0020-0255(75)90036-5

Zairulazha, Z. (2005). Modeling and Vibration control of a Gantry crane (Master's report). University Teknologi Malaysia.

Chapter 15
Finite Element Analysis-Based Thermo-Mechanical Performance Study of Heavy Vehicle Medium Duty Transmission Gearbox

Ashwani Kumar
Graphic Era University, India

Yatika Gori
Graphic Era University, India

Pravin P. Patil
Graphic Era University, India

ABSTRACT

The main objective of this chapter is to investigate the performance of automobile transmission gearbox under the influence of load, rotational speed, and lubrication on multi speed gearbox gear surface. Gear oil SAE 80W-90 was used as gearbox lubricant, for cooling of transmission gearbox for high performance. An assumption has been made at the air-gear oil mist within transmission is under steady state condition, in isothermal equilibrium with the transmission gear oil bath of lubricant. The lubrication in multi speed transmission is subjected to thermo-elastohydrodynamic lubrication. The present chapter deals with the thermo-mechanical performance study of multi speed transmission (4 speed, excluding reverse gear) system, which combines transient structure analysis of the gear train assembly. The engaged gear teeth pairs transmit torque subjected to thermo-elastohydrodynamic arrangements of lubrication. The study here analyzed transmission in second gear pair.

DOI: 10.4018/978-1-7998-4939-1.ch015

INTRODUCTION

In vehicle transmission gearbox assembly gear oil is used as lubricant for cooling of gear train assembly and release the heat to cooling fluid of radiator. Gear oil can perform the cooling function at (-18 to 100)^{0}C temperature. Gear oil provides good sealing, friction durability, antifoam and non-corrosive to gear train parts and internal components of gearbox casing. 4-speed transmission gearbox excluding reverse gear was studied for transient structural and thermal characteristics. Researchers have performed various types of studies on transmission gearbox casing, gear train assembly, gear oil properties, role of gear oil for transmission efficiency and flow of lubricant in mini-micro channels heat exchangers.

(Kumar and Patil, 2014) have studied the dynamic vibration problem of heavy vehicle truck transmission system. The simulation results were obtained for different looseness condition of transmission casing. Zero displacement constraint based boundary condition was applied by constraining all 37 bolt holes. The loose casing frequency varies from (1311-3110) Hz. (Karamavruc et al., 2011) have studied the heat transfer and fluid-flow characteristics of an automatic transmission fluid (ATF). They have investigated the thermo-physical properties of transmission fluid. Specific heat capacity, density, thermal conductivity and viscosity are the thermo-physical properties that highly depend on temperature. The viscosity of ATF is very important as it change in exponential manner with change in temperature but other thermo-physical properties shows linear relation with temperature. (Bozca and Fietkau, 2010) have studied the rattle noise problem in vehicle transmission by optimization of gearbox geometric design. Rattle noise is caused by torsional vibration transmitted from engine to the shaft of transmission. In experimental work the rattle noise were calculated by varying various parameters. The numerical, analytical work was also performed. (Zimparov, 2002) has studied the optimal design problem of the heat exchangers. The objective was to design a heat exchanger with minimum system losses. The entropy generation and irreversible friction flow problem was studied. (Salah et al., 2007) have investigated radiator heat exchanger, which is used for transmission oil cooling. The aim of study was to find a relation between transmission cooling and vehicle fuel economy. The results were in favor that if the transmission gear train cooling is proper and efficient, it increases the vehicle fuel economy. The results shows that improved use of ATF can reduce the fuel use by 1.0-1.4%.

(Quaiyum, 2012) has studied the Minichannel Heat Exchanger (MICHX) application. He studied the ATF cooling and flow behaviour. The wind tunnel based experimental setup was used for the micro channel analysis. ATF flow on a mini channel heat exchanger was studied for varying convective heat transfer coefficient and Reynold's no. (Pennestri et al., 2012) have calculated the efficiency of epicyclic gear trains used in hybrid vehicles. They have also studied the power-flow in the transmission by using graph theory approach. (Kumar et al., 2015) have studied the transmission casing for different material application. A series of four materials were studied for vibration based suitability analysis. Vehicle transmission power-train thermo-elastohydrodynamic lubrication, dynamics, thermal balance, gear interaction and gear oil were studied (Cruz et al., 2012). Convective heat transfer co-efficient (h) measure how faster heat transfer takes place. The correlation shows that the h changes linearly with Reynold's No. an exponent of 1.044 (Quaiyum, 2012). Transmission gear train components material properties were considered for dynamic analysis (Qin-man, 2011). In this study the effect of gear oil on thermo-mechanical performance was measured for high performance of multi speed transmission gearbox. The inclusion of convective heat transfer coefficient is essential in the study of multi speed transmission when dealing with increasing the transmission thermo-mechanical performance. Under isothermal conditions the temperature around transmission gear train was assumed throughout same as bath gear oil. Steady state, isothermal condition reduces the complexity of simulation when considering average convective heat transfer coefficient.

MODELLING OF GEARBOX ASSEMBLY

Multi speed transmission gearbox has been investigated here. The transmission gear train assembly comprises input shaft, output shaft and lay shaft. Noise, vibration and harness (NVH) with thermo-mechanical performance are the two important design parameters of compact multi speed transmission. Transmission gearbox consists of arrangement of gears and gearbox casing. All designing parameters were obtained from measurements and drawing sheets. The casing encloses the gear assembly, input, output and lay shaft. In this study only direct drive 4-speed gears were considered. Figure 1 shows the full assembly of transmission gearbox of vehicle. The isometric view of vehicle shows casing designed in three parts. The main part covers the gear assembly and other two parts used to enclose the transmission casing. Figure 2 shows 4-speed gear assembly and shafts for deformation analysis. The numerical simulation was performed for the transient structural, thermal analysis of gear train under the influence of load, rotational speed and convection heat transfer coefficient. Gear oil works as gearbox lubricant to cool the gears and transfer the heat through convective process. The overheating of gears reduces the efficiency and life span of gears. The transmission gearbox assembly replacement is costly, so this study focus the thermal prospectus of gearbox so that the life span of a transmission gear train can be increased. FEA simulation works on meshing of objects. The meshed model of gear assembly is shown in Figure 3. The gear assembly is meshed using 5, 75, 383 nodes and 3, 39,898 elements (Linear Tetrahedron Element, Tet 4). This research work presents a strong base for the thermal-mechanical analysis of gearbox to increase the performance by understanding the role of gear oil and loading parameters such as torque, rotational speed and average convection heat transfer coefficient. Transmission gearbox components were assigned materials (Qin-man [60]) for accurate analysis of temperature on gearbox surface.

Figure 1. Heavy vehicle medium duty transmission gearbox assembly

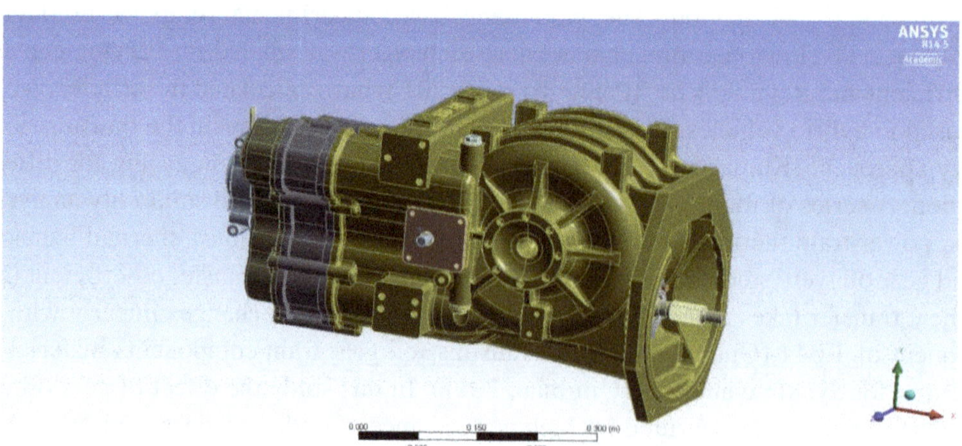

Figure 2. Gearbox assembly of 4-speed transmission gearbox

Figure 3. FEA meshed model of gearbox

MATERIAL PROPERTIES AND BOUNDARY CONDITIONS

The teeth in contact causes thin film thickness lubrication, thus a thermo-elastohydrodynamic arrangement of lubrication is expected. The thin film thickness is due to heat generated between meshing teeth pairs, reduces the viscosity of lubricant. The simulation study of transmission model is reported here. The noted point is the isothermal temperature on gear train is assumed to be the temperature of bath of oil. FEA simulation mechanical strength was measured using structural analysis and steady state thermal analysis was performed by applying isothermal temperature of gear oil bath lubricant on gear train. By applying uniform isothermal temperature on gear train, the temperature variation on gears was measured under the effect of heat, friction and gear oil temperature. The dynamic response in terms of stress, strains and deformation was evaluated when vehicle was running on second gear at 1500 rpm and having 245 Nm torque. When vehicle was running on full loading and high rpm the heat generation due to engaged gear pair meshing is very high.

High internal temperature around gear train changes the thermo-physical properties of gear oil which effects the vehicle performance and service life. In this research work gear oil thermal properties (convective heat transfer coefficient (h) and gear oil temperature) effect was studied on gears surface. In first step 1500 rpm with 245 Nm torque load was applied on second gear assuming average convection heat transfer coefficient varies (100-600) w/m^2 k and gear oil temperature varies in range of $(70\text{-}80)^0$C. In second step for same loading conditions gear oil temperature was kept in higher range of $(80\text{-}100)^0$C, by applying these conditions the effect was measured on gear surface in terms of temperature.

DYNAMIC STRUCTURAL ANALYSIS

FEA based Ansys 14.5 evaluate the results of structural and steady state thermal analysis. Structural results evaluate the performance of transmission gear train on strength point. Inertia and damping effects was not considered for simulation. When vehicle was running at full loading and 1500 rpm and generates 245 Nm torque. Torque and rotational velocity were applied for full dynamic loading. At maximum torque condition the gear meshing causes heat generation in large value that was studied in this research work. Von-mises stress and strain were mentioned in Figure 4. Figure 4 (a) the simulation results shows that the equivalent (Von-Mises) stress distribution is within safe limit at its minimum value. The shear elastic strain due to dynamic load shows deformation. This deformation is safe (Figure 4 (b)). No high deformation regions were identified.

Figure 4 structural results show that the gear train design is safe to sustain the applied dynamic load. Transient structural analysis was performed at 1500 rpm when vehicle was on high gearing (third gear selection) and producing 119.31 Nm torque. The time period of full loading was 9 seconds and dynamic behaviour of 4-speed transmission gear box was evaluated. Figure 4 (c and d) shows transient results. Figure 4 (c) explains the total deformation variation in gears. When gears are in meshing at full loading the red hues region shows stress concentration where deformation is maximum. The blue hues signify the minimum deformation level and light green hues shows the regions where deformation is average. Maximum value of total deformation (red hues) is 0.126 mm within limit. Figure 4 (d) shows maximum principal stress variation on 4-speed transmission gearbox. Maximum principal stress available on gears is 5.346e8 Pa. Blue hues show the principal stress variation, which is uniform in nature. A uniform stress signifies that failure chances are less.

Figure 4. Dynamic Structural Analysis of transmission gearbox: (a) Von-Mises stress distribution in gearbox surface. (b) Shear elastic strain distribution in XY plane. (c)Total Deformation on gearbox (transient). (d) Maximum Principal Stress variation on gears (transient).

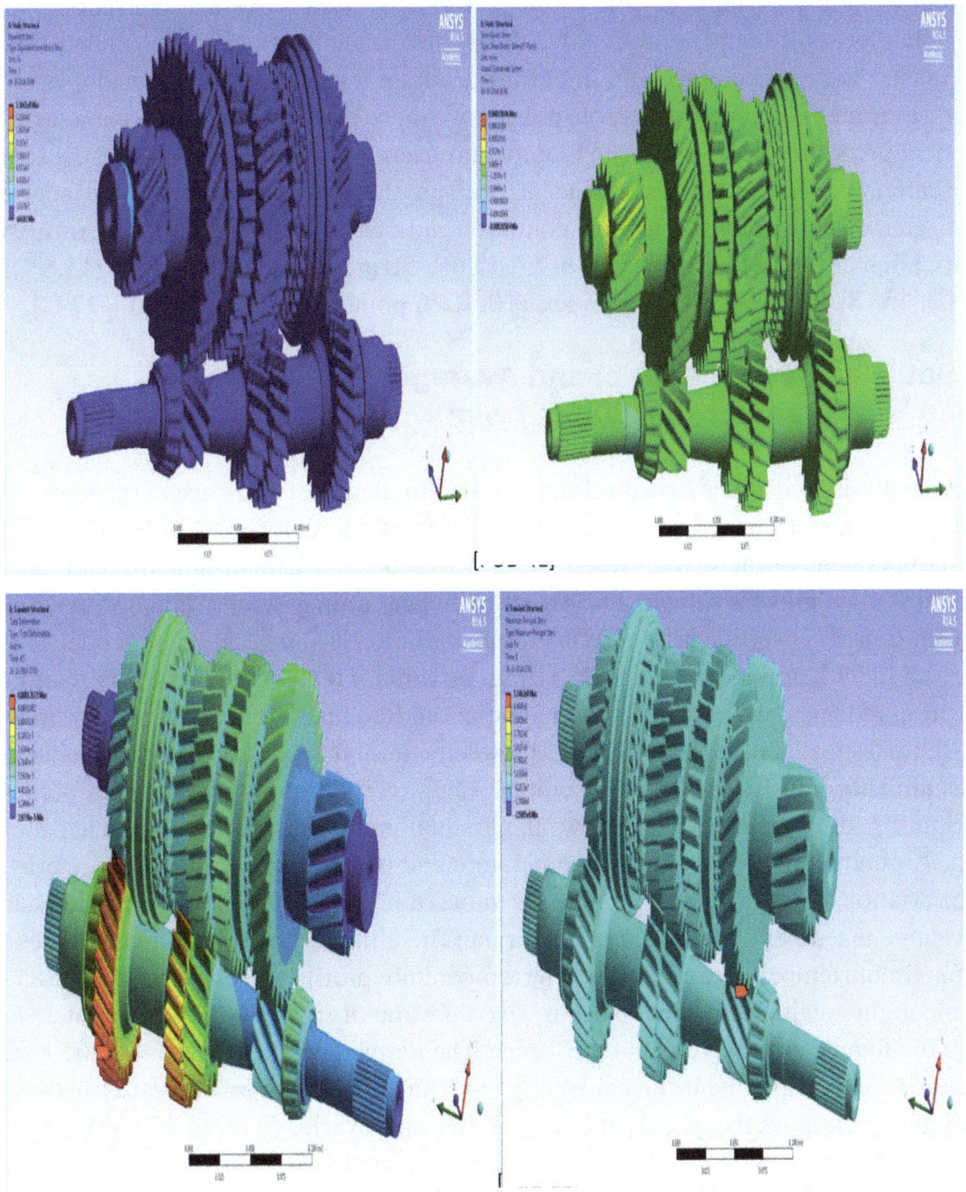

STEADY STATE THERMAL ANALYSIS

The present research work concerns with the transmission assembly of medium duty trucks. In earlier study the authors have considered only gear tooth or simple geometry of transmission but here we have simulated the full assembly to highlight the effects of different operating conditions (load, rotational speed, lubrication). In steady state thermal analysis the effect of gear oil heating was considered. The gear oil temperature varies between (-18 to 100)^{0}C. When vehicle is running at 1500 rpm and 245 Nm loading the gear oil temperature may vary (70-100)^{0}C. Gear oil temperature variation is highly nonlinear problem totally depends on operating condition and loading. To simulate the gear oil heating a constant isothermal temperature was applied with varying (100-600) w/m^{2}k average convective heat transfer coefficient. When gear oil temperature is 80^{0}C the performance of transmission fluid is high, and its life span increases. Gear oil cooling is an important medium for release of frictional heat from casing. The studies of gear oil play an important role to improve the efficiency in automotive industry. Gear oil thermo-physical properties depend on temperature and frictional environment inside the transmission.

In this research work efforts have been made to measure the surface temperature variation of gearbox surface due to gear oil temperature, frictional condition and dynamic loading under the influence of SAE 80W-90 lubricating gear oil. The properties of SAE 80W-90 are- density 887 kg/m^3 (15.6^{0}C), viscosity 139cSt (40^{0}C), 15 cSt (100^{0}C), Viscosity Index 110, flash point 218^{0}C, pour point (-27^{0}C).

Gear Oil Bath Temperature (80^{0}c) and Average Convection Heat Transfer (h=100-600) w/m^{2}k

Finite element analysis used for the structural and thermal simulation of transmission gearbox. The internal temperature of gearbox has an influence on thermal stresses and deformations leads to failure. Figure 5 shows the temperature variation at different point of gearbox surface for gear oil bath temperature of 80^{0}C. Figure 5(a) highlights the temperature variation in gear train assembly at the value of h_{100w/m^2k} for convective heat transfer co-efficient (h). Temperature variation is shown in colour code. The minimum temperature is 342.94 k and maximum is 350.01 k. Maximum temperature effect is found on counter shaft right end in red hues, which was fixed in simulation. Red hues signify the maximum temperature change and thermal stress generation portion. In between minimum and maximum temperature 8 another temperatures are shown in figure. The temperature varies very gradually at different points, it can be seen from Figure 7. In Figure.7, h_{100} (1) shows the gradual increment in temperature at different points of gear train. Around second gear pair the temperature and deformation is high. Figure 5(b) shows the temperature variation in gear train assembly at the value of h_{200w/m^2k}. As the value of h increases there is very small change in temperature profile of gear train. The minimum temperature decreases 0.09 k and increase in maximum temperature of 1.58 K. The temperature profile varies between (342.85-351.59) k. The high temperature region is at same place of fixed portion of right side end of counter shaft. Figure 7.5(c) shows the temperature increment on 3rd gear. The temperature reached to 350.01 k. On 2nd gear the temperature is 349k under the influence of $h_{300w/m^2k.}$ The temperature profile varies between (342.93-352.03) k. Figure 5(d) shows the gear train temperature profile varies (342.92-352.38) k. First gear is in loose meshing and its temperature is 347.12 k and 2nd, 3rd gear shows temperature increment and have a value of (349.23-351.33) k. The lay/counter shaft also shows variation of temperature (346.07-352.38) k for $h_{400w/m^2k.}$ The 3rd gear profile shows temperature increment in red hues and this area is prone for thermal stresses more for ($h_{500w/m^2k,}$ h_{600w/m^2k}) convective heat transfer coefficient values. In Figure 7, h_{400} and h_{500} shows the gear train temperature profile variation.

Figure 5. Gearbox surface temperature variation- Isothermal gear oil bath temperature (80^0C). (a) Temperature variation at h = 300 W/m² k (b) Temperature variation at h = 400 W/m² k (c) Temperature variation at h = 500 W/m² k (d) Temperature variation at h = 600 W/m² k

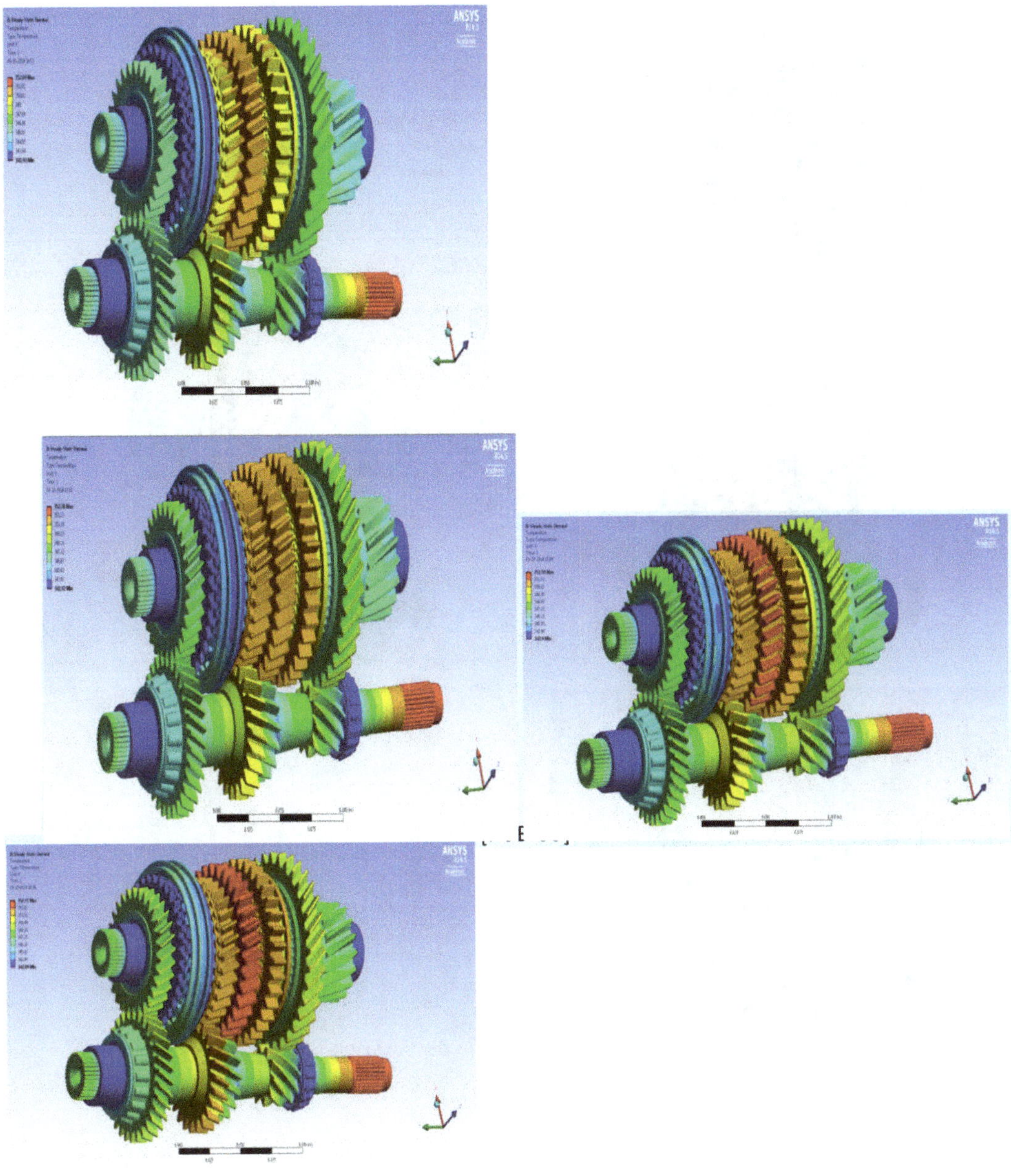

Gear Oil Bath Temperature (100^0c) and Average Convection Heat Transfer (h=100-600) w/m²k

For second part of study the gear oil temperature is increased to 100^0 C. Figure 6 shows the temperature profile of gearbox surface at gear oil temperature of 100^0C. Figure 6 (a) shows the temperature profile of gear train assembly at h_{100w/m^2k} convective heat transfer coefficient (h) value. The minimum tempera-

Figure 6. Temperature profile variation on gearbox surface- Isothermal gear oil bath temperature (100^{0}C). (a) Temperature variation at h = 200 W/m^2 k (b) Temperature variation at h = 300 W/m^2 k (c) Temperature variation at h = 400 W/m^2 k (d) Temperature variation at h = 500 W/m^2 k

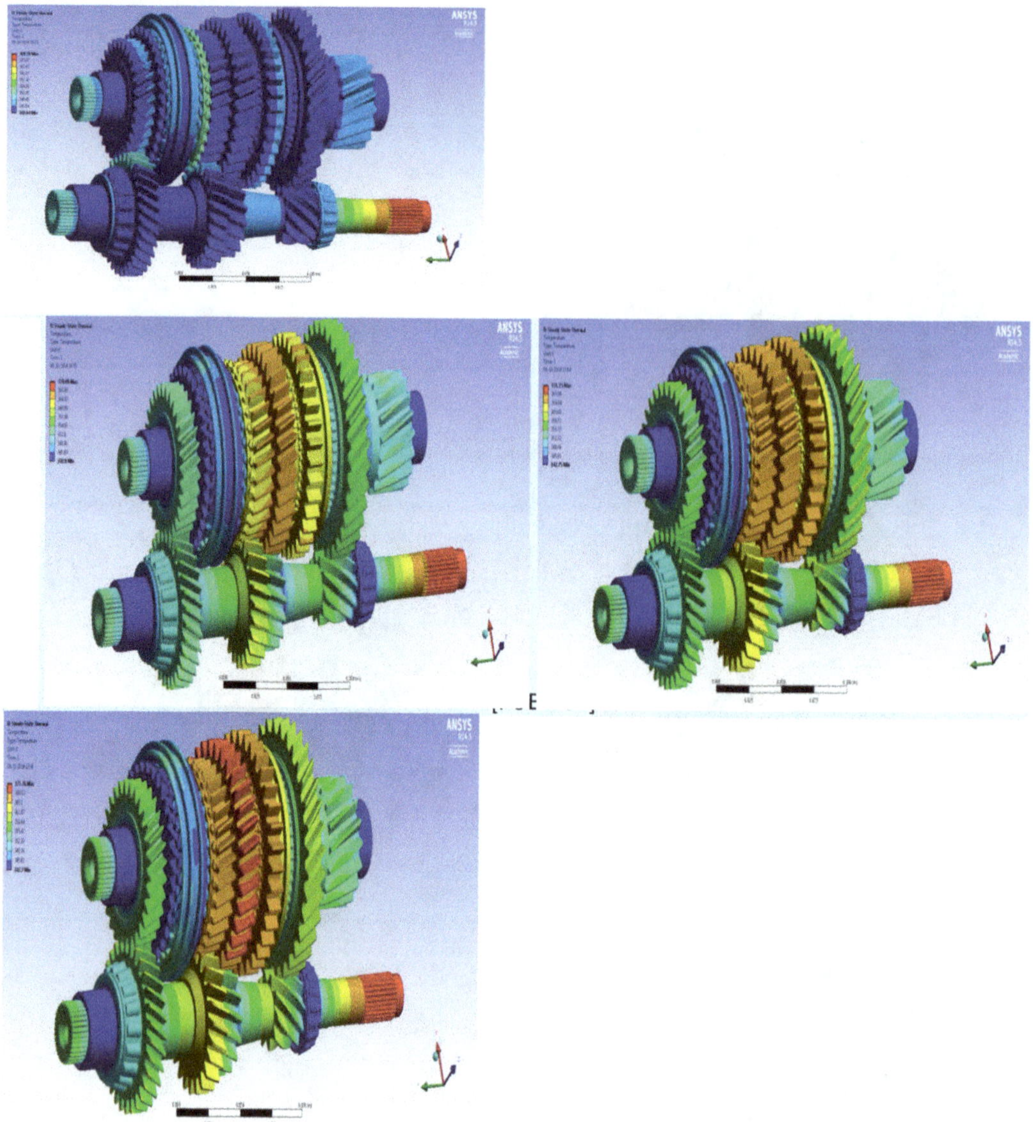

ture is 342.81 k and maximum is 364.02 k. Maximum temperature effect is found on same place as for $h_{100w/m^2 k}$ at 80^{0}C on counter shaft right end in red hues. Red hues shows hot areas subjected to thermal stresses and deformation. Figure 8, h_{100} shows the gradual increment in temperature profile of gear train at different points. Figure 6 (b) explains the temperature variation in transmission gear train at $h_{200 w/m^2 k}$. As the value of h increases there is increase in temperature of gear train by 4.76 k, mention by red hues. The high hues temperature region is found at fixed portion of right side end of counter shaft. Figure 6 (c) shows dark yellow hues on 3rd gear. The dark yellow hues designate increment in temperature. The maximum value of temperature reached to 370.09 k. The right side fixed counter shaft end temperature is maximum for h_{300w/m^2k}. The temperature profile varies between (342.8-370.09) k. In Figure 7.8, h_{300} (3) shows the temperature linear temperature variation at different point on gear train. Figure 6 (d) shows gradual change in temperature. Gear profile of 2nd and 3rd shows red hues of hot areas. In Figure 8, h_{600} shows the gear train temperature profile variation.

FEA RESULTS AND VALIDATION

Multi speed transmission gearbox and gear oil analysis is highly nonlinear problem. From FEA analysis (Figure 5 and 6), when h is 100 w/m²k and gear oil temperature is constant at 80⁰C, the gear train profile temperature varies very gradually and linearly. As we increase the value of convective heat transfer coefficient to 200 W/m² k the difference in maximum temperature is only 1.58 k. Further as there is increment in value of h (300,400,500 & 600) the difference in gearbox surface temperature profile varies only (1.58-2.7) k (Figure 7).

Figure 7. Temperature variations on gearbox surface at oil bath temperature 80⁰C with varying h.

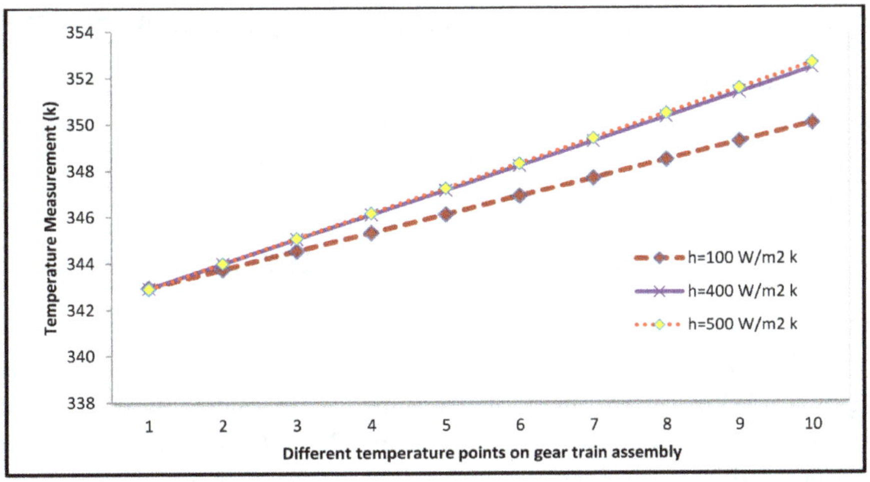

Figure 8. Temperature variations on gearbox surface at oil bath temperature of 100⁰C with varying h.

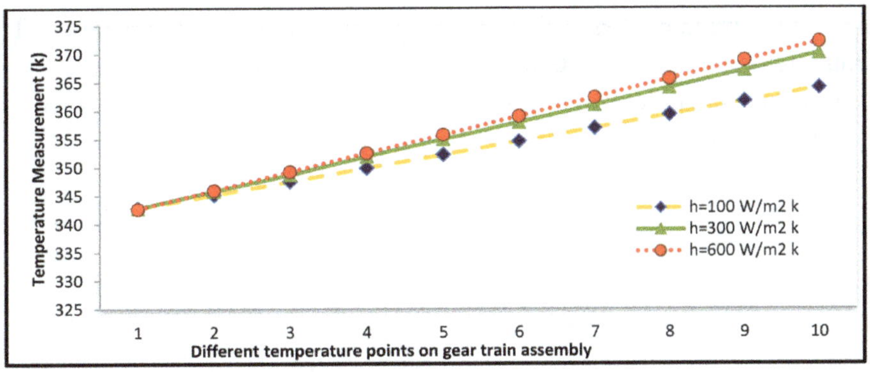

Same temperature profile as Figure 7 is generated when gear oil temperature is constant at 100⁰C and value of h increases from (100-600) W/m^2 k, the difference in temperature is (1-8) k (Figure 7.8). From Figure 7 and Figure 8 it can be concluded that if gear oil temperature is constant (isothermal) the gear train thermal stresses at each convective heat transfer coefficient value is within permissible limits. The temperature profile of gear train shows that the lower temperature is approximate constant (Figure 5 & 6) and maximum temperature is varying by (1-5) k (Figure 7 & 8). Convective heat transfer co-efficient (h) refer to transfer of heat with fluid movement. In general for the increase value of h the rate of heat transfer increases. It signifies that the gearbox surface temperature should be reduced with increase in h value. The numerical simulation results show the same results as per thermal concept. In this research work for thermal analysis gear oil temperature was varied for 80⁰C and 100⁰C and h varied for (100-600) W/m^{2}k, for these loading conditions the temperature at 10 different gear assembly points was measured.

Figure 9. FEA simulation result validation using experimental results, Long et al. [38]

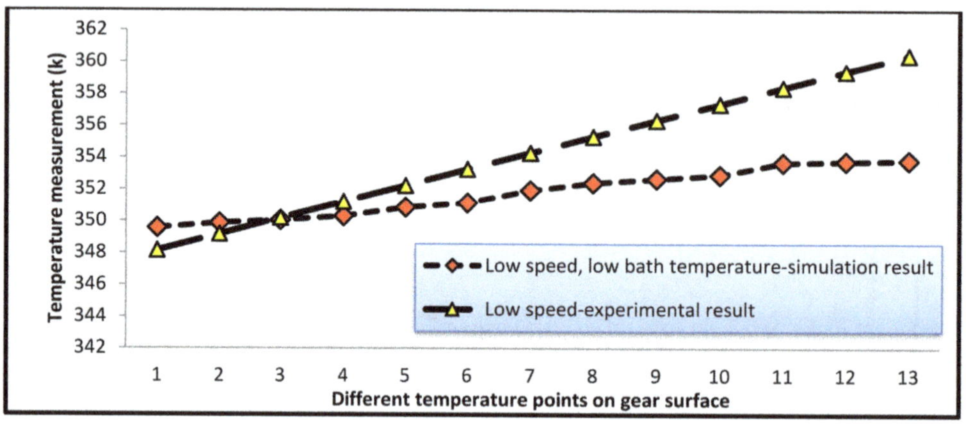

FEA based numerical simulation result was validated with experimental results available in literature (Figure 9). Long et al. [38] have performed the experimental thermal analysis of single gear tooth of vehicle gearbox. The experimental study was performed at 2000, 6000 and 10000 rpm with combined loading. At different rotational speed and loading conditions the gear surface temperature was evaluated. The gear temperature is mainly governed by loading condition, rotational speed has less effect. In this study when vehicle was running at low speed (1500-2000) rpm at 245Nm load value, the gearbox surface temperature varies (349.56-353.82)k and the experimental results varies (348.14-360.42)k. FEA simulation temperature range lies within experimental gear surface temperature. The deviation in results for minimum temperature is less than 1% and for maximum temperature distribution deviation is less than 2%. It indicates that deviation in result is less than 5%, it shows satisfactory measurement of gearbox surface temperature.

CONCLUSION

The analysis of multi speed transmission gearbox with gear oil has analytical and theoretical significance for thermo-mechanical performance improvement of gearbox. The FEA simulation results of medium duty transmission gearbox concludes that-

1. Gradual increase in temperature around engaged gear pairs were reported due to effect of heat generation on surface of gears under the influence of gear meshing, frictional heat, average heat transfer and gear oil bath temperature. Increment in temperature is very low, in resultant thermal stress generation due to meshing action of gear will be very less.
2. It was found that varying convective heat transfer (h) method reduces thermal stresses generation (Figure 5 & 6). All stresses like equivalent von-mises stresses and thermal strain are in permissible range for the transmission gearbox.
3. The most important outcome of this analysis is that by varying convective heat transfer coefficient (h) phenomena overheating of gear oil is less, which refer cooling will be more and it increase the thermo-mechanical performance of multi speed transmission system.
4. Increase in thermo-mechanical performance of multi speed transmission system signifies higher fuel economy.

REFERENCES

Abouel-Seoud, S. S., Mohamed, E. S., Abdel-Hamid, A. A., & Abdallah, A. S. (2013). Analytical Technique for Predicting Passenger Car Gearbox Structure Noise Using Vibration Response Analysis. *British Journal of Applied Science and Technology*, *3*(4), 860–883. doi:10.9734/BJAST/2013/3277

Ansys R 14.5. (2013). Academic, Structural analysis Guide.

Cho, D., & Lee, D. (1998). Manufacturing of co-cured aluminum composite shafts with compression during co-curing operation to reduce residual thermal stresses. *Journal of Composite Materials*, *32*, 1221–1241. doi:10.1177/002199839803201205

Cho, D., Lee, D., & Choi, J. (1997). Manufacture of one-piece automotive drive shafts with aluminum and composite. *Composite Structures*, *38*(1-4), 309–319. doi:10.1016/S0263-8223(97)00065-2

Chowdhury, M. A., & Helali, M. M. (2008). The Effect of Amplitude of Vibration on the Coefficient of Friction. *Tribology International*, *41*(4), 307–314. doi:10.1016/j.triboint.2007.08.005

Chowdhury, M. A., & Helali, M. M. (2010). The Influence of Natural Frequency of the Experimental Set-up on the Friction Coefficient of Stainless Steel-304. *Tribology in Industry*, *32*, 19–24.

Crowther, A., Singh, R., Zhang, N., & Chapman, C. (2007). Impulsive response of an automatic transmission system with multiple clearances: Formulation, Simulation and experiment. *Journal of Sound and Vibration*, *306*(3-5), 444–466. doi:10.1016/j.jsv.2007.05.053

Czech, P. (2012). Diagnosis of industrial gearboxes Condition by vibration and time frequency, Scale-frequency, frequency-frequency analysis. *Metalurgija*, *51*, 521–524.

De-gang, L., Weixing, H., & Fengzhou, W. (2004). Fatigue Life Analysis of a Component Based on the Finite Elements Technology. *Journal of The China Railway Society*, *26*, 47–51.

Dogan, S. N. (1999). Loose part vibration in vehicle transmissions Gear rattle. *Transactions Journal of Engineering and Environmental Science*, *23*, 439–454.

Gordon, T.J., & Bareket, Z. (2007). *Vibration Transmission from Road Surface Features– Vehicle Measurement and Detection.* Technical Report for Nissan Technical Center North America, Inc. UMTRI- 2007-4.

Ingle, R., & Ahuja, B. (2006). An experimental investigation on dynamic analysis of high speed carbon epoxy shaft in aerostatic conical journal bearings. *Composites Science and Technology*, *66*(3-4), 604–612. doi:10.1016/j.compscitech.2005.03.021

Kar, C., & Mohanty, A. R. (2006). Monitoring gear vibrations through motor current signature analysis and wavelet transform. *Mechanical Systems and Signal Processing*, *20*(1), 158–187. doi:10.1016/j.ymssp.2004.07.006

Karamavruc, A., Shi, Z., & Gunther, D. (2011). *Determination of Empirical Heat Transfer Coefficient via CFD to Predict the Interface Temperature of Continuously Sleeping Clutches.* SAE Technical Paper 2011- 01-0313.

Kostic, S. C., & Ognjanovic, M. (2007). The Noise Structure of Gear Transmission Units and the Role of Gearbox Walls. *FME Transactions*, *35*, 105–112.

Kumar, A., Dwivedi, A., Jaiswal, H., & Patil, P. P. (2015). Material Based Vibration Characteristic Analysis of Heavy Vehicle Transmission Gearbox Casing Using Finite Element Analysis (FEA). *Advances in Intelligent Systems and Computing*, *308*, 527–533. doi:10.1007/978-81-322-2012-1_56

Kumar, A., Jaiswal, H., Mishra, A., & Patil, P. P. (2014). Mode Shape Vibration Analysis of Truck Transmission Housing Based on FEA. *Proc. of the Intl. Conf. on Advances in Engineering and Technology*, 334-338. DOI: 10.15224/ 978-1-63248-028-6-03-66

Kumar, A., Joshi, H., Saini, A., & Patil, P. P. (2014). Vibration Diagnosis of Heavy Vehicle Truck Transmission Gearbox Casing Using FEA. Composite Materials-Proc. of Intl. Conf. on Mechanical Engineering, 169-177.

Kuo, K. (2011). Simulation and Analysis of the Shift Process for an Automatic Transmission. *World Academy of Science, Engineering and Technology*, *52*, 341–347.

Li, Y., & Guagnqiang, W. (2006). Analysis on Fatigue Life of Rear Suspension Based on Virtual Test rig. *Computer-Aided Engineering*, *15*, 128–130.

Long, H., Lord, A. A., Gethin, D. T., & Roylance, B. J. (2003). Operating temperatures of oil-lubricated medium-speed gears: Numerical models and experimental results, Proceedings of the I MECH E part G. *Journal of Aerospace Engineering*, *217*(2), 87–106.

Minfeng, H., & Yingchun, J. (2008). Fatigue life analysis of automobile component based on FEM. *Mechanical Research and Application*, *21*, 57–60.

Miyasato, H. H., Simionatto, V. G. S., & Junior, M. D. (2011). Study of the gear rattle phenomena in automotive powertrain systems. *Proceedings of COBEM 2011. 21st International Congress of Mechanical Engineering*, 1-10.

Nacib, L., Pekpe, K. M., & Sakhara, S. (2013). Detecting gear tooth cracks using cepstral analysis in gearbox of helicopters. *International Journal of Advances in Engineering and Technology*, *5*, 139–145.

Pennestrì, E., Mariti, L., Valentini, P., & Mucino, V. H. (2012). Efficiency evaluation of gearboxes for parallel hybrid vehicles: Theory and applications. *Mechanism and Machine Theory*, *49*, 157–176. doi:10.1016/j.mechmachtheory.2011.10.012

Pro-E 5.0. (2013). *Designing guide manual.*

Qin-man, F. (2011). Modal analysis of a truck transmission based on ANSYS. *Fourth International conference on Information and Computing*. doi:10.1109/ICIC.2011.75

Quaiyum, A. (2012). *Experimental investigation of automatic transmission fluid (ATF) in an air cooled mini-channel heat exchanger* (M.S. Thesis). Department Mechanical, Automotive, and Materials Engineering, University of Windsor.

Saada, A., & Velex, P. (1995). An extended model for the analysis of the dynamic behavior of planetary trains. *ASME Journal of Mechanical Design*, *117*(2A), 241–247. doi:10.1115/1.2826129

Sayer, F., Bürkner, F., Blunk, M., Wingerde, A. M., & Busmann, H. G. (2009). Influence of Loads and Environmental Conditions on Material Properties over the Service Life of Rotor Blades. *DEWI Magazin*, *34*, 24–31.

Sevkat, E., & Tumer, H. (2013). Residual torsional properties of composite shafts subjected to impact loadings. *Materials & Design*, *51*, 956–967. doi:10.1016/j.matdes.2013.05.004

Singh, R. (2000). Dynamic design of automotive systems-Engine mounts and structural joints. *Dynamic Design of Automotive Systems*, *25*(3), 319–330. doi:10.1007/BF02703548

Solidedge. (2006). *Version 19.0.*

The Metals Databook. (2008). (4th ed.). Tata: McGraw- Hill.

Tuma, J. (2009). Gearbox Noise and Vibration Prediction and Control. *International Journal of Acoustics and Vibration*, *14*(2), 1–11. doi:10.20855/ijav.2009.14.2242

Vandi, G., Cavina, N., Corti, E., Mancini, G., Moro, D., Ponti, F., & Ravaglioli, V. (2014). Development of a software in the loop environment for automotive powertrain systems. *Energy Procedia*, *45*, 789–798. doi:10.1016/j.egypro.2014.01.084

Velex, P., & Flamand, L. (1996). Dynamic response of planetary trains to mesh parametric excitations. *Journal of Mechanical Design*, *118*(1), 7–14. doi:10.1115/1.2826860

Wang, J., Zheng, J., & Yang, A. (2012). An Analytical Study of Bifurcation and Chaos in a Spur Gear Pair with Sliding Friction. *Procedia Engineering*, *31*, 563–570. doi:10.1016/j.proeng.2012.01.1068

Yu, F., Li, Y., Sun, D., Shen, W., & Xia, W. (2013). Analysis for the Dynamic Characteristic of the Automobile Transmission Gearbox. *Research Journal of Applied Sciences, Engineering and Technology*, *5*(4), 1449–1453. doi:10.19026/rjaset.5.4886

Yulong, L., Xingzhong, L., Weipeng, L., & Hanyong. (2011). Hydraulic System Optimization and Dynamic Characteristic Simulation of Double Clutch Transmission. *Procedia Environmental Sciences*, *10*, 1065–1070. doi:10.1016/j.proenv.2011.09.170

Chapter 16
Pulsed Laser Micro-Forming of Thin Sheets

Kuntal Maji
https://orcid.org/0000-0002-5602-8948
National Institute of Technology, Patna, India

ABSTRACT

This chapter presents a study on the deformation behavior of thin sheets in pulsed laser micro-forming. Deformation mechanics of thin sheets under different operating conditions under pulsed laser irradiation are described and explained. Modeling of pulsed laser micro-forming for analyzing and optimizing the process parameters using different techniques mentioned in the published literature are discussed. The estimation of temperature and deformation fields in pulsed laser micro-forming of thin sheets of different shapes considering the effects of different process parameters such as laser power, spot diameter, pulse on time, pulse frequency, and sample size are presented. Response surface modeling and finite element analysis of the pulsed laser micro-forming process are described and explained. Experimental procedures for conducting pulsed laser micro-forming tests are described and explained. Recent investigations on pulsed laser micro-forming published elsewhere are presented, and possible scopes for future investigations are highlighted.

INTRODUCTION

Laser micro-forming is a non-contact technique to deform a metal sheet at micro-scale using the controlled applications of laser energy to produce a shock wave in a medium or to induce thermal stresses in the workpiece. A precise deformation or adjustment of thin sheet metal components by continuous and pulsed laser irradiations has potential applications in micro-manufacturing, micro-machines and devices, actuators, hard disks, microelectronics components and others. Pulsed laser micro-forming has advantages of very low laser irradiation time in the order of nanosecond, pico-second, femto-second etc. to generate high peak power compared to that of continuous laser irradiation. The pulsed laser micro-forming can be conducted by utilizing a stationary or a moving laser beam depending on the sample size and the desired shape to be formed. In pulsed laser micro-forming (PLμF), thin sheet metal parts are

DOI: 10.4018/978-1-7998-4939-1.ch016

heated in a discontinuous manner. The process parameters in pulsed laser forming are laser power, spot diameter, pulse frequency and duty cycle. The pulse on time depends on the combination of laser pulse frequency and duty cycle. As the pulse frequency increased at a constant duty cycle, the pulse on-time decreased and consequently, the energy per pulse decreased. The deformation mechanism in pulsed laser micro-forming depends on the process route, i.e., thermal, non-thermal, hybrid etc., combination of different process parameters, material properties and workpiece geometry. Pulsed laser micro-forming can be classified based on the deformation mechanism as shown in Fig. 1.

Figure 1. Classification of pulsed laser micro-forming (PLμF) processes

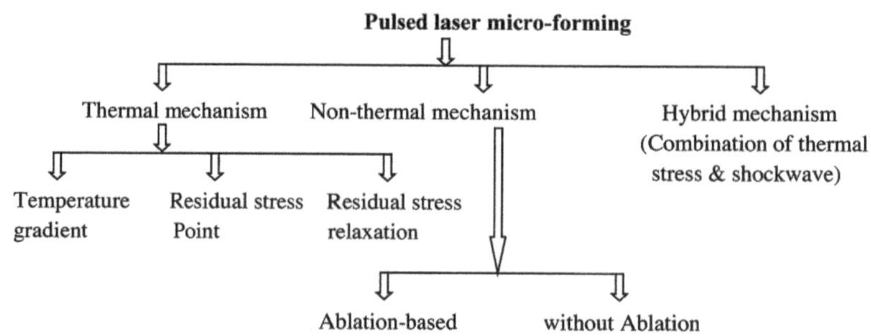

The deformation mechanism in pulsed laser thermal micro-forming can be explained by considering the temperature gradient, residual stress point and residual stress relaxation mechanisms. Laser power, duration of laser irradiation and size of plastic zone produced by the thermal stress decide the amount of deformation. Temperature gradient-based mechanism of thermal PLμF is shown in Fig. 2.

Figure 2. Deformation mechanism in thermal PLμF process

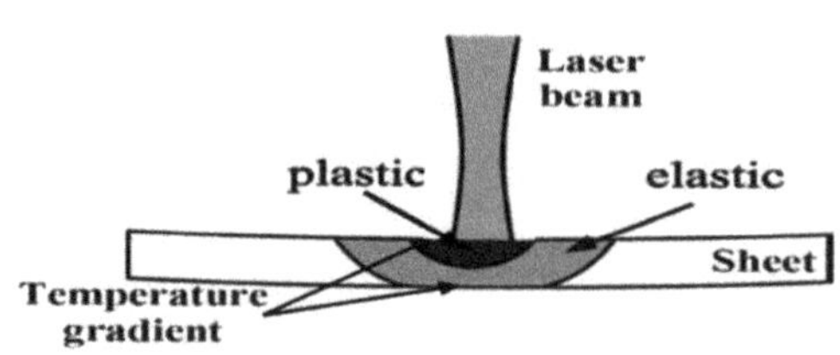

In case of multiple laser pulses irradiation at constant laser power and duty cycle, the pulse-on time and time between two pulses got reduced with the increase in frequency. If the time interval between the two pulses becomes more than the characteristic thermal diffusion time, then the heating at a given point will remain unaffected by the previous pulses. Otherwise, heating due to the previous pulse will have some effect on the next pulse. The magnitude of deformation depends on temperature gradient and peak surface temperature reached which is decided by the Fourier number.

There are various mechanisms of laser forming process depending on the processing conditions or specific combinations of the process variables, i.e., component geometries and laser processing conditions. In laser forming, though the deformation takes place through the heating and cooling cycles, there are 25 variations of thermal forming mechanisms had been investigated in the Center for Laser Technologies of Metals (CLTM) of the Polish Academy of Sciences (PASc.) and Kielce Technical University, Poland. The mechanisms of laser forming process can be classified based on the existence of thermal gradient. The thermal forming mechanisms with the existence of temperature gradient are the Temperature Gradient Mechanism (TGM), Residual Stress Point Mechanism (RSPM) and Martensite Expansion Mechanism (MEM); and those without temperature gradient are the Buckling Mechanism (BM), Upsetting Mechanism (UM) and Residual Stress Relaxation Mechanism (RSRM). However, the three main mechanisms of laser forming process mostly reported in literature are TGM, BM and UM. Now, among these three mechanisms, the TGM is the most common one and widely used mechanism because of its ease of control. The details of different mechanisms and their comparisons are discussed below.

The temperature gradient mechanism (TGM) is the most widely reported laser-forming mechanism, also called the bending mechanism because an out of plane bend is produced. Due to the rapid heating of the surface by the defocused laser beam and slow heat conduction into the sheet, a steep temperature gradient along the thickness sets in and thus, results in a differential thermal expansion. To create a high temperature gradient, laser beam must traverse the workpiece at such a speed that the thermal diffusion depth is small compared to the workpiece thickness. The thermal diffusion depth is approximately given by $d_{th} = \sqrt{\kappa\tau} = \sqrt{\kappa d / v}$, where κ, τ, d and v are thermal diffusivity, interaction time, laser spot diameter and laser scan speed, respectively. For establishing TGM, $d_{th} < s, or \sqrt{\kappa d / v} < s, or\ \kappa d / s^2 v < 1$, where s is sheet thickness and $\kappa d / s^2 v$ is Fourier's number. Therefore, the TGM become dominant, if the Fourier number is very small, less than one. In TGM, as the material is heated, the thermal expansion of the top surface becomes greater than that of the bottom surface, initially. This results in a bending opposite to the laser beam, which is called counter-bending. Due to this bending moment, a small amount of plastic tensile strain occurs at the heated top surface. With further heating, the bending moment opposes the counter-bending away from the laser beam, and the yield stress of the material is reduced with the temperature rise. Once the thermal stress reaches the temperature-dependent flow stress of the material, any additional thermal expansion is converted into a plastic compressive strain, because of the fact that the free expansion is restricted by the surrounding material. During cooling, the material contracts again in the upper layer, and as it has been compressed, there is a local shortening of the upper layer, and a bending angle develops that finally bends the specimen towards the laser beam. Generally, the TGM may be used for bending thick sheets along straight lines towards the laser beam. The irradiation of the surface may be repeated in order to increase the bending angle. The amount of bending per pass is approximately 1–3°. It remains constant for the first few passes for a given material, laser power and beam size. Thereafter, the bending angle per pass starts to fall owing to work hardening and thickening of the material at the bent edge.

BACKGROUND

Investigations on pulsed laser micro-forming of sheet metal were carried out in past by different researchers considering different processing conditions utilizing numerical simulations and experiments as discussed in the following. Vollertsen and Sakkiettibutra (2010) presented a review on different techniques of forming of sheet metal by a laser beam. Process design and optimization of thermal laser micro-forming processes by scaling rules were also discussed. Qi et al. (2010) carried out experimental investigations and finite element analysis to study the effects of parameters on the laser adjustment of a rod. Griffith et al. (2010) carried out finite element simulation to study the deformation behavior in laser forming at macro- and micro-scales. Jamil et al. (2011) performed finite element simulations of laser forming of stainless steel sheets and tubes under both buckling and upsetting mechanisms using a high power diode laser. The process mechanisms were explained and the simulations results were validated by published experimental data. Bellisario and Quadrini (2012) investigated formability of 5005 aluminium alloy sheet by laser beam. Parametric effects were studied through numerical simulation and experiments, and significant process parameters for controlling the process were also identified. Laser thermal adjustments of actuators made of stainless steel and aluminum alloy were studied by investigators (Shen & Yao, 2011; Shen et al. 2012; Shen at al., 2015). Actuators with different cut-out geometries was investigated by Shen et al. 2012 considering the effects of different process parameters, and it was concluded that the effects of cut-out geometries must be considered in optimizing the process parameters. Zheng et al. (2013) investigated the effect of confining overlay in micro-scale bulge forming of copper foil by laser induced shock wave mechanism. The confining layer observed to have significant effect on the deformation and surface morphology. Sagasika et al. (2014) investigated pulse laser micro-forming of thin titanium sheet utilizing ultra-short-pulse laser radiation for manufacturing of complex shaped micro-parts. Ismail et al. (2016) developed a three dimensional thermo-mechanical finite element model to study thermal deformation of thin stainless steel sheet in laser micro-welding. The developed model could predict the thermal deformation with good accuracy.

Experimental study on laser thermal adjustment of sheet metal micro-parts was carried out by Griffith et al. (2014) using pulsed modes of laser irradiations with different process parameters. It was observed that ablation of the laser irradiated sheet surface for short duration pulse was less compared to the long duration pulse heating. It was also found that type and amount of deformation was significantly influenced by both sample geometry and material properties. Li et al. (2016) investigated formability of micro-gears by laser dynamic flexible punching in brass foil by employing short pulse laser beam. Folkersma et al. (2016) proposed an algorithm for alignment of micro tube through laser micro forming using statistical analysis of the previous deformation data. The proposed algorithm was validated by micro tube laser bending experiments and found to predict alignment with good accuracy. Laser forming of shaped surfaces were investigated using finite element analysis experiments and other methods by different researchers (Jovic et al., 2017; Maji et al., 2018). Deformation of the samples with the different process parameters was studied and the most significant variables were identified to affect bending and thickening in laser forming of shaped surfaces. Effects of process parameters on pulsed laser thermal micro-forming of thin sheets were carried out by Maji (2019) employing response surface methodology and finite element simulations. The developed models were able to explain the parametric effects and optimization results were validated by experiments. Response surface and soft computing techniques were utilized by modeling and optimization of different complex manufacturing processes (Maji et al., 2010; Maji et al., 2014; Maji and Kumar, 2019). Maji (2019) investigated spatial and temporal-variations of pulsed laser forming of thin sheets and showed that pulsed mode of laser forming could produce greater deformation compared to that generated by continuous-wave mode. Micro-scale laser dynamic forming on metallic surfaces to create micro-coining features was investigated by Shen et al. (2019) utilizing numerical simulations and surface morphology was affected through void nucleation, growth, and coalescence by repeated laser induced shock loadings. Shen et al. (2019) studied the effects of plasticine layer on laser dynamic flexible forming of thin sheet material and could reduce the rebound effects. Semmar et al.

(2020) carried out experimental investigations on micro-spikes formation on porous silicon materials by applying picosecond laser pulses. Size and uniformity of the formed spikes were found to enhance with the number of laser pulses applied.

The micro-scale thermal forming and adjustments of thin sheet metal parts by stationary pulsed laser beam irradiation is investigated in this paper. The deformation behavior is studied under different processing conditions and optimal set of parameters are identified to achieve maximum deformation.

METHODOLOGY

Finite Element Analysis

For analysis of deformation in laser forming process using FEM, a time dependent or transient thermal structural analysis is performed in some time steps. The equilibrium states at all the time steps starting from 0 to t can be obtained using the principle of virtual work or minimum potential energy method. The equilibrium equation at time $t+\Delta t$ can be obtained from the virtual work principle as given in equation (1).

$$\int_V \sigma.\delta\varepsilon.dV = \int_V q.\delta u.dV + \int_S p.\delta u.dS, \tag{1}$$

where σ and ε are the stress and strain tensors in updated Lagrange configuration descriptions, respectively; q and p are the body force and surface pressure, respectively; δu is the virtual displacement. Discretization of this problem is done using the standard procedure and a group of non-linear equations are obtained. Newton-Raphson method is used to linearize the non-linear equations.

$$[K]\{\Delta u\} = \{F^a\} - \{F^{nr}\}, \tag{2}$$

where $[K] = \int_V [B]^T [D^{ep}][B] dV$ is the tangential stiffness matrix, $[B]$ is the general geometric matrix, $[D^{ep}]$ is the elasto-plastic stress-strain matrix, $\{\Delta u\}$ is the displacement incremental vector at the element nodes, $\{F^a\}$ is the applied force vector, and $\{F^{nr}\} = \int_V [B]^T \{\sigma\} dV$ is the Newton-Raphson restored force vector.

Laser forming is a complex, transient, non-linear, structural thermal problem and its numerical simulation can be performed in two ways. It can be modeled and solved in coupled or decoupled manner using commercial FEM packages like ANSYS. In the decoupled approach, transient thermal analysis for scanning by a laser beam is performed first. The transient or time dependent problems are solved in time steps and the obtained results at various sub-steps are recorded and accessible. Now, the solution giving temperature at various sub-steps (time steps) can be used as an input thermal loading in terms of temperature to the subsequent static structural analysis to find deformation induced due to that temperature loading at that sub-step. Therefore, the effect of stress-strain history on the deformation cannot be captured by this kind of analysis and this technique has not been used in this thesis. The simulation process adopted in this thesis is explained below.

Thermo-mechanical FE analysis of PLμF is carried out to study the deformation behavior of thin sheet metal micro-parts. Laser thermal micro-forming is considered as a weakly-coupled thermo-mechanical process and the analysis is performed using the finite element code ANSYS. In this simulation process, first temperature distribution is calculated and then the deformation field is obtained. The thermal equilibrium equation for heat transfer analysis of an isotropic material can be written as given in equation (3).

$$k\nabla^2 T = \rho c \frac{\partial T}{\partial t} \tag{3}$$

where ρ, c and k are the density, specific heat and thermal conductivity of the material, respectively, and the rate of heat generation due to plastic deformation is neglected. The basic finite element equation for the indirectly-coupled thermo-structural analysis for PLμF process can be written as given in equation (4).

$$\begin{bmatrix} [0] & [0] \\ [0] & [C] \end{bmatrix} \begin{Bmatrix} \{\dot{u}\} \\ \{\dot{T}\} \end{Bmatrix} + \begin{bmatrix} [K] & [0] \\ [0] & [K_T] \end{bmatrix} \begin{Bmatrix} \{u\} \\ \{T\} \end{Bmatrix} = \begin{Bmatrix} \{F\} \\ \{Q\} \end{Bmatrix} \tag{4}$$

where $[C] = \int_V \rho c [N][N]^T dV$ is the heat capacity matrix; $[K] = \int_V [B]^T [D^{ep}][B] dV$ is the tangential stiffness matrix; $[K_T] = \int_V k[B][B]^T dV$ represents the heat conduction matrix. $\{u\}$ and $\{T\}$ are the nodal displacement and temperature vector, respectively. $[N]$, $[B]$ and $[D^{ep}]$ denote the shape function matrix, general geometric matrix and elasto-plastic stress-strain matrix, respectively. $\{F\}$ and $\{Q\}$ are the nodal force vector and heat flux vector, respectively. A computer simulation of the laser-forming process is complex and the main steps of the analysis are the thermal and structural analyses. In these analyses, the finite element code: ANSYS has been used to calculate the thermo-structural coupled field with a large plastic deformation. An indirect coupled-field analysis (i.e., two sequential analyses) has been performed. The flow-chart of the FEM simulation is shown in Fig. 3. In this method, the results from the transient thermal analysis for nodal temperatures are applied as loads for the structural analysis. Therefore, the calculation of the temperature distribution is very important for the calculation of the stress–strain distribution.

The sheet material is assumed as isotropic, flat and free of residual stresses. Gaussian energy distribution of laser beam and free convection to air are considered. Temperature dependent properties, i.e., thermal conductivity, specific heat, density, young's modulus, Poisson's ratio etc. were considered in the analysis and taken from Ueda et al., 1991. Von-mises yield criterion and bilinear isotropic hardening model are used in the simulation process. In this simulation method, the results from the transient thermal analysis for nodal temperatures are applied as loads for the structural analysis. The calculated temperature field is used for the determination of the deformation field. For the thermal and structural analyses, the elements: solid70 and solid 185 were used respectively.

Figure 3. Flow-chart of FEM simulation of pulsed laser micro-forming process

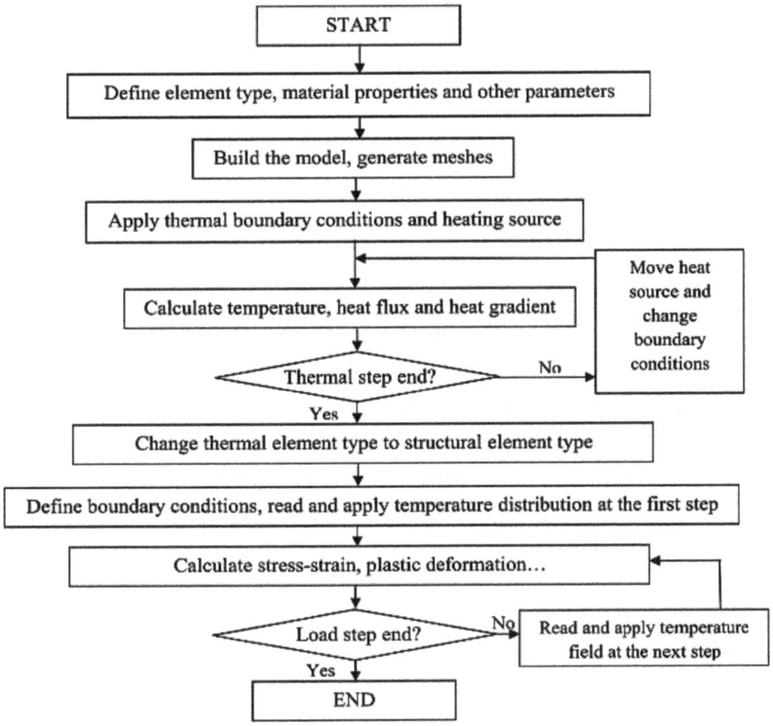

Response Surface Methodology

Response surface methodology was developed by Box and Wilson in 1951 to aid the improvement of manufacturing processes in the chemical industry. Response surface methodology (RSM) is a group of statistical techniques that are useful for the modeling and optimization of a response of interest, which is influenced by several independent variables. In response surface modeling, outputs of a process are assumed to be continuous, while the inputs may be either discrete or continuous in nature. The empirical model formed is a second order one (usually a quadratic model), which is used to develop a response surface. A response surface is a graphical representation of relationship between the dependent (responses or outputs) and independent variables (input factors). The response surfaces provide information regarding direct effects, interaction effects and curvilinear variable effects. For representing the curvilinear effects, the response surface models may have cubic terms also apart from the main, quadratic and interaction effect terms. But, it has been observed that the quadratic models are sufficient for most of the applications. The response surface can be expressed as shown in equation (5), when all the independent variables are measurable.

$$y = f(x_1, x_2, ... x_n) + \varepsilon, \tag{5}$$

where x_1, x_2, ..., x_n are the independent variables, and ε is the observed error or noise of the response y.

Response surface methodology (RSM) is a group of statistical techniques that are useful for the modeling and optimization of a response of interest, which is influenced by several independent variables (Montgomery, 2001). Generally, a second-order polynomial (refer to equation (6)) is used for establishing an approximate model of the response in terms of the measured independent variables kept within their respective ranges.

$$y = \beta_0 + \sum_{i=1}^{k} \beta_i x_i + \sum_{i=1}^{k} \beta_{ij} x_i^2 + \sum_{i=1}^{k-1} \sum_{j=2}^{k} \beta_{ij} x_i x_j \quad (6)$$

where the β parameters are called the regression coefficients. The coefficients of the approximating polynomial are obtained using the method of least square for error minimization. A response surface graphically represents the relationships between the outputs or responses and the input factors, and it also provides information regarding direct, interaction and curvilinear variable effects. The response optimization to determine the optimum operating conditions in PLMF was carried out using the desirability function method which starts with multiple solutions and finally obtains the optimal solution by using the reduced gradient algorithm.

EXPERIMENTATION

Laser micro-forming experiments were conducted on a fiber laser (wavelength = 1.07 μm, Make: SPI) having a maximum laser power of 400 W (refer to Fig. 4) with a 5 mm diameter collimated laser beam and minimum available spot diameter of 25 μm at the focal plane. Argon gas was used for shielding to protect the optical system. The work-pieces are AISI 304 stainless steel sheet was cut into the required size and shape in WEDM machine and cleaned by acetone.

Figure 4. Experimental Set up for PLμF of thin sheet

Deformation of the laser formed samples were measured using a laser displacement sensor (Make: Micro-Eplison, Model: OPTO NCDT 1320) as shown in Figs. 5 and 6, respectively.

Figure 5. Laser formed circular and rectangular samples by pulsed laser micro-forming

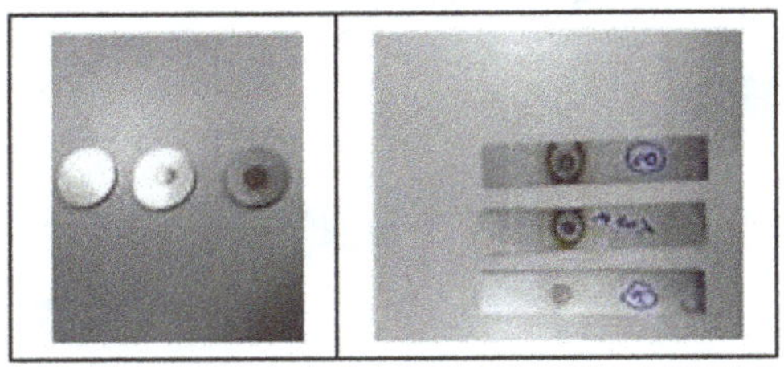

Figure 6. Measurement of deformation in laser formed samples

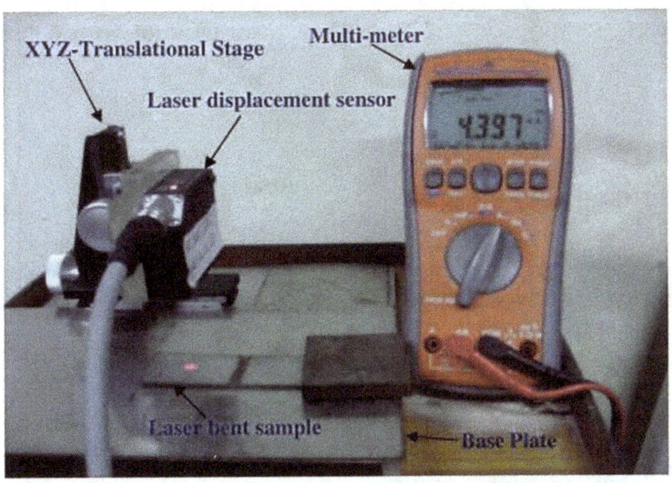

Deformation in terms of bending angle was calculated by the triangulation method. Results of the finite element simulations were compared to the experimental ones as discussed in the following section.

RESULTS AND DISCUSSION

This section discuss and explains the simulation results of deformation in pulsed laser micro forming of stainless steel micro-sheet metal samples of different sizes and shapes considering different process parameters.

Analysis of Rectangular Sheet Samples

Laser micro forming process of the rectangular metal sheet ($10 \times 0.5 \times 0.25$ mm^3) of AISI 304 Stainless steel was simulated as thermal-mechanical coupling process by using ANSYS program. Fig. 7 shows the meshed model with SOLID 70 element for thermal analysis having number of elements and nodes of 4800 and 6795, respectively. Size of the smallest element is 0.0001 m. During the structural analysis the element type was changed from SOLID 70 to SOLID 185.

Figure 7. Meshed model of rectangular metal sheet

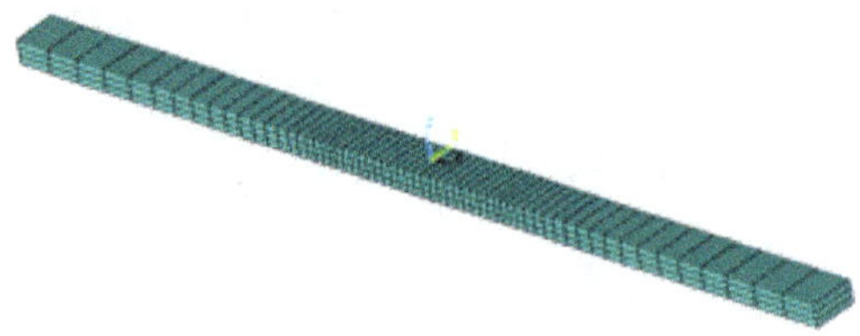

The end of the bar is fixed, and the other end was kept free. The middle position of the bar surface was irradiated by the laser beam. Fig. 8 shows the temperature and deformation fields for process parameters of 100 W laser power, 0.5 mm spot diameter, 1ms laser irradiation time and frequency of 50 Hz.

Figure 8. (a) Temperature and (b) deformation field of the rectangular sample

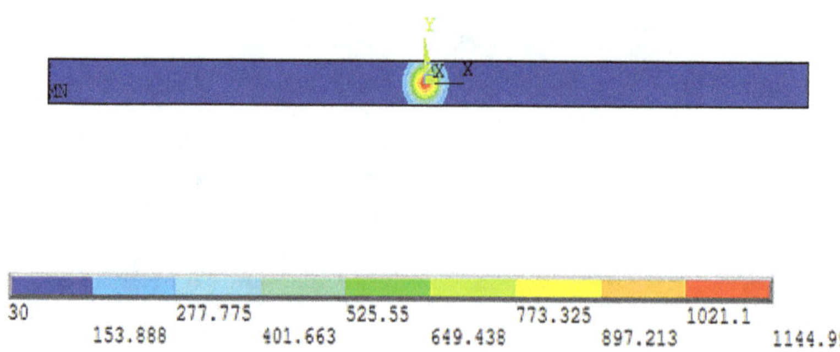

(a)

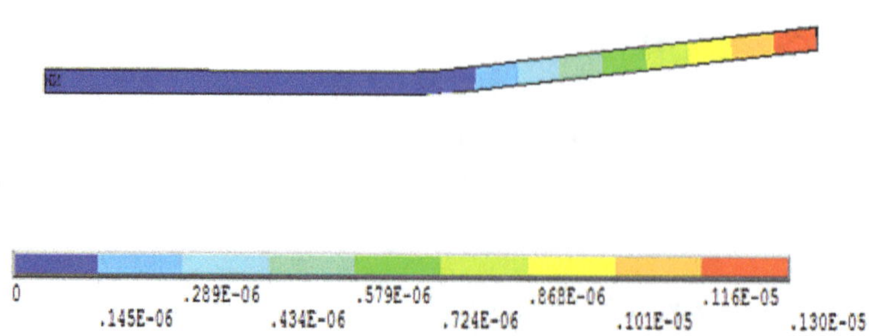

(b)

To study the effects of process parameters, the ranges of process parameter in PLμF of thin metal sheets are decided as given in Table 1.

Table 1. Ranges of input process parameters

	Minimum	Intermediate	Maximum
Laser Power, P_L (W)	50	65	80
Spot diameter, d (mm)	0.4	0.5	0.6
Pulse time, t_p (ms)	0.5	1	1.5

Total fifteen sets of different combinations according to central composite design method and the corresponding deformation values in terms of bending angle are given in Table 2. The response bending angle was expressed in terms of the process parameters through the following equation (7).

$$\begin{aligned}\alpha_b &= 0.296 + 0.00157P_L - 0.0928d + 0.1422t_p + 0.000007P_L^2 \\ &+0.961d^2 - 0.0376t_p^2 - 0.00475P_L d + 0.001883P_L t_p - 0.2275dt_p\end{aligned} \tag{7}$$

The variations of the deformation in terms of bending angle are shown in Fig. 9 through the surface plots.

Table 2. Deformation values for the designed set of parameters

SL. No.	Power, P_L (W)	Spot Diameter, d (mm)	Pulse Time, t_p (ms)	Bending Angle, αb (mrad)	
				Rectangular Sample	Circular Sample
1	80	0.4	1.5	0.327	0.190
2	65	0.5	1	0.175	0.102
3	50	0.4	1.5	0.208	0.126
4	65	0.4	1	0.220	0.133
5	65	0.5	1.5	0.181	0.109
6	65	0.5	0.5	0.121	0.091
7	65	0.6	1	0.120	0.089
8	50	0.6	0.5	0.070	0.053
9	50	0.4	0.5	0.145	0.109
10	80	0.4	0.5	0.185	0.139
11	80	0.6	0.5	0.104	0.078
12	50	0.6	1.5	0.110	0.082
13	80	0.6	1.5	0.178	0.133
14	80	0.5	1	0.182	0.137
15	50	0.5	1	0.142	0.107

Figure 9. Variation of bending angle for rectangular metal sheet

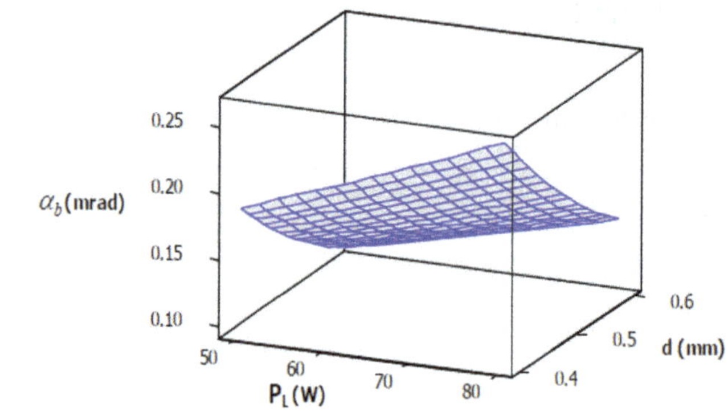

(a)

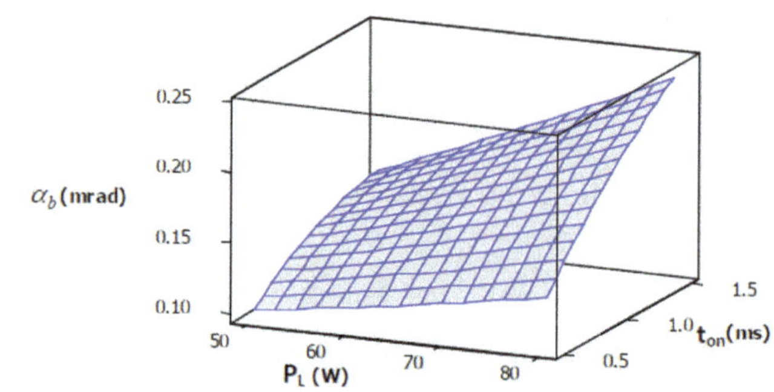

(b)

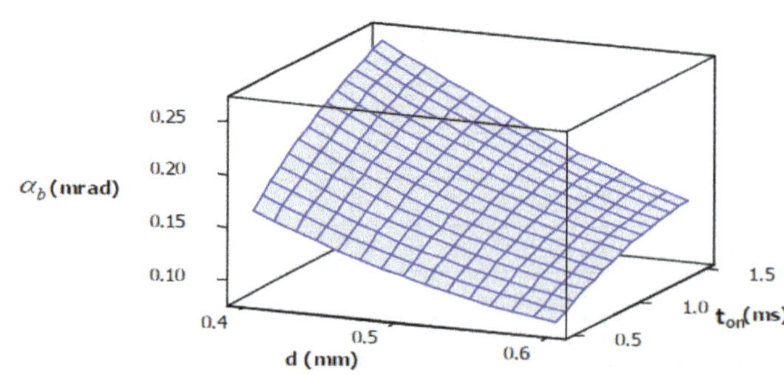

(c)

Bending angle was observed to increase with the increase of laser power and pulse on time, because pulse energy increased which produced more thermal stress and more deformation. Deformation was reduced with the increase of spot diameter. Smaller spot diameter produces higher energy density and peak surface temperature which generates larger deformation. Maximum bending angle was obtained as 0.315 mrad and the corresponding input parameters, i.e., laser power, spot diameter and pulse on time were 80 W, 0.4 mm, and 1.5 ms, respectively. An error of 0.012 mrad was found when compared with the experimental result. As the laser power and pulse duration are increased, the energy per pulse increases and hence, the enhanced thermal energy input produces the higher thermal stress and larger bending. As the laser irradiation time increases, the dimension of the spot increases. With this, the energy input per unit volume decreases and consequently, the deformation decreases. Therefore, there is an optimum value of spot diameter for the maximum bending angle within the ranges of parameters considered in this study.

Analysis of Circular Sheet Samples

Finite element analysis was also carried out to study the deformation behavior of circular sheet sample of size 5 mm diameter and 0.25 mm thick in PLµF. The meshed model and contour plots of temperature and deformation fields for the circular sample is shown in Figs. 10 and 11, respectively for laser power of 100 W, spot diameter of 0.5 mm and pulse on time of 1 ms.

Figure 10. Meshed 3-D circular model

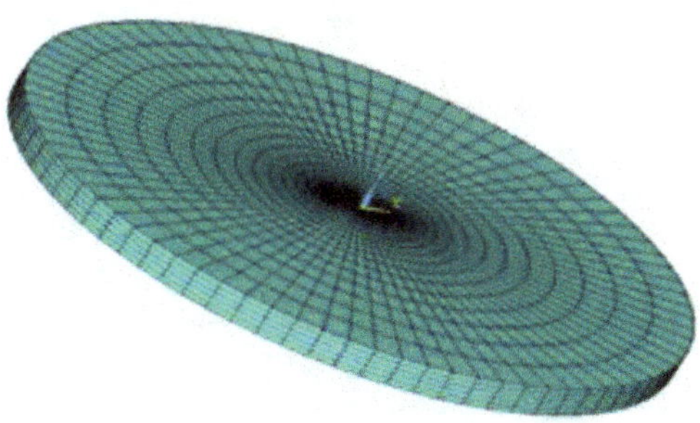

Deformation generated for circular sample is less compared to that for rectangular sample because of lower peak temperature reached for circular sample due more material surrounding the irradiated zone works as heat sink. Finite element simulated bending angles for different input parameters combination are shown in Table 2. The response out-of-plane deformations or bending angle for circular sample was expressed as given in the equation (8).

$$\alpha_b = 0.424 - 0.00651P_L - 0.432d + 0.0158t_p + 0.000058P_L^{\,2} + 0.224d^2 - 0.0341t_p^{\,2} - 0.00148P_L d + 0.000999P_L t_p + 0.0431dt_p \quad (8)$$

The variations of deformations for the circular sample in PLMF are shown in Fig. 12.

Figure 11. (a) Temperature and (b) Deformation contour plot of the circular sample in PLμF.

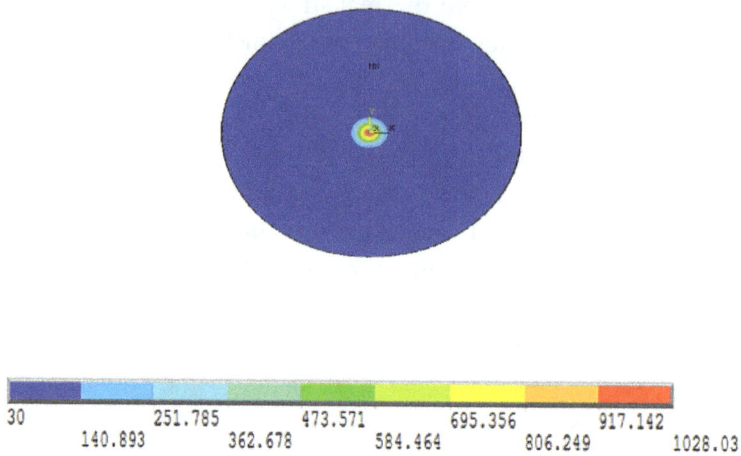

(a)

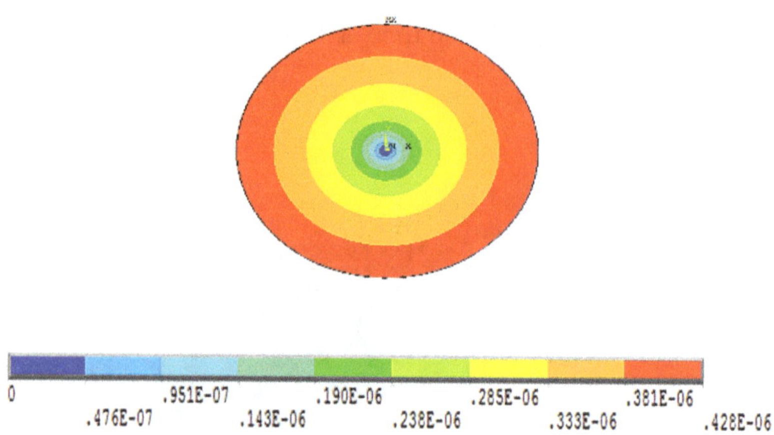

(b)

The maximum bending angle obtained by response optimization method from the equation (5) was 0.183 mrad and an error of 0.007 mrad was seen compared to the experimental value. The input process parameters corresponding to the maximum bending angle were found as laser power of 80 W, pulse on time of 1.5 ms and spot diameter of 0.4 mm. The maximum deformation was found for maximum laser power and pulse on time and minimum spot diameter with the operating window considered. The effects of multiple laser pulses and sample size are discussed in section 5.3.

Figure 12. Surface plots of deformation in terms of bending angle in thin circular metal sheet

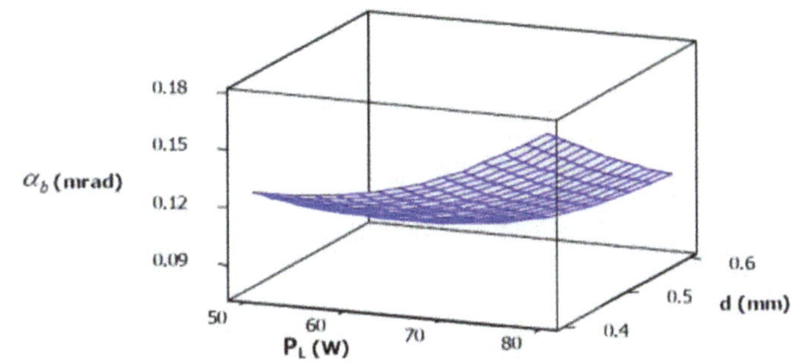

(a)

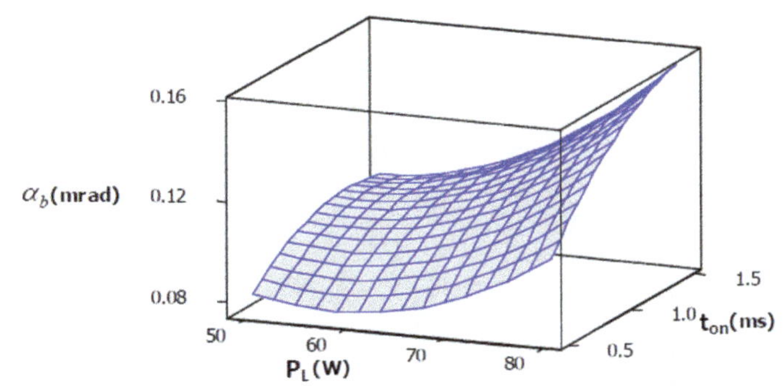

(b)

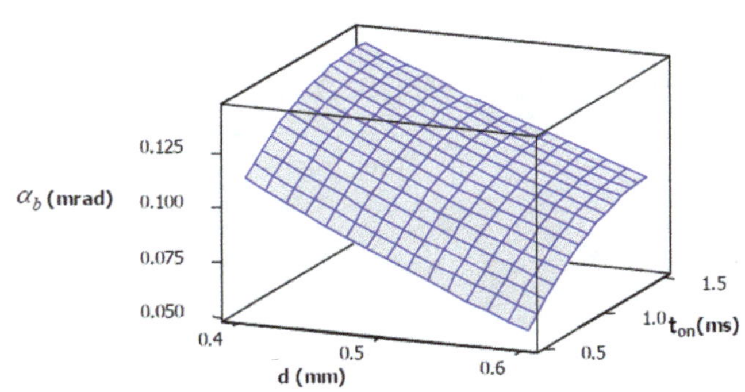

(c)

Figure 13. Transient-temperature distribution on top surface at the center of laser beam spot for (a) f= 50 Hz (b) f = 100 Hz

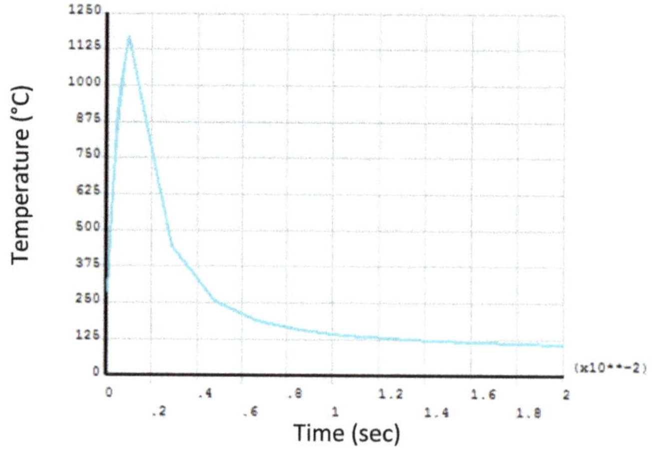

(a)

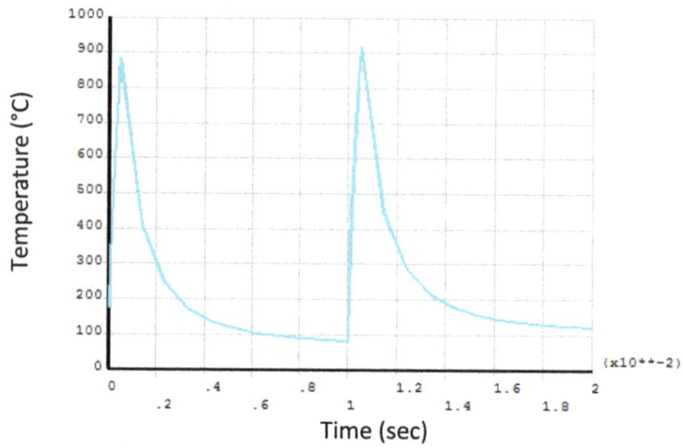

(b)

Effects of Pulse Frequency and Sample Size

The effects of pulse frequency and sample size are described and explained below. The transient temperature values for two values of pulse frequency, i.e., 50 Hz and 100 Hz keeping other parameters at constant values of 100 W laser power, 0.5 mm spot diameter and 1 ms pulse on time as shown in Fig. 13. The deformation was found more for 50 Hz frequency because peak surface temperature was more though the total energy input was same as shown in Fig. 14.

The rise in surface temperature (ΔT) can be expressed in terms of laser energy density (H) and pulse on duration (t_p) as $\Delta T \propto H\sqrt{t_p}$ assuming 1-D heat conduction[19]. Peak surface temperature decreased because pulse-on-time decreases with the increase in frequency at constant duty cycle. Therefore, micro-scale thermo-elasto-plastic deformation of thin sheet metallic micro-parts depends not only on the laser pulse energy input, but also on how fast the laser energy is coupled to the workpiece, i.e., which determines the rise in temperature.

Figure 14. Deformation in thin metal sheet in PLμF for pulse frequencies (a) f = 50Hz (b) f = 100Hz

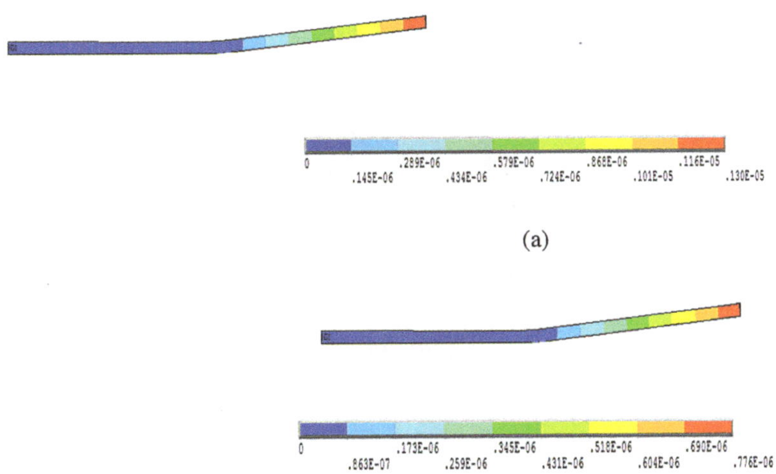

(a)

(b)

The amount of deformation was found to increase with increasing sample size keeping the energy density constant due to more energy input to induce more thermal stress as shown in Fig. 15.

FUTURE RESEARCH DIRECTIONS

The present described the experimental investigations and modeling of pulsed laser micro-forming processes for different materials applications and using techniques like finite element method, statistical regression analysis and mechanical and metallurgical characterization of the laser processes work surface. Hybrid micro-forming technique combining both laser heating and laser induced shock wave can be investigated in future to increase the efficiency of the micro-feature forming on high strength

Figure 15. Deformation contour plot of the circular sample for single laser pulse for sample diameter of (a) 3 mm (b) 7 mm

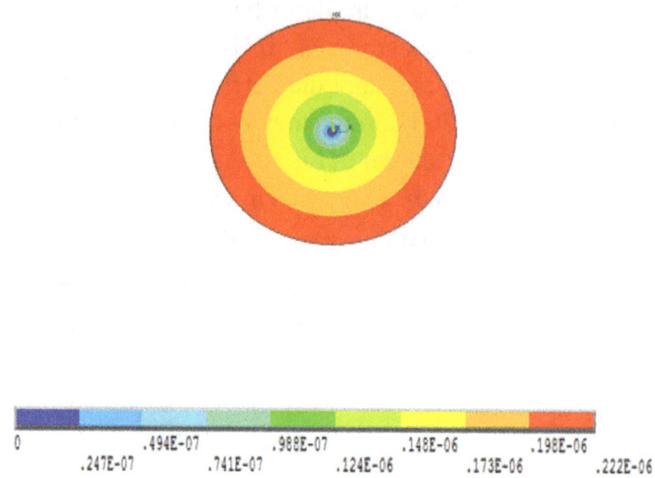

(a)

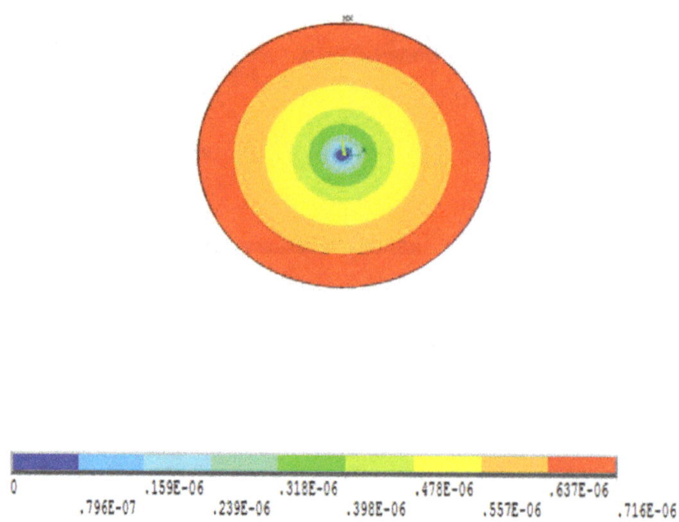

(b)

thin sheet materials. Process parameters optimization and control to obtain the desired morphology of the pulsed laser micro-formed samples can be studied in future.

CONCLUSION

Finite element analysis on laser thermal forming of sheet metal micro-parts of AISI 304 stainless steel thin sheet was carried out to investigate the deformation behavior. Response surface models of deformation were developed based on the finite element simulation data. Deformation values for both rectangular and circular shaped samples were observed to increase with the increase in laser power and pulse on time, and decreased with the increasing spot diameter. Maximum deformation was obtained for both cases for maximum laser power and pulse on time and minimum spot diameter. The finite element simulation results were validated by experiments. The deformation was found to decrease with increasing pulse frequency, and increased with the increasing sample size for constant heat energy input.

ACKNOWLEDGMENT

The author gratefully acknowledges the financial support of Department of Science and Technology, SERB, India for funding this research work under Early Carrier Research award with project File No. ECR/2016/001134.

REFERENCES

Bellisario, D., & Quadrini, F. (2012). Laser bending of 5005 aluminum alloy sheet. *International Journal of Manufacturing, Materials, and Mechanical Engineering*, *2*(3), 1–15. doi:10.4018/ijmmme.2012070101

Chakraborty, S. S., More, H., & Nath, A. K. (2018). Laser forming of a bowl shaped surface with a stationary laser beam. *Optics and Lasers in Engineering*, *77*, 126–136. doi:10.1016/j.optlaseng.2015.08.006

Folkersma, K. G. P., Brouwer, D. M., Römer, G. R. B. E., & Herder, J. L. (2016). Robust precision alignment algorithm for micro tube laser forming. *Precision Engineering*, *46*, 301–308. doi:10.1016/j.precisioneng.2016.05.011

Griffiths, J., Edwardson, S. P., Dearden, G., & Watkins, K. G. (2010). Finite element modeling of laser forming at macro and micro scales. *Physics Procedia*, *5*, 371–380. doi:10.1016/j.phpro.2010.08.064

Griffiths, J., Edwardson, S. P., Dearden, G., & Watkins, K. G. (2012). Thermal laser micro-adjustment using picosecond pulse durations. *Applied Surface Science*, *258*(19), 7639–7643. doi:10.1016/j.apsusc.2012.04.109

Griffiths, J., Sheikholeslami, G., & Dearden, G. (2014). Optimization of beam mode for high efficiency laser thermal forming within metallurgical constraints. *Physics Procedia*, *56*, 1392–1399. doi:10.1016/j.phpro.2014.08.069

Ismail, M. I. S., Okamoto, Y., & Okada, A. (2016). Thermo-mechanical analysis on thermal deformation of thin stainless steel in laser micro-welding. *International Journal of Manufacturing, Materials, and Mechanical Engineering*, *6*(2), 51–66. doi:10.4018/IJMMME.2016040104

Jamil, M. S. C., Sheikh, M. A., & Li, L. (2011). A finite element study of buckling and upsetting mechanisms in laser forming of plates and tubes. *International Journal of Manufacturing, Materials, and Mechanical Engineering*, *1*(1), 1–17. doi:10.4018/ijmmme.2011010101

Jovic, S., Makragic, S., & Jovanovic, M. (2017). Parameters influence of laser forming on shaped surface by soft computing technique. *Optik (Stuttgart)*, *142*, 451–454. doi:10.1016/j.ijleo.2017.04.089

Li, J., Liu, H., Shen, Z., Qian, Q., Zhang, H., & Wang, X. (2016). Formability of micro-gears fabrication in laser dynamic flexible punching. *Journal of Materials Processing Technology*, *234*, 131–142. doi:10.1016/j.jmatprotec.2016.03.018

Maji, K. (2019). Parametric study and optimization of pulsed laser thermal micro-forming of thin sheets. *International Journal of Manufacturing, Materials, and Mechanical Engineering*, *9*(2), 47–61. doi:10.4018/IJMMME.2019040103

Maji, K. (2019). Experimental investigations on spatial and temporal laser irradiations in pulsed laser forming of thin metal sheets. *Journal of Laser Applications*, *31*(4), 0420101–0420115. doi:10.2351/1.5055842

Maji, K., & Kumar, G. (2019). Inverse analysis and multi-objective optimization of single-point incremental forming of AA5083 aluminum alloy sheet. *Soft Computing*, 1–17. doi:10.100700500-019-04211-z(.0123456789().,-volV)(0123456789().,-volV)

Maji, K., Pratihar, D. K., & Nath, A. K. (2014). Analysis of pulsed laser bending of sheet metal using neural networks and neuro-fuzzy system. *Proceedings of the Institution of Mechanical Engineers. Part B, Journal of Engineering Manufacture*, *228*(9), 1015–1026. doi:10.1177/0954405414522212

Maji, K., Pratihar, D. K., & Nath, A. K. (2018). Forward and Inverse predictions of deformations in coupling mechanism based laser forming of shaped surfaces. *Journal of Laser Applications*, *30*(3), 03201101–03201110. doi:10.2351/1.5033450

Maji, K., Pratihar, D. K., & Patra, S. (2010). Modelling of electrical discharge machining process using regression analysis, adaptive neuro-fuzzy inference system and genetic algorithm. *International Journal of Data Mining. Modelling and Management*, *2*(1), 75–94.

Qi, L., & Namba, Y. (2011). Precision laser adjustment using CW diode laser. *Precision Engineering*, *35*(1), 126–132. doi:10.1016/j.precisioneng.2010.08.002

Qin, Y., Chen, Y., Ni, X., Shen, Z., Bi, J., & Zhang, X. (2010). Axisymmetric numerical simulation of plastic damage in aluminum alloy induced by long pulsed laser. *Optics and Lasers in Engineering*, *48*(3), 361–367. doi:10.1016/j.optlaseng.2009.10.006

Sagisaka, Y., Yamashita, K., Yanagihara, W., & Ueta, H. (2014). Micro parts processing using laser cutting and ultra-short-pulse laser peen forming. *Procedia Engineering*, *81*, 2367–2372. doi:10.1016/j.proeng.2014.10.335

Semmar, N., Talbi, A., Mikikian, M., Stolz, A., Melhem, A., & Meneses, D. S. (2019, November). ((Accepted). Micro-spikes formed on mesoporous silicon by UV picosecond laser irradiation. *Applied Surface Science*, *20*. doi:10.1016/j.apsusc.2019.144820

Shen, H., Gong, C., Hu, J., & Yao, Z. (2012). Numerical and experimental study on bi-direction deformations in laser micro forming of two-bridge actuators. *International Journal of Machine Tools & Manufacture*, *54*, 66–72. doi:10.1016/j.ijmachtools.2011.11.006

Shen, H., Hu, J., & Yao, Z. (2015). The interactive effects on deformation behavior in laser thermal adjustment of two-bridge actuators. *International Journal of Machine Tools & Manufacture*, *90*, 1–7. doi:10.1016/j.ijmachtools.2014.12.001

Shen, H., Hu, J., Yu, X., & Yao, Z. (2012). The effects of cut-out on stress distribution and deformations in laser thermal adjustment. *International Journal of Machine Tools & Manufacture*, *62*, 46–52. doi:10.1016/j.ijmachtools.2012.06.006

Shen, H., & Yao, Z. (2011). Numerical and experimental investigation of cooling time in laser micro-adjustment of two-bridge actuators. *Applied Thermal Engineering*, *31*(8-9), 1447–1456. doi:10.1016/j.applthermaleng.2011.01.013

Shen, Z., Liu, H., Wang, X., & Wang, C. (2019). Surface degradation of micro-mold in micro-scale laser dynamic forming and its effects on workpiece. *Optics & Laser Technology*, *117*, 114–125. doi:10.1016/j.optlastec.2019.03.047

Shen, Z., Zhang, J., Liu, H., Wang, X., & Ma, Y. (2019). Reducing the rebound effect in micro-scale laser dynamic flexible forming through using plasticine as pressure-carrying medium. *International Journal of Machine Tools & Manufacture*, *141*, 1–18. doi:10.1016/j.ijmachtools.2019.03.004

Ueda, Y., Lida, K., Saito, M. & Okamoto, A. (1991) Finite element model and residual stress calculation for multi-pass welded joint between a plate and penetrating pipe. *Modeling of casting welding and advanced solidification processes*, 219-227.

Vollertsen, F., & Sakkiettibutra, J. (2010). Different types to use laser as a forming tool. *Physics Procedia*, *5*, 193–203. doi:10.1016/j.phpro.2010.08.044

Zheng, C., Sun, S., Zhang, G., Song, L., & Ji, Z. (2013). Effect of confining overlay in micro scale laser bulge forming. *Applied Surface Science*, *285*, 477–482. doi:10.1016/j.apsusc.2013.08.080

Zhou, J., Shen, H., Yu, X., Hu, J., & Yao, Z. (2013). On the competition between in-plane and out-of-plane deformations in laser thermal adjustment. *Optics & Laser Technology*, *45*, 689–696. doi:10.1016/j.optlastec.2012.05.010

Chapter 17
Additive Manufacturing and Its Need, Role, Applications in the Automotive Industry

Dhinakaran V.
Chennai Institute of Technology, India

Varsha Shree M.
Chennai Institute of Technology, India

Swapna Sai M.
Chennai Institute of Technology, India

Rishiekesh Ramgopal
National Institute of Technology, Tiruchirappalli, India

ABSTRACT

Additive manufacturing (AM) emerged from rapid prototyping to relinquish sustainable industrial production. The role of AM in the industrial field is to diminish manufacturing pace, functioning cost, and assembling the AM lightweight particles together, which enhances the malleable fabrication of personalized user defined components without symbolic concussion. The automotive manufacturing industry plays a chief role in the aggressive trade field where time to market declines. The engraving design structures with weight reduction materials are the demands faced by the automotive industry that can be ominously resolved by additive manufacturing technology. This research work provides a better understanding of AM technology and its role in automotive sector to enhance modern vehicle designs and enduring features and augments the knowledge of both researchers and industrialists to overcome the efficacy in manufacturing process by fabricating relatively high strength geometries with reduced weight.

DOI: 10.4018/978-1-7998-4939-1.ch017

INTRODUCTION

Additive Manufacturing (AM) is most commonly known as 3D Printing that favours design configuration of 2D data into construction of 3D prototypes. Nowadays, Additive Manufacturing technology deploys its place in every industry including food, textile, medicine, aerospace and automobile. It has tremendous applications around the world which upturns the quench of researchers to develop additive manufacturing technologies for individualized processing and to create a revolution in multiple manufacturing and engineering industries (Lewis 2014). Additive Manufacturing is greatly prioritized in Automotive industry as it enables the fast production of components and saves time, money and materials. The prototyping of robust and flexible assembly aids is achieved at a faster rate through Additive Manufacturing due to its speed and flexibility of designing accurate components with precise materials. This technology enhances the production requirements to satisfy the quality of end-use component (Chee et al. 2017). The vast designing process is turned into an easy to do method with the expansion of design freedom to create complex geometric structures and to fabricate with the appropriate material in precise quantity by a single machine in a one step process within the stimulated time. The Additive Manufacturing consists of seven methodologies in which each individualized method is suitable for various materials and works accordingly with its environmental conditions.

AM IN AUTOMOTIVE INDUSTRY

Additive Manufacturing Technique

The Additive manufacturing technologies familiarize many innovative and monetary gains when compared to conservative subtractive manufacturing methods in rapid prototyping (RP). The Additive Manufacturing (AM) technology is developed from the RP technology to enhance the quality of the output product. Additive Manufacturing acts as a basic principle of creating three dimensional (3D) objects generated through Computer Aided Design (CAD) systems. In AM technology, the components are made from CAD data and slicing software to create specified part geometry rather than complex tooling and additional fixtures that are used in conventional manufacturing methodologies (Costas et al. 2018). The structures are built in a layer-by-layer fashion with a specified cross section and which is not only used in manufacturing industries to fabricate automobile components and dynamic mechanical structures but also enlists its application in tissue engineering with the capacity of bioprinting to create biomedical implants, artificial human organs and drug delivery systems. The AM technology acts as a key for solving environmental and engineering issues since it has Free-Form Fabrication (FFF) that facilitates to produce user defined geometries with all classes of raw materials without any limitations unlike metals, non-metals, alloys, and synthetic polymers with no wastage of materials. The development of this technology improves with its increasing applications all over the engineering industry. Figure 1 explain the additive manufacturing technology.

Figure 1. Additive manufacturing technology

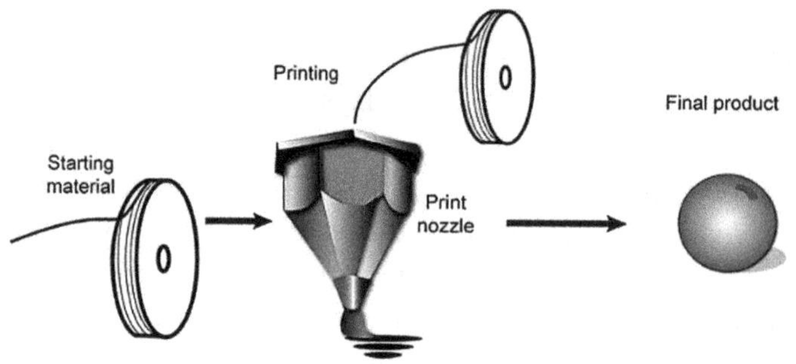

ADDITIVE MANUFACTURING METHODOLOGY

Computer-aided design (CAD) is the method that employs computers or workstations to assist in the process of, designing, modelling, creating, adaptation, simulation, or in optimising design structures. The great increase in productivity of design with improved quality through documentation and communication can be achieved by CAD software. The result from CAD software is obtained as electronic data files that are used for machining, and other manufacturing procedures. Now, the CAD model is converted into STL (stereolithographic) file which displays the peripheral closed surface of the CAD geometry and used for performing slice calculation by a slicing software. The build parameters comprising support structure, layer thickness, part orientation are customized according to the designer are done using slicer program (van Duin 2016). The Additive Manufacturing machine is set with materials and appropriate process parameters to generate the component with the stimulated time. The 3D printing machine requires high maintenance to deliver enhanced calibration of the component to meet the required quality. After proper completion of the product, the interaction of the part with the machine is cut down by adjusting the machine temperature and then detached. In post processing the part is cleaned before use and they are treated mechanically such as reducing, polishing, and pressurised cleansing for smoother surface finish and required texture.

NEED FOR AM IN AUTOMOTIVE INDUSTRY

Additive Manufacturing helps us to build up components with an extensive array of uneven, elevated temperature materials and hence the shipping companies are just abrading the facade of Additive Manufacturing for their vehicles. Figure 2 summarizes the need of AM in automotive industry. The major requirement of automotive industry is to obtain less weighted structures with new engineering materials such as polymers and metal alloys for the advancement of vehicle design create optimised components to minimize the fuel consumption in automobile industry thereby, reducing the cost of production and construction of design structures promoting monetary gains for the manufacturing. The design freedom in AM technology greatly helps in creating parts with internal channels which allows various features like cooling, free flow of fluids, fine meshed structures with curved surfaces and thin walls that has both stability and less weight. The another compatibility of Additive Manufactured components is to withstand high temperatures as well as low temperature which is an engraving requirement in automotive industries. The testing of engineering materials such as light weight polymers, metal alloys and other new material to identify the materials with high moisture resistant property can be done through AM which acts as a great boon for automotive parts to use for long term duration (Wong 2016). The AM components are not only tested but also post processed desirably according the needs and application of the component to improvise superior properties enduring resistance to water, corrosion and to barricade dust, moisture thereby developing the adaptability to discrete environmental conditions. Being a single step fabrication technology, Additive Manufacturing has the ability to reconstruct the multi-structured automotive parts into a single component with complex features that reduces the use of materials and benefits the wearability and easy replacement of components.

Figure 2. Need for AM in automotive industry

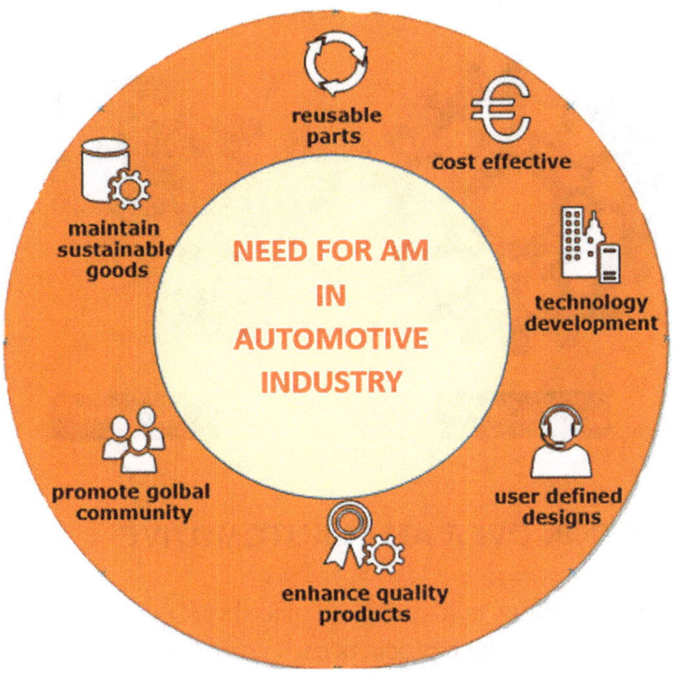

ROLE OF AM IN AUTOMOTIVE INDUSTRY

The significant advancement in Additive Manufacturing evolved as an eye opener for Automotive Industry to attain momentous development over manufacturing to give superior upturn for automobiles around the mounting modern world. Figure 3 explains the role of AM in automotive industry. Additive Manufacturing excels by accelerating the demand for easy to do structures as an alternate for conventional complex structures by implementing new materials for light weight productions with specific mechanical properties and upgraded design freedom for creating complex multifaceted designs in a single geometry paving new brands in automobile industry. The main custom of a vehicle is attained by its design assembly with high quality and durability via prototyping (Kassou 2018). The high scale models are manufactured through Material jetting and Selective Laser Sintering that consent the integration of design community. The Fused Deposition Modelling is the cost effective among the other AM technologies finds a communal place in automotive industry for the investigation of the performances of prototypes in a detailed manner to study the definite mechanical properties and their applications. The manufacture of fluid handling systems relays on Selective Laser Melting and Electron Beam Melting to produce fuel pumps and valves. The interior and seating arrangements of a car or any four is made through polymer material in order to reduce the weight and cost of production with effective durability of the product and also the analysis and testing of different polymeric materials to find its strength and durability through simulations. Metal alloys like Titanium, aluminium are being manipulated to create engine components, hubcaps, suspension springs, and tyres with resistance to wear and tear that can be highly achieved through Additive Manufacturing.

Figure 3. Role of AM in automotive industry

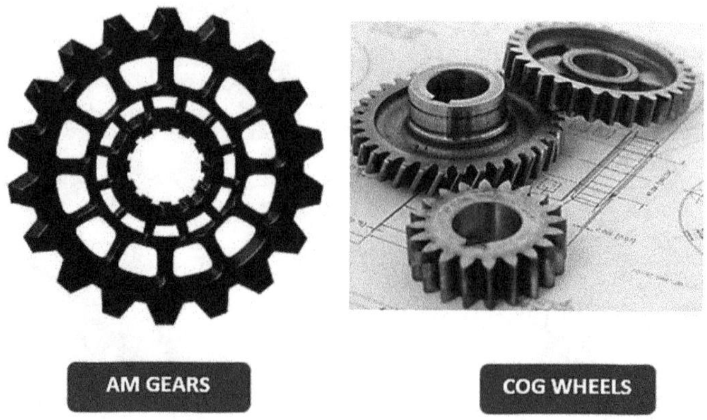

EFFECT OF PRODUCTION OVER AM IN AUTOMOTIVE

Additive Manufacturing predominantly holds a distinct place in production technology. The production of automotive industry achieves over one lakh products each year; AM technology being used for prototyping is comparatively higher than the complete part production, the main constrain for this shortcoming is the size of printers. The expansion in the size of industrial printers must be made to cope with industries inorder to maintain a balance in large scale production to make revolution especially in high end automotive industries. Additive Manufacturing acts as an effective tool for production as it integrates multifaceted tools in a short duration with cheaper composition development are made relatively quickly, in view of the fact that the expertise does not oblige for upscale assembling for production (Tillman 2019). Figure 4 shows the Additive manufactured components. The existence in the swift path requires durability to threatening environmental conditions such as intense speeds and warmth. The shipping management requires Additive Manufacturing tools to withstand ruthless road conditions and trivial parts ensuring safety for human being to avoid unnecessary drag. AM developed and reinforced frivolous components that guided to establish additional well-organized vehicles which are very much needed by the automotive companies. Some of the applications that have transmuted the industry involve parts made from intricate work which are not able to be fabricated with current industrialized methods, flexible prototypes, anastomotic exemplaries, frameworks, ritual characteristics and great filling. Additive Manufacturing contributes to Hybrid Manufacturing by providing opportunities to reproduce aftermarket parts for renovating classic cars and bring a new market and quality to them.

Figure 4. Additive manufactured components

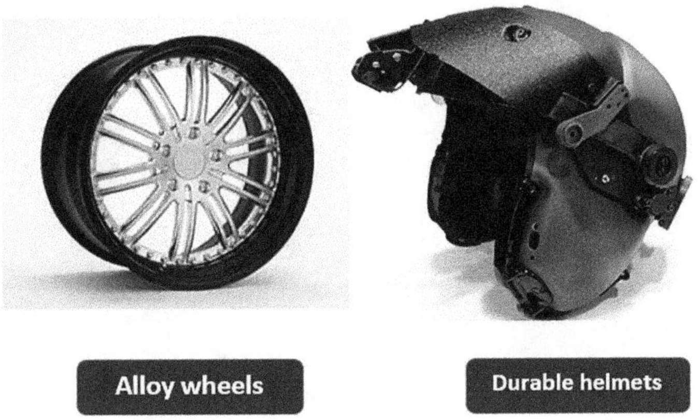

AUTOMOTIVE APPLICATIONS OF AM

Being capable to fabricate a composite of weight less band instantaneously has become a marque of the Additive Manufacturing methodology. Figure 5 shows the automotive applications of AM. AM method including smaller amount of input from a machinist to swiftly take a design from a computer and assemble the parts in a diminutive period exclusively by using Selective Laser Sintering (SLS) which performs as semi-functional holler fragments where elasticity is obligatory in mating or assembly. In general, the applications of the AM part are resolved to perform monotonous flexing movements that need substantial flexing, and former Polyethylene based SLS materials such as Duraform “Flex” are improved. AM can be recycled recurrently to fabricate automotive machineries that depend on aesthetics more than utility resulting in the production of faction mirrors and radiance housings for map-reading the wheels and complete internal control panel arrangement being created. The generally employed methods for producing parts of aesthetic prototypes from a photo-activated resin are Material jetting and SLA printing (Jan 2014). Tremendous feature, silky and exact 3D printed replica are engaged in the automotive industry to exhibit the various designs and perception of innovative vehicles. AM is made use of in automotive industry to mould the parts to explicit vehicles (making them ritual and weight less components) or even driver (e.g. spaces for racing cars). This is principally practical when the cost of exclusive machineries is vindicated by a substantial development in vehicle presentation. The polymeric techniques of stereo-lithography and selective Laser Sintering made the dashboards and seat frames manufacturing process in short duration.

Figure 5. Automotive applications of AM

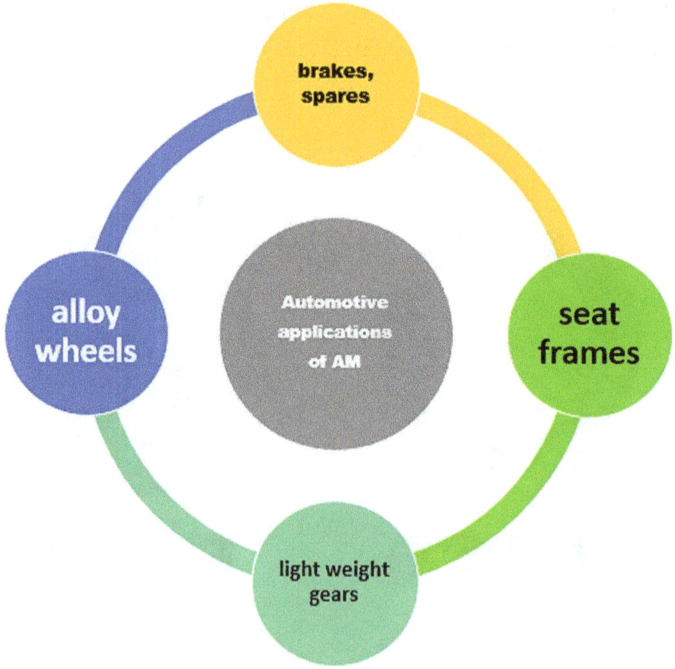

CONCLUSION

This research work provides a better understanding of AM technology and its role in automotive sector and the following conclusions are drawn.

- The transition from analog to digital processing in the areas involving communication, imaging, designing, and engineering collectively created a digital revolution known as "Additive Manufacturing" that has specialised digital flexibility for smart manufacturing.
- AM acts as a simplified manufacturing process which can perform different operations in a single process hence it is used over various industries.
- The automotive industry is intensely contingent on quantity as well as sequence production. AM in helps to develop new models and facelifts in short term, requiring new tools in the automotive industry.
- The 3D printing technology can be viewed as an alternate process for conventional manufacturing methodologies.
- Nevertheless, large-scale production in additive manufacturing is an essential and emergent area of exploration with technologies that can sustain superior build sizes, such as Wire Arc Additive Manufacturing (WAAM) and Big Area Additive Manufacturing (BAAM), actively being researched and developed. The supplementary investment into emerging AM-specific skills should be raised worldwide.
- The design for AM and its operation, engineering materials and post-processing are the dynamic aids that should be industrialized and cherished to yield remarkable applications.

REFERENCES

Azman, Vignat, & Villeneuve. (2018). CAD tools and file format performance evaluation in designing lattice structures for additive manufacturing. *Jurnal Teknologi, 80*(4).

Bikas, Stavropoulos, & Chryssolouris. (2016). Additive manufacturing methods and modelling approaches: a critical review. *The International Journal of Advanced Manufacturing Technology 83*(1-4), 389-405.

Böckin, D., & Tillman, A.-M. (2019). Environmental assessment of additive manufacturing in the automotive industry. *Journal of Cleaner Production, 226*, 977–987. doi:10.1016/j.jclepro.2019.04.086

Campbell, T., Williams, C., Ivanova, O., & Garrett, B. (2011). Could 3D printing change the world. Technologies, Potential, and Implications of Additive Manufacturing, Atlantic Council.

Compton, B. G., & Lewis, J. A. (2014). 3D-printing of lightweight cellular composites. *Advanced Materials, 26*(34), 5930–5935. doi:10.1002/adma.201401804 PMID:24942232

Ding, D., Shen, C., Pan, Z., Cuiuri, D., Li, H., Larkin, N., & van Duin, S. (2016). Towards an automated robotic arc-welding-based additive manufacturing system from CAD to finished part. *Computer Aided Design, 73*, 66–75. doi:10.1016/j.cad.2015.12.003

Dwivedi, G., Srivastava, S. K., & Srivastava, R. K. (2017). Analysis of barriers to implement additive manufacturing technology in the Indian automotive sector. *International Journal of Physical Distribution & Logistics Management, 47*(10), 972–991. doi:10.1108/IJPDLM-07-2017-0222

Han, H. (2016). *A Study of Additive Manufacturing Technology as a Manufacturing Tool of the Automotive Industry*. Academic Press.

Hettesheimer, Hirzel, & Roß. (2018). Energy savings through additive manufacturing: an analysis of selective laser sintering for automotive and aircraft components. *Energy Efficiency, 11*(5), 1227-1245.

Kerbrat, O., Mognol, P., & Hascoët, J.-Y. (2011). A new DFM approach to combine machining and additive manufacturing. *Computers in Industry, 62*(7), 684–692. doi:10.1016/j.compind.2011.04.003

Khajavi, Partanen, & Holmström. (2014). Additive manufacturing in the spare parts supply chain. *Computers in Industry, 65*, 150-63.

Leal, R., Barreiros, F. M., Alves, L., Romeiro, F., Vasco, J. C., Santos, M., & Marto, C. (2017). Additive manufacturing tooling for the automotive industry. *International Journal of Advanced Manufacturing Technology, 92*(5-8), 1671–1676. doi:10.100700170-017-0239-8

Lee, J.-Y., An, J., & Chua, C. K. (2017). Fundamentals and applications of 3D printing for novel materials. *Applied Materials, 7*, 120–133. doi:10.1016/j.apmt.2017.02.004

Lim, Le, Lu, & Wong. (2016). An overview of 3-D printing in manufacturing, aerospace, and automotive industries. *IEEE Potentials, 35*, 418-22.

MacDonald, E., & Wicker, R. (2016). Multiprocess 3D printing for increasing component functionality. *Science, 353*(6307), aaf2093. doi:10.1126cience.aaf2093 PMID:27708075

Makdoun, I., Mezzour, G., Carley, K. M., & Kassou, I. (2018). Analyzing the Needs of the Automotive Job Market in Morocco. In *2018 13th International Conference on Computer Science & Education* (pp. 1-6). IEEE. 10.1109/ICCSE.2018.8468854

Ngo, T. D., Kashani, A., Imbalzano, G., Nguyen, K. T. Q., & Hui, D. (2018). Additive manufacturing (3D printing): A review of materials, methods, applications and challenges. *Composites. Part B, Engineering, 143*, 172–196. doi:10.1016/j.compositesb.2018.02.012

Paritala, P. K., Manchikatla, S., & Yarlagadda, P. K. D. V. (2017). Digital manufacturing-applications past, current, and future trends. *Procedia Engineering, 174*, 982–991. doi:10.1016/j.proeng.2017.01.250

Patalas-Maliszewska, J., Topczak, M., & Kłos, S. (2020). The Level of the Additive Manufacturing Technology Use in Polish Metal and Automotive Manufacturing Enterprises. *Applied Sciences (Basel, Switzerland), 10*(3), 735. doi:10.3390/app10030735

Richardson & Haylock. (2012). Designer/maker: the rise of additive manufacturing, domestic-scale production and the possible implications for the automotive industry. *Computer-Aided Design & Applications, 2*, 33-48.

Sarvankar, S. G., & Yewale, S. N. (2019). Additive Manufacturing in Automotive Industry. *Internationl Journal of Research in Aeronautical and Mechanical Engineering, 7*(4), 1–10.

Shree, Varsha, Dhinakaran, Rajkumar, Bupathi Ram, Vijayakumar, & Sathish. (2019). Effect of 3D printing on supply chain management. *Materials Today: Proceedings.*

Straub, B., & Daimler, A. G. (2018). Automotive Displays-Increasing and Challenging Market. In *2018 25th International Workshop on Active-Matrix Flatpanel Displays and Devices* (pp. 1-4). IEEE. 10.23919/AM-FPD.2018.8437443

Tofail, Koumoulos, Bandyopadhyay, Bose, O'Donoghue, & Charitidis. (2018). Additive manufacturing: scientific and technological challenges, market uptake and opportunities. *Materials Today, 21*, 122-37.

Zadpoor & Malda. (2017). *Additive manufacturing of biomaterials, tissues, and organs*. Academic Press.

KEY TERMS AND DEFINITIONS

Additive Manufacturing: It can be described as a "freedom fabrication" as it can construct any complex geometry with diverse materials by addition of elements in a layer by layer process.

Direct Metal Deposition: This method is not based on other types of powder bed, but it uses a feed nozzle to extrude the powder into the laser beam.

Direct Metal Laser Sintering: This technique is similar to selective laser melting but has greater efficiency on comparing with SLM technique.

Electron Beam Melting: A beam of electron is used to heat the raw material under vacuum and fused together.

Hybrid Manufacturing: The term "hybrid manufacturing" designates the process of combining additive manufacturing addition of materials in layers frame a component.

Laser Metal Fusion: In this method a 3D part formed by selectively melting the powder from the power bed and fusing it in layers with their support structures.

Selective Laser Melting: In this method the metallic powders are fused together using high density lasers.

Chapter 18
Structure Properties Relationship Studies of Vinyl Ester Hybrid Syntactic Foam

Pravin R. Kubade
KIT's College of Engineering (Autonomous), India

Amol N. Patil
KIT's College of Engineering (Autonomous), India

Hrushikesh B. Kulkarni
https://orcid.org/0000-0002-0580-4572
Symbiosis Skills and Open University, India

ABSTRACT

Syntactic foam is the porous composite produced by mixing prefabricated hollow spherical particle into the matrix. Syntactic foams are used as energy absorption sandwich core for several applications like marine, automotive, and aerospace. In this work, low density hollow glass microspheres are hybridized with fly ash cenosphere in Bisphenol-A epoxy-based vinyl ester matrix. Hybrid syntactic foams is created with 60% total filler content. Within these hybrid systems internal composition of two fillers were varied in a step of 25 vol% with respect to each other. Hybrid syntactic foams are prepared by the hand lay-up (molding) method. The physical characterization parameter contains density and matrix porosity whereas tensile, quasi-static compression, flexural (3-point bending), Izod impact, and micro Vickers hardness are grouped as mechanical characterization parameters. Scanning electron microscopy was performed on fractured surfaces to examine deformation and fracture mechanisms related with each loading condition.

DOI: 10.4018/978-1-7998-4939-1.ch018

INTRODUCTION

Development in technology is always assisted by the development of associated materials. The best technological design in marine, space, structural, aerospace or automotive industry is of no use in the absence of appropriate materials to withstand the service loads and the service conditions. Humans are advancing very fast in technology development for the transportation sector in the 21st century with increasing performance demands (fuel economy, increased payload capacity, reduction in pollution, etc.) that calls for the development of lightweight engineering materials. However, it is difficult to attain such elevated and stringent performance standards by only some single conventional materials, hence fresh materials are engineered by means of merging two or more predictable materials termed as composite materials, which gives a distinctive amalgamation of properties that cannot be achieved from any single conventional materials.

A formal definition of composite materials given by The American Society of Metals (ASM) Handbook is "macroscopic combination of two or more distinct materials, having a recognizable interface between them" (M. Gauthier, 2016) Composites are generally prepared by integrating some reinforcement such as particles or fibers in a matrix. This reinforcement provides a balance of structural properties that is superior to either constituent material alone. The major benefits of composite materials are an increase in strength, modulus, bending stiffness and chemical resistance, low moisture uptake, etc. Properties of composites could be altered according to particular design necessities and special properties. At the space-age, composite materials have their widespread applications in marine, aeronautics, automobiles and the space sector because of their lightweight and high strength. Almost all means of transportation and at some extent oil and gas industry, wind energy sector and sports equipment use a considerable amount of composites (M. Biron, 2013) Low cost and simple fabrication methods make polymer matrix composites very popular. Low level of mechanical properties limits the usage of non-reinforced polymers as structure materials therefore, composites with diverse reinforcements provides the greater options with elevated properties for various applications.

Syntactic Foams

Syntactic Foams were buoyancy aid materials developed for deep-sea applications in the 1960s. Currently, syntactic foams are used in aerospace applications, airplane and ship structures. Syntactic foams are porous particulate composites fabricated by mixing hollow particles called micro-balloons into a matrix material such as metal, polymer, ceramic etc. Fillers i.e. micro balloons are used to decrease the amount of more expensive matrix materials and/or to enhance or tailor some properties of matrix materials. Typical examples of micro balloons are hollow glass microspheres, fly ash cenospheres, polymeric micro balloons, carbon micro balloons, metal oxides, etc. The use of hollow particles, such as glass micro balloons, fly ash cenospheres, phenolic micro balloons, etc. is considerably increased in recent years. These hollow particles are used for the production of low-density core materials for sandwich structures. These fillers can be available in different densities and particle sizes. These low-density composites are termed as closed-cell structured foams and named as syntactic foams. The low density is a major property of syntactic foams. The different variations in density and properties of syntactic foam could be obtained by changing the material, volume fraction and/ or density of micro balloons (A. K. Kaw,2006) . Syntactic foams are familiar for their high specific compressive strength, less moisture absorption and better damping characteristics, higher thermal and chemical stability, high damage tolerance and ability to keep damage localized. On account of these properties, syntactic foams are used as core materials for sandwich composites for thermal insulation and conduction, weight-susceptive structures, sports equipment's, body armor, etc.

Figure 1(a) shows the basic two phase structure of syntactic foam, whereas figure 1(b) is showing three phase structure of syntactic foam. Three phase structure shows the matrix porosities present in syntactic foam which are basically the voids formed during fabrication of syntactic foam. These voids are inherent and cannot be eliminated completely. Syntactic foams are further categorized into different types according to different types of fillers and matrix used to synthesize the particular syntactic foam. Amongst these all ceramic syntactic foam is widely used variety of syntactic foams as a core material for sandwich composites due to their ease of fabrication and good mechanical properties compare to types of syntactic foams. Some of the advantages, disadvantages and applications of ceramic syntactic foam are listed below.

Figure 1. (a) Two phase structure of syntactic foam, (b) Three phase structure of syntactic foam *(M. Gilbert, 2005)*

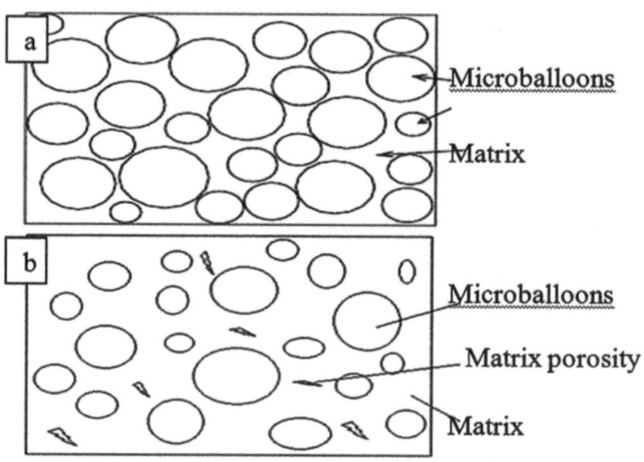

Advantages of Syntactic Foam

- Low density
- High specific compressive strength
- Less moisture and water absorption
- Good damping characteristics
- Good acoustic Characteristics
- Higher thermal stability
- Higher chemical stability
- High damage tolerance keeps damage localized
- syntactic foams can be made as electrically conductive or insulating according to need

Disadvantage

- Fabrication methods for syntactic foam are limited
- Surface characteristics are non-uniform
- Ceramic syntactic foams are brittle in nature

Application

Syntactic foams are used as core materials of sandwich composites for various applications such as:

- Deep-sea applications such as deep sea exploration vehicles
- Thermal insulation and conduction
- Sound insulation
- Aeronautical applications
- Weight-susceptive structures
- Sports equipment's
- Body armor, etc.

Hybrid Syntactic Foams

Hybrid syntactic foams are generally synthesized by mixing two or more different kinds of hollow particles in a matrix. Hybridization in syntactic foam further improves or tailors the particular properties of syntactic foams. Use of two or more different kinds of hollow particulates with different properties and surface structure will produce different deformation and fracture behavior; therefore hybridization may improve particular mechanical characteristics of hybrid syntactic foam.

This particular study focuses on hybridization two types of ceramic microballons in vinyl ester matrix. The primary filler 'hollow glass microspheres' are highly engineered materials comprising of silica and alumina and possess good properties such as good specific strength, uniformity of their surface, resistivity to wetting, low viscosity of the resin-micro balloons mixture and simplicity of production (N. Gupta, 2013). The secondary filler used in this study is fly ash cenospheres which is basically a waste by- product of combustion process. Flyash cenospheres are lightweight, idle, and empty circles comprising of silica and alumina, are loaded with air or gasses, and possess characteristics such as good specific strength, irregularity of their surface, average-wetting resistivity, good energy absorption properties, very low cost thus showing good values to replace conventional fillers and thereby lowering composite cost and reducing land pollution (C. Swetha, 2011).

BACKGROUND

An extensive Literature Survey on the syntactic foam was carried out to highlight the present status of different polymer matrix materials, filler materials, micro-scale and nano-scale reinforcement, their manufacturing/ processing method and their effects on the properties of syntactic foams with deformation and fracture mechanics.

(Breunig, P. et al. 2020) have studied the mechanical response and damage mechanisms associated with syntactic foam core sandwich composites subjected to dynamic impact loading, influence of varying cenosphere volume fraction in syntactic foam core sandwich composites subjected to varying dynamic impact loading and further elucidates the extent and diversity of corresponding damage mechanisms. The syntactic foam cores are first fabricated using epoxy resin as the matrix and cenospheres as the reinforcement with four cenosphere volume fractions of 0% (pure epoxy), 20%, 40%, and 60%. Dynamic impact tests are performed on the sandwich composite specimens at two energy levels of 80 J and 160 J, upon which the data are post-processed to gain a quantitative understanding of the impact response and damage mechanisms incurred by the specimens.

(Ding, Junjie, et al. 2020) has prepared co-continuous hollow glass microspheres (HGMs)/epoxy resin syntactic foams by filling the HGM preforms with epoxy resin. The effects of HGM wall thickness and solid loading on the density, open porosity, and morphology of HGM preform and HGM/epoxy resin syntactic foams were studied in detail. However, the density of preforms increased with the true density of HGMs. The density of HGM/epoxy resin syntactic foams decreased with the contents of HGMs, and the measured density was essentially equal to the theoretical density. The compressive strength decreased with the contents of HGM, regardless of the HGMs wall thickness. The co-continuous HGM/epoxy resin syntactic foams possessed low density (0.567–1.002 g/cm3), and excellent compressive strengths (30.32–131.60 MPa). These properties were due to the elimination of matrix pores, and are a type of promising material in deep sea exploration area.

(Carolan, D, et al. 2020) studied a combined numerical-experimental method that enables accurate prediction of not only the elastic moduli and tensile failure strengths of syntactic foams, but also accounts for the experimentally observed scatter in these measurements is presented. In general, for the systems studied, an increase in microsphere content resulted in an increase in tensile modulus and a decrease in tensile strength. For the first time, a direct link between the underlying microstructure and the experimentally observed scatter in fracture strength is drawn and substantiated with modelling.

(Labella et al. 2016) reported that elastic energy absorption of fly ash cenosphere reinforced vinyl ester syntactic foam was higher in high strain rate compression than quasi-static compression. They observed that compressive strength was decreased by 12-19% and modulus was increased by 50-70% in quasi-static compression than the neat vinyl resin. They reported decreased flexural strength and increased flexural modulus with increased vol.% . At 60 vol.% flexural strength and flexural modulus were high as 73% and 47%, respectively. The decrease in flexural strength is ascribed to stress concentration in materials due to cenospheres and also defects in cenospheres. The coefficient of thermal expansion decreased with increased vol % of fly ash cenospheres.

(Ghamsari et al. 2016) developed bucky epoxy matrix syntactic foam utilizing imidazolium-based ionic liquid fillers (IL, wt. filler%=0.5, 1, 7.5, 15) with dispersing single-walled carbon nanotube (CNT, wt. filler%=0.25, 1, 2). They reported degradation of yield strength, elastic modulus and energy absorption in compression loading at high CNT and IL content however at low content of CNT (0.25 wt.%) and IL (0.5 wt.%), improvement in mechanical properties was noticed. They reported a marginal increase in electrical conductivity by addition of CNT and IL hybrid gel. They also reported controverting effects on mechanical and electrical properties.

(Colloca et al. 2014) reported a 10-47% increase in tensile strength and modulus for 50 vol.% of glass microballoon epoxy syntactic foams, with the reinforcement of carbon nano-fiber in the range of 0.30 to 0.42 vol.%. However, for 30 vol.% of glass microballoon 8-15% decrease in tensile strength and modulus is observed for same the carbon nano-fiber reinforcement level.

(Zhang and Ma, 2014) investigated the effect of carbon Nano-fiber (CNF, Filler% = 0.5, 1.0, 1.5, 2.0) addition on the compressive, flexural and fracture toughness properties of hollow carbon microspheres filled phenolic resin syntactic foam. They observed that no major improvement in compressive strength is observed with the addition of CNFs as crushing of microsphere was the main failure mechanism, while flexural strength and fracture toughness was improved considerably by the addition of CNFs up to 1.5 vol.% and after which it was decreased. The decrease was attributed to agglomeration and clustering of CNFs. They reported crack deflection, microsphere deboning, microsphere crushing and deformation, step structure (CNF pull-out with deboning from the matrix) as the failure mechanism.

(Desai et al. 2013) investigated an effect of surface modification of cenospheres on cenospheres filled acrylonitrile butadiene styrene (ABS) matrix syntactic foam (SFs). They showed the effect of surface modification on thermal, electrical, fire retardancy (M. Colloca, 2013) and dynamic mechanical [8] properties of syntactic foam. SFs were manufactured by PIM. They reported low heat deflection temperature (HDT) i.e. distortion temperature in treated cenospheres to that of untreated cenosphere SFs at a given concentration and it was increased with the addition of both type of cenosphere. Glass transition temperature and decomposition temperature was nearly same as of ABS matrix thus thermal stability is detected. They found 20 wt.% cenosphere as critical fraction up to which dielectric constant increased but above that, it was decreased and it was higher for untreated cenosphere for all samples. They observed a small difference between treated and untreated cenosphere and fire retardancy increases with an increase in cenosphere content due to their ceramic nature (M. Colloca, 2013). In the dynamic mechanical analysis, they found that addition of the cenosphere increases storage modulus and simultaneously decreases loss modulus while the loss factor shows fluctuating trend and highest loss factor was observed at 15 wt.% cenosphere content at room temperature (B. B. Kumar, 2016).

(Zegeye et al. 2013) reported the effect of pristine graphene platelets (filler% = 0.1, 0.3, 0.5) reinforcements on tensile and compressive properties of HGM epoxy SFs. They reported 3.6% and 2.4% increase in compressive strength, 15.9% and 14.7% increase in tensile strength, 4.57% and 14.70% increase in tensile modulus at 0.1% and 0.3% volume fraction of graphene platelets respectively. They found 0.3% volume fraction as critical volume fraction beyond which mechanical properties deteriorated. They also observed an increase in fracture strain which increases the toughness for all graphene reinforced foams. They relate enhancement in properties to the heavy interaction amongst the graphene platelets layer and polymer matrix, which in turn restrains the motion of polymer chains and thus delayed crack initiation and growth.

(Poveda et al. 2013) worked to study the effect of carbon nano-fiber (CNF, wt.filler%=1,2) reinforcement on the visco-elastic properties (storage and loss modulus, loss factor, max. use and glass transition temperature) of HGM/epoxy-based syntactic foam in the temperature range of -750C to 1500C. Experimentation was done with two types of HGM fillers with densities of 0.22, 0.46 g/cc, respectively, as well as filler% was taken as 30, 40. They reported a maximum increase in storage modulus and increase loss modulus at room temperature. They observed a decline in storage modulus and an enhancement in loss modulus at room temperature with the addition of carbon nano-fibers to syntactic foam. Max. Use and glass transition temperature increased through max. of 27.1% and 25%, respectively to neat epoxy resin.

(Huang, 2015) analyzed the consequence of HGM volume fraction on elastic behavior and fracture mechanics of epoxy syntactic foam in compression loading. They developed a FEM model with representative elementary volume for different volume fractions. They found that localized stresses may create vertical splitting fracture of micro balloons and that was responsible for the macro crack formation in a matrix as fracture mechanisms. Also, micro-cracks get coupled with neighboring micro-crack to form a macro crack, either in a longitudinal direction for the low volume fraction or in a diagonal direction for the high volume fraction. They reported an increase in young's modulus and critical strain with an increase in volume fraction regarding to above-stated failure mechanisms.

(Kumar et al. 2016) worked on the effects of surface modification of cenospheres for the high-density polyethylene (HDPE) matrix syntactic foam at cenosphere filler% of 0, 20, 40 and 60. They also studied the effect of blending methods (mechanical mixing and Brabender mixer) and surface modification on the tensile (R. L. Poveda, 2014) and flexural (R. Huang, 2015) properties of polymer injection molded (PIM) syntactic foams (SFs). They treated cenospheres with a 3Aminopropyltriethoxysilane coupling agent and HDPE was functionalized with Dibutylmaleate (DBM). They reported an increase in density, tensile modulus and tensile strength with an increment of treated cenosphere content. Highest strength and modulus were observed for treated cenosphere (filler%=40)/ functionalized HDPE and untreated cenosphere (filler%=60)/ virgin HDPE SFs respectively. They highlighted the use of Brabender mixing process over mechanical mixing process. The highest tensile strength and modulus were obtained in Brabender mixing due to effectiveness in particle cluster braking which leads to high matrix-particle surface area generation (R. L. Poveda, 2014). Also, they stated that density, flexural strength and flexural modulus was improved with an increase in treated cenosphere content and functionalized HDPE SFs and were dependent on mixing methods. Brabender mixing yields the highest strength and modulus for syntactic foam with treated cenosphere (Filler%=60) and functionalized HDPE with 41% and 70% improvement than the neat HDPE. They employed Gupta and Bardella-Genna model to predict flexural properties in which later was predicted better at high filler content.

Kumar et al. (B. R. B. Kumar, 2016) worked on the strain rate sensitivity of syntactic foam (SFs). The cenosphere fillers are blended with high-density polyethylene (HDPE) by using a mechanical mixing and afterward feed into polymer injection molded (PIM) machine with quasi-static (10-4/s, 10-3/s, 10-2/s) and high (3430/s, 2700/s, 1810/s) strain rate compression. They reported that compressive strength and modulus get improved with increasing cenosphere content than the neat HDPE for the same quasi-static strain rate, as well as the average strength at high strain rate is 2-fold of the strength at quasi-static strain rate for same cenosphere content.

(Kumar et al. 2016) worked on the tensile characterization of cenosphere filled high-density polyethylene (HDPE) matrix syntactic foam produced by optimized polymer injection molding (PIM) machine. They had used cenospheres with a true particle density of 0.8 g/cc with vol.% of 0, 20, 40 and 60. The tensile test results showed comparatively brittle failure with a minute plastic deformation. They observed matrix as the main crack propagation carrier while flaws in cenosphere, non-spherical shape and lack of particle-matrix adhesion also contributed to failure mechanisms. Theoretical modeling was used to estimate the properties of the cenospheres.

(Gupta et al. 2016) reported an increased specific compressive strength and modulus of glass micro balloons reinforced vinyl ester syntactic foams with a decrease in radius ratio of particle. They also stated that the specific compressive strength and modulus were increased with an increase in a volume fraction of same radius ratio particle. They also reported that the tensile modulus was 15-30% higher than the compressive modulus. They observed particle deboning and microsphere fracture as the main fracture mechanisms in tensile and compressive loading, respectively.

(Q. Tian and D. Yu, 2016) worked on the effect of interfacial adhesion between epoxy matrix and hollow/solid poly-divinylbenzene-co-glycidyl methacrylate P[DVB-GMA] polymer microsphere. They studied the effect of interfacial adhesion on the tensile, electrical and optical properties of syntactic foam. Hollow and solid polymer microspheres/epoxy composite films were obtained through the powder mixing method and UV-cured polymerization. An interfacial adhesion was produced by a chemical reaction between the epoxy faction of the microspheres and epoxy matrix. They reported higher tensile strength and higher modulus for hollow polymer microsphere to that of solid microsphere syntactic foam. Also, there was an increase in tensile strength and tensile modulus with an increase in filler fraction to critical limit and then properties were decreased. The decrease was attributed to a shift in failure mechanism from shear yielding deformation and crack blowing to the particle deboning. They reported a slight decrease in electrical resistivity and UV transmittance with hollow microsphere content increased.

(Rutz B. and Berg J. 2010) reviewed the possibility of reducing the weight of structural polymeric composites by creating voids and by uncompromising mechanical characteristics. They discussed the efficacy of making lightweight composites by two methods, creating gas bubbles and inclosing voids within shell i.e. micro balloons. They highlighted the fracture mechanics of syntactic foam under different loading conditions. Also, they highlighted key issues of syntactic foam such as micro balloons deboning, their dispersion, size effects, agglomeration of nano-fillers, etc.

(Dimchev M. et al. 2010) discussed efficiency of CNF ($_{filler\%}$= 0.25) addition into HGM (ρ_{TP}=0.22, 0.32, 0.37, 0.46 g/cc, $\Phi_{filler\%}$=30, 40, 50)/ epoxy matrix syntactic foam (SFs) under tensile and compressive loading. They reported increased tensile strength, tensile modulus and unaffected compressive strength with the addition of CNF compared to plain SFs. They observed micro balloons deboning and matrix-CNF interphase fracture in tensile loading and particle crushing in compression loading as fracture mechanisms.

Literature Summary

- The above studies in the literature on polymer matrix syntactic foam composites was largely done on thermoset polymers such as epoxy, vinyl ester, cyanate ester, phenolic resins and thermoplastic polymers such as polyethylene, polystyrene resin system and fillers such as hollow glass microsphere, cenosphere, phenolic/carbon microballoons.
- In addition, few researchers have used modified polymer resin systems. A very few researchers did surface modification of micro balloons. Few studies incorporated nano-clay, carbon nano Fiber, Carbon nano Tube as nano reinforcing materials, while some used glass and aramid fibers as micro reinforcement.
- Most studies have compared their results to previous studies done by other researchers, which used different resin-hardener systems, curing cycles, manufacturing methods, etc., which affect obtained results.
- There are very few researchers who attempted to hybrid two different material micro balloons at time.

PROBLEM IDENTIFICATION

Increased interest of transportation sector in the lightweight composites has built considerable interest for exploring new techniques to improve the properties of syntactic foams. In hybrid syntactic foam, two different types of micro balloons i.e. glass micro balloons and fly ash cenospheres used, have different properties and surface structures can build up different deformation and fracture behavior.

As these syntactic foams might be used as a sandwich core in a variety of weight-sensitive applications makes it necessary to study and predict the deformation and fracture behavior of such hybrid syntactic foams.

Objectives of Research Work

The prime objective of this dissertation is to develop a 'Hybrid Syntactic Foam' and understanding the deformation as well as fracture behavior of syntactic foams subjected to various loading conditions. Sub-objective are as follows:

- To fabricate syntactic foams by hand lay-up method.
- To investigate the effects of micro balloons volume fraction and density, micro balloons material type, on:
    - Physical properties -Density, Void Content.
    - Mechanical properties -Tensile, Compression, Flexural, Impact.
- To understand the deformation behavior and fracture behavior using scanning electron microscopy.
- To study the structural relationship of a composite.

Motivation

According to world energy council, India will become a 2nd largest coal-consuming country by the year 2020 with expected fly ash generation of 350 MT. in India only 3% fly ash is utilized to a commercial product, whereas in Germany it's around 80%. Today's technology strives to reduce pollution, make effective use of industrial by waste. The residual fly ash from coal thermal power plants, which is 80% of coal ash, must usually be disposed of by landfilling since it has hardly any commercial value.

The landfill space is rapidly fading in many regions and the creation of new landfills is very costly. In this way, much endeavor has been directed at ascertaining utilization for fly ash remains with the objective that the fly ash does not require to be landfilled and any functioning item from this waste material is praiseworthy.

EXPERIMENTAL DESIGN

Design of Experiment

The full factorial design of experiments is used with two control factors. The total volume content of fillers (i.e. glass micro balloons + fly ash cenospheres) with respect to the volume of composite is the first control factor with one level. The individual volume fraction of flyash cenospheres and hollow glassmicrospheres in total filler content is the second control factor with 5 levels. Table 1 gives details of control factors and their levels.

Hybrid systems was created with 60% total filler content in a matrix, i.e. the total volume fraction of mixture of two fillers was fixed at 60% and rest 40% is matrix, whereas, within these hybrid systems an internal composition of two fillers were varied in a step of 25 vol% with respect to each other. The following specimen coding is used as HSF-YYY-ZZZ. HSF is an abbreviation of hybrid syntactic foam.

Table 1. Control factors and their levels

Factors	Levels				
Total volume content of fillers (glass micro balloons + fly ash cenospheres) with respect to volume of composite	60%				
Individual volume fraction of flyash cenospheres in total filler content	0	25	50	75	100

Table 2. Hybrid syntactic foam nomenclature with their filler composition for 60% total filler

Material	Total 60% Filler Content in Composite (vol.%)		Hybrid Syntactic Foam Nomenclature
	Glass Microballoons	Flyash Cenosphere	
Neat Vinyl Ester	-	-	Neat Vinyl Ester
HSF with FAC only	0	100	HSF-00-100
HSF with K20 glass microballoons only	100	0	HSF-100-00
HSF with K20 glass microballoons and FAC	75	25	HSF-75-25
HSF with K20 glass microballoons and FAC	50	50	HSF-50-50
HSF with K20 glass microballoons and FAC	25	75	HSF-25-75

The abbreviation YYY represents K20 Hollow glass micro balloons with volume fraction and ZZZ represents the volume fraction fly ash cenospheres, respectively in total filler volume fractions. For example, HSF-75-25 represents hybrid syntactic foam with 75 vol% of K20 glass micro balloons and 25 vol% of fly ash cenospheres in 60% total filler volume fraction. Table 2 represents the nomenclature with their filler composition for all hybrid syntactic foam samples.

Fabrication Process

Hybrid syntactic foams were fabricated by hand stir mixing (Hand Layup) method (see Figure 2). It is an inexpensive method has low productivity. The raw material used in the study also limits the fabrication technique to hand layup method because other techniques may cause breakage of micro balloons during fabrication. The volume percentage (fraction) of fillers within the syntactic foam is the limitation of the stirring method. Generally, the matrix is 2- 8 times heavier than fillers which cause fillers to float below 30 vol%. On another side, mixing beyond 60 vol% of fillers in syntactic foam make breakage and clustering of fillers. In this study we have selected maximum possible filler level in resin (i.e. 60%) so as to keep density of foam as low as possible.

The vinyl ester resin to curing agent ratio was fixed at 97:3 according to the manufacturers guidelines. A measured quantity of vinyl ester resin taken in a glass beaker and catalyst was added at 3% by volume and mixed on a magnetic stirrer (see Figure 4 (a)). Then measured the quantity of glass micro balloons and fly ash cenospheres were added to resin and mixture was gently stirred with glass dowel until homogeneous mixture obtained. Care should be taken in mixing process not to damage micro balloons. Appendix (I) gives details about the procedure to obtain the weight of constituent materials

Figure 2. Hand layup method

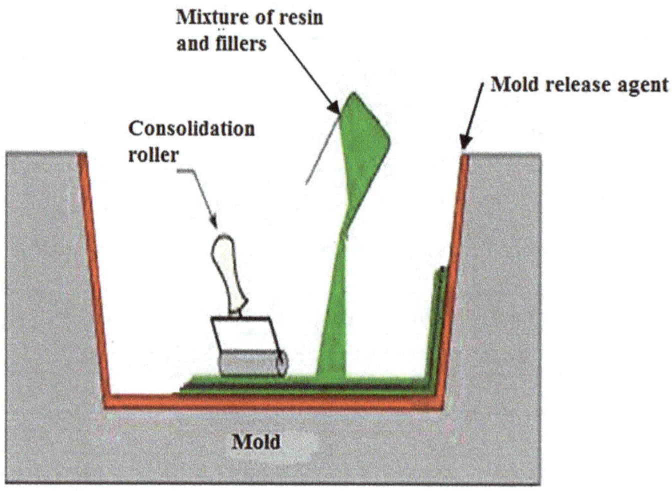

and calculated weight for each test specimens. Weight of constitute materials was calculated for each mold separately. The slurry cast in an aluminum mold (see Figure 4 (b)) of required dimensions of test specimens. This mold left for 24 h at room temperature for curing and then 3 h at 70 °C in an air oven for post-curing. A thin layer of mold release wax and polyvinyl alcohol was coated on the mold for easy removal of cast specimens. Figure 3 depicts a schematic illustration of the fabrication process.

Figure 3. Flow chart of fabrication process

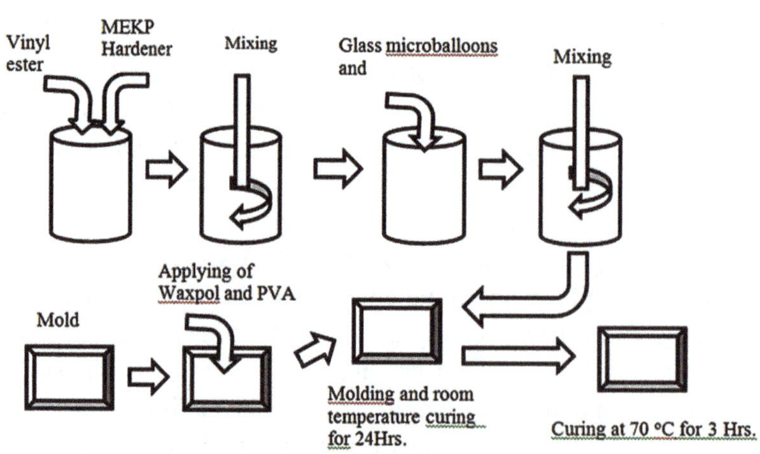

Figure 4. (a) Heating on magnetic stirrer and mixing with glass dowel (b) Pouring slurry into the aluminum mold

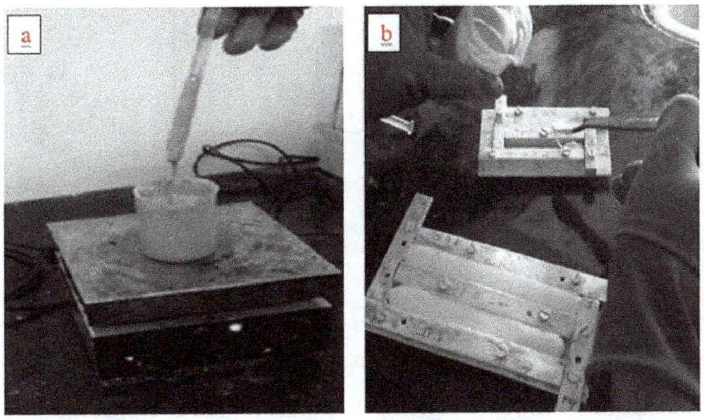

RESULTS AND DISCUSSIONS

Density and Matrix Porosity

Theoretical density, experimental density and matrix porosity of various compositions are represented in Table 3. Neat vinyl ester resin has more density than the all compositions of hybrid syntactic foam. The hybrid syntactic foam contains low-density fillers which are hollow in nature which causes reduces the density. Thus the density of hybrid syntactic foams is a function of density and volume fraction of fillers. Theoretical and experimental densities vary because of air entrapped during the fabrication process. Experimental density has a lower value than theoretical density due to voids formed during processing. Figure 6 shows an SEM image of fabricated 60HSF-25-75 hybrid syntactic foam which confirms the uniform distribution of fillers into the matrix.

Matrix porosity increased from 1.2% to 2% by addition of fly ash cenosphere in hybrid syntactic foams compared to their plain syntactic foams (HSF-100-00). Matrix porosity increased with fly ash cenosphere content, due to surface irregularities and cracked micro balloons present on fly ash cenosphere. Surface irregularities of fly ash cenosphere lead to air entrapment during mixing with a matrix. Matrix porosity was determined with respect to the total volume of hybrid syntactic foam. Matrix porosity of all hybrid syntactic foams is in the range of 7-10% of the total volume of hybrid syntactic foam.

Table 3. Effect of composition on density and Matrix porosity

Specimen Nomenclature	Theoretical Density (g/cm^3)	Experimental Density (g/cm^3)	Matrix Porosity (%)
Neat Vinyl Ester	1.22505	1.22136	2.33
HSF-00-100	0.72402	0.660885	9.33
HSF-100-00	0.61002	0.557314	8.64
HSF-75-25	0.63852	0.582586	8.76
HSF-50-50	0.66702	0.606254	9.11
HSF-25-75	0.69552	0.630211	9.39

Figure 5. Density and matrix porosity of various hybrid syntactic foams

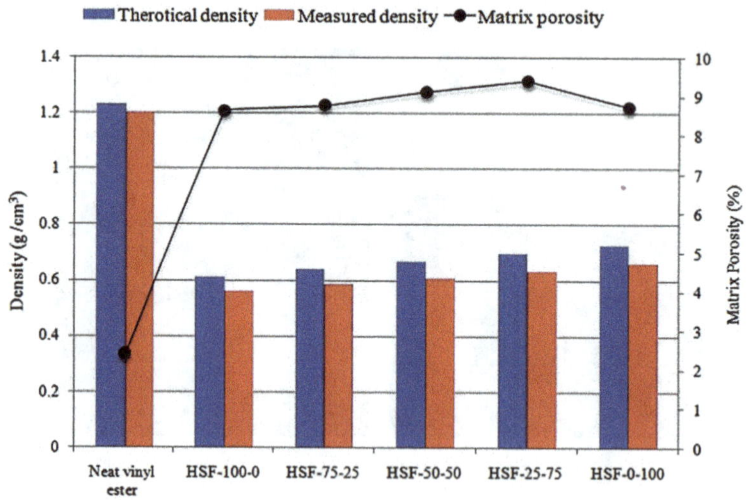

Figure 6 (a) and (b) shows the matrix porosity as well as air entrapment surrounding the fly ash cenosphere. Figure 6 (a) shows a microscopic image of hybrid syntactic foam at 132x magnification which shows the voids present in the composite. These voids are also called as matrix porosities which are formed during fabrication process. Improper mixing and pouring during fabrication of syntactic also increases void space. These void contents can be reduced by proper mixing and pouring during the fabrication but cannot be eliminated completely. Figure 6 (b) shows the air entrapment surrounding to cenospheres. The rough surface of fly ash cenosphere affects the adhesion of matrix at the surface which may cause air entrapment. Matrix porosity is also attributed to fabrication process. The matrix porosities present in composite may reduce the density of composite but adversely effects on strength of composite.

Figure 6. SEM images showing (a) matrix porosity (b) air entrapment of hybrid syntactic foams

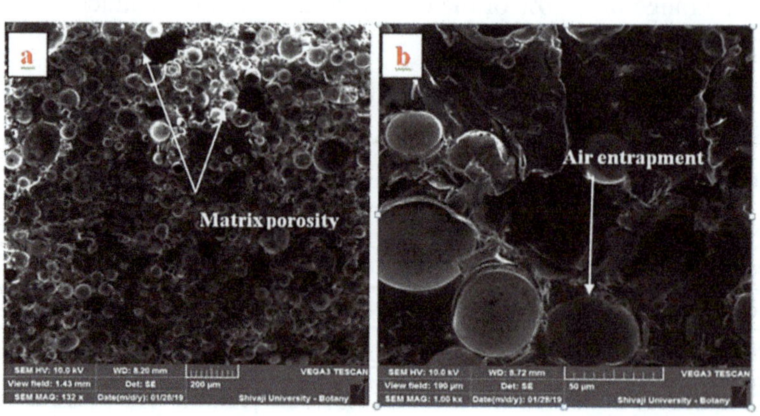

Tensile Characteristics of Hybrid Syntactic Foam

Tensile strength and specific tensile strength of hybrid syntactic foam with 60% total filler volume are presented in Figure 8. The tensile strength of all compositions of hybrid syntactic foams shows lower values than the neat vinyl ester resin sample. In hybrid syntactic foams, all hybrid compositions showed higher values of tensile strength than plane syntactic foam with only K20 glass micro balloons. Plane Flyash cenosphere (i.e. HSF-00-100) based syntactic foams also has lower tensile strength than all combinations of hybrid syntactic foam. In hybrid syntactic foam system, HSF-25-75 showed maximum 135% increase in tensile strength than plain syntactic foam with K20 glass microballoon (i.e. HSF-100-00). It has been observed that tensile strength increases with increase in fly ash cenosphere volume fraction in filler content of hybrid syntactic foam system. As fly ash cenosphere content are increased by 25, 50 and 75 vol% in hybrid fillers, tensile strength increased by 64, 54 and 135% respectively.

The specific tensile strength of neat vinyl ester resin has lower values than all compositions of hybrid syntactic foams. In hybrid syntactic foams, all combinations of hybrid syntactic foam showed higher values of specific tensile strength than the plane syntactic foam with only fly ash cenosphere (i.e.HSF-00-100). Plane syntactic foam with only K20 glass micro balloons (i.e. HSF-100-00) showed

Figure 7. Fractured specimens of hybrid syntactic foams in tensile test

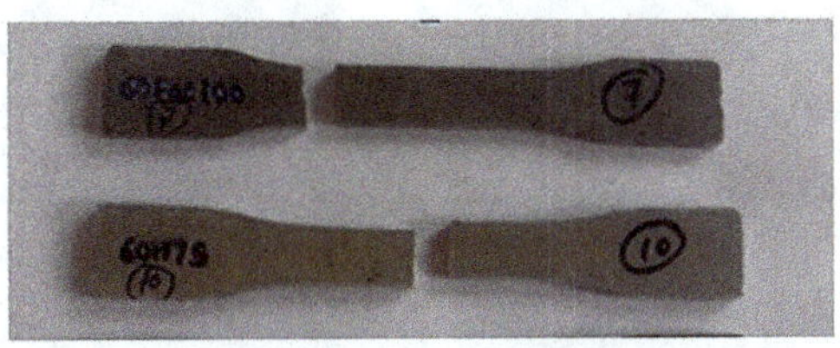

Figure 8. Tensile characteristics of hybrid syntactic foams

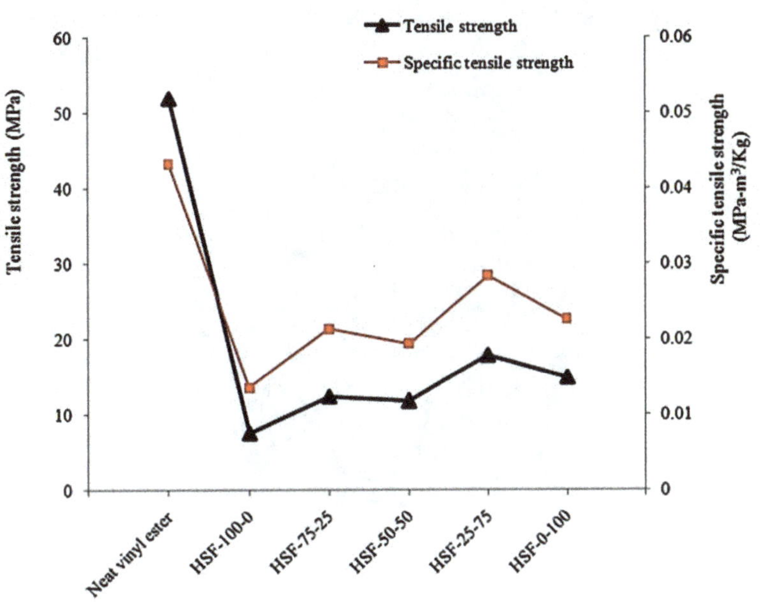

lower specific tensile strength than all Hybrid syntactic foams. In hybrid syntactic foam system, HSF-25-75 showed a maximum of 108% increase in specific tensile strength than plain syntactic foam with K20 micro balloons (i.e.HSF-100-00). Increase in Specific tensile strength is also noticed in hybrid syntactic foam system with an increase in fly ash cenosphere percentage. With an increase in fly ash cenosphere content in filler by 25, 50 and 75 vol% net increase of 56, 42 and 108% in specific tensile strength is observed with respect to plane syntactic foam with K20 glass micro balloons.

Figure 9 (a), (b) and (c) shows a microscopic image of fractured specimens of hybrid syntactic foams in the tensile test. Brittle failure of the test specimen is also observed for hybrid syntactic foam. Microballoon cracking is also observed with matrix cracking (see Figure 9 (a) & (c)). The Figure 9 (b) showed a deboning at both types of fillers, however, in hybrid syntactic foams with 60% total fillers showed micro balloons crushing at some places. This will be attributed to the high filler quantity in the total composition of hybrid syntactic foam. Flyash cenosphere in composition causes to increase in microballoon cracking and matrix-micro balloons deboning due to surface flaws of fly ash cenosphere. This matrix-micro balloons deboning hampers the tensile properties of hybrid syntactic foam.

Figure 9. Microscopic images of fractured specimens in tensile test for 60% total filler hybrid syntactic foam test at (a) 117x magnification, (b) 482x magnification, (c) 1.00kx magnification

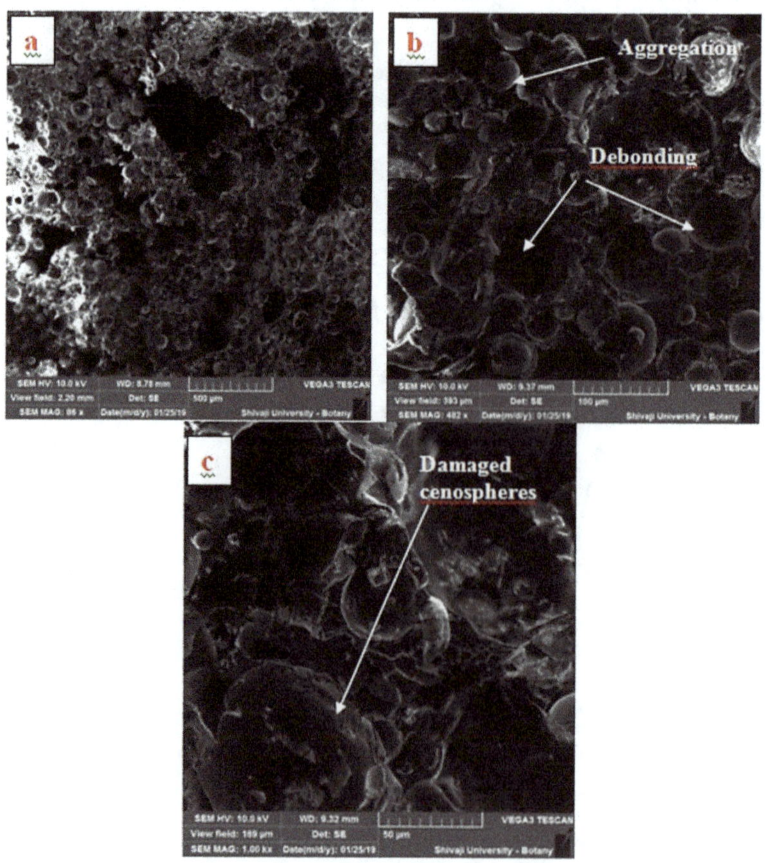

Compressive Characteristics of Hybrid Syntactic Foams

Compressive strength and specific compressive strength of hybrid syntactic foam are presented in Figure 10. The compressive strength of all compositions of hybrid syntactic foam shows lower values than the neat vinyl ester resin. In plain syntactic foams, syntactic foam containing K20 glass micro balloons (i.e. HSF-100-00) showed the lowest value than all combinations of hybrid syntactic foam and fly ash cenosphere (i.e.HSF-00-100) based plane syntactic foams. In hybrid syntactic foam system, HSF-50-50 showed a maximum of 12% increase than plain syntactic foam with K20 glass micro balloons (i.e. HSF-100-00). As fly ash cenosphere fraction in filler content of in hybrid syntactic foam increased by 25 vol%, 50 vol% and 75 vol% increase in compressive strength by 2%, 12% and 8% is observed. Plane syntactic foam with only fly ash cenosphere (i.e. HSF-00-100) showed highest compressive strength.

The specific compressive strength of all compositions of hybrid syntactic foams shows near about similar values that of neat vinyl ester resin specimen. Graph of specific compressive strength follows a similar trend as a graph of compressive strength for hybrid syntactic foams. The increase in fly ash cenosphere content in hybrid syntactic foam system had shown a gradual increase in compressive strength and specific compressive strength. Specific compressive strength increased by 12%, 10% and 11% as fly ash cenosphere in filler content increased by 25, 50 and 75 vol% respectively.

Figure 10. Compressive characteristics of hybrid syntactic foams

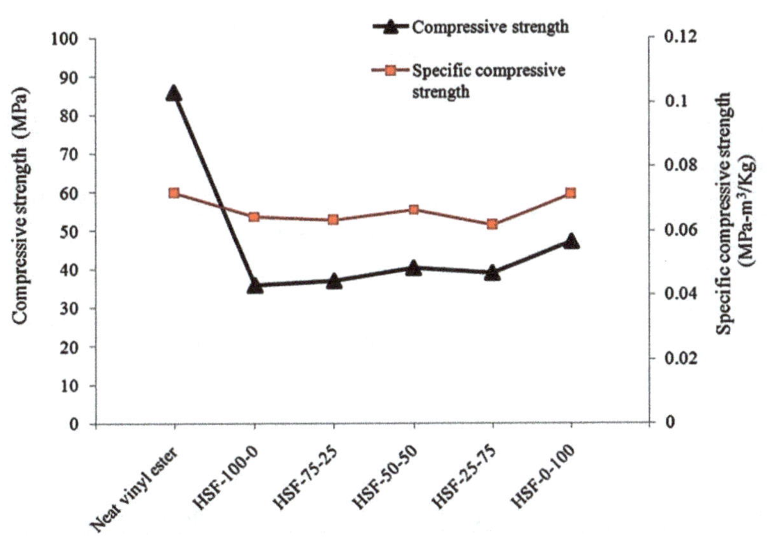

The hybridization of fly ash cenosphere in glass micro balloon based system, showed increased compressive strength while there was a horizontal line type graph for specific compressive strength for hybrid syntactic foam system is observed. This phenomenon can be attributed to hard alumina-silicate composition of fly ash cenosphere. The increase in density because of high-density fly ash cenosphere is compensated by hard alumina-silicate composition of fly ash cenosphere.

Impact Properties of Hybrid Syntactic Foams

For 60% total filler configuration; Impact strength of plane hollow glass microsphere syntactic foam (HSF-100-0) is higher than all hybrid syntactic foams. In this hybrid syntactic foam system as the volume fraction of fly ash cenospheres in filler content of foam increases by 25% sudden drop in impact strength was observed. This behavior can be attributed to the stirring process and surface flaws of fly ash cenospheres. Impact strength further increased gradually as fly ash cenospheres vol% in hybrid fillers increased to 50% and 75% respectively; however, the increased impact strength is less than plane hollow glass microsphere syntactic foam. The net decrease in impact strength observed was 27%, 19.7% and 19.7% for an increase of fly ash cenospheres volume fraction in hybrid fillers of syntactic foam with 25, 50 and 75% respectively in comparison to plane hollow glass microsphere syntactic foam. The decrease in strength is to be credited to the irregular surface morphology of fly ash cenospheres and mechanism of fracture at a high volume fraction. Specific impact strength also follows a similar trend; decrease in specific impact strength observed is 30%, 26%, 29% for an increase in a volume fraction of fly ash cenospheres in filler by 25, 50, 75 vol% respectively in hybrid syntactic foam system.

Figure 11. Impact strength and specific impact strength of hybrid syntactic foams

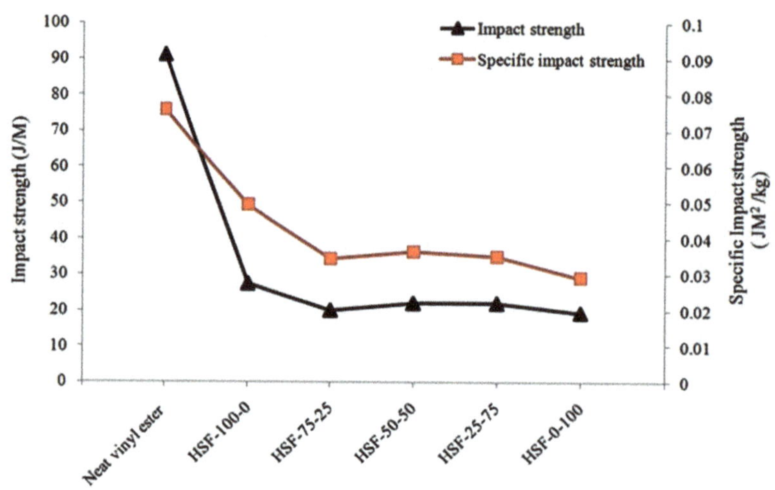

Absorbed impact energy for hybrid syntactic foam system with was decreased by 28%, 21% and 21% with an increase of fly ash cenospheres volume fraction by 25, 50, 75 vol% respectively in filler content of hybrid syntactic foam. Plane fly ash cenospheres based syntactic foam (i.e. HSF-0-100) shows less strength than all hybrid syntactic foam system. Absorbed Impact energy and specific impact energy for hybrid syntactic foam is presented in figure 12.

Figure 13 (a) and (b) represents the microscopic observations of unnotched Izod impact tests of HSF-50-50 type syntactic foam. figure 13 (a) shows the fractured surface of hybrid syntactic foam. while figure 13 (b) shows microballoon crushing as the failure mechanisms under the impact. At some places aggregation of microballoons and cenosphere is observed, this also may cause a decrease in impact strength. This aggregation of microballoons and cenosphere may be caused because of improper

Figure 12. Absorbed Impact energy and specific impact energy of hybrid syntactic foams

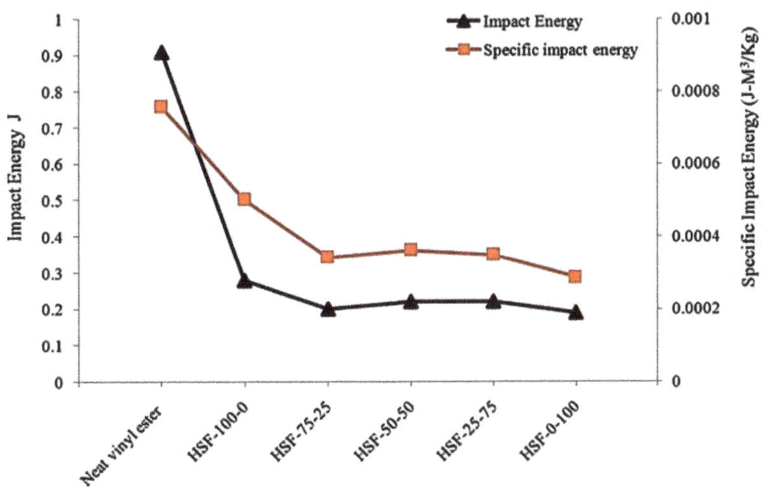

Figure 13. Microscopic images of unnotched izod impact tested HSF-50-50 type syntactic foam. (a) 100x magnification, (b) 513x magnification.

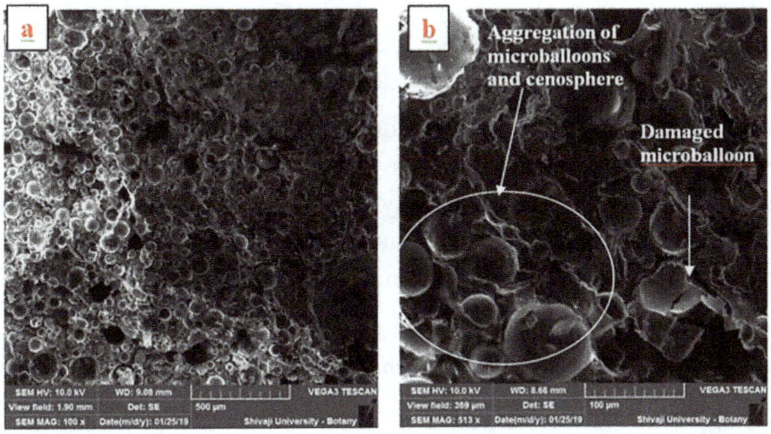

blending during fabrication. In hybrid syntactic foam low resin content also leads to weak bonding amongst fillers and matrix. The poor bond among microballoons and matrix causes the debonding of microballoon and cenospheres.

Flexural Characteristics of Hybrid Syntactic Foam

Effect of hybridization on flexural strength and specific flexural strength of hybrid syntactic is presented in Figure 14. In plain syntactic foams, syntactic foam containing only K20 glass micro balloons (i.e. HSF-100-0) had the lowest flexural strength of 22.70 MPa. The plane fly ash cenospheres syntactic foam (i.e. HSF-0-100) had a higher value than all hybrid syntactic foam system. The flexural strength of plane fly ash cenospheres syntactic foam (i.e. HSF-0-100) was 44.22 MPa. As vol% of fly-ash cenospheres in

Figure 14. Flexural strength and specific flexural strength of hybrid syntactic foams

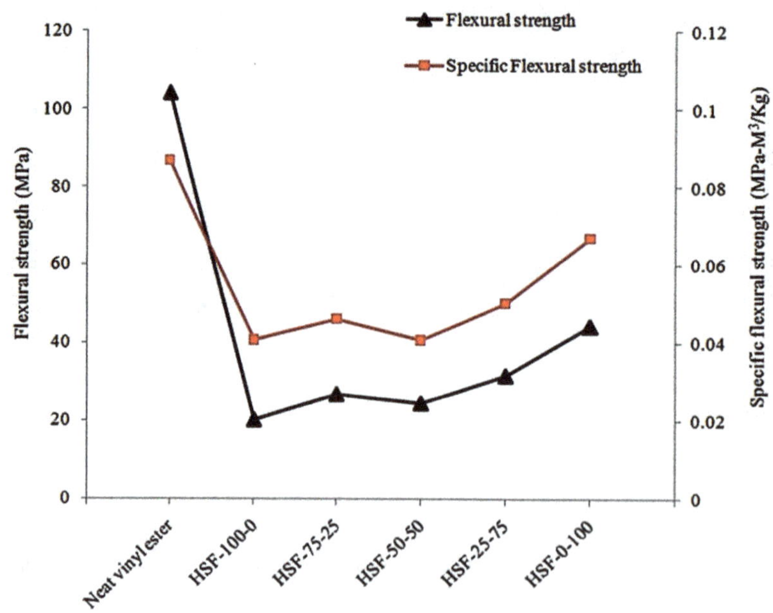

hybrid fillers increases by 25%, 50%, 75% the flexural strength varies 18%, 9% and 39% respectively in comparison to plane hollow glass microballoon syntactic foam (i.e. HSF-100-0).

Effect of hybrid compositions over flexural modulus and specific flexural modulus of hybrid syntactic foam is presented in Figure 15. Flexural modulus of hybrid syntactic foam increased linearly with increase in fly ash cenospheres content in the hybrid filler system of hybrid syntactic foam. The increase of fly ash cenospheres content with higher density may cause an increase in the flexural modulus but reduces flexural strength because of the stress concentration at defect present in cenospheres.

Figure 15. Flexural modulus and specific flexural modulus of hybrid syntactic foams

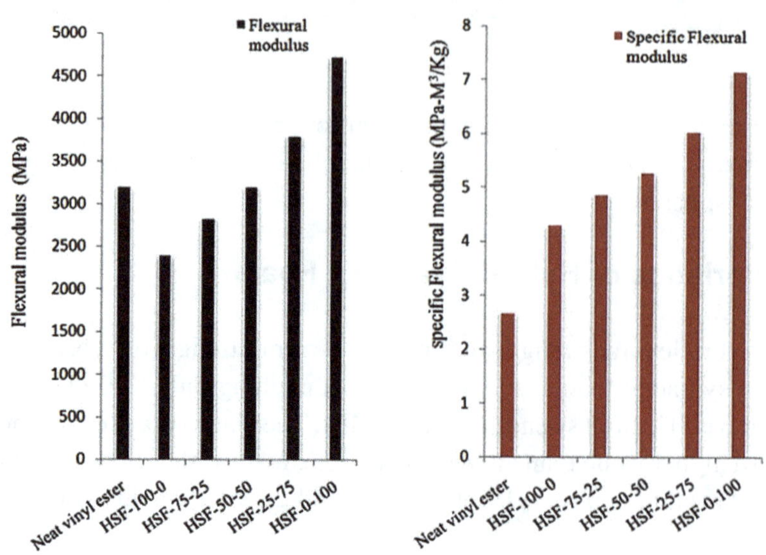

Specific flexural modulus for hybrid syntactic foam system is higher than neat resin. Flexural modulus and specific flexural modulus increased with increase in fly ash cenospheres content. Flexural modulus increased by 17%, 33%, and 58% as fly ash cenospheres percentage in fillers increased by 25%, 50%, 75% in comparison to plane hollow glass microsphere syntactic foam. Also increase in Specific flexural modulus observed by 13%, 22%, and 40% as percentage of fly ash cenospheres in fillers of hybrid syntactic foam increased by 25%, 50%, 75% in comparison to plane hollow glass microsphere syntactic foam.

Figure 16. Flexural Stress - strain curve of hybrid syntactic foams

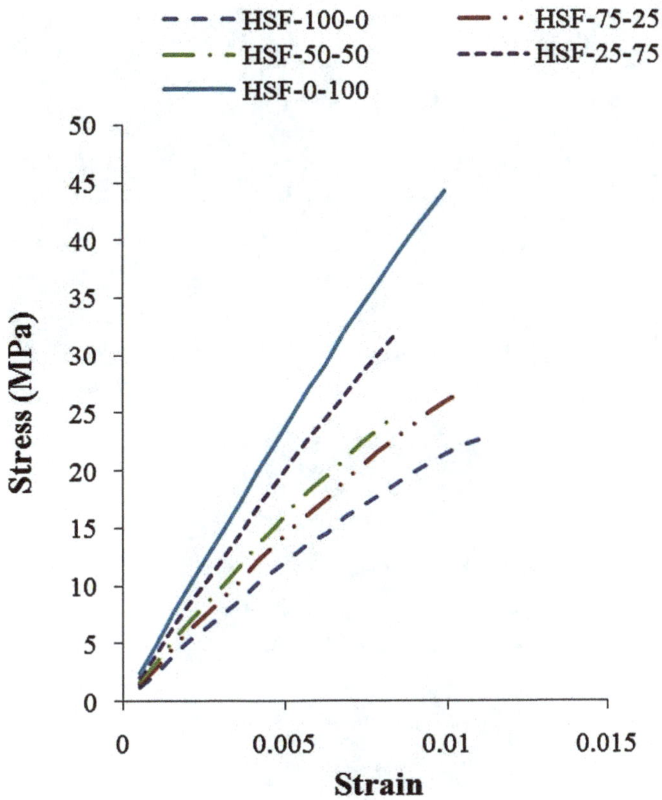

Figure 18 (a), (b), (c) shows SEM images of fractured surfaces of HSF-50-50 hybrid syntactic foams in 3-point bending configuration. SEM images reveal matrix-micro balloons deboning, matrix cracking, as deformation mechanisms. Very less fractured micro balloons are observed in microscopic images. From Figure 18 (b), it has been observed that an elevated volume fraction of micro balloons in composite leads to deboning. This deboning is basically attributed to poor wetting of cenospheres and hollow glass micro balloons, which cause weak interface between resin and fillers. Failure of syntactic foam at low loading is observed because of easy deboning between matrix and fillers.

Figure 17. Fractured specimens of hybrid syntactic foams in Flexural test.

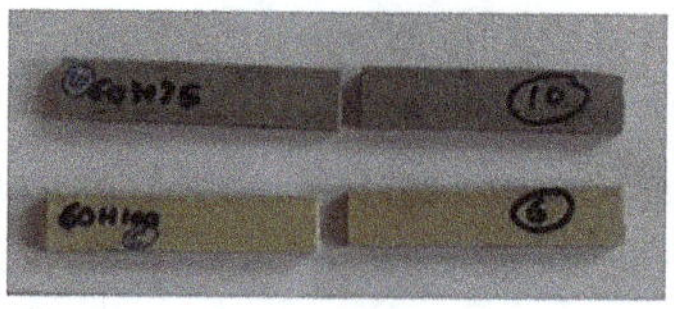

Figure 18. SEM images of fractured surface of hybrid syntactic foam in flexural testing test at (a) 132x magnification, (b) 483x magnification, (c) 1.00kx magnification

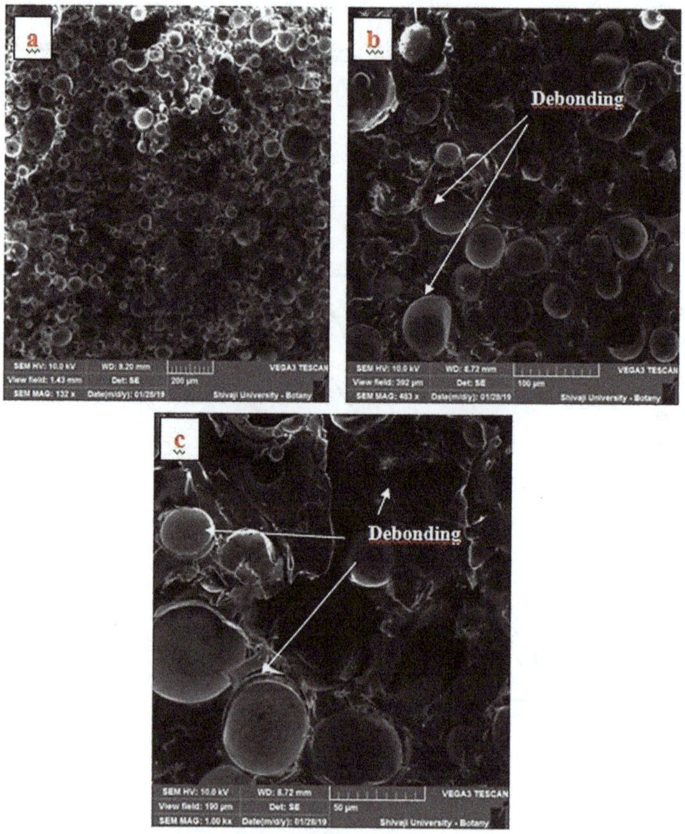

Micro Vickers's Hardness Properties of Hybrid Syntactic Foam

Effect of hybridization in syntactic foam on the hardness of hybrid syntactic foam is presented in Figure 19. The hardness of each composition of hybrid syntactic foams shows higher values than the neat (VER) specimen. In plain syntactic foams, syntactic foam containing fly ash cenosphere only (i.e. HSF-00-100) showed a higher value than all combinations of hybrid syntactic foam. In hybrid syntactic foam system, HSF-25-75 showed 8.67% increase than their plain syntactic foam containing only K20 micro balloons (i.e. HSF-100-00). There was a gradual increase found in hardness values with increase in fly ash cenosphere in filler content of hybrid syntactic foam. Micro Vickers hardness values ware increased by -3%, 0% and 8.67% for an increase in vol% of fly ash cenosphere in hybrid fillers by 25, 50, 75 vol% respectively with reference to plane syntactic foam with only K20 glass micro balloons.

An increase of fly ash cenosphere in hybrid syntactic foam caused increase into Micro Vicker's hardness due to the hard nature of alumina silicate composition present in fly ash cenosphere.

Figure 19. Micro Vickers Hardness properties of hybrid syntactic foams

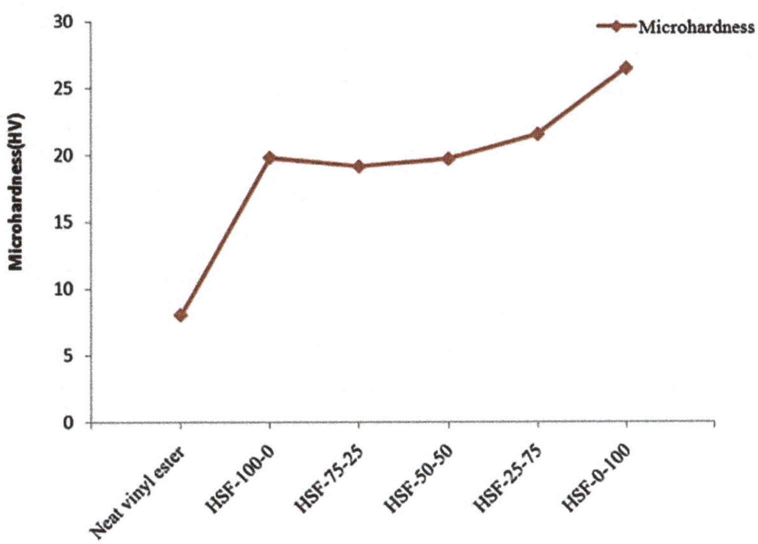

CONCLUSION

The hybrid glass micro balloons/ fly ash cenosphere filled syntactic foams are successfully prepared by hand lay-up method. The physical characterization parameter includes density and matrix porosity whereas tensile, compression, flexural, impact, and hardness grouped as mechanical characterization parameters. Scanning electron microscopy was performed on fractured surfaces to find out deformation and fracture mechanisms. The main findings are as follows:

1. The density of each hybrid syntactic foam specimen was lower than the density of neat vinyl ester resin specimen and it is a function of density and volume content of micro balloons.
2. Matrix porosity was increased with the content of flyash cenosphere in hybrid syntactic foams due to surface flaws present in flyash cenosphere.
3. Tensile properties are much affected by the addition of flyash cenosphere into fillers of hybrid syntactic foams. Tensile strength is successfully increased by hybridization. Matrix cracking, matrix-micro balloons debonding are characterized as deformation and fracture behavior.
4. Maximum of 135% increase in tensile strength is observed as flyash cenosphere content increased to 75%.
5. Compressive strength is not much affectedly the addition of flyash cenosphere; however, some combinations showed an increase in compressive strength. Maximum of 12% improvement was done by adding fly ash cenosphere. This increase in strength is attributed to hard alumina-silicate

composition of flyash cenosphere. Matrix cracking and micro balloons crushing are characterized as deformation and fracture behavior.

6. Flexural modulus was increased with the hybridization of flyash cenosphere in fillers of hybrid syntactic foams due to the hard nature of flyash cenosphere. 40% increase in specific flexural modulus was observed in hybrid syntactic foam as compared to plane syntactic foam. Hybrid syntactic foams with 60% total filler showed a gradual increase in flexural modulus with increase in secondary fillers.
7. Increase in flexural strength is observed for hybrid compositions. About 9 to 39% increase in flexural strength is observed for hybrid compositions. Matrix cracking, matrix-micro balloons debonding are characterized as deformation and fracture behavior.
8. Impact energy and specific impact energy gets decreased gradually with the increase of flyash cenosphere content due to surface flaws of flyash cenosphere and its deformation and fracture mechanism are characterized by matrix failure, matrix-micro balloon debonding and aggregation of fillers.
9. Micro-Hardness of hybrid syntactic foams increased with flyash cenosphere content due to the hardness of alumina silicate composition.
10. Therefore, it can be possible to use flyash cenosphere with glass micro balloons in various applications without much hindrance to properties. Hence, reducing land pollution and saving money can be possible.

FUTURE SCOPE

The present investigation focuses on the hybridization of glass micro balloons and flyash cenosphere at total filler content with 60 vol% in the vinyl ester polymer resin. Fabrication is done by hand layup process. Apart from this, the present study leaves wide scope for future investigation. This is listed below:

1. Apart from vinyl ester resin other thermoset and thermoplastic resin can be investigated for hybridization of glass micro balloons and flyash cenospheres.
2. Apart from the hand lay-up process, other fabrication processes can be employed and their comparison can be done.
3. The thermal and dynamic study can be done to evaluate composite properties.
4. Environmental effect study should be done to understand the deformation and fracture behaviour under various conditions.
5. Natural fiber can be utilized as reinforcement in these composites.
6. MADM methods can be used for ranking and their comparative study can be done.
7. Surface treatments of fillers can be done to investigate the performance and compared to the untreated one.

ACKNOWLEDGMENT

Dr. Pravin R. Kubade (Corresponding author) would like to acknowledge Shivaji University, Kolhapur (MH) India for providing research fund under 'Research Initiation Scheme (RIS) 2017-18'. The Author is also thankful to KIT's College of Engineering (Autonomous), Kolhapur for providing infrastructure and facility to carry out research.

REFERENCES

American Society for Testing and Materials. (2008a). ASTM D792-08, Standard Test Methods for Density and Specific Gravity (Relative Density) of Plastics by Displacement.

ASTM. (1998b). Standard test method for flatwise compressive properties of sandwich cores. ASTM International.

ASTM. (2007). *Standard test methods for flexural properties of unreinforced and reinforced plastics and electrical insulating materials.* ASTM D790-07.

ASTM. (2014). 638-14. *Standard test method for tensile properties of plastics.* ASTM.

Bharath Kumar, B. R., Zeltmann, S. E., Doddamani, M., Gupta, N., Gurupadu, S., & Sailaja, R. R. N. (2016). Effect of cenosphere surface treatment and blending method on the tensile properties of thermoplastic matrix syntactic foams. *Journal of Applied Polymer Science, 133*(35). doi:10.1002/app.43881

Biron, M. (2013). *Thermosets and composites: material selection, applications, manufacturing and cost analysis.* Elsevier.

Breunig, P., Damodaran, V., Shahapurkar, K., Waddar, S., Doddamani, M., Jeyaraj, P., & Prabhakar, P. (2020). Dynamic impact behavior of syntactic foam core sandwich composites. *Journal of Composite Materials, 54*(4), 535–547. doi:10.1177/0021998319885000

Carolan, D., Mayall, A., Dear, J. P., & Fergusson, A. D. (2020). Micromechanical modelling of syntactic foam. *Composites. Part B, Engineering, 183*, 107701. doi:10.1016/j.compositesb.2019.107701

Chanda, B., Kumar, R., Kumar, K., & Bhowmik, S. (2015). Optimisation of Mechanical Properties of Wood Dust-reinforced Epoxy Composite Using Grey Relational Analysis. In *Proceedings of fourth international conference on soft computing for problem solving* (pp. 13-24). Springer. 10.1007/978-81-322-2220-0_2

Colloca, M., Gupta, N., & Porfiri, M. (2013). Tensile properties of carbon nanofiber reinforced multiscale syntactic foams. *Composites. Part B, Engineering, 44*(1), 584–591. doi:10.1016/j.compositesb.2012.02.030

Desai, J. R., Shit, S. C., & Jain, S. K. (2014). Studies on thermal, electrical and flame properties of surface modified cenosphere filled ABS composites. *International Journal of Plastics Technology, 18*(3), 390–396. doi:10.100712588-014-9099-3

Desai, J. R., Shit, S. C., & Jain, S. K. (2016). Analysis of 3-methacryloxypropyl trimethoxysilane treated cenosphere inclusion on dynamic mechanical properties of ABS composites. *International Journal of Plastics Technology*, *20*(2), 241–248. doi:10.100712588-016-9152-5

Dimchev, M., Caeti, R., & Gupta, N. (2010). Effect of carbon nanofibers on tensile and compressive characteristics of hollow particle filled composites. *Materials & Design*, *31*(3), 1332–1337. doi:10.1016/j.matdes.2009.09.007

Ding, J. (2020). Compressive properties of co-continuous hollow glass microsphere/epoxy resin syntactic foams prepared using resin transfer molding. *Journal of Reinforced Plastics and Composites*, *39*(3-4), 132-143.

Gauthier, M. M. (Ed.). (1995). *Engineered materials handbook* (Vol. 1). ASM International.

Ghamsari, A. K., Wicker, S., & Woldesenbet, E. (2014). Bucky syntactic foam; multi-functional composite utilizing carbon nanotubes-ionic liquid hybrid. *Composites. Part B, Engineering*, *67*, 1–8. doi:10.1016/j.compositesb.2014.06.030

Gilbert, M. (Ed.). (2016). *Brydson's plastics materials*. William Andrew.

Gupta, N., & Paramsothy, M. (2014). Metal-and polymer-matrix composites: Functional lightweight materials for high-performance structures. *JOM*, *66*(6), 862–865. doi:10.100711837-014-0969-0

Gupta, N., Pinisetty, D., & Shunmugasamy, V. C. (2013). *Reinforced polymer matrix syntactic foams: effect of nano and micro-scale reinforcement*. Springer Science & Business Media. doi:10.1007/978-3-319-01243-8

Gupta, N., Ye, R., & Porfiri, M. (2010). Comparison of tensile and compressive characteristics of vinyl ester/glass microballoon syntactic foams. *Composites. Part B, Engineering*, *41*(3), 236–245. doi:10.1016/j.compositesb.2009.07.004

Hossain, M. M., & Shivakumar, K. (2011). Compression fatigue performance of a fire resistant syntactic foam. *Composite Structures*, *94*(1), 290–298. doi:10.1016/j.compstruct.2011.05.035

Huang, R., & Li, P. (2015). Elastic behaviour and failure mechanism in epoxy syntactic foams: The effect of glass microballoon volume fractions. *Composites. Part B, Engineering*, *78*, 401–408. doi:10.1016/j.compositesb.2015.04.002

Kaw, A. K. (2005). *Mechanics of composite materials*. CRC Press. doi:10.1201/9781420058291

Kazimieras Zavadskas, E., Antucheviciene, J., Adeli, H., & Turskis, Z. (2016). Hybrid multiple criteria decision making methods: A review of applications in engineering. *Scientia Iranica*, *23*(1), 1–20. doi:10.24200ci.2016.2093

Kubade, P. R., Kulkarni, H. B., & Gavali, V. C. (2020). Property Enhancement of Carbon Fiber-Reinforced Polylactic Acid Composites Prepared by Fused-Deposition Modeling. In *Handbook of Research on Recent Developments in Electrical and Mechanical Engineering* (pp. 455–478). IGI Global. doi:10.4018/978-1-7998-0117-7.ch017

Kumar, B. B., Doddamani, M., Zeltmann, S. E., Gupta, N., Gurupadu, S., & Sailaja, R. R. N. (2016). Effect of particle surface treatment and blending method on flexural properties of injection-molded cenosphere/HDPE syntactic foams. *Journal of Materials Science*, *51*(8), 3793–3805. doi:10.100710853-015-9697-2

Kumar, B. B., Doddamani, M., Zeltmann, S. E., Gupta, N., & Ramakrishna, S. (2016). Data characterizing tensile behavior of cenosphere/HDPE syntactic foam. *Data in Brief*, *6*, 933–941. doi:10.1016/j.dib.2016.01.058 PMID:26937472

Kumar, B. B., Doddamani, M., Zeltmann, S. E., Gupta, N., Ramesh, M. R., & Ramakrishna, S. (2016). Processing of cenosphere/HDPE syntactic foams using an industrial scale polymer injection molding machine. *Materials & Design*, *92*, 414–423. doi:10.1016/j.matdes.2015.12.052

Kumar, B. B., Singh, A. K., Doddamani, M., Luong, D. D., & Gupta, N. (2016). Quasi-static and high strain rate compressive response of injection-molded cenosphere/HDPE syntactic foam. *JOM*, *68*(7), 1861–1871. doi:10.100711837-016-1912-3

Labella, M., Zeltmann, S. E., Shunmugasamy, V. C., Gupta, N., & Rohatgi, P. K. (2014). Mechanical and thermal properties of fly ash/vinyl ester syntactic foams. *Fuel*, *121*, 240–249. doi:10.1016/j.fuel.2013.12.038

Naffakh, M., Diez-Pascual, A. M., Marco, C., Ellis, G. J., & Gomez-Fatou, M. A. (2013). Opportunities and challenges in the use of inorganic fullerene-like nanoparticles to produce advanced polymer nanocomposites. *Progress in Polymer Science*, *38*(8), 1163–1231. doi:10.1016/j.progpolymsci.2013.04.001

Njuguna, J. (Ed.). (2016). *Lightweight composite structures in transport: design, manufacturing, analysis and performance*. Woodhead Publishing.

Patil, A. N., & Kubade, P. R. (n.d.). *A Review on Modifications and Hybridization in Syntactic Foam.* Academic Press.

Patil, A. N., Kubade, P. R., & Kulkarni, H. B. (n.d.). *Effect of Composition on Impact and Flexural Properties of Hybrid Glass Microballoons/Fly Ash Cenosphere Filled Vinyl Ester Matrix Syntactic Foams*. Academic Press.

Poveda, R. L., Achar, S., & Gupta, N. (2014). Viscoelastic properties of carbon nanofiber reinforced multiscale syntactic foam. *Composites. Part B, Engineering*, *58*, 208–216. doi:10.1016/j.compositesb.2013.10.079

Rutz, B. H., & Berg, J. C. (2010). A review of the feasibility of lightening structural polymeric composites with voids without compromising mechanical properties. *Advances in Colloid and Interface Science*, *160*(1-2), 56–75. doi:10.1016/j.cis.2010.07.005 PMID:20692640

Sankar, B. R., Umamaheswarrao, P., Srinivasulu, V., & Chowdari, G. K. (2015). Optimization of milling process on jute polyester composite using Taguchi based grey relational analysis coupled with principle component analysis. *Materials Today: Proceedings*, *2*(4-5), 2522–2531. doi:10.1016/j.matpr.2015.07.197

Shunmugesh, K., & Panneerselvam, K. (2017). Grey Relational Analysis Based Optimization Of Multiple Responses in Drilling Of Carbon Fiber-Epoxy Composites. *Materials Today: Proceedings*, *4*(2), 2861–2870. doi:10.1016/j.matpr.2017.02.166

Swetha, C., & Kumar, R. (2011). Quasi-static uni-axial compression behaviour of hollow glass microspheres/epoxy based syntactic foams. *Materials & Design*, *32*(8-9), 4152–4163. doi:10.1016/j.matdes.2011.04.058

Tagliavia, G., Porfiri, M., & Gupta, N. (2010). Analysis of flexural properties of hollow-particle filled composites. *Composites. Part B, Engineering*, *41*(1), 86–93. doi:10.1016/j.compositesb.2009.03.004

Thomas, C. R. (1973). Syntactic carbon foams. *Materials Science and Engineering*, *12*(5-6), 219–233. doi:10.1016/0025-5416(73)90033-5

Tian, Q., & Yu, D. (2016). Preparation and properties of polymer microspheres filled epoxy composite films by UV-curable polymerization. *Materials & Design*, *107*, 221–229. doi:10.1016/j.matdes.2016.06.035

Zegeye, E., Ghamsari, A. K., & Woldesenbet, E. (2014). Mechanical properties of graphene platelets reinforced syntactic foams. *Composites. Part B, Engineering*, *60*, 268–273. doi:10.1016/j.compositesb.2013.12.040

Zhang, L., & Ma, J. (2013). Effect of carbon nanofiber reinforcement on mechanical properties of syntactic foam. *Materials Science and Engineering A*, *574*, 191–196. doi:10.1016/j.msea.2013.03.028

Chapter 19
Microstructure and Mechanical Properties of Silicon Carbide–Treated Ferritic Stainless Steel Welds

Muhammed O. H. Amuda
University of Lagos, Nigeria

Taiwo Folorunsho Lawal
University of Lagos, Nigeria

Shahjahan Mridha
University of Strathclyde, UK

ABSTRACT

To further previous works on grain refinement in ferritic stainless steel (FSS) welds via elemental powder addition, microstructure evolution and mechanical properties in non-metallic carbide treated FSS weld is discussed. Silicon carbide (SiC) powder preplaced on medium chromium FSS bevelled joint was fusion melted using tungsten inert gas (TIG) torch at different combination of process parameters. The incorporation of the powder in the weld pool constricted the weld geometry in addition to grain modification from columnar to equiaxed grain structure. Whilst significant improvement was obtained in hardness characteristics in powder treated welds, those of ductility and tensile strength were marginal. The study also established that welds mechanical properties are controlled by a combination of phenomena such as phase fraction and grain morphology.

INTRODUCTION

FSSs possess body-centered cubic structure of ferrite and contain chromium in the range 11-30% with smaller amount of other elements such as molybdenum, nickel, titanium, and niobium (Banerjee, 2017).

DOI: 10.4018/978-1-7998-4939-1.ch019

The near zero or insignificant amount of nickel in the alloy confers better economic value relative to the austenitic variety (Amuda et al., 2017; Cortie and Du Toit, 2016). As such, they find suitable applications in construction, transportation, chemical and petrochemical industries (Amuda et al., 2016). Other than this, they exhibit superior stress corrosion cracking and are particularly credited with excellent radiation shielding characteristics relative to the austenitic counterpart (Amuda and Mridha, 2012). Thus, they equally find ready application as nuclear reactor components in power generation.

Whereas the attractive attributes in FSS had generated growing interest in its deployment for structural application, its after-weld properties are often poor relative to the base metal (Delgado et al., 2016; Anbazhagan and Nagalakshmi, 2002). These unacceptable after-weld characteristics are connected to undesirable phenomena such as precipitation of chromium carbide and more importantly grain coarsening (Shahid et al., 2018; Amuda, 2011). Thus, most often, such welds exhibit reduced ductility, strength, hardness, impact properties as well as intra-granular corrosion resistance (Delgado et al., 2016; Jozwik and Bojar, 2007).

Numerous techniques targeted at improving weldability in FSS welds have been reported and a detailed review of these is available in Amuda and Mridha (2010). One of such techniques involves addition of elements with similar atomic size as iron. For instance, Villafuerte et al. (1990) accomplished grain refinement in FSS welds via treatments with varying amounts of aluminum (Al) and titanium (Ti) under varying welding process conditions. Though, the work successfully generated equiaxed grains in the welds particularly at higher welding speeds, it was limited to experimental FSS with constitutive presence of Al and Ti. This requires bulk alloying, which is rather expensive and time consuming. The phenomenon of grain coarsening is limited to the weld region but not the bulk of the material. Therefore, grain refinement treatment should be restricted to the weld region (Amuda, 2011)

Mohandas et al. (1999) developed a non-bulk approach for the introduction of alloying elements into the weld pool of base metal by incorporating the elements in powder form in electrode flux coatings using consumable electrode. The incorporated metal powder in the electrode coatings not only control the energy input into the base metal during fusion melting, it equally serves as nucleation sites for the resolidification of the molten weld pool. This enables the formation of equiaxed grains post-resolidification of the weld pool. However, electrode flux coating for the incorporation of elemental powder for grain refinement is not appropriate for autogenous non-filler process like the TIG welding process.

Amuda et al. (2018a) and Amuda (2011) developed a simple yet innovative technique for the incorporation of metal powder into the weld pool of FSS for grain refinement. In this technique, specific concentration of Al and Ti metal powder was individually preplaced onto the substrate of the material before welding for grain refinement. These works reported appreciable grain refinement in the resolidified weld of the material but there were substantial cracks in welds treated with Ti powder. Amuda et al. (2020) consequently attempted the conditions that approximate those used in Villafuerte et al. (1990). They incorporated specific concentration of a mixture of Al and Ti powder in the weld pool of FSS to assist in the achievement of crack free grain refinement. The investigation established that the mixture of Al and Ti powder produced a defect free grain refined structure in the FSS welds with marginal improvement in mechanical properties.

The potential of non-metallic carbide such as SiC for grain refinement was investigated by Schiffl and Easton (2009) as well as Easton et al. (2006). Though the research was conducted on magnesium alloy; the authors established that SiC effectively grain-refined the alloy via formation of binary and ternary carbides. Further, it was equally established that the size of the SiC particles is significant to the grain refinement. Although, this treatment was a bulk alloying approach, powders of this ceramic could be employed for a non-bulk treatment to investigate its potential in producing a defect-free grain-refined medium chromium FSS weld. This has not been attempted before now. Thus, the current investigation sought to produce defect free grain refined equiaxed weld in medium chromium FSS by incorporating SiC powder in the weld pool in autogenous TIG welding process.

MATERIALS AND METHOD

Materials Preparation

The medium chromium FSS used in the study was commercially procured as AISI 430 grade in 1000 mm x 1000 mm sheet with the chemical composition as presented in Table 1. The grain size and some mechanical properties are also included. The 1.5 mm thick sheet was sectioned into coupons of dimensions 150 mm x 90 mm preparatory TIG torch arc melting after pre-cleaning in acetone in a Branson 2500 ultrasonic cleaner for 5 minutes. The surfaces of the coupons were prepared for metal injection by roughening with 300 grit size emery paper. Two specifications of powder with 99.9% purity and average particle sizes, 9 and 10 µm; mixture of Al + Ti powder and that of SIC, respectively were used as grain refining agents. Pastes of pre-mixed Ti and Al, and SiC powders were prepared in a solution of polyvinyl alcohol (PVA) organic binder. Amuda and Mridha (2017) elaborately provided the recipe for the PVA binder solution in previous work. The different powder pastes in different amounts were preplaced on the coupons substrate using preplacement technique and subsequently oven-dried at 60°C for 24 hours to drive off moisture.

Experimental Procedure

Autogenous single pass full penetration bead-on-plate welds were produced on the prepared coupons using direct current straight polarity (DCEN) arc torch from Miller-Telwin TIG welding machine. This was shielded and stabilized with argon gas flowing at a rate 0.72 l/min. Details of the experimental set-up and the specific process parameter combination for the mixture of Al and Ti which was adapted in the present investigation were discussed in a previous publication (Amuda et al., 2018a). However, specific process parameter combination for SiC is presented in Table 2 whilst the generic melting conditions are presented in Table 3.

Table 1. Chemical composition of AISI 430 FSS validated with EDXRF spectroscopy

Material	Composition							Mechanical Properties			Grain Size (µm)	KFF	Cr_{eq}/Ni_{eq}
AISI 430 FSS	C	Cr	Si	Mn	P	S	Fe	UTS (MPa)	EL (%)	Hardness (VHN)	5	14.7	3.85
	0.12	16.19	0.75	1	0.04	0.3	Bal.	432	26	167			

Table 2. Process parameter combination for TIG torch melting of SiC powder onto AISI 430 FSS

Weld Run	I (A)	υ (mm/s)	V (V)	EI (J/mm)	Ar Flow Rate (L/min)	Type of Powder	Powder Content (mg/mm^2)
1	90	2.5	30	518	0.72	SiC	1.2
2	70	3.0	30	336	0.72	SiC	1.2
3	110	3.0	30	528	0.72	SiC	1.2
4	90	3.5	30	370	0.72	SiC	1.2
5	70	2.5	30	403	0.72	SiC	1.2
6	110	2.5	30	634	0.72	SiC	1.2
7	90	3.0	30	432	0.72	SiC	1.2
8	70	3.5	30	288	0.72	SiC	1.2
9	110	3.5	30	453	0.72	SiC	1.2
10	90	2.5	30	518	0.72	SiC	1.2
11	90	3.5	30	370	0.72	SiC	1.2
12	110	3.0	30	528	0.72	SiC	1.2
13	70	3.0	30	336	0.72	SiC	1.2

The coupons were prepared for metallographic examination using electrical discharge machining (EDM) in direction normal to the weld interface. Microstructural examination was conducted using the conventional approaches provided in literature (Amuda et al., 2018b). Nikon Epiphot 200 light optical microscope incorporated with IMT image analysis software was employed for a more comprehensive examination of the microstructure. Fractography examination was conducted using JEOL JSM 5600 scanning electron microscope. Phase Identification in both the base metal and weld section was undertaken with Shimadzu XRD 6000 diffractometer. Microhardness examination of the resolidified weld was done with a Mitutoyo MVK-H2 Vickers digital microhardness tester. The examination was conducted at a test load of 500gf for a dwell time of 10s in both transverse and thickness sections of the weld. The test was repeated thrice and the average value recorded. Specimens for tensile testing were prepared from the resolidified welds using EDM wirecut as shown in Figure 1. Transverse tensile test was conducted

Table 3. Generic melting conditions (Amuda et al., 2018)

Process	DCEN straight polarity full bead on plate penetration TIG weld
Position	Flat
Melting Conditions Current Voltage Speed Arc length Elemental Powder Torch orientation Electrode configuration Electrode stick-out	 70, 90, 110A 30V 2.5-3.5 mm/s 1.5 mm Al + Ti, SiC Vertical 2.44 mm W-2 pct. Th., 60^0 cone included angle 3mm
Shielding environment	99.9 pct. argon at a flow rate of 0.72L/min

thrice on the samples at room temperature on Shimadzu Trapezium AGX-250 electro-mechanically controlled universal testing machine at a cross head speed of 5 mm/min.

Figure 1. Specimen for tensile testing (a) prepared specimen from weld section at the middle; (b) dimension of test specimen (Amuda et al., 2018a)

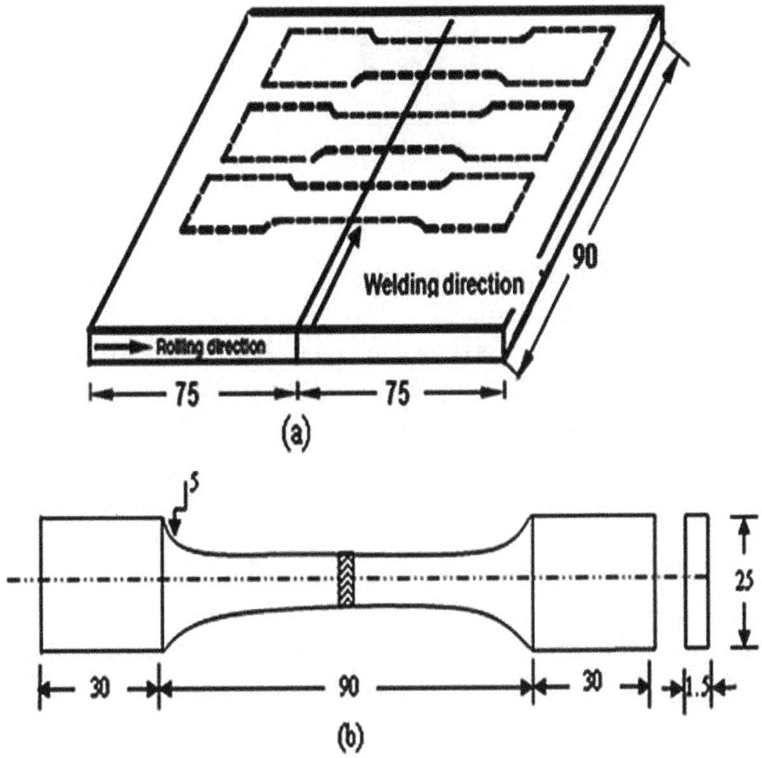

RESULTS AND DISCUSSION

Macrostructural Features of the Welds

The topographies of the welds produced with elemental metal and non-metallic carbide powder addition are related to that of the conventional weld in Figure 2. The figure reveals that all the welds produced with the powders exhibited narrower width compared to the conventional weld. Further, the non-metallic carbide (SiC) treated weld showed continuous ridged track in contrast to the smooth blister free track in mixture of Al + Ti powder treated weld. The well-printed track profile in the SiC-treated weld relative to the mixture of Al + Ti powder treated welds demonstrated that ceramic compound rather than elemental powder produces quality surface track devoid of defect. This is presumed to be facilitated by the likelihood that ceramic compound enhances arc stability in weld metal than elemental powder addition. This presumption is further supported by the findings of Amuda et al. (2018a) and Amuda and Mridha (2012). In these works, it was established that Al and Ti powder addition in the weld pool of medium chromium FSS produced different outcomes. In Al treated welds, the surface track was con-

Figure 2. Track surface profiles in: (a) conventional weld (b) mixture of Al + Ti powder treated weld and (c) SiC treated weld

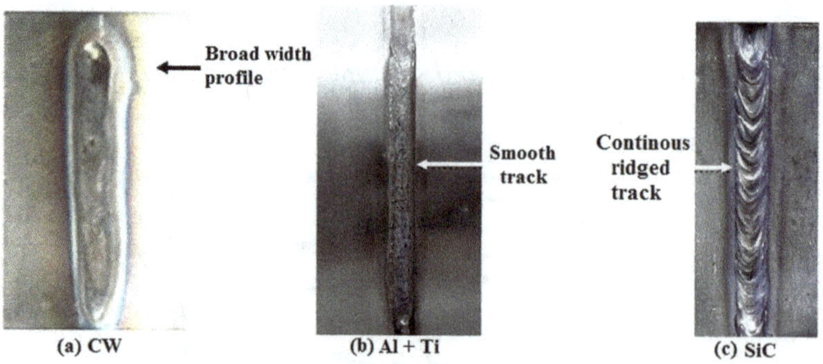

tinuous, smooth and defect free but in Ti treated welds, the tracks were discontinuous with cracks and blisters indicating non-smoothness of the tracks.

The macrostructures of SiC-treated welds shown in Figure 3 in relation to the conventional weld show a narrower width and shallower depth. The constriction in bead profile is similar to that in mixture of powder treated welds (Figure 3b). As obtained in previous studies (Amuda et al., 2018a; Amuda et al. 2020), the macro-grains orientation in the conventional welds appears to project in the lateral direction suggesting the direction of heat dissipation in the welds. No such clear macro-grain orientation is visible in both mixture of Al + Ti powder-treated and SiC-treated welds (Figure 3). A further examination of the influence of heat input on the penetration in the SiC-treated welds in Figure 4 indicates that the degree of dilution as well as the depth of the weld bead is deeper with increasing heat input. It however appears that the amount of powder injected in the weld pool was not adequate to fully fill the prepared grooves (Figure 4).

Figure 3. Macrostructure of the conventional weld and powder treated welds (a) CW, (b) mixture Al + Ti powder and (c) SiC powder

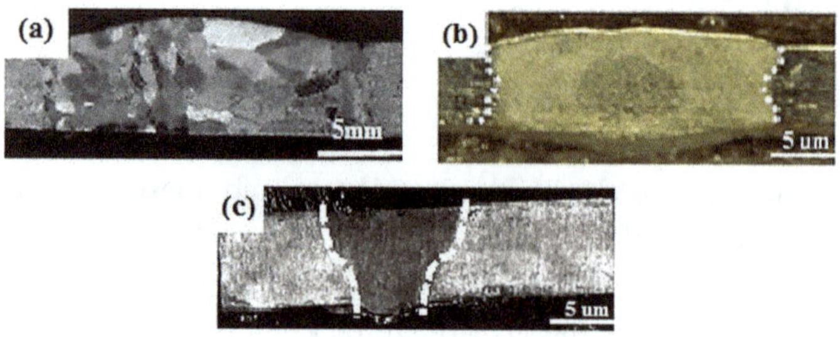

Figure 4. Macroprofile of SiC treated weld showing varying degree of dilution and depth profile at different heat inputs: (a) 370J/mm and (b) 452 J/mm

Weld Dimensions

Figure 5 presents the width – depth dimensions of the welds acquired with Nikon MM-400L trinocular head measuring microscope. The figure reveal an increase in weld dimension with increasing energy input for conventional weld, mixture of Al + Ti powder treated weld and non-metallic carbide treated weld. This suggests that the use of higher current and transverse speed results in a higher weld dimension. Amuda et al. (2019) explained that higher current delivers more energy to the weld and hence melts more volume of materials in the surface while the melting speed influences the duration of interaction of the melting torch with the substrate; hence the depth of the resolidified weld. As such, the width of the weld bead is determined by the arc current while the depth is determined by the traverse speed during melting. There is constriction in the geometry of the powder treated welds with the mixture of powder treated welds exhibiting about 43% constriction at the heat affected zone (HAZ) while SiC treated welds exhibit about 37% constriction in the HAZ. A sudden increase in the weld dimension is noticed at the energy input of 432 J/mm at the fusion zone (FZ) for all the welds whereas there is fluctuation in weld dimension at this energy input at the HAZ which suggests irregularities in the arc current and transverse speed relating to this energy input.

Figure 5. Weld dimensions of conventional weld in relation to mixture of Al + Ti powder treated and SiC treated welds at varying heat inputs: (a) FZ and (b) HAZ

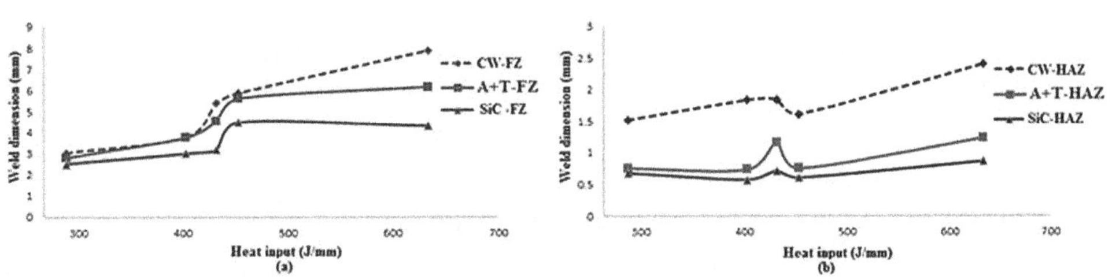

Phase Identification in Powder Treated Welds

Figure 6 presents the phase characterisation of the resolidified melt pool carried out via XRD analysis. The XRD spectra of the welds treated with mixture of Al + Ti powder and carbide were superimposed on the spectrum of the conventional weld. The figure indicates the phases in the spectra characterised using powder diffraction file (PDF) card searching and matching. The phases identified in the powder treated welds are quite distinct from those of the conventional weld. The phases present in the powder treated welds are as obtained in previous studies (Amuda et al. 2020, Amuda et al., 2019; Amuda et al. 2018a) except for the presence of Fe_5Cr_3Si identified in the SiC treated welds.

Microstructural Analysis of the Powder Treated Welds

Figure 7 presents the microstructure of the as-received FSS base metal revealing equiaxed grains elongated in the rolling direction. As reported by Amuda et al. (2018a), the microstructures of the conventional welds comprises ferrite ($\alpha + \delta$), martensite (α') and chromium carbide ($Cr_{23}C_6$) precipitates. The

Figure 6. XRD spectra in conventional welds and powder treated welds: (a) Al+Ti powder treated weld, (b) SiC treated weld and (c) conventional weld

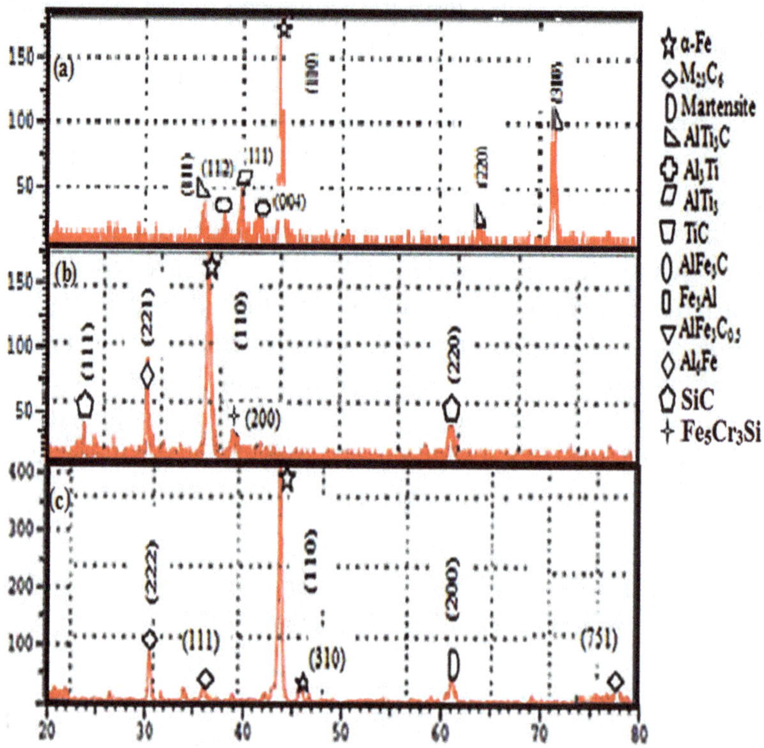

grain structure (Figure 8) in the conventional welds reveals coarse and columnar grains elongated in the direction of solidification. This in relation to the as-received base metal grain structure established that

Figure 7. Optical micrograph of ferritic stainless steel base metal; insert shows ferrite and carbide phases (Amuda et al., 2018)

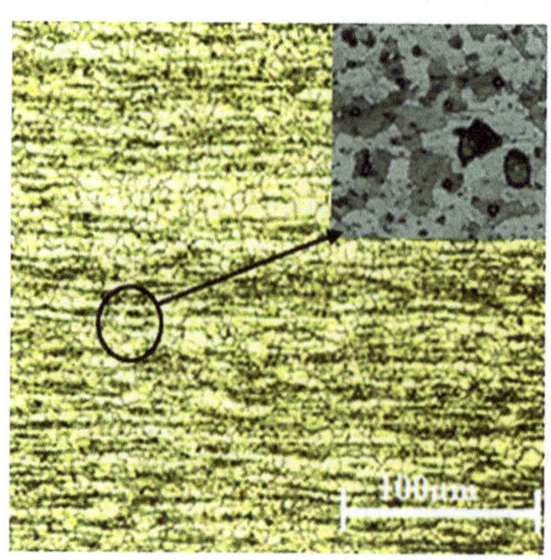

the apparent grain coarsening and directional orientation in the conventional welds is predicated by the incident heat from the moving TIG torch.

The comparison of the microstructure of the welds post-powder treatment with conventional welds shown in Figure 8 further illustrated that the powder treated welds are free of defects. Additionally, the grain structures are equiaxed and refined relative to the conventional weld. The grain distribution in the SiC-treated welds is observed to be homogenous with random orientation of SiC with shorter arm. However, that in the mixture of Al + Ti powder treated welds is inhomogeneous which has been ascribed to the flow pattern of the preplaced powder into the melt pool (Figure 8d). The pattern has equally been reported in previous works by Mridha *et al.* (2001) and Mridha and Baker (1994) in titanium-aluminide coating deposited on titanium substrate as well as laser nitriding of commercial titanium, respectively. This flow pattern has been attributed to the influence of marangoni convectional fluid flow (Kou, 2003; Lancaster, 1993). The flow pattern obtained in the mixture of Al + Ti powder treated welds is absent in SiC treated welds. Furthermore, weld treated with the mixture of powder exhibits about 20 percent increase in delta ferrite while that in metal carbide was about 16 percent increase in delta ferrite (Figure 9). The formation of equiaxed grains in the powder treated welds is promoted by powder preplacement because a greater part of the energy input was converted into the melting of the preplaced metal powder thus reducing the effective energy available to melt the substrate. In addition, the intermetallic formed by the powders during in-situ reaction acts as nucleation sites for the resolidified melt pool and also pinned subsequent growth of the grains at the grain boundary. Other than this, the increase in delta ferrite distribution observed in the mixture of Al +Ti powder and SiC-treated welds is expected to result in a decrease in mechanical properties (Amuda, 2011).

Figure 8. Microstructure of welds at energy input of 452J/mm, and that of the melt pool: (a) CW, (b) mixture of Al + Ti treated weld, (c) SiC treated weld and (d) melt pool showing the flow loop

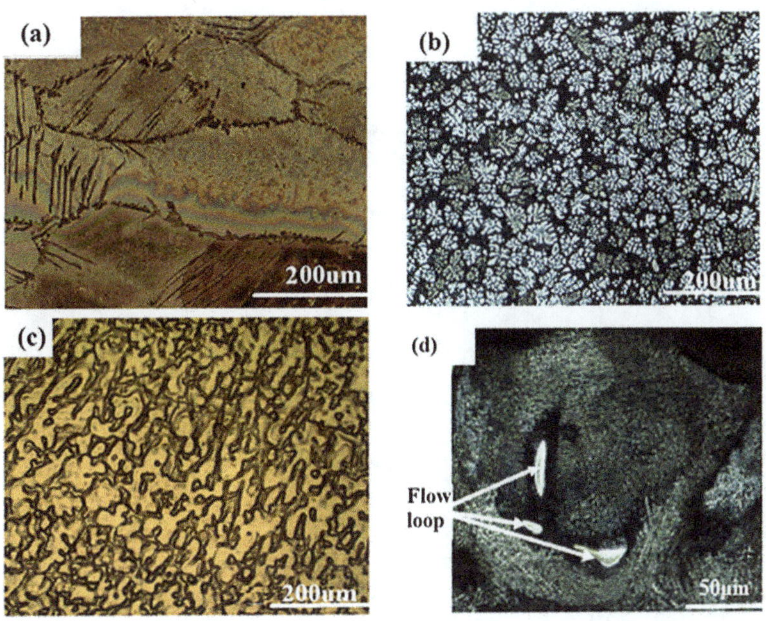

Figure 9. Delta ferrite content in melt pool of conventional welds and those of powder treated welds.

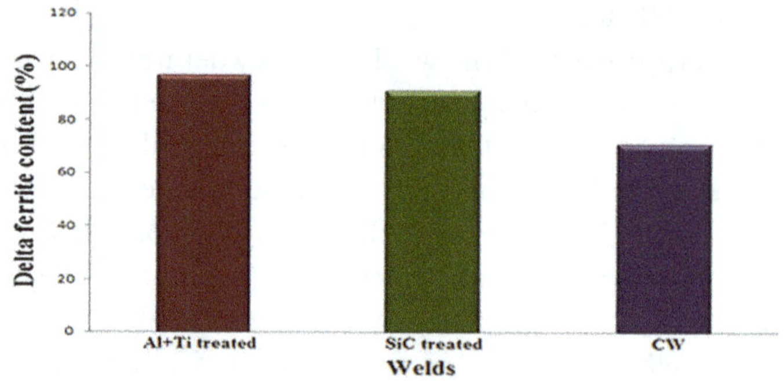

Grain Size Analysis of the Powder Treated Welds

Figure 10 relates the grain size of the welds treated with different powders to that of the conventional welds. The grain sizes were measured using quantitative microscopy based on Abrams Three Circle procedure with IMT image analysis software. The figure reveals that the grain size in the powder treated welds is smaller relative to the conventional weld across all energy inputs and the weld treated with mixture of Al + Ti powder generated the smallest grain size. In particular, grain size in the mixture of powder treated weld is less than 10 μm but it ranged 27-31 μm in SiC treated weld while it ranged 33-40 μm in conventional weld. As such, in relation to the previous works (Amuda et al., 2018a; Amuda et al., 2019), the mixture of Al + Ti powder treated still presents the highest grain refinement. For instance, in

Figure 10. Combined effects of metal powder and metal carbide with heat input on weld grain size

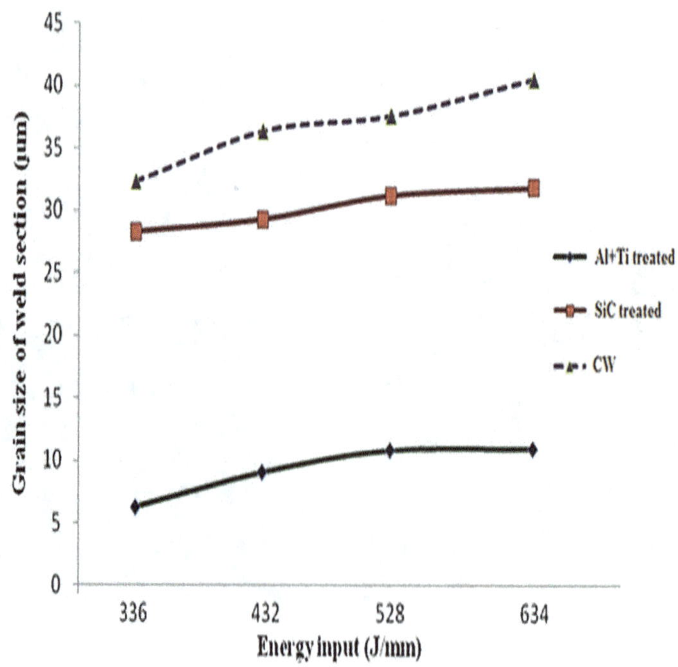

the previous works of Amuda et al. (2018a), grain refinement in Ti powder treated weld was 70 percent. It was 15 percent in Al powder treated welds; whereas in the current work, SiC powder treated welds generates 24 percent grain refinement whilst the mixture of Al + Ti powder treated welds generate about 85 percent. The improvement in grain refinement in the powder treated welds is attributable to the creation of more sites for heterogeneous nucleation of grains.

Mechanical Analysis

Microhardness Analysis

Hardness characteristic of the welds was conducted by measuring the hardness distribution of the re-solidified melt pool at 1 mm interval in the transverse direction and 0.25mm in the thickness direction. Figure 11 reveals that the hardness profile of the powder treated welds in the transverse direction are conspicuously higher at the centerline relative to the conventional weld. The SiC treated weld exhibits the highest hardness value of about 612VHN at the centerline while the maximum hardness exhibited by the mixture of Al + Ti powder treated weld is 579VHN at the centerline. The hardness values of all the welds converge to a minimum at positions away from the weld centerline suggesting that carbide formation during metallurgical reactions in the melt pool are restricted to the melting zone.

Figure 11. Transverse hardness distribution in welds treated with mixture of Al+Ti powder and SiC powder in relation to conventional welds.

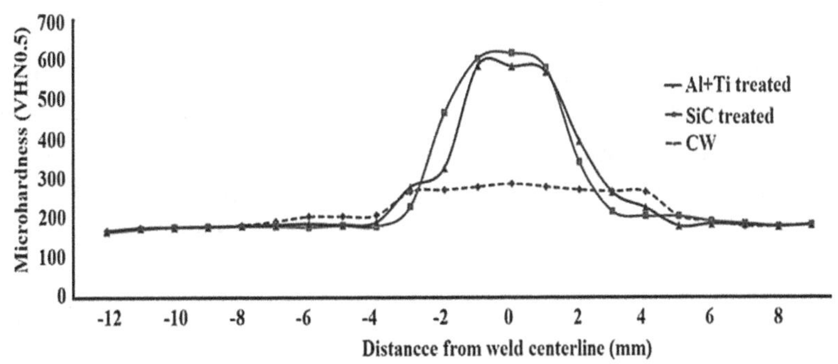

Figure 12 reveals that the hardness profile of the welds in the thickness direction increases with increasing distance from the welds surface. The increase in hardness with increasing depth from the surface is presumed to be due to the combined effect of second phase particle as well as increasing cooling rate. The figure further indicates that powder treated welds exhibit higher hardness value than the conventional weld. The mixture of Al + Ti powder treated weld exhibited the highest value of hardness near the weld's surface up to a depth of about 0. 45 mm. Beyond this position, the SiC treated weld exhibited the highest hardness profile with increasing distance from the surface. Maximum hardness value of 782VHN is obtained in SiC treated weld while in the mixture of Al + Ti powder treated welds, the peak obtainable hardness is 621VHN in the thickness direction. The increased hardness in the SiC treated welds is attributable to formation of $M_xC_ySi_z$ precipitates of the form Fe_5Cr_3Si and undissolved

SiC powder. The increased in the hardness value of mixture of Al + Ti powder treated welds could be ascribed to the presence of both ternary and binary iron-alumino carbide as well as titanium carbide in the resolidified melt pool.

Figure 12. Hardness distribution in the thickness direction in welds treated with mixture of Al+Ti powder and SiC powder in relation to conventional welds.

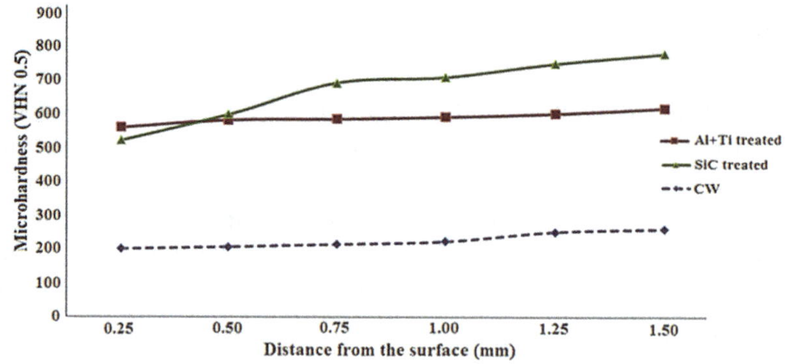

Analysis of Tensile Strength

Figure 13 relates the tensile strength characteristics of the mixture of powder and SiC-powder treated welds to the conventional welds. The figure reveals that the strength values in the weld are influenced by energy input. The strength value in the SiC-treated weld is higher relative to the conventional weld across all energy input. However, the strength value decreases with increasing energy input with the highest value of 468 MPa been exhibited at the energy input 336 J/mm. The relatively high value of tensile strength exhibited at low energy input can be related to undissolved carbide precipitates in the resolidified welds pinning the grains at the grain boundaries. The tensile strength behavior for welds

Figure 13. Combined effects of metal powder and metal carbide with energy input on weld tensile strength

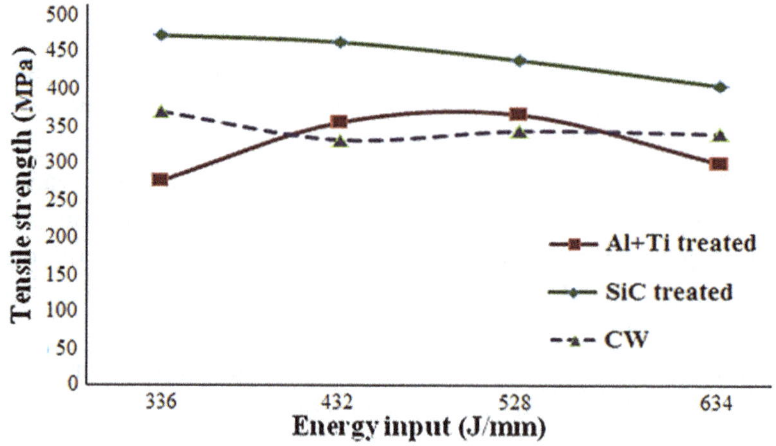

treated with mixture of Al + Ti powder indicates that the strength value is strongly influenced by energy input. In relation to the conventional weld, the mixture of powder treated weld exhibits higher tensile strength distribution in the energy input range 432-528J/mm.

Outside this range, the conventional weld exhibited higher tensile strength. The maximum tensile strength exhibited by the mixture powder treated weld is 363 MPa. Incomplete melting of the preplaced powder coupled with incomplete homogenisation of the melt pool might have caused low tensile strength at low energy input. On the other hand, dissolution of grain pinning precipitates might have accounted for the reduction in tensile strength at high energy input.

Influence of Grain Refinement on Weld Ductility

The ductility of the powder treated welds as well as that of the conventional welds was conducted by evaluating the strain of the weld at fracture during tensile testing. Figure 14 reveals that the ductility of the mixture of Al + Ti powder treated weld in relation to the conventional weld is energy input dependent. At an energy input lower than 432 J/mm, the ductility is higher than that in conventional weld while at an energy input greater than 432 J/mm; the ductility is lower than in the conventional weld. The relative ductility of the conventional weld averaged 40% of the base metal; that in the mixture of powder treated welds is about 48% of the base metal whilst those in SiC treated welds is about 24% of the base metal.

Figure 14. Combine effect of metal powder and metal carbide with energy input on weld ductility

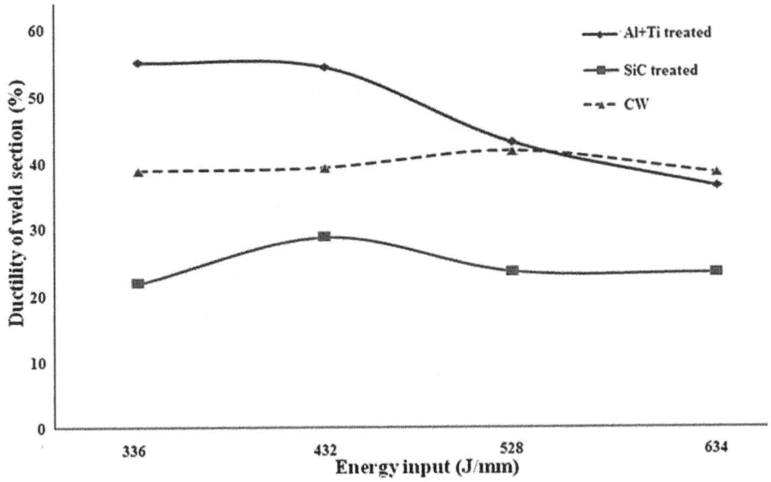

The variations in the mechanical properties of mixture of Al + Ti powder treated welds and those of SiC-treated welds suggest that, in addition to grain morphology, mechanical properties of medium chromium FSS welds are equally affected by other phenomena such as the presence and distribution of precipitate and delta ferrite.

Fractography of the Welds

The fractographs of the powder treated welds in relation to the conventional weld is presented in Figure 15. The figure demonstrates that the fracture mode in the conventional weld is quasi cleavage elongated in the director of fracture mode but that in the mixture of Al + Ti powder treated weld exhibited a mix of brittle cleavage and dimple fracture consisting of fibrous facets.

Figure 15. Tensile fractographs of the metal powder and metal carbide powder in relation to the conventional weld: (a) CW, (b) mixture of Al + Ti powder treated welds and (c) SiC treated welds

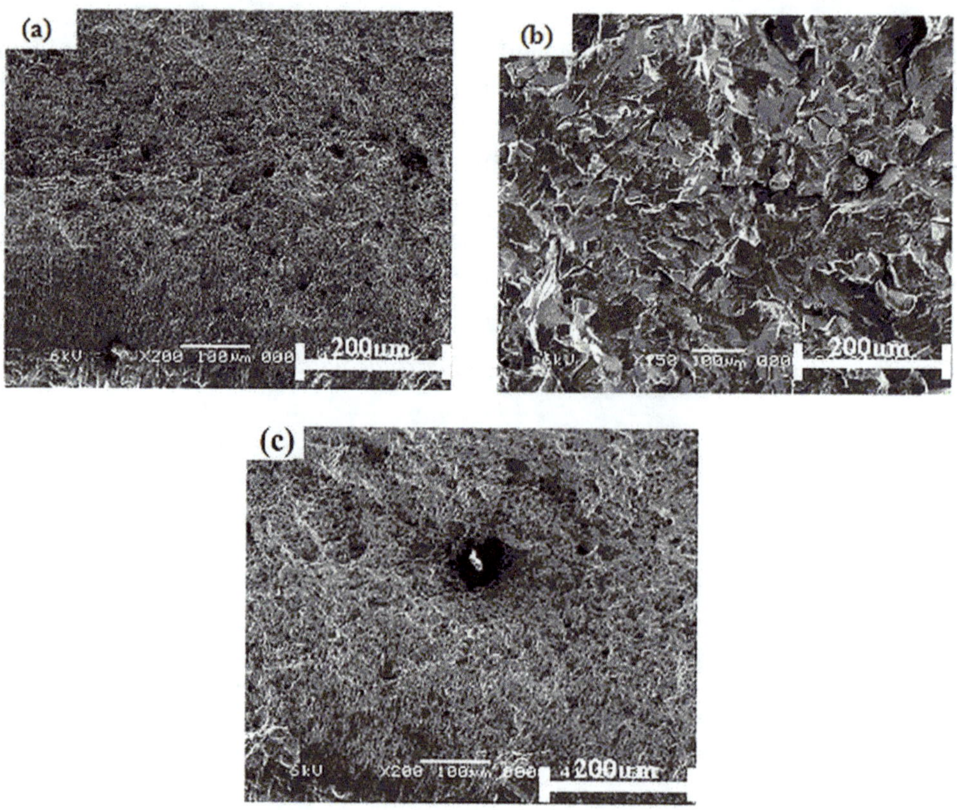

The fracture mode in the SiC treated welds exhibited brittle cleavage and the presence of undissolved SiC particle is apparent in the fractograph (Figure 15c). The facets in the powder treated welds are essentially quasi equiaxed unlike the elongated columnar facet in the conventional welds. The injection of metal powder into the weld pool reduces the facet size and changed the elongated morphology in the welds to equiaxed structure. The fracture points occurred at the interface between the FZ and the HAZ in all the powder treated welds whilst those in the conventional welds were located at the FZ.

CONCLUSION

The microstructural evolution in medium chromium FSS welds conforming to AISI 430 grade treated with mixture of Al + Ti powder as well as SiC powders for grain refinement was characterized in relation to mechanical properties. The following deductions emerged from the study:

1. The injection of mixture of Al + Ti powder or SiC powder into the weld pool of medium chromium FSS enhanced the production of quality defect-free welds. These powders also constricted the weld geometry relative to the conventional weld. Constriction generated in mixture of Al + Ti powder treated welds was about 43 percent whilst SiC treated welds generated about 37 percent constriction.
2. The injection of mixture of Al + Ti powder or SiC powder into the melt pool generated equiaxed and refined grain structure in the weld. The grain size in the conventional welds is between 33-40 μm, but that in SiC-treated weld is 27-31 μm while in mixture of Al + Ti powder treated weld, it is about 10 μm. The percentage grain refinements obtained in these various powder treated welds are: 24 percent in SiC treated welds and 85 percent in the mixture of Al + Ti powder treated welds. Further, the addition of the powders increased the distribution of delta ferrite in the melt pool.
3. Additions of mixture of Al + Ti metal powders and SiC ceramic powders improve the weld hardness and tensile strength. But the improvement in hardness was confined to the weld centerline. The relative ductility in the mixture of Al + Ti powder treated welds averaged 48 percent and that in SiC-treated weld averaged 24 percent of the base metal.
4. Mechanical properties of welds appear to be influenced by several phenomena other than grain morphology. For instance, in the present work, the presence and distribution of delta ferrite was established as additional phenomenon in addition to grain morphology.

REFERENCES

Amuda, M. O. H. (2011). *Microstructural features and properties of TIG melted AISI 430 ferritic stainless steel welds* (Doctoral Thesis). International Islamic University Malaysia, Kuala Lumpur, Malaysia.

Amuda, M. O. H., Akinlabi, E. T., & Mridha, S. (2016). Ferritic stainless steels: Metallurgy, Application and Weldability. In *Reference Module in Materials Science and Materials Engineering*. New York: Elsevier. doi:10.1016/B978-0-12-803581-8.04010-8

Amuda, M. O. H., Akinlabi, E. T., & Mridha, S. (2017). Influences of energy input and metal powder addition on carbide precipitation in AISI 430 ferritic stainless steel welds. *Materials Today: Proceedings*, *4*(2), 234–243. doi:10.1016/j.matpr.2017.01.017

Amuda, M. O. H., Lawal, F. T., Daramola, O., Awarun, A. B., & Subair, W. O. (2019). Wear and corrosion characteristics of silicon carbide surface modified mild steel. *UNILAG Journal of Medicine. Science and Technology.*, *6*(1), 129–146.

Amuda, M. O. H., Lawal, F. T., & Mridha, S. (2020). Microstructure evolution in AISI 430 ferritic stainless steel welds treated with aluminum + titanium powder mixture. Journal of Engineering Research. (in press)

Amuda, M. O. H., Lawal, F. T., Onitiri, M. A., Akinlabi, E. T., & Mridha, S. (2018a). Microstructure and Mechanical Properties of Metal Powder Treated AISI-430 FSS Welds. *International Journal of Manufacturing, Materials, and Mechanical Engineering*, *8*(4), 63–83. doi:10.4018/IJMMME.2018100104

Amuda, M. O. H., Lawal, T. F., Enumah, K. S., Ezeonu, L. L., & Onitiri, M. A. (2018b). Influence of thermal treatments on sensitization in Cr-Mn-Cu austenitic stainless steel welds. *UNILAG Journal of Medicine. Science and Technology*, *6*(2), 91–106.

Amuda, M. O. H., & Mridha, S. (2010). Grain refinement in ferritic stainless steel welds: The journey so far. *Advanced Materials Research*, *83-86*, 1165–1172. doi:10.4028/www.scientific.net/AMR.83-86.1165

Amuda, M. O. H., & Mridha, S. (2012). Comparative evaluation of grain refinement in AISI 430 FSS welds by elemental metal powder addition and cryogenic cooling. *Materials & Design*, *35*, 609–618. doi:10.1016/j.matdes.2011.09.066

Anbazhagan, V., & Nagalakshmi, R. (2002). Metallurgical studies in ferritic stainless steel. *Welding Research Institute Journal*, *23*(3), 25–37.

Banerjee, M. K. (2017). Surface and heat treatment processes. In *Comprehensive Materials Finishing* (Vol. 2). New York: Elsevier.

Cortie, M., & Du Toit, M. (2016). Stainless steels, Ferritic. In *Reference Module in Materials Science and Materials Engineering*. New York: Elsevier. doi:10.1016/B978-0-12-803581-8.02501-7

Delgado, A. J., Ambriz, R. R., Cuenca-Alvarez, R., Alatorre, N., & López, F. F. (2016). Heat input effect on the microstructural transformation and mechanical properties in GTAW welds of a 409L ferritic stainless steel. *Revista de Metalurgia*, *52*(2), 68.

Easton, M. A., Schiffl, A., Yao, J. Y., & Kaufmann, H. (2006). Grain refinement of Mg–Al(–Mn) alloys by SiC additions. *Scripta Materialia*, *55*(4), 379–382. doi:10.1016/j.scriptamat.2006.04.014

Jozwik, P., & Bojar, Z. (2007). Analysis of grain size effect on tensile properties of Ni3Al-based intermetallic strips. *Archives of Metallurgy and Materials*, *2*(2), 321–327.

Kou, S. (2003). *Welding metallurgy* (2nd ed.). John-Wiley and Sons.

Lancaster, J. F. (1993). *Metallurgy of welding* (5th ed.). London: Chapman and Hall.

Mohandas, T., Reddy, G. M., & Naveed, M. (1999). A comparative evaluation of gas tungsten and shielded metal arc welds of a "ferritic" stainless steel. *Journal of Materials Processing Technology*, *94*(2-3), 133–140. doi:10.1016/S0924-0136(99)00092-8

Mridha, S., & Baker, T. N. (1994). Crack-free hard surfaces produced by laser nitriding of commercial purity titanium. *Materials Science and Engineering A*, *188*(1-2), 229–239. doi:10.1016/0921-5093(94)90376-X

Mridha, S., Ong, H. S., Poh, L. S., & Cheang, P. (2001). Intermetallic coatings produced by TIG surface melting. *Journal of Materials Processing Technology*, *113*(1-3), 516–520. doi:10.1016/S0924-0136(01)00609-4

Schiffl, A., & Easton, M. A. (2009). Influence of SiC particles on the grain refinement of an Mg-Al alloy. *Materials Science Forum*, *618-619*, 445–448. doi:10.4028/www.scientific.net/MSF.618-619.445

Shahid, Z., Muhammad, A., Khattak, M., & Nasir, T. (2018). Effects of welding on the microstructural properties of AISI 430 Ferritic Stainless Steel. *Journal of Advanced Research in Materials Science*, *44*(1), 25–32.

Villafuerte, J. C., Pardo, E., & Kerr, H. W. (1990). The effect of alloy composition and welding conditions on columnar-equiaxed transitions in ferritic stainless steel gas-tungsten arc welds. *Metallurgical Transactions. A, Physical Metallurgy and Materials Science*, *21*(7), 2009–2019. doi:10.1007/BF02647249

Compilation of References

Abouel-Seoud, S. S., Mohamed, E. S., Abdel-Hamid, A. A., & Abdallah, A. S. (2013). Analytical Technique for Predicting Passenger Car Gearbox Structure Noise Using Vibration Response Analysis. *British Journal of Applied Science and Technology*, *3*(4), 860–883. doi:10.9734/BJAST/2013/3277

Abouridouane, M., Klocke, F., & Döbbeler, B. (2016). Analytical temperature prediction for cutting steel. *CIRP Annals - Manufacturing Technology*, *65*, 77-80. doi:10.1016/j.cirp.2016.04.039

Akgöz, B., & Civalek, Ö. (2013). Free vibration analysis of axially functionally graded tapered Bernoulli–Euler microbeams based on the modiðed couple stress theory. *Composite Structures*, *98*, 314–322. doi:10.1016/j.compstruct.2012.11.020

Akgöz, B., & Civalek, Ö. (2015). Bending analysis of FG microbeams resting on winkler elastic foundation via strain gradient elasticity. *Composite Structures*, *134*, 294–301. doi:10.1016/j.compstruct.2015.08.095

Akinnuli, B.O., Dahunsi, O.A., Ayodeji S.P. & Bodunde, O.P. (2018) Whole-body vibration exposure on earthmoving equipment operators in construction industries, *Cogent Engineering*, *5*(1).

Akturk, M. S., & Avci, S. (1996). An integrated process planning approach for CNC machine tools. *International Journal of Advanced Manufacturing Technology*, *12*(3), 221–229. doi:10.1007/BF01351201

Ali, S., Hassan, A., Hassan, G., Bae, J., & Lee, C. H. (2016). All-printed humidity sensor based on graphene/methyl-red composite with high sensitivity. *Carbon*, *105*, 23–32. doi:10.1016/j.carbon.2016.04.013

Al-Shebeeb, O. A. (2019). *Analysis of the Integration of DFM Techniques and Effective Machining Parameter Selection in Metal Parts Manufacturing*. Academic Press.

Al-Shebeeb, O., & Gopalakrishnan, B. (2016). *Computer Aided Process Planning Approach for Cost Reduction and Increase in Throughput*. Academic Press.

American Society for Testing and Materials. (2008a). ASTM D792-08, Standard Test Methods for Density and Specific Gravity (Relative Density) of Plastics by Displacement.

Amuda, M. O. H. (2011). *Microstructural features and properties of TIG melted AISI 430 ferritic stainless steel welds* (Doctoral Thesis). International Islamic University Malaysia, Kuala Lumpur, Malaysia.

Amuda, M. O. H., Lawal, F. T., & Mridha, S. (2020). Microstructure evolution in AISI 430 ferritic stainless steel welds treated with aluminum + titanium powder mixture. Journal of Engineering Research. (in press)

Amuda, M. O. H., Akinlabi, E. T., & Mridha, S. (2016). Ferritic stainless steels: Metallurgy, Application and Weldability. In *Reference Module in Materials Science and Materials Engineering*. New York: Elsevier. doi:10.1016/B978-0-12-803581-8.04010-8

Amuda, M. O. H., Akinlabi, E. T., & Mridha, S. (2017). Influences of energy input and metal powder addition on carbide precipitation in AISI 430 ferritic stainless steel welds. *Materials Today: Proceedings*, *4*(2), 234–243. doi:10.1016/j.matpr.2017.01.017

Amuda, M. O. H., Lawal, F. T., Daramola, O., Awarun, A. B., & Subair, W. O. (2019). Wear and corrosion characteristics of silicon carbide surface modified mild steel. *UNILAG Journal of Medicine. Science and Technology.*, *6*(1), 129–146.

Amuda, M. O. H., Lawal, F. T., Onitiri, M. A., Akinlabi, E. T., & Mridha, S. (2018a). Microstructure and Mechanical Properties of Metal Powder Treated AISI-430 FSS Welds. *International Journal of Manufacturing, Materials, and Mechanical Engineering*, *8*(4), 63–83. doi:10.4018/IJMMME.2018100104

Amuda, M. O. H., Lawal, T. F., Enumah, K. S., Ezeonu, L. L., & Onitiri, M. A. (2018b). Influence of thermal treatments on sensitization in Cr-Mn-Cu austenitic stainless steel welds. *UNILAG Journal of Medicine. Science and Technology*, *6*(2), 91–106.

Amuda, M. O. H., & Mridha, S. (2010). Grain refinement in ferritic stainless steel welds: The journey so far. *Advanced Materials Research*, *83-86*, 1165–1172. doi:10.4028/www.scientific.net/AMR.83-86.1165

Amuda, M. O. H., & Mridha, S. (2012). Comparative evaluation of grain refinement in AISI 430 FSS welds by elemental metal powder addition and cryogenic cooling. *Materials & Design*, *35*, 609–618. doi:10.1016/j.matdes.2011.09.066

Anand, G., Alagumurthi, N., Elansezhian, R., Palanikumar, K., & Venkateshwaran, N. (2018). Investigation of drilling parameters on hybrid polymer composites using grey relational analysis, regression, fuzzy logic, and ANN models. *Journal of the Brazilian Society of Mechanical Sciences and Engineering*, *40*(4), 214. doi:10.100740430-018-1137-1

Anbazhagan, V., & Nagalakshmi, R. (2002). Metallurgical studies in ferritic stainless steel. *Welding Research Institute Journal*, *23*(3), 25–37.

An, S., Lee, M. W., Kim, N. Y., Lee, C., Al-Deyab, S. S., James, S. C., & Yoon, S. S. (2014). Effect of viscosity, electrical conductivity, and surface tension on direct-current-pulsed drop-on-demand electrohydrodynamic printing frequency. *Applied Physics Letters*, *105*(21), 214102–214105. doi:10.1063/1.4902241

Ansys R 14.5. (2013). Academic, Structural analysis Guide.

ArcelorMittal. (2014). Driving Advanced Automotive Steel solutions. *ArcelorMittal North America*.

Arefi, M., & Zenkour, A. M. (2017). Analysis of wave propagation in a functionally graded nanobeam resting on visco-Pasternak's foundation. *Theoretical & Applied Mechanics Letters*, *7*(3), 145–151. doi:10.1016/j.taml.2017.05.003

Arefi, M., & Zenkour, A. M. (2017). Wave propagation analysis of a functionally graded magneto-electro-elastic nanobeam rest on Visco-Pasternak foundation. *Mechanics Research Communications*, *79*, 51–62. doi:10.1016/j.mechrescom.2017.01.004

Arntz, K., Wegener, M., & Liu, Y. (2015). Computer aided manufacturing supported process planning of additive manufacturing by laser deposition welding. *Journal of Laser Applications*, *27*(S1), S14002. doi:10.2351/1.4823748

Arpaci, H., & Ozguven, O. F. (2011). ANFIS and PID controller design and comparison for overhead cranes. *Indian Journal of Engineering and Materials Sciences*, *18*(3), 191–203.

Arsecularatne, J. A., Zhang, L. C., & Montross, C. (2006). Wear and tool life of tungsten carbide, PCBN and PCD cutting tools. *International Journal of Machine Tools & Manufacture*, *46*(6), 482–491. doi:10.1016/j.ijmachtools.2005.07.015

Asadi, P., Mahdavinejad, R. A., & Tutunchilar, S. (2011). Simulation and experimental investigation of FSP of AZ91 magnesium alloy. *Materials Science and Engineering A*, *528*(21), 6469–6477. doi:10.1016/j.msea.2011.05.035

Asad, S., Salahat, M., & Zalata, A., & Rawashdeh, A.A. (2011). Design of fuzzy PD controlled overhead crane system with anti-swinging compensation. *Journal of Engineering and Computer Innovations*, *2*(3), 51–58.

Ashby, M. F., & Lu, T. (2003). Metal Foams: A Survey. *Science in China. Series B, Chemistry*, *46*, 521–532. doi:10.1360/02yb0203

Ashley, S. (2010). Shedding Pounds on a Magnesium Diet. *Automotive Engineering International*, 34–36.

Ashok Kumar Reddy, Y. (2013). *Study of Performance, Combustion, Emission and Vibration analysis of an Indirect Injection (IDI) Diesel Engine using preheated Jatropha Methyl Ester* (Unpublished doctoral dissertation). Andhra University, Visakhapatnam, India.

ASTM International. (2013). ASTM F2792-12a. *Rapid Manufacturing Association, 1–3*. doi:10.1520/F2792-12A.2

ASTM. (1998b). Standard test method for flatwise compressive properties of sandwich cores. ASTM International.

ASTM. (2007). *Standard test methods for flexural properties of unreinforced and reinforced plastics and electrical insulating materials.* ASTM D790-07.

ASTM. (2014). 638-14. *Standard test method for tensile properties of plastics*. ASTM.

Atzeni, E., Barletta, M., Calignano, F., Iuliano, L., Rubino, G., & Tagliaferri, V. (2016). Abrasive Fluidized Bed (AFB) finishing of AlSi10Mg substrates manufactured by Direct Metal Laser Sintering (DMLS). *Additive Manufacturing*, *10*, 15–23. doi:10.1016/j.addma.2016.01.005

Ausloos, M., & Berman, D. H. (1985). A Multivariate Weierstrass-Mandelbrot Function. *Proceedings of the Royal Society A: Mathematical, Physical and Engineering Sciences, 400*(1819), 331–350. 10.1098/rspa.1985.0083

Awais, M. N., Kim, H. C., Doh, Y. H., & Choi, K. H. (2013). ZrO_2 flexible printed resistive (memristive) switch through electrohydrodynamic printing process. *Thin Solid Films*, *536*, 308–312. doi:10.1016/j.tsf.2013.04.003

Azman, Vignat, & Villeneuve. (2018). CAD tools and file format performance evaluation in designing lattice structures for additive manufacturing. *Jurnal Teknologi, 80*(4).

Back, S. Y., Song, C. H., Yu, S., Lee, H. J., Kim, B. S., Yang, N. Y., ... Ahn, H. (2012). Drop-on-Demand Printing of Carbon Black Ink by Electrohydrodynamic Jet Printing. *Journal of Nanoscience and Nanotechnology*, *12*(1), 446–450. doi:10.1166/jnn.2012.5370 PMID:22524000

Badkar, D. S., Pandey, K. S., & Buvanashekaran, G. (2011). Parameter optimization of laser transformation hardening by using Taguchi method and utility concept. *International Journal of Advanced Manufacturing Technology*, *52*(9-12), 1067–1077. doi:10.100700170-010-2787-z

Bagci, E., & Aykut, S. (2006). A study of Taguchi optimization method for identifying optimum surface roughness in CNC face milling of cobalt-based alloy (stellite 6). *International Journal of Advanced Manufacturing Technology*, *29*(9-10), 940–947. doi:10.100700170-005-2616-y

Bai, W., Sun, R., Roy, A., & Silberschmidt, V. V. (2017). Improved analytical prediction of chip formation in orthogonal cutting of titanium alloy Ti6Al4V. *International Journal of Mechanical Sciences*, *133*, 357–367. doi:10.1016/j.ijmecsci.2017.08.054

Ballagh, K. O. (1996). Acoustical properties of wool. *Applied Acoustics*, *48*(2), 101–120. doi:10.1016/0003-682X(95)00042-8

Balogun, V. A., Edem, I. F., Adekunle, A. A., & Mativenga, P. T. (2016). Specific energy based evaluation of machining efficiency. *Journal of Cleaner Production*.

Banerjee, M. K. (2017). Surface and heat treatment processes. In *Comprehensive Materials Finishing* (Vol. 2). New York: Elsevier.

Baohai, W., Di, C., Xiaodong, H., Dinghua, Z., & Kai, T. (2016). Cutting tool temperature prediction method using analytical model for end milling. *Chinese Journal of Aeronautics*, *29*(6), 1788–1794. doi:10.1016/j.cja.2016.03.011

Bara, A., Dale, S., & Nagy, Z. T. (2009). Comparative real-time experimental study case for control algorithms, from implementation point of view. *Proceedings of the 8th WSEAS International Conference on System Science and Simulation in Engineering*, 302-306.

Bartarya, G., & Choudhury, S. K. (2011). State of the art in hard turning. *International Journal of Machine Tools & Manufacture*, *53*(1), 1–14. doi:10.1016/j.ijmachtools.2011.08.019

Baseri, H. (2011). Simulated annealing based optimization of dressing conditions for increasing the grinding performance. *International Journal of Advanced Manufacturing Technology*, *59*(5-8), 531–538. doi:10.100700170-011-3518-9

Batish, A., Bhattacharya, A., Kaur, M., & Cheema, M. S. (2014). Hard turning: Parametric optimization using genetic algorithm for rough/finish machining and study of surface morphology. *Journal of Mechanical Science and Technology*, *28*(5), 1629–1640. doi:10.100712206-014-0308-y

Bauer, W. (2011). *Hydropneumatic suspension system*. Springer. doi:10.1007/978-3-642-15147-7

Baumers, M., Dickens, P., Tuck, C., & Hague, R. (2016). The cost of additive manufacturing: Machine productivity, economies of scale and technology-push. *Technological Forecasting and Social Change*, *102*, 193–201.

Baykasoglu, C., Akyildiz, O., Candemir, D., Yang, Q., & To, A. C. (2018). Predicting Microstructure Evolution During Directed Energy Deposition Additive Manufacturing of Ti-6Al-4V. *Journal of Manufacturing Science and Engineering*, *140*(5), 051003. doi:10.1115/1.4038894

Bellisario, D., & Quadrini, F. (2012). Laser bending of 5005 aluminum alloy sheet. *International Journal of Manufacturing, Materials, and Mechanical Engineering*, *2*(3), 1–15. doi:10.4018/ijmmme.2012070101

Belunce, A., Pandolfo, V., Roozbahani, H., & Handroos, H. (2015). Novel control method for Overhead cranes load stability. *Procedia Engineering*, *106*, 108–125. doi:10.1016/j.proeng.2015.06.014

Benevento, M. (2016). *Adhesives and Sealants for Automotive Weight Reduction, Safety, and Performance, Industrial Market Insight*. https://www.adhesives.org/resources/knowledge-center/aggregate-single/adhesives-and-sealants-for-automotive-weight-reduction-safety-and-performance

Berardi, U., & Iannace, G. (2015). Acoustic characterization of natural fibers for sound absorption applications. *Building and Environment*, *94*, 840–852. doi:10.1016/j.buildenv.2015.05.029

Berry, M. V., & Lewis, Z. V. (1980). On the Weierstrass-Mandelbrot Fractal Function. *Proceedings of the Royal Society A: Mathematical, Physical and Engineering Sciences, 370*(1743), 459–484. 10.1098/rspa.1980.0044

Bharath Kumar, B. R., Zeltmann, S. E., Doddamani, M., Gupta, N., Gurupadu, S., & Sailaja, R. R. N. (2016). Effect of cenosphere surface treatment and blending method on the tensile properties of thermoplastic matrix syntactic foams. *Journal of Applied Polymer Science*, *133*(35). doi:10.1002/app.43881

Bikas, Stavropoulos, & Chryssolouris. (2016). Additive manufacturing methods and modelling approaches: a critical review. *The International Journal of Advanced Manufacturing Technology* *83*(1-4), 389-405.

Billiet, T., Vandenhaute, M., Schelfhout, J., Van Vlierberghe, S., & Dubruel, P. (2012). A review of trends and limitations in hydrogel-rapid prototyping for tissue engineering. *Biomaterials*, *33*(26), 6020–6041. doi:10.1016/j.biomaterials.2012.04.050 PMID:22681979

Birch, S. (2015). New-generation Audi R8: The Race Driver's View. *SAE International.* Available from: http://articles.sae.org/14469

Biron, M. (2013). *Thermosets and composites: material selection, applications, manufacturing and cost analysis*. Elsevier.

Böckin, D., & Tillman, A.-M. (2019). Environmental assessment of additive manufacturing in the automotive industry. *Journal of Cleaner Production*, *226*, 977–987. doi:10.1016/j.jclepro.2019.04.086

Boney, T., & Richman, D. (2014). *The Aluminum Association: 2015 Ducker Worldwide Automaker Survey.* Available from http://www.aluminumintransportation.org/research-resources/PDF/Webinars/2014/2015-ducker-worldwide-survey-of-automakers

Bonny, K., De Baets, P., Perez, Y., Vleugels, J., & Lauwers, B. (2010). Friction and wear characteristics of WC–Co cemented carbides in dry reciprocating sliding contact. *Wear*, *268*(11–12), 1504–1517. doi:10.1016/j.wear.2010.02.029

Bray, L. (2019). Mineral Commodity Summaries - Magnesium Metal 2019. *U.S. Geological Survey.* https://www.usgs.gov/centers/nmic/magnesium-statistics-and-information

Brennan, J. M. (2011). *Intelligent Simulation Technology Delivering Weight Efficient Vehicles.* CAR Management Briefing Seminars 2011, Traverse City, MI.

Breunig, P., Damodaran, V., Shahapurkar, K., Waddar, S., Doddamani, M., Jeyaraj, P., & Prabhakar, P. (2020). Dynamic impact behavior of syntactic foam core sandwich composites. *Journal of Composite Materials*, *54*(4), 535–547. doi:10.1177/0021998319885000

Brooke, L. (2012, November). Creating the 54.5 mpg Car, SAE. *Automotive Engineering International*, 32–36.

Buczkowski, R., Kleiber, M., & Starzynski, G. (2014). Normal contact stiffness of fractal rough surfaces. *Archives of Mechanics*, *66*(6), 411–428. doi:10.24423/AOM.1286

Burstein, L. (2015). *MATLAB in quality assurance sciences*. Amsterdam: Elsevier - Woodhead Publishing.

Burstein, L. (2020). *A MATLA Primer for Technical Programming for Materials Science and Engineering*. Amsterdam: Elsevier.

Buzio, R., Boragno, C., & Valbusa, U. (2003). Contact mechanics and friction of fractal surfaces probed by atomic force microscopy. *Wear*, *254*(9), 917–923. doi:10.1016/S0043-1648(03)00247-3

Caffaro, F., Cremasco, M. M., Preti, C., & Cavallo, E. (2016). Ergonomic analysis of the effects of a telehandler's active suspended cab on whole body vibration level and operator comfort. *International Journal of Industrial Ergonomics*, *53*, 19–26. doi:10.1016/j.ergon.2015.10.009

Cai, W., Wang, P. C., & Yang, W. (2005). Assembly Dimensional Prediction for Self-Piercing Riveted Aluminum Panels. *International Journal of Machine Tools & Manufacture*, *45*(6), 695–704. doi:10.1016/j.ijmachtools.2004.09.023

Cai, Z. Z., Li, H. M., & An, G. (2012). Vibration Simulation of Tracked Vehicle Suspension and Damper Consumption Analysis. *Applied Mechanics and Materials*, *184-185*, 748–751. doi:10.4028/www.scientific.net/AMM.184-185.748

Campbell, T., Williams, C., Ivanova, O., & Garrett, B. (2011). Could 3D printing change the world. Technologies, Potential, and Implications of Additive Manufacturing, Atlantic Council.

Carolan, D., Mayall, A., Dear, J. P., & Fergusson, A. D. (2020). Micromechanical modelling of syntactic foam. *Composites. Part B, Engineering*, *183*, 107701. doi:10.1016/j.compositesb.2019.107701

Cebon, D. (2000). *Handbook of Vehicle-Road Interaction*. Swets & Zeitlinger.

CEI-IEC 61156, 2002-12. Multicore and symmetrical pair/quad cables for digital communications, Parts1, 5, 6, and 7. International standard.

Chakraborty, S. S., More, H., & Nath, A. K. (2018). Laser forming of a bowl shaped surface with a stationary laser beam. *Optics and Lasers in Engineering*, *77*, 126–136. doi:10.1016/j.optlaseng.2015.08.006

Chakraborty, S., Kar, S., Ghosh, S. K., & Dey, V. (2017). Parametric optimization of electric discharge coating on Aluminium-6351 alloy with green compact silicon carbide and copper tool: A Taguchi coupled utility concept approach. *Surfaces and Interfaces*, *7*, 47–57. doi:10.1016/j.surfin.2017.02.001

Chanda, B., Kumar, R., Kumar, K., & Bhowmik, S. (2015). Optimisation of Mechanical Properties of Wood Dust-reinforced Epoxy Composite Using Grey Relational Analysis. In *Proceedings of fourth international conference on soft computing for problem solving* (pp. 13-24). Springer. 10.1007/978-81-322-2220-0_2

Chang, R., Emami, K., Wu, H., & Sun, W. (2010). Bio-fabrication of a Three-Dimensional Liver Micro-organ as an In-Vitro Drug Metabolism Model. *Bio-Fabrication, 2*(4).

Chang, C. Y., & Chiang, T. C. (2009). Overhead cranes fuzzy control design with deadzone compensation. *Neural Computing & Applications*, *18*(7), 749–757. doi:10.100700521-009-0264-0

Chang, C., Hsu, S., & Chiang, K. (2005). A practical fuzzy controllers scheme of overhead crane. *Journal of Control Theory and Applications*, *3*(3), 266–270. doi:10.100711768-005-0047-9

Chartrain, N. A., Williams, C. B., & Whittington, A. R. (2018). A review on fabricating tissue scaffolds using vat photopolymerization. *Acta Biomaterialia*, *74*, 90–111. doi:10.1016/j.actbio.2018.05.010 PMID:29753139

Chatterjee, B., & Sahoo, P. (2013). Finite Element based Parametric Study of Elastic-plastic Contact of Fractal Surfaces. *Procedia Engineering*, *64*, 552–561. doi:10.1016/j.proeng.2013.09.130

Chatterjee, B., & Sahoo, P. (2014). Finite Element Based Contact Analysis of Fractal Surfaces – Effect of Varying Elastic Modulus. *Procedia Engineering*, *90*, 116–122. doi:10.1016/j.proeng.2014.11.823

Chen, C. H., Saville, D. A., & Aksay, I. A. (2006). Scaling laws for pulsed electrohydrodynamic drop formation. *Applied Physics Letters*, *89*(2), 124103–3. doi:10.1063/1.2356891

Chen, C., Saville, D. A., & Aksay, I. A. (2006). Electrohydrodynamic "drop-and-place" particle deployment. *Applied Physics Letters*, *88*(25), 154104–3. doi:10.1063/1.2191733

Chen, D., Yang, J., & Kitipornchai, S. (2015). Elastic buckling and static bending of shear deformable functionally graded porous beam. *Composite Structures*, *133*, 54–61. doi:10.1016/j.compstruct.2015.07.052

Chen, Z. M., Meng, W. J., & Zhang, J. G. (2012). Intellihent anti-swing control for bridge crane. *Journal of Central South University*, *19*(10), 2774–2781. doi:10.100711771-012-1341-6

Chethan, Y. D., Ravindra, H. V., & Krishnegowda, Y. T. (2019). Optimization of machining parameters in turning Nimonic-75 using machine vision and acoustic emission signals by Taguchi technique. *Measurement*, *144*, 144–154. doi:10.1016/j.measurement.2019.05.035

Cho, D., & Lee, D. (1998). Manufacturing of co-cured aluminum composite shafts with compression during co-curing operation to reduce residual thermal stresses. *Journal of Composite Materials*, *32*, 1221–1241. doi:10.1177/002199839803201205

Cho, D., Lee, D., & Choi, J. (1997). Manufacture of one-piece automotive drive shafts with aluminum and composite. *Composite Structures*, *38*(1-4), 309–319. doi:10.1016/S0263-8223(97)00065-2

Cho, H. S., Kim, K. S., Choi, D. S., Kim, Y. C., Park, S. H., Oh, B. S., ... Kook, J. H. (2013). Structure analysis on automotive seat recliner housing with high tension steel plate. *Journal of the Korea Academia-Industrial Cooperation Society*, *14*(8), 3644–3649. doi:10.5762/KAIS.2013.14.8.3644

Choi, H. K., Park, J.-U., Park, O. O., Ferreira, P. M., Georgiadis, J. G., & Rogers, J. A. (2008). Scaling laws for jet pulsations associated with high-resolution electrohydrodynamic printing. *Applied Physics Letters*, *92*(12), 123109–3. doi:10.1063/1.2903700

Chowdhury, M. A., & Helali, M. M. (2008). The Effect of Amplitude of Vibration on the Coefficient of Friction. *Tribology International*, *41*(4), 307–314. doi:10.1016/j.triboint.2007.08.005

Chowdhury, M. A., & Helali, M. M. (2010). The Influence of Natural Frequency of the Experimental Set-up on the Friction Coefficient of Stainless Steel-304. *Tribology in Industry*, *32*, 19–24.

Citroen. (1977). Hydraulic course notes, Man. 008440. *Historic Archive Citroen.*

Colloca, M., Gupta, N., & Porfiri, M. (2013). Tensile properties of carbon nanofiber reinforced multiscale syntactic foams. *Composites. Part B, Engineering*, *44*(1), 584–591. doi:10.1016/j.compositesb.2012.02.030

Compton, B. G., & Lewis, J. A. (2014). 3D-printing of lightweight cellular composites. *Advanced Materials*, *26*(34), 5930–5935. doi:10.1002/adma.201401804 PMID:24942232

Coral-Enriquez, H., Pulido-Guerrero, S. P., & Cortes-Romero, J. (2019). Robust disturbance rejection based control with extended state resonant observer for sway reduction in uncertain tower cranes. *International Journal of Automation and Computing*, *16*(6), 812–827. doi:10.100711633-019-1179-6

Cortie, M., & Du Toit, M. (2016). Stainless steels, Ferritic. In *Reference Module in Materials Science and Materials Engineering*. New York: Elsevier. doi:10.1016/B978-0-12-803581-8.02501-7

Costello, G. A. (1997). Theory of wire rope. Springer.

Creese, R., & Adithan, M. (1992). *Estimating and costing for the metal manufacturing industries*. CRC Press.

Crowther, A., Singh, R., Zhang, N., & Chapman, C. (2007). Impulsive response of an automatic transmission system with multiple clearances: Formulation, Simulation and experiment. *Journal of Sound and Vibration*, *306*(3-5), 444–466. doi:10.1016/j.jsv.2007.05.053

Czech, P. (2012). Diagnosis of industrial gearboxes Condition by vibration and time frequency, Scale-frequency, frequency-frequency analysis. *Metalurgija*, *51*, 521–524.

D'Acunto, A., Le Coz, G., Moufki, A., & Dudzinski, D. (2017). Effect of cutting edge geometry on chip flow direction – analytical modelling and experimental validation. *16th CIRP Conference on Modelling of Machining Operations*, *58*, 353-357. 10.1016/j.procir.2017.03.327

Dai, M., Tang, D., Xu, Y., & Li, W. (2015). Energy-aware integrated process planning and scheduling for job shops. *Proceedings of the Institution of Mechanical Engineers, Part B: Journal of Engineering Manufacture, 229*(1_suppl), 13–26. 10.1177/0954405414553069

Das, R. (2018). Optimization of E-Jet Based Micro-manufacturing Process Using Desirability Function Analysis. In S. Bhattacharyya, S. Sen, M. Dutta, P. Biswas, & H. Chattopadhyay (Eds.), Industry Interactive Innovations in Science, Engineering and Technology, Lecture Notes in Networks and Systems (Vol. 11, pp. 477-484). Singapore: Springer. doi:10.1007/978-981-10-3953-9_46

Deb, K., Pratap, A., Agarwal, S., & Meyarivan, T. (2002). A fast and elitist multiobjective genetic algorithm: NSGA-II. *IEEE Transactions on Evolutionary Computation*, *6*(2), 182–197. doi:10.1109/4235.996017

De-gang, L., Weixing, H., & Fengzhou, W. (2004). Fatigue Life Analysis of a Component Based on the Finite Elements Technology. *Journal of The China Railway Society*, *26*, 47–51.

R. Del Rey, J. Alba, & A. S. V. Sanch (Eds.). (2007). Proposal of an Empirical Model for Absorbent Acoustical Materials Based in Kenaf. *Proceedings of the 19th International Congress on Acoustics*, 3361-3366.

Delgado, A. J., Ambriz, R. R., Cuenca-Alvarez, R., Alatorre, N., & López, F. F. (2016). Heat input effect on the microstructural transformation and mechanical properties in GTAW welds of a 409L ferritic stainless steel. *Revista de Metalurgia*, *52*(2), 68.

Deng, J.L. (1989). Introduction to Grey System Theory. *J Grey Syst, 1*(1), 1-24.

Deng, H., Chen, K. D., Cheng, W., & Zhao, S. G. (2017). Vibration and buckling analysis of double-functionally graded Timoshenko beam system on Winkler-Pasternak elastic foundation. *Composite Structures*, *160*, 152–168. doi:10.1016/j.compstruct.2016.10.027

Deprez, K., Moshou, D., Anthonis, J., De Baerdemaeker, J., & Ramon, H. (2005). Improvement of vibrational comfort on agricultural vehicles by passive and semi-active cabin suspensions. *Computers and Electronics in Agriculture*, *49*(3), 431–440. doi:10.1016/j.compag.2005.08.009

Derby, B., & … . (2010). Inkjet Printing of Functional and Structural Materials: Fluid Property Requirements, Feature Stability, and Resolution. *Annual Review of Materials Research*, *40*(1), 395–414. doi:10.1146/annurev-matsci-070909-104502

Desai, J. R., Shit, S. C., & Jain, S. K. (2014). Studies on thermal, electrical and flame properties of surface modified cenosphere filled ABS composites. *International Journal of Plastics Technology*, *18*(3), 390–396. doi:10.100712588-014-9099-3

Desai, J. R., Shit, S. C., & Jain, S. K. (2016). Analysis of 3-methacryloxypropyl trimethoxysilane treated cenosphere inclusion on dynamic mechanical properties of ABS composites. *International Journal of Plastics Technology*, *20*(2), 241–248. doi:10.100712588-016-9152-5

Dickrell, D. J., Dugger, M. T., Hamilton, M. A., & Sawyer, W. G. (2007). Direct contact-area computation for MEMS using real topographic surface data. *Journal of Microelectromechanical Systems*, *16*(5), 1263–1268. doi:10.1109/JMEMS.2007.901120

Dilberoglu, U. M., Gharehpapagh, B., Yaman, U., & Dolen, M. (2017). The Role of Additive Manufacturing in the Era of Industry 4.0. *Procedia Manufacturing*, *11*(June), 545–554. doi:10.1016/j.promfg.2017.07.148

Dimchev, M., Caeti, R., & Gupta, N. (2010). Effect of carbon nanofibers on tensile and compressive characteristics of hollow particle filled composites. *Materials & Design*, *31*(3), 1332–1337. doi:10.1016/j.matdes.2009.09.007

Ding, J. (2020). Compressive properties of co-continuous hollow glass microsphere/epoxy resin syntactic foams prepared using resin transfer molding. *Journal of Reinforced Plastics and Composites*, *39*(3-4), 132-143.

Ding, D., Shen, C., Pan, Z., Cuiuri, D., Li, H., Larkin, N., & van Duin, S. (2016). Towards an automated robotic arc-welding-based additive manufacturing system from CAD to finished part. *Computer Aided Design*, *73*, 66–75. doi:10.1016/j.cad.2015.12.003

Dixon, R. K., McGowan, E., Onysko, G., & Scheer, R. M. (2010). US energy conservation and efficiency policies: Challenges and opportunities. *Energy Policy*, *38*(11), 6398–6408. doi:10.1016/j.enpol.2010.01.038

Dogan, S. N. (1999). Loose part vibration in vehicle transmissions Gear rattle. *Transactions Journal of Engineering and Environmental Science*, *23*, 439–454.

Dong, Y. W., & Yan, L. (2012). Sound absorption performance of natural fibers and their composites. *Science China. Technological Sciences*, *55*(8), 2278–2283. doi:10.100711431-012-4943-1

Dubuc, B., Zucker, S. W., Tricot, C., Quiniou, J. F., & Wehbi, D. (1989). Evaluating the Fractal Dimension of Surfaces. *Proceedings of the Royal Society A: Mathematical, Physical and Engineering Sciences, 425*(1868), 113–127. 10.1098/rspa.1989.0101

Ducker Worldwide LLC. (2014). *2015 North American Light Vehicle Aluminum Content Study, June, 2014*. Author.

Dwivedi, G., Srivastava, S. K., & Srivastava, R. K. (2017). Analysis of barriers to implement additive manufacturing technology in the Indian automotive sector. *International Journal of Physical Distribution & Logistics Management*, *47*(10), 972–991. doi:10.1108/IJPDLM-07-2017-0222

Easton, M. A., Schiffl, A., Yao, J. Y., & Kaufmann, H. (2006). Grain refinement of Mg–Al(–Mn) alloys by SiC additions. *Scripta Materialia*, *55*(4), 379–382. doi:10.1016/j.scriptamat.2006.04.014

Ebrahimi, F., & Salari, E. (2015). Thermal buckling and free vibration analysis of size dependent Timoshenko FG nanobeams in thermal environments. *Composite Structures*, *128*, 363–380. doi:10.1016/j.compstruct.2015.03.023

Ebrahimi, F., & Zia, M. (2015). Large amplitude nonlinear vibration analysis of functionally graded Timoshenko beams with porosities. *Acta Astronautica*, *116*, 117–125. doi:10.1016/j.actaastro.2015.06.014

Ellmann, S., & Löser, F. (2002). *Elektromechanisches Leveling System*. Forum Thyssen Krupp.

Emmerich, M. T. M., & Deutz, A. H. (2018). A tutorial on multiobjective optimization: Fundamentals and evolutionary methods. *Natural Computing*, *17*(3), 585–609. doi:10.100711047-018-9685-y PMID:30174562

Equbal, M. I., Talukdar, P., Kumar, V., & Ohdar, R. K. (2014). Deformation Behavior of Micro-alloyed Steel by Using Thermo Mechanical Simulator and Finite Element Method. *Procedia Materials Science, 6*(Icmpc), 674–681. doi:10.1016/j.mspro.2014.07.083

Erriquez, M. (2017). What did we learn from a teardown and benchmarking of ten EV models? *McKinsey Center for Future Mobility*. https://www.mckinsey.com/industries/automotive-and-assembly/our-insights/trends-in-electric-vehicle-design

Eskandari, B., Davoodi, B., & Ghorbani, H. (2018). Multi-objective optimization of parameters in turning of N-155 iron-nickel-base superalloy using gray relational analysis. *Journal of the Brazilian Society of Mechanical Sciences and Engineering*, *40*(4), 233. doi:10.100740430-018-1156-y

Eulenbach, D. (2003). *Nivomat – Stand und Entwicklungstrends hydropneumatischer Niveauregelsysteme. Tagung: Kraftfahrzeugstoßdämpfer und ihr Einfluss auf das Fahrwerk*. Komponenten und Systeme.

European Agency for Safety and Health at Work. (2008). *Workplace exposure to vibration in Europe: an expert review*. Author.

European Aluminium Association (EAA). (2015). EAA Aluminium Automotive Manual – Joining. Brussels: Author.

European Council Directive. (1989). *On the Introduction of measures to encourage improvements in the safety and health of workers.* Author.

European Parliament and European Council. (2002). *On the minimum health and safety requirements regarding the exposure of workers to the risks arising from physical agents (vibration).* Author.

Evans, R. W., & Scharning, P. J. (2001). Axisymmetric compression test and hot working properties of alloys. *Materials Science and Technology*, *17*(8), 995–1004. doi:10.1179/026708301101510843

Fan, C., Peng, Y., Yang, H., Zhou, W., & Yan, H. (2017). Hot deformation behavior of Al–9.0Mg–0.5Mn–0.1Ti alloy based on processing maps. [English Edition]. *Transactions of Nonferrous Metals Society of China*, *27*(2), 289–297. doi:10.1016/S1003-6326(17)60033-5

Fedoriková, A., Kvačkaj, T., Kočiško, R., Bidulský, R., Petroušek, P., Bidulská, J., & Domovcová, L. (2016). Hot compression test of 9 Cr-1 mo steel – Numerical simulation. *Acta Metallurgica Slovaca*, *22*(2), 102–110. doi:10.12776/ams.v22i2.616

Fedtke, S., & Boysen, N. (2017). Gantry crane and shuttle car scheduling in modern rail-rail transshipment yards. *OR-Spektrum*, *39*(2), 473–503. doi:10.100700291-016-0461-z

Ferraresi, C., Quaglia, G., & Sorli, M. (1997). Force control laws for semi-active vehicular suspensions. *European Journal of Mechanical and Environmental Engineering*, *42*(3), 145–151.

Fiat Chrysler Automobiles (FCA). (2019). *2019 Jeep® Cherokee: Authentic, Premium Jeep Design and New Advanced Global Turbo Engine Enhance Evolution of Renowned Mid-size SUV.* https://media.fcanorthamerica.com/newsrelease.do?id=18756&mid=426

Folkersma, K. G. P., Brouwer, D. M., Römer, G. R. B. E., & Herder, J. L. (2016). Robust precision alignment algorithm for micro tube laser forming. *Precision Engineering*, *46*, 301–308. doi:10.1016/j.precisioneng.2016.05.011

Fouladi, M. H., Ayub, M., & Nor, M. J. M. (2011). Analysis of coir fiber acoustical characteristics. *Applied Acoustics*, *72*(1), 35–42. doi:10.1016/j.apacoust.2010.09.007

Fuh, J., Chang, C., & Melkanoff, M. (1992). A logic-based integrated manufacturing planning system. *Computers in Engineering*, *1*, 391–391.

Funakawa, Y., & Nagataki, Y. (2019). *High Strength Steel Sheets for Weight Reduction of Automobiles*. JFE Technical Report, No. 24. https://www.jfe-steel.co.jp/en/research//report/024/pdf/024-02.pdf

Fung, C. P. (2003). Manufacturing process optimization for wear property of fiber-reinforced polybutylene terephthalate composites with grey relational analysis. *Wear*, *254*(3-4), 298–306. doi:10.1016/S0043-1648(03)00013-9

Fu, Z., Chen, X., Mao, J., & Xiong, T. (2018). An analytical force mode applied to three-dimensional turning based on a predictive machining theory. *International Journal of Mechanical Sciences*, *136*, 94–105. doi:10.1016/j.ijmecsci.2017.12.021

Ganesan, H., & Mohankumar, G. (2013). Optimization of machining techniques in CNC turning centre using genetic algorithm. *Arabian Journal for Science and Engineering*, *38*(6), 1529–1538. doi:10.100713369-013-0539-8

Ganesan, P., & Karthik, T. (2016). Development of acoustic nonwoven materials from kapok and milkweed fibers. *Journal of the Textile Institute*, *107*(4), 477–482. doi:10.1080/00405000.2015.1045251

Gao, Y., Sun, R., & Leopold, J. (2018). An analytical force model for modulation-assisted turning. *Journal of Manufacturing Processes*, *31*, 712–730. doi:10.1016/j.jmapro.2017.12.024

Gauthier, M. M. (Ed.). (1995). *Engineered materials handbook* (Vol. 1). ASM International.

Gehm, R. (2016). Composite Bonding Adhesive. *SAE International*. Available from: http://articles.sae.org/14862/

Genta, G. (2009). *Vibration Dynamics and Control*. Springer. doi:10.1007/978-0-387-79580-5

Genta, G., & Morello, L. (2008). *The automotive chassis*. Springer.

George, D. (1988). *Mechanical Metallurgy. McGraw-Hill Book Company.*

Gerdemeli, I., Kurt, S., & Anil, A. S. (2013). Fatigue Life Analysis of Wire Rope Strands with Finite Element Method. *Key Engineering Materials*, *572*(Sep), 513–516. doi:10.4028/www.scientific.net/KEM.572.513

Ghamsari, A. K., Wicker, S., & Woldesenbet, E. (2014). Bucky syntactic foam; multi-functional composite utilizing carbon nanotubes-ionic liquid hybrid. *Composites. Part B, Engineering*, *67*, 1–8. doi:10.1016/j.compositesb.2014.06.030

Ghayesh, M. H., & Farokhi, H. (2018). Bending and vibration analyses of coupled axially functionally graded tapered beams. *Nonlinear Dynamics*, *91*(1), 17–28. doi:10.100711071-017-3783-8

Gibson, I., Rosen, D., & Stucker, B. (2015). *Additive Manufacturing Technologies*. New York, NY: Springer New York; doi:10.1007/978-1-4939-2113-3

Gies, P. E. (1996). *Development of a hydropneumatic suspension system for an earthmoving machine*. SAE Technical Papers.

Gilbert, M. (Ed.). (2016). *Brydson's plastics materials*. William Andrew.

Giliomee, C. L., & Els, P. S. (1998). Semi-active hydropneumatic spring and damper system. *Journal of Terramechanics*, *35*(2), 109–117. doi:10.1016/S0022-4898(98)00016-0

Gokuldoss, P. K., Kolla, S., & Eckert, J. (2017). Additive Manufacturing Processes: Selective Laser Melting, Electron Beam Melting and Binder Jetting—Selection Guidelines. *Materials (Basel)*, *10*(6), 672. doi:10.3390/ma10060672

Gong, G., Deng, Q., Chiong, R., Gong, X., Huang, H., & Han, W. (2019). Remanufacturing-oriented process planning and scheduling: Mathematical modelling and evolutionary optimisation. *International Journal of Production Research*, 1–19. doi:10.1080/00207543.2019.1634848

Gopalakrishnan, B. (1989). Computer integrated machining parameter selection in a job shop using expert systems. *Journal of Mechanical Working Technology*, *20*, 163–170. doi:10.1016/0378-3804(89)90027-2

Gopalakrishnan, B., & Pandiarajan, V. (1991). Materials and manufacturing processes selection system for product designs in concurrent engineering. *Journal of Materials Processing Technology*, *28*(1), 93–103. doi:10.1016/0924-0136(91)90209-W

Gopalakrishnan, B., Plummer, R., & Alkadi, N. (2001). Analysis of energy conservation opportunities in glass manufacturing facilities. *Energy Engineering*, *98*(6), 27–49. doi:10.1080/01998590109509330

Gorabal, S. V., Surbet, S. N., & Appukuttan, K. K. (2011). Design and evaluation of hydro-pneumatic friction damper suspension system. *International Journal of Manufacturing, Materials, and Mechanical Engineering*, *1*(4), 65–74. doi:10.4018/ijmmme.2011100105

Gordon, T.J., & Bareket, Z. (2007). *Vibration Transmission from Road Surface Features– Vehicle Measurement and Detection.* Technical Report for Nissan Technical Center North America, Inc. UMTRI- 2007-4.

Graf, P. G. (2011). *A 2nd generation electrohydrodynamic jet (e-jet) printing system, parametric studies of e-jet nozzles and integrated electrode e-jet deposition* (M.S. Thesis). University of Illinois at Urbana-Champaign. https://www.ideals.illinois.edu/handle/2142/24288

Griffiths, J., Edwardson, S. P., Dearden, G., & Watkins, K. G. (2010). Finite element modeling of laser forming at macro and micro scales. *Physics Procedia*, *5*, 371–380. doi:10.1016/j.phpro.2010.08.064

Griffiths, J., Edwardson, S. P., Dearden, G., & Watkins, K. G. (2012). Thermal laser micro-adjustment using picosecond pulse durations. *Applied Surface Science*, *258*(19), 7639–7643. doi:10.1016/j.apsusc.2012.04.109

Griffiths, J., Sheikholeslami, G., & Dearden, G. (2014). Optimization of beam mode for high efficiency laser thermal forming within metallurgical constraints. *Physics Procedia*, *56*, 1392–1399. doi:10.1016/j.phpro.2014.08.069

Grott, M. (2010). *Design of Suspension Systems and Control Algorithms for Heavy Duty Vehicles* (PhD thesis). University of Trento.

Grzesik, W., & Kwiatkowska, E. (1997). An energy approach to chip-breaking when machining with grooved tool inserts. *International Journal of Machine Tools & Manufacture*, *37*(5), 569–577. doi:10.1016/S0890-6955(96)00076-4

Guignard, J. C. (1971). Human sensitivity to vibration. *Journal of Sound and Vibration*, *15*(1), 11–16. doi:10.1016/0022-460X(71)90354-3

Gupta, D., & Gopalakrishnan, B. (2010). Energy sensitive machining parameter optimisation. *International Journal of Industrial and Systems Engineering*, *5*(4), 405–423. doi:10.1504/IJISE.2010.032964

Gupta, N., & Paramsothy, M. (2014). Metal-and polymer-matrix composites: Functional lightweight materials for high-performance structures. *JOM*, *66*(6), 862–865. doi:10.100711837-014-0969-0

Gupta, N., Pinisetty, D., & Shunmugasamy, V. C. (2013). *Reinforced polymer matrix syntactic foams: effect of nano and micro-scale reinforcement*. Springer Science & Business Media. doi:10.1007/978-3-319-01243-8

Gupta, N., Ye, R., & Porfiri, M. (2010). Comparison of tensile and compressive characteristics of vinyl ester/glass microballoon syntactic foams. *Composites. Part B, Engineering*, *41*(3), 236–245. doi:10.1016/j.compositesb.2009.07.004

Han, H. (2016). *A Study of Additive Manufacturing Technology as a Manufacturing Tool of the Automotive Industry*. Academic Press.

Han, S. W., & Jung, H. S. (2011). Weight reducing of aluminum extrusion profiles of a railway-car body based on topology and size optimization. *Transactions of the Korean Society of Mechanical Engineers-A*, *25*(2), 213–221. doi:10.3795/KSME-A.2011.35.2.213

Hansson, P. A. (1995). Optimization of agricultural tractor cab suspension using the evolution method. *Computers and Electronics in Agriculture*, *12*(1), 35–49. doi:10.1016/0168-1699(94)00036-P

Han, Y., Sun, Y., Zhang, W., & Chen, H. (2017a). Hot deformation and processing window optimization of a 70MnSiCrMo carbide-free bainitic steel. *Materials (Basel)*, *10*(3), 318. doi:10.3390/ma10030318 PMID:28772678

Hao, G. L., Xu, Q. P., Han, F. S., Li, W. D., & Bai, S. M. (2009). Processing and Damping Behaviour of Porous Copper. *Powder Metallurgy*, *52*(2), 145–150. doi:10.1179/003258908X356885

Hartley, P., & Pillinger, I. (2006). Numerical simulation of the forging process. *Computer Methods in Applied Mechanics and Engineering*, *195*(48–49), 6676–6690. doi:10.1016/j.cma.2005.03.013

Hegab, H. A., Darras, B., & Kishawy, H. A. (2018). Towards sustainability assessment of machining processes. *Journal of Cleaner Production*, *170*, 694–703. doi:10.1016/j.jclepro.2017.09.197

Hemmatnezhad, M., Ansari, R., & Rahimi, G. H. (2013). Large-amplitude free vibrations of functionally graded beams by means of a finite element formulation. *Applied Mathematical Modelling*, *37*(18-19), 8495–8504. doi:10.1016/j.apm.2013.03.055

Heo, U., Kim, S. K., Song, M. J., Yang, I. Y., & Im, K. H. (2011). Effectiveness Evaluation for Seat Parts by Seat Belt Anchorage Strength Analysis. *IJPEM*, *12*(6), 1031–1034. doi:10.100712541-011-0137-1

Hess, D. P., & Soom, A. (1991). Normal Vibrations and Friction Under Harmonic Loads: Part I—Hertzian Contacts. *Journal of Tribology*, *113*(1), 80–86. doi:10.1115/1.2920607

Hess, D. P., & Soom, A. (1993). Normal vibrations and friction at a hertzian contact under random excitation: Perturbation solution. *Journal of Sound and Vibration*, *164*(2), 317–326. doi:10.1006/jsvi.1993.1216

Hess, D. P., Soom, A., & Kim, C. H. (1992). Normal vibrations and friction at a Hertzian contact under random excitation: Theory and experiments. *Journal of Sound and Vibration*, *153*(3), 491–508. doi:10.1016/0022-460X(92)90378-B

Hessenberger. (n.d.). Strength Analysis of Seat Belt Anchorage According to ECE R14 and FMVSS. *4th Europeon LS-DYNA Users Conferrnce: Crash / Automotive Applications,* (2), 15-20.

Hettesheimer, Hirzel, & Roß. (2018). Energy savings through additive manufacturing: an analysis of selective laser sintering for automotive and aircraft components. *Energy Efficiency, 11*(5), 1227-1245.

Hilton, D. J., & Moran, P. (1975). Experiments in reducing tractor ride vibration with a cab suspension. *Journal of Agricultural Engineering Research*, *20*(4), 433. doi:10.1016/0021-8634(75)90079-7

Ho, C. Y., & Lin, Z. C. (2003). Analysis and Application of Grey Relation and ANOVA in Chemical-Mechanical Polishing Process Parameters. *International Journal of Advanced Manufacturing Technology*, *21*(1), 10–14. doi:10.1007001700300001

Holmes, L. (2015). Cadillac CT6 Aluminum Construction Saves 198 Pounds Compared to Steel. *Automobile Magazine*. Available from: http://www.automobilemag.com

Ho, M. P., Wanga, H., Lee, J. H., Ho, C. K., Lau, K. T., Leng, J., & Hui, D. (2012). Critical factors on manufacturing processes of natural fiber composites. *Composites. Part B, Engineering*, *43*(8), 3549–3562. doi:10.1016/j.compositesb.2011.10.001

Honda North America. (2014). *The Second-Generation NSX: 2017 Model Media Information*. https://hondanews.com/releases/2017-acura-nsx-press-kit-overview

Hossain, M. M., & Shivakumar, K. (2011). Compression fatigue performance of a fire resistant syntactic foam. *Composite Structures*, *94*(1), 290–298. doi:10.1016/j.compstruct.2011.05.035

Hsia, S. Y., & Chou, Y. T. (2015). Fabrication improvement of cold forging hexagonal nuts by computational analysis and experiment verification. *Mathematical Problems in Engineering*, *2015*, 1–11. doi:10.1155/2015/835038

Huang, R., & Li, P. (2015). Elastic behaviour and failure mechanism in epoxy syntactic foams: The effect of glass microballoon volume fractions. *Composites. Part B, Engineering*, *78*, 401–408. doi:10.1016/j.compositesb.2015.04.002

Huang, Y., & Luo, Q. (2011). A simple method to determine the critical buckling loads for axially inhomogeneous beams with elastic restraint. *Computers & Mathematics with Applications (Oxford, England)*, *61*(9), 2510–2517. doi:10.1016/j.camwa.2011.02.037

Huang, Y., Wang, T., Zhao, Y., & Wang, P. (2018). Effect of axially functionally graded material on whirling frequencies and critical speeds of a spinning Timoshenko beam. *Composite Structures*, *192*, 355–367. doi:10.1016/j.compstruct.2018.02.039

Huang, Y., Yang, L., & Luo, Q. (2013). Free vibration of axially functionally graded Timoshenko beams with non-uniform cross-section. *Composites. Part B, Engineering*, *45*(1), 1493–1498. doi:10.1016/j.compositesb.2012.09.015

Huang, Z., Peng, X., Li, G., & Hao, L. (2019). Response of a Two-Degree-of-Freedom Vibration System with Rough Contact Interfaces. *Shock and Vibration*, *2019*, 1–13. doi:10.1155/2019/1691582

Hudaa, S., & Yang, Y. (2009). A novel approach of manufacturing light-weight composites with polypropylene web and mechanically split cornhusk. *Industrial Crops and Products*, *30*(1), 17–23. doi:10.1016/j.indcrop.2008.12.007

Huetter, J. (2016). *New BMW 5 Series loses 137 lbs.; has aluminum, HS steel, magnesium — but no carbon fiber*. https://www.repairerdrivennews.com/2016/10/14/new-bmw-5-series-loses-137-lbs-has-aluminum-hs-steel-magnesium-but-no-carbon-fiber/

Hwang, Y. J., Kang, S. K., Kim, J. B., Lee, S. S., Choi, C. G., & Son, J. H. (2003). Topology optimal design for lightweight shape of the vehicle mechanical component. *Journal of the Korean Society for Precision Engineering*, *20*(7), 177–184.

Hyland, B. (2012). An Improved and Simple Cable Simulation Model. *Maxim Integrated Application Note*, *5141*, 1–12.

Hyun, S., Pei, L., Molinari, J. F., & Robbins, M. O. (2004). Finite-element analysis of contact between elastic self-affine surfaces. *Physical Review E: Statistical Physics, Plasmas, Fluids, and Related Interdisciplinary Topics*, *70*(2), 12. doi:10.1103/PhysRevE.70.026117 PMID:15447555

Ilo, S., Just, C., & Xhiku, F. (2012). Optimisation of multiple quality characteristics of hardfacing using grey-based Taguchi method. *Materials & Design*, *33*, 459–468. doi:10.1016/j.matdes.2011.04.050

Ingle, R., & Ahuja, B. (2006). An experimental investigation on dynamic analysis of high speed carbon epoxy shaft in aerostatic conical journal bearings. *Composites Science and Technology*, *66*(3-4), 604–612. doi:10.1016/j.compscitech.2005.03.021

InterRegs Ltd. (2012). Federal Motor Vehicle Safety Standard: FMVSS 202 Head Restraints. 77(4), 1-5.

Ismail, M. I. S., Okamoto, Y., & Okada, A. (2016). Thermo-mechanical analysis on thermal deformation of thin stainless steel in laser micro-welding. *International Journal of Manufacturing, Materials, and Mechanical Engineering*, *6*(2), 51–66. doi:10.4018/IJMMME.2016040104

ISO/ASTM. (2015). International standard ISO / ASTM 52900 Additive manufacturing — General principles — Terminology. *International Organization for Standardization*, *5*, 1–26. doi:10.1520/ISOASTM52900-15

Jackson, R. L., & Green, I. (2005). A Finite Element Study of Elasto-Plastic Hemispherical Contact Against a Rigid Flat. *Journal of Tribology*, *127*(2), 343–354. doi:10.1115/1.1866166

Jain, A. K. (1973). Monte Carlo simulation of crosstalk in communication tables. *WSC '73 Proceedings of the 6th conference on Winter simulation*, 844-857.

Jamil, M. S. C., Sheikh, M. A., & Li, L. (2011). A finite element study of buckling and upsetting mechanisms in laser forming of plates and tubes. *International Journal of Manufacturing, Materials, and Mechanical Engineering*, *1*(1), 1–17. doi:10.4018/ijmmme.2011010101

Jana, T., Mitra, A., & Sahoo, P. (2019). Dynamic analysis of elastically and plastically graded spherical and cylindrical contact. *Proceedings of the Institution of Mechanical Engineers, Part J: Journal of Engineering Tribology*. 10.1177/1350650119841756

Jana, T., Mitra, A., & Sahoo, P. (2017). Dynamic contact interactions of fractal surfaces. *Applied Surface Science*, *392*, 872–882. doi:10.1016/j.apsusc.2016.09.025

Jang, I. S., & Min, B. J. (2006). A Study on the Weight Optimization for the Passenger Car Seat Frame Part. *Transaction of the Korean Society of Automotive Engineers*, *14*(5), 155–163.

Jang, J. Y., Yeom, E. S., & Lim, D. B. (Eds.). (2018). *Moving the center of the light weight of automobiles, lightening the material*. Samsung KPMG Issue Monitor.

Jasper, S. P. F. C., & Dautzenberg, J. H. (2002). Material behaviour in conditions similar to metal cutting: Flow stress in the primary shear zone. *Journal of Materials Processing Technology*, *122*(2-3), 322–330. doi:10.1016/S0924-0136(01)01228-6

Jayamani, E., Hamdan, S., Rahman, M. R., Heng, S. K., & Bakri, M. K. B. (2014). Processing and Characterization of Epoxy/Luffa Composites: Investigation on Chemical Treatment of Fibers on Mechanical and Acoustical Properties. *BioResources*, *9*(3), 5542–5546. doi:10.15376/biores.9.3.5542-5556

Jayasinghe, S. N., Qureshi, A. N., & Eagles, P. A. M. (2005). Electrohydrodynamic Jet Processing: An Advanced Electric-Field-Driven Jetting Phenomenon for Processing Living Cells. *Small*, *2*(2), 216–219. doi:10.1002mll.200500291 PMID:17193023

Jeon, D., & Jeon, E. S. (2017). Lightweight Optimization of Headrest Modules Including Flat-Plate Sliding Brackets. *International Journal of Applied Engineering Research*, *12*(8), 1719–1724.

Jeong, C. H., Oh, H. S., & Cheon, S. S. (2016). Strength characterization of automotive composite seat frame by FE analysis based on FMVSS 207. *Journal of Korean Society Precision Engineering*, *16*(12), 158–159.

Jeon, S. J., Kim, M. H., Lee, G. H., & Lee, M. Y. (2010). Experimental Study on Side Impact Characteristics for Automotives Door module. *Journal of the Korean Society for Technology of Plasticity*, *10*(10), 391–394.

Jeysingh, J., Rao, B., & Reddy, A. (2008). Development of a ductile fracture criterion in cold forming. *Materials Science Research Journal*, *2*(3), 4.

Jiang, J., Xu, X., & Stringer, J. (2019). Optimization of process planning for reducing material waste in extrusion based additive manufacturing. *Robotics and Computer-integrated Manufacturing*, *59*, 317–325. doi:10.1016/j.rcim.2019.05.007

Jiang, Z., Jiang, Y., Wang, Y., Zhang, H., Cao, H., & Tian, G. (2019). A hybrid approach of rough set and case-based reasoning to remanufacturing process planning. *Journal of Intelligent Manufacturing*, *30*(1), 19–32. doi:10.100710845-016-1231-0

Jia, X. L., Ke, L. L., Feng, C. B., Yang, J., & Kitipornchai, S. (2015). Size effect on the free vibration of geometrically nonlinear functionally graded micro-beams under electrical actuation and temperature change. *Composite Structures*, *133*, 1137–1148. doi:10.1016/j.compstruct.2015.08.044

Jin, C., & Wang, X. (2015). Accurate free vibration analysis of Euler functionally graded beams by the weak form quadrature element method. *Composite Structures*, *125*, 41–50. doi:10.1016/j.compstruct.2015.01.039

Jin, L., Zhang, C., & Fei, X. (2019). Realizing energy savings in integrated process planning and scheduling. *Processes (Basel, Switzerland)*, *7*(3), 120. doi:10.3390/pr7030120

Jolevski, D., & Bego, O. (2015). Model predictive control of gantry/bridge crane with anti-sway algorithm. *Journal of Mechanical Science and Technology*, *29*(2), 827–834. doi:10.100712206-015-0144-8

Jovic, S., Makragic, S., & Jovanovic, M. (2017). Parameters influence of laser forming on shaped surface by soft computing technique. *Optik (Stuttgart)*, *142*, 451–454. doi:10.1016/j.ijleo.2017.04.089

Jozwik, P., & Bojar, Z. (2007). Analysis of grain size effect on tensile properties of Ni3Al-based intermetallic strips. *Archives of Metallurgy and Materials*, *2*(2), 321–327.

Jung, H. J., Lee, D. S., Kwon, Y. D., & Han, D. H. (2010). A study of optimum design and analysis with D.O.E for automotive seat frame. *Transaction of the Korean Society of Automotive Engineerss*, *10*(11), 2598–2603.

Jung, S. H., Seo, B. T., & Kim, T. H. (2016). A Study of safe and lightweight car seat by applying multi material design. *Proceedings of the Korea Auto-Vehicle Safety Association autumn conference except collection*, 4-5.

Ju, Y. J., & Jeon, E. S. (2017). Lightweight Optimization of Joint Type Seatback including Plastic Materials. *International Journal of Applied Engineering Research*, *12*(15), 5173–5180.

Ju, Y. J., & Jeon, E. S. (2019). Topology Optimization of Seat Cushion Frame With Dissimilar Materials for Lightweight Strength Optimization. *International Journal of Manufacturing, Materials, and Mechanical Engineering*, *9*(1), 1–12. doi:10.4018/IJMMME.2019010101

Kabir, M. M., Wang, H., Lau, K. T., & Cardona, F. (2012). Chemical treatments on plant-based natural fiber reinforced polymer composites: An overview. *Composites. Part B, Engineering*, *43*(7), 2883–2892. doi:10.1016/j.compositesb.2012.04.053

Kadoli, R., Akhtar, K., & Ganesan, N. (2008). Static analysis of functionally graded beams using higher order shear deformation theory. *Applied Mathematical Modelling*, *32*(12), 2509–2525. doi:10.1016/j.apm.2007.09.015

Kajtoch, J. (2007). Strain in the upsetting process. *Metallurgy and Foundry Engineering*, *33*(1), 1–11. doi:10.7494/mafe.2007.33.1.51

Kamyshny, A., & Magdassi, S. (2014). Conductive Nanomaterials for Printed Electronics. *Small*, *10*(17), 3515–3535. doi:10.1002mll.201303000 PMID:25340186

Kanani, A. S., Niknam, H., Ohadi, A. R., & Aghdam, M. M. (2014). Effect of nonlinear elastic foundation on large amplitude free and forced vibration of functionally graded beam. *Composite Structures*, *115*, 60–68. doi:10.1016/j.compstruct.2014.04.003

Kang, K., Yang, D., Park, J., Kim, S., Cho, I., Yang, H.-H., ... Park, I. (2017). Micropatterning of metal oxide nanofibers by electrohydrodynamic (EHD) printing towards highly integrated and multiplexed gas sensor applications. *Sensors and Actuators. B, Chemical*, *250*, 574–583. doi:10.1016/j.snb.2017.04.194

Kanneganti, H., Gopalakrishnan, B., Crowe, E., Al-Shebeeb, O., Yelamanchi, T., Nimbarte, A., ... Abolhassani, A. (2017). Specification of energy assessment methodologies to satisfy ISO 50001 energy management standard. *Sustainable Energy Technologies and Assessments*, *23*, 121–135. doi:10.1016/j.seta.2017.09.003

Karamavruc, A., Shi, Z., & Gunther, D. (2011). *Determination of Empirical Heat Transfer Coefficient via CFD to Predict the Interface Temperature of Continuously Sleeping Clutches*. SAE Technical Paper 2011- 01-0313.

Kar, C., & Mohanty, A. R. (2006). Monitoring gear vibrations through motor current signature analysis and wavelet transform. *Mechanical Systems and Signal Processing*, *20*(1), 158–187. doi:10.1016/j.ymssp.2004.07.006

Kaur, A., Priyahansha, Kumari, S., & Singh, T. (2014). Position control of overhead cranes using fuzzy controllers. *International Journal of Advanced research in Electrical Electronics and Instrumentation Engineering*, *3*(5), 9341–9350.

Kaw, A. K. (2005). *Mechanics of composite materials*. CRC Press. doi:10.1201/9781420058291

Kazimieras Zavadskas, E., Antucheviciene, J., Adeli, H., & Turskis, Z. (2016). Hybrid multiple criteria decision making methods: A review of applications in engineering. *Scientia Iranica*, *23*(1), 1–20. doi:10.24200ci.2016.2093

Kerbrat, O., Mognol, P., & Hascoët, J.-Y. (2011). A new DFM approach to combine machining and additive manufacturing. *Computers in Industry*, *62*(7), 684–692. doi:10.1016/j.compind.2011.04.003

Ketabchi, M., Mohammadi, H., & Izadi, M. (2012). Finite-Element Simulation and Experimental Investigation of Isothermal Backward Extrusion of 7075 Al Alloy. *Arabian Journal for Science and Engineering*, *37*(8), 2287–2296. doi:10.100713369-012-0320-4

Khajavi, Partanen, & Holmström. (2014). Additive manufacturing in the spare parts supply chain. *Computers in Industry*, *65*, 150-63.

Khidir, E. A., Ali, M. O. M., Ali, M. M., Tahir, M. F. M., & Zulkifli, R. (2014). Analysis of Sound Absorption of Date Palm Fibers Based on Flow Resistivity. *Applied Mechanics and Materials*, *471*, 285–290. doi:10.4028/www.scientific.net/AMM.471.285

Kien, N. (2013). Large displacement response of tapered cantilever beams made of axially functionally graded material. *Composites. Part B, Engineering*, *55*, 298–305. doi:10.1016/j.compositesb.2013.06.024

Kim, Y. J. (2009). On-demand electrohydrodynamic jetting with meniscus control by a piezoelectric actuator for ultra-fine patterns. *J. Micromech. Microeng.*, *19*(10). doi:10.1088/0960-1317/19/10/107001

Kim, D. S., Khan, A., Rahman, K., Khan, S., Kim, H. C., & Choi, K. H. (2016). Drop-on-Demand Direct Printing of Colloidal Copper Nanoparticles by Electrohydrodynamic Atomization. *Materials and Manufacturing Processes*, *26*(9), 1196–1201. doi:10.1080/10426914.2011.551956

Kim, H. S., Lee, Y. S., Yang, S. M., & Kang, H. Y. (2016). Structural analysis on variable characteristics of automotive seat frame by FEA. *International Journal of Precision Engineering and Manufacturing-Green Technology*, *3*(1), 75–79. doi:10.100740684-016-0010-x

Kim, H. Y., & Lee, S. K. (1999). Prediction of Impact Characteristics for Automobile Seat. *Transaction of the Korean Society of Automotive Engineers*, *7*(3), 196–204.

Kim, J. S., Roh, H. S., Min, Y. J., Hui, S. H., Shin, S. J., & Jang, I. G. (2017). Lightweight design of a bulk trailer using topology optimization. *Transaction of the Korean Society of Automotive Engineers*, *25*(5), 548–554. doi:10.7467/KSAE.2017.25.5.548

Kim, J., Oh, H., & Kim, S. S. (2008). Electrohydrodynamic drop-on-demand patterning in pulsed cone-jet mode at various frequencies. *Journal of Aerosol Science*, *39*(9), 819–825. doi:10.1016/j.jaerosci.2008.05.001

Kim, K. S., Cho, H. S., Kim, Y. C., & Cho, J. U. (2014). Structural study of automotive seat frame with high tension steel plate using analysis and experiment. *Journal of the Korea Academia-Industrial Cooperation Society*, *15*(1), 27–31. doi:10.5762/KAIS.2014.15.1.27

Kim, K., Lee, B. U., Hwang, G. B., Lee, J. H., & Kim, S. (2010). Drop-on-Demand Patterning of Bacterial Cells Using Pulsed Jet Electrospraying. *Analytical Chemistry*, *82*(5), 2109–2112. doi:10.1021/ac9027966 PMID:20143835

Kim, S. S., Kim, K. S., Choi, D. S., Park, S. H., Kim, S. H., & Cho, J. U. (2012). Structural strength analysis at cushion frame and back frame of automotive seat. *Journal of the Korea Academia-Industrial Cooperation Society*, *13*(11), 4956–4962. doi:10.5762/KAIS.2012.13.11.4956

Kim, Y. C., Hong, J. K., & Jang, G. W. (2011). Lightweight crane design by using topology and shape optimization. *Transactions of the Korean Society of Mechanical Engineers-A*, *35*(7), 821–826. doi:10.3795/KSME-A.2011.35.7.821

Kingkam, W., Li, N., Zhang, H. X., & Zhao, C. Z. (2017). Hot Deformation Behavior of High Strength Low Alloy Steel by Thermo Mechanical Simulator and Finite Element Method. *IOP Conference Series. Materials Science and Engineering*, *205*(1), 012001. doi:10.1088/1757-899X/205/1/012001

Kittusamy, N. K., & Buchholz, B. (2004). Whole-body vibration and postural stress among operators of construction equipment: A literature review. *Journal of Safety Research*, *35*(3), 255–261. doi:10.1016/j.jsr.2004.03.014 PMID:15288559

Klinger, J. F. (2012). *Automotive Body Structure Assembly - Mass & Cost Saving Potential of Laser Welding Compared to Spot Welding* (Master's thesis). KTH Royal Institute of Technology.

Koenigsberger, F. (2013). *Design principles of metal-cutting machine tools*. Elsevier.

Koizumi, M. (1997). FGM activities in Japan. *Composites. Part B, Engineering*, *28B*(1-2), 1–4. doi:10.1016/S1359-8368(96)00016-9

T. Koizumi (Ed.). (2002). The Development of Sound Absorbing Materials Using Natural Bamboo Fibers and Their Acoustic Properties. *Proceedings of the Inter-Noise.*

Koizumi, T., Tsujiuchi, N., & Adachi, A. (2002). The Development of Sound Absorbing Materials Using Natural Bamboo Fibers. *High Performance Structures and Composites*, *59*, 157–166.

Kolakoti, A., & Appa Rao, B. V. (2017). Performance and emission analysis of a naturally aspirated and supercharged IDI diesel engine using palm methyl ester. *Biofuels*, 1–12. doi:10.1080/17597269.2017.1374770

Kolakoti, A., & Appa Rao, B. V. (2019). Effect of fatty acid composition on the performance and emission characteristics of an IDI supercharged engine using neat palm biodiesel and coconut biodiesel as an additive. *Biofuels*, *10*(5), 591–601. doi:10.1080/17597269.2017.1332293

Komvopoulos, K., & Ye, N. (2000). Three-Dimensional Contact Analysis of Elastic-Plastic Layered Media With Fractal Surface Topographies. *Journal of Tribology*, *123*(3), 632–640. doi:10.1115/1.1327583

Kostic, S. C., & Ognjanovic, M. (2007). The Noise Structure of Gear Transmission Units and the Role of Gearbox Walls. *FME Transactions*, *35*, 105–112.

Kotadi, H. R. (2019). Challenges and Opportunities in Remote Laser Welding of Steel to Aluminium. *MATEC Web of Conferences*, *269*, 02012. doi:10.1051/matecconf/201926902012

Kou, S. (2003). *Welding metallurgy* (2nd ed.). John-Wiley and Sons.

Kovačević, M., Madić, M., Radovanović, M., & Rančić, D. (2014). Software prototype for solving multi-objective machining optimization problems: Application in non-conventional machining processes. *Expert Systems with Applications*, *41*(13), 5657–5668. doi:10.1016/j.eswa.2014.03.037

Krog, L., & Sperling, K. (2019). A comprehensive framework for strategic energy planning based on Danish and international insights. *Energy Strategy Reviews*, *24*, 83–93. doi:10.1016/j.esr.2019.02.005

Kubade, P. R., Kulkarni, H. B., & Gavali, V. C. (2020). Property Enhancement of Carbon Fiber-Reinforced Polylactic Acid Composites Prepared by Fused-Deposition Modeling. In *Handbook of Research on Recent Developments in Electrical and Mechanical Engineering* (pp. 455–478). IGI Global. doi:10.4018/978-1-7998-0117-7.ch017

Kübler, F., Böhner, J., & Steinhilper, R. (2015). Resource efficiency optimization of manufacturing processes using evolutionary computation: A turning case. *The 22nd CIRP conference on Life Cycle Engineering*, *29*, 822-827. doi: 10.1016/j.procir.2015.02.208

Ku, H., Wang, H., Pattarachaiyakoop, N., & Trada, M. (2011). A review on the tensile properties of natural fiber reinforced polymer composites. *Composites. Part B, Engineering*, *42*(4), 856–873. doi:10.1016/j.compositesb.2011.01.010

Kuijt-Evers, L. F. M., Krause, F., & Vink, P. (2003). Aspects to improve cabin comfort of wheel loaders and excavators according to operators. *Applied Ergonomics*, *34*(3), 265–271. doi:10.1016/S0003-6870(03)00032-2 PMID:12737927

Kukuryk, M. (2012). Analysis of deformation and damage evolution in hot elongation forging. *Archives of Metallurgy and Materials*, *57*(2), 417–424. doi:10.2478/v10172-012-0041-4

Kukuryk, M. (2018). Numerical analysis of strains and stresses in the hot cogging process. *Journal of Applied Mathematics and Computational Mechanics*, *17*(3), 45–52. doi:10.17512/jamcm.2018.3.04

Kumar, A., Jaiswal, H., Mishra, A., & Patil, P. P. (2014). Mode Shape Vibration Analysis of Truck Transmission Housing Based on FEA. *Proc. of the Intl. Conf. on Advances in Engineering and Technology*, 334-338. DOI: 10.15224/ 978-1-63248-028-6-03-66

Kumar, A., Joshi, H., Saini, A., & Patil, P. P. (2014). Vibration Diagnosis of Heavy Vehicle Truck Transmission Gearbox Casing Using FEA. Composite Materials-Proc. of Intl. Conf. on Mechanical Engineering, 169-177.

Kumar, S., Mitra, A., & Roy, H. (2015) Geometrically nonlinear free vibration analysis of axially functionally graded taper beams. *Engineering Science and Technology, an International Journal, 18*, 579-593.

Kumar, A., Dwivedi, A., Jaiswal, H., & Patil, P. P. (2015). Material Based Vibration Characteristic Analysis of Heavy Vehicle Transmission Gearbox Casing Using Finite Element Analysis (FEA). *Advances in Intelligent Systems and Computing*, *308*, 527–533. doi:10.1007/978-81-322-2012-1_56

Kumar, B. B., Doddamani, M., Zeltmann, S. E., Gupta, N., Gurupadu, S., & Sailaja, R. R. N. (2016). Effect of particle surface treatment and blending method on flexural properties of injection-molded cenosphere/HDPE syntactic foams. *Journal of Materials Science*, *51*(8), 3793–3805. doi:10.100710853-015-9697-2

Kumar, B. B., Doddamani, M., Zeltmann, S. E., Gupta, N., & Ramakrishna, S. (2016). Data characterizing tensile behavior of cenosphere/HDPE syntactic foam. *Data in Brief*, *6*, 933–941. doi:10.1016/j.dib.2016.01.058 PMID:26937472

Kumar, B. B., Doddamani, M., Zeltmann, S. E., Gupta, N., Ramesh, M. R., & Ramakrishna, S. (2016). Processing of cenosphere/HDPE syntactic foams using an industrial scale polymer injection molding machine. *Materials & Design*, *92*, 414–423. doi:10.1016/j.matdes.2015.12.052

Kumar, B. B., Singh, A. K., Doddamani, M., Luong, D. D., & Gupta, N. (2016). Quasi-static and high strain rate compressive response of injection-molded cenosphere/HDPE syntactic foam. *JOM*, *68*(7), 1861–1871. doi:10.100711837-016-1912-3

Kumar, P., Barua, P. B., & Gaindhar, J. L. (2000). Quality optimization (multi-characteristics) through Taguchi's technique and utility concept. *Quality and Reliability Engineering International*, *16*(6), 475–485. doi:10.1002/1099-1638(200011/12)16:6<475::AID-QRE342>3.0.CO;2-0

Kumar, S., Meenu, & Satsangi, P. S. (2013). Multiple-response optimization of turning machining by the taguchi method and the utility concept using uni-directional glass fiber-reinforced plastic composite and carbide (k10) cutting tool. *Journal of Mechanical Science and Technology*, *27*(9), 2829–2837. doi:10.100712206-013-0731-5

Kunchala, P., & Kappagantula, K. (2018). 3D printing high density ceramics using binder jetting with nanoparticle densifiers. *Materials & Design*, *155*, 443–450. doi:10.1016/j.matdes.2018.06.009

Kuo, K. (2011). Simulation and Analysis of the Shift Process for an Automatic Transmission. *World Academy of Science, Engineering and Technology*, *52*, 341–347.

La Fé Perdomo, I., Quiza, R., Haeseldonckx, D., & Rivas, M. (2020). (in press). Sustainability-focused multi-objective optimization of a turning process. *International Journal of Precision Engineering and Manufacturing-Green Technology*. doi:10.100740684-019-00122-4

La Fé, I., Quiza, R., Rivas, M., & Ramtahalsing, V. (2018). Hybrid modelling and optimization of the oblique cutting of AISI 1045 steel. *International Journal of Manufacturing, Materials, and Mechanical Engineering*, *8*(4). doi:10.4018/IJMMME.2018100101

Labella, M., Zeltmann, S. E., Shunmugasamy, V. C., Gupta, N., & Rohatgi, P. K. (2014). Mechanical and thermal properties of fly ash/vinyl ester syntactic foams. *Fuel*, *121*, 240–249. doi:10.1016/j.fuel.2013.12.038

Lalwani, D. I., Mehta, N. K., & Jain, P. K. (2009). Extension of Oxley's predictive machining theory for Johnson and Cook flow stress model. *Journal of Materials Processing Technology*, *209*(12-13), 5305–5312. doi:10.1016/j.jmatprotec.2009.03.020

Lancaster, J. F. (1993). *Metallurgy of welding* (5th ed.). London: Chapman and Hall.

Lauwers, B., Klocke, F., Klink, A., Tekkaya, A. E., Neugebauer, R., & Mcintosh, D. (2014). Hybrid processes in manufacturing. *CIRP Annals-Manufacturing Technology*, *63*(2), 561–583. doi:10.1016/j.cirp.2014.05.003

Leal, R., Barreiros, F. M., Alves, L., Romeiro, F., Vasco, J. C., Santos, M., & Marto, C. (2017). Additive manufacturing tooling for the automotive industry. *International Journal of Advanced Manufacturing Technology*, *92*(5-8), 1671–1676. doi:10.100700170-017-0239-8

Lee, D. J., Park, C. S., Lee, K. T., Kim, S. B., & Kim, H. Y. (2005). Evaluation of Static/Dynamic Structural Strength for Automotive Round Recliner. *Transaction of the Korean Society of Automotive Engineers*, *13*(1), 140–146.

Lee, H. H., & Cho, S. K. (2001). A new fuzzy logic anti-swinging control for Industrial three dimensional overhead cranes. *Proceedings of IEEE International conference on Robotics and Automation*, 2956-2961. doi: 10.1109/ROBOT.2001.933070

Lee, H. H., Liang, Y., & Segura, D. (2006). A sliding-mode anti-swing trajectory control for overhead cranes with high-speed load hoisting. *Journal of Dynamic Systems, Measurement, and Control*, *128*(4), 842–846. doi:10.1115/1.2364010

Lee, I. C., Lee, T. K., & Jang, D. H. (2015). Development Process of Side Impact Beam for Automotive Light- Weighting Door using Sheet Type. *Transactions of Materials Processing*, *24*(2), 130–137. doi:10.5228/KSTP.2015.24.2.130

Lee, J.-Y., An, J., & Chua, C. K. (2017). Fundamentals and applications of 3D printing for novel materials. *Applied Materials*, *7*, 120–133. doi:10.1016/j.apmt.2017.02.004

Lee, K. S. (2014). *A study on improvement of the correlation between frontal sled FEA and test of vehicle seatbelt system.* Department of Mechanical Engineering, Graduate School of Engineering, Ajou University.

Lee, K. T. (2004). *CAE for Development of Round Recliner in Automotive seat System. Department of Mechanical Engineering, Graduate School Kangwon National University.*

Lee, M. W., Kang, D. K., Kim, N. Y., Kim, H. Y., James, S. C., & Yoon, S. S. (2012). A study of ejection modes for pulsed-DC electrohydrodynamic inkjet printing. *Journal of Aerosol Science*, *46*, 1–6. doi:10.1016/j.jaerosci.2011.11.002

Lee, Y. S., Lee, S. U., Van Tyne, C. J., Joo, B. D., & Moon, Y. H. (2011). Internal void closure during the forging of large cast ingots using a simulation approach. *Journal of Materials Processing Technology*, *211*(6), 1136–1145. doi:10.1016/j.jmatprotec.2011.01.017

Leonesio, M., Tosatti, L. M., Pellegrinelli, S., & Valente, A. (2013). An integrated approach to support the joint design of machine tools and process planning. *CIRP Journal of Manufacturing Science and Technology*, *6*(3), 181–186. doi:10.1016/j.cirpj.2013.03.002

Le, V. A., Le, H. X., Nguyen, L., & Phan, M. X. (2019). An efficient adaptive hierarchical sliding mode control strategy using neural networks for 3D overhead cranes. *International Journal of Automation and Computing*, *16*(5), 614–627. doi:10.100711633-019-1174-y

Li, B., & Zhou, Z. (2016). Test method for coupling attenuation of screened twisted pair cable based on mixed-mode S-parameters. In *Proceedings of 2016 Asia-Pacific International Symposium on Electromagnetic Compatibility (APEMC)*. Shenzhen, China: IEEE.

Li, J. L. (2006). On the meniscus deformation when the pulsed voltage is applied. *Journal of Electrostatics*, *64*(1), 44–52. doi:10.1016/j.elstat.2005.04.005

Li, J., Liu, H., Shen, Z., Qian, Q., Zhang, H., & Wang, X. (2016). Formability of micro-gears fabrication in laser dynamic flexible punching. *Journal of Materials Processing Technology*, *234*, 131–142. doi:10.1016/j.jmatprotec.2016.03.018

Li, L., Haghighi, A., & Yang, Y. (2018). A novel 6-axis hybrid additive-subtractive manufacturing process: Design and case studies. *Journal of Manufacturing Processes*, *33*, 150–160. doi:10.1016/j.jmapro.2018.05.008

Lim, J. Y. (2017). *Lightweight Design of the Automotive Seat Back Frame using FEM.* Department of Mechanical Engineering, Graduate School of Chonbuk National University.

Lim, Le, Lu, & Wong. (2016). An overview of 3-D printing in manufacturing, aerospace, and automotive industries. *IEEE Potentials, 35*, 418-22.

Lin, C. L., Lin, J. L., & Ko, T. C. (2002). Optimisation of the EDM Process Based on the Orthogonal Array with Fuzzy Logic and Grey Relational Analysis Method. *International Journal of Advanced Manufacturing Technology*, *19*(4), 271–277. doi:10.1007001700200034

Lines, J. A., Whyte, A. T., & Stayner, A. M. (1989). *Agricultural vehicle suspensions - suspensions for tractor cabs.* Report Presented to the third International Symposium of the International section of the ISSA for research on prevention of Occupational Risks.

Ling, F. F. (1990). Fractals, engineering surfaces and tribology. *Wear, 136*(1), 141–156. doi:10.1016/0043-1648(90)90077-N

Lin, Y. C., Chen, M. S., & Zhong, J. (2008). Numerical simulation for stress/strain distribution and microstructural evolution in 42CrMo steel during hot upsetting process. *Computational Materials Science*, *43*(4), 1117–1122. doi:10.1016/j.commatsci.2008.03.010

Lin, Y. C., & Lee, H. S. (2009). Optimization of machining parameters using magnetic-force-assisted EDM based on gray relational analysis. *International Journal of Advanced Manufacturing Technology*, *42*(11-12), 1052–1064. doi:10.100700170-008-1662-7

Lin, Y., Wang, K., Wu, J., Sun, O., & Chen, K. (2018). Using Deep Learning Algorithm to Modify the Design Parameters for Extending the Operation Life of the Flexible Cables. *Proceedings 67th International Cable and Connectivity Symposium (IWCS 2018). Providence, Rhode Island. Leadership, Innovation and Professional Development*, 139-150.

Liu, D., Yi, J., Zhao, D., & Wang, W. (2004). Swing free transporting of two dimensional overhead crane using sliding mode fuzzy control. *Proceedings of the American control conference*, 1764-1769.

Liu, F.-H., Lee, R.-T., Lin, W.-H., & Liao, Y.-S. (2013). Selective Laser Sintering of Bio-Metal Scaffold. *Procedia CIRP*, *5*, 83–87. doi:10.1016/j.procir.2013.01.017

Liu, X., Yan, X., Li, L., & Zhang, H. (2015). Sound-Absorption Properties of Kapok Fiber Nonwoven Fabrics at Low Frequency. *Journal of Natural Fibers*, *12*(4), 311–322. doi:10.1080/15440478.2014.919891

Li, X. F., Kang, Y. A., & Wu, J. X. (2013). Exact frequency equations of free vibration of exponentially functionally graded beams. *Applied Acoustics*, *74*(3), 413–420. doi:10.1016/j.apacoust.2012.08.003

Li, Y. P., Onodera, E., Matsumoto, H., & Chiba, A. (2009). Correcting the stress-strain curve in hot compression process to high strain level. *Metallurgical and Materials Transactions. A, Physical Metallurgy and Materials Science*, *40*(4), 982–990. doi:10.100711661-009-9783-7

Li, Y., & Guagnqiang, W. (2006). Analysis on Fatigue Life of Rear Suspension Based on Virtual Test rig. *Computer-Aided Engineering*, *15*, 128–130.

Lohar, H., Mitra, A., & Sahoo, S. (2016) Free vibration analysis of axially functionally graded linearly taper beam on elastic foundation. *IOP Conf. Series: Materials Science and Engineering*, *149*. 10.1088/1757-899X/149/1/012130

Lohar, H., Mitra, A., & Sahoo, S. (2016). Geometric nonlinear free vibration of axially functionally graded non-uniform beams supported on elastic foundation. *Curved and Layered Structures*, *3*(1), 223–239. doi:10.1515/cls-2016-0018

Lohar, H., Mitra, A., & Sahoo, S. (2016). Natural frequency and mode shapes of exponential tapered AFG beams on elastic foundation. *International Frontier Science Letters*, *9*, 9–25. doi:10.18052/www.scipress.com/IFSL.9.9

Lohar, H., Mitra, A., & Sahoo, S. (2018). Free vibration of initially deflected axially functionally graded non-uniform Timoshenko beams on elastic foundation. *Romanian Journal of Acoustics and Vibration*, *15*(2), 75–89.

Lohar, H., Mitra, A., & Sahoo, S. (2018). Large amplitude forced vibration analysis of an axially functionally graded tapered beam resting on elastic foundation. *Materials Today: Proceedings*, *5*(2), 5303–5312. doi:10.1016/j.matpr.2017.12.114

Lohar, H., Mitra, A., & Sahoo, S. (2019). Nonlinear response of axially functionally graded Timoshenko beams on elastic foundation under harmonic excitation. *Curved and Layered Structures*, *6*(1), 90–104. doi:10.1515/cls-2019-0008

Long, H., Lord, A. A., Gethin, D. T., & Roylance, B. J. (2003). Operating temperatures of oil-lubricated medium-speed gears: Numerical models and experimental results, Proceedings of the I MECH E part G. *Journal of Aerospace Engineering*, *217*(2), 87–106.

Luan, J., Sun, C., Li, X., & Zhang, Q. (2014). Constitutive model for AZ31 magnesium alloy based on isothermal compression test. *Materials Science and Technology*, *30*(2), 211–219. doi:10.1179/1743284713Y.0000000341

Lu, S. Q., Li, X., Wang, K. L., Liu, S. B., & Fu, M. W. (2013). A method for prediction of unstable deformation in hot forging process by simulation. *Transactions of Nonferrous Metals Society of China*, *23*(12), 3739–3747. doi:10.1016/S1003-6326(13)62924-6

MacDonald, E., & Wicker, R. (2016). Multiprocess 3D printing for increasing component functionality. *Science*, *353*(6307), aaf2093. doi:10.1126cience.aaf2093 PMID:27708075

Mahfouf, M., Kee, C. H., Abbod, M. F., & Linkens, D. A. (2000). Fuzzy logic-based anti-swing control design for overhead cranes. *Neural Computing & Applications*, *9*(1), 38–43. doi:10.1007005210070033

Maji, K. (2019). Experimental investigations on spatial and temporal laser irradiations in pulsed laser forming of thin metal sheets. *Journal of Laser Applications*, *31*(4), 0420101–0420115. doi:10.2351/1.5055842

Maji, K. (2019). Parametric study and optimization of pulsed laser thermal micro-forming of thin sheets. *International Journal of Manufacturing, Materials, and Mechanical Engineering*, *9*(2), 47–61. doi:10.4018/IJMMME.2019040103

Maji, K., & Kumar, G. (2019). Inverse analysis and multi-objective optimization of single-point incremental forming of AA5083 aluminum alloy sheet. *Soft Computing*, 1–17. doi:10.100700500-019-04211-z(.0123456789().,-volV)(0123456789().,-volV)

Maji, K., Pratihar, D. K., & Nath, A. K. (2014). Analysis of pulsed laser bending of sheet metal using neural networks and neuro-fuzzy system. *Proceedings of the Institution of Mechanical Engineers. Part B, Journal of Engineering Manufacture*, *228*(9), 1015–1026. doi:10.1177/0954405414522212

Maji, K., Pratihar, D. K., & Nath, A. K. (2018). Forward and Inverse predictions of deformations in coupling mechanism based laser forming of shaped surfaces. *Journal of Laser Applications*, *30*(3), 03201101–03201110. doi:10.2351/1.5033450

Maji, K., Pratihar, D. K., & Patra, S. (2010). Modelling of electrical discharge machining process using regression analysis, adaptive neuro-fuzzy inference system and genetic algorithm. *International Journal of Data Mining. Modelling and Management*, *2*(1), 75–94.

Majumdar, A., & Bhushan, B. (1990). Role of fractal geometry in roughness characterization and contact mechanics of surfaces. *Journal of Tribology*, *112*(2), 205–216. doi:10.1115/1.2920243

Majumdar, A., & Tien, C. L. (1990). Fractal characterization and simulation of rough surfaces. *Wear*, *136*(2), 313–327. doi:10.1016/0043-1648(90)90154-3

Makdoun, I., Mezzour, G., Carley, K. M., & Kassou, I. (2018). Analyzing the Needs of the Automotive Job Market in Morocco. In *2018 13th International Conference on Computer Science & Education* (pp. 1-6). IEEE. 10.1109/ICCSE.2018.8468854

Malik, A., & Manna, A. (2018). Multi-response optimization of laser-assisted jet electrochemical machining parameters based on gray relational analysis. *Journal of the Brazilian Society of Mechanical Sciences and Engineering*, *40*(3), 148. doi:10.100740430-018-1069-9

Mandelbrot, B. B., & Blumen, A. (1989). Fractal Geometry: What is it, and What Does it do? [and Discussion]. *Proceedings of the Royal Society A: Mathematical, Physical and Engineering Sciences*, *423*(1864), 3–16. 10.1098/rspa.1989.0038

Mandelbrot, B. (1967). How long is the coast of Britain? Statistical self-similarity and fractional dimension. *Science*, *156*(3775), 636–638. doi:10.1126cience.156.3775.636 PMID:17837158

Ma, Q. L., Kahraman, A., Perret-Liaudet, J., & Rigaud, E. (2007). An investigation of steady-state dynamic response of a sphere-plane contact interface with contact loss. *Journal of Applied Mechanics, Transactions ASME*, *74*(2), 249–255. 10.1115/1.2190230

Marko, H., Simon, K., Tomaz, I., Matej, P., Joze, B., & Miran, B. (2014). Turning Parameters Optimization using Particle Swarm Optimization. *Procedia Engineering*, *69*, 670–677. doi:10.1016/j.proeng.2014.03.041

Mårtensson, P., Zenkert, D., & Åkermo, M. (2015). Cost and Weight Efficient Partitioning of Composite Automotive Structures. *Polymer Composites*, *38*(10), 2174–2181. doi:10.1002/pc.23795

Martinsen, K., Hu, S.J., & Carlson, B.E. (2015). Joining of dissimilar materials. *CIRP Annals - Manufacturing Technology, 64*(2), 679-699. doi:10.1016/j.cirp.2015.05.006

Matthews, J., & Talamo, J. D. C. (1965). Ride comfort for tractor operators. *Journal of Agricultural Engineering Research*, *10*(2), 93–108. doi:10.1016/0021-8634(65)90057-0

Mazda News Release. (2003). *Mazda Develops World's First Aluminum Joining Technology Using Friction Heat*. Available from: https://www.mazda.com/publicity/release/200302/0227e.html

McCoy, G., & Douglass, J. G. (2000). *Energy management for motor driven systems*. Department of Energy's Motor Challenge Program.

Mcmaster.com. (n.d.). www.mcmaster.com

Mehtonen, S. V., Karjalainen, L. P., & Porter, D. (2014). Modeling of the high temperature flow behavior of stabilized 12–27wt% Cr ferritic stainless steels. *Materials Science and Engineering A*, *607*, 44–52. doi:10.1016/j.msea.2014.03.124

Meisel, N. A., Elliott, A. M., & Williams, C. B. (2015). A procedure for creating actuated joints via embedding shape memory alloys in PolyJet 3D printing. *Journal of Intelligent Material Systems and Structures*, *26*(12), 1498–1512. doi:10.1177/1045389X14544144

Minfeng, H., & Yingchun, J. (2008). Fatigue life analysis of automobile component based on FEM. *Mechanical Research and Application*, *21*, 57–60.

Mironov, V., Visconti, R. P., Kasyanov, V., Forgacs, G., Drake, C. J., & Markwald, R. R. (2009). Organ Printing: Tissue Spheroids as Building Blocks. *Biomaterials*, *30*(12), 2164–2174. doi:10.1016/j.biomaterials.2008.12.084 PMID:19176247

Mishra, S. (2010). High-speed and drop-on-demand printing with a pulsed electrohydrodynamic jet. *J. Micromech. Microeng.*, *20*(9). doi:10.1088/0960-1317/20/9/095026

Mittal, V., Saini, R., & Sinha, S. (2016). Natural fiber-mediated epoxy composites A review. *Composites. Part B, Engineering*, *99*, 425–435. doi:10.1016/j.compositesb.2016.06.051

Miyasato, H. H., Simionatto, V. G. S., & Junior, M. D. (2011). Study of the gear rattle phenomena in automotive powertrain systems. *Proceedings of COBEM 2011. 21st International Congress of Mechanical Engineering*, 1-10.

Mocera, F., & Nicolini, A. (2017). Multibody simulation of a small size farming tracked vehicle. *Procedia Structural Integrity*, *8*, 118–125. doi:10.1016/j.prostr.2017.12.013

Modi, S. (2016). Material Qualification in the Automotive Industry. Center for Automotive Research.

Mohandas, T., Reddy, G. M., & Naveed, M. (1999). A comparative evaluation of gas tungsten and shielded metal arc welds of a "ferritic" stainless steel. *Journal of Materials Processing Technology*, *94*(2-3), 133–140. doi:10.1016/S0924-0136(99)00092-8

Mohapatra, S. K., & Maity, K. P. (2016). Parametric Optimization of Simulated Extrusion of Square to Square Section Through Linear Converging Die. *IOP Conference Series. Materials Science and Engineering*, *115*(1), 012031. doi:10.1088/1757-899X/115/1/012031

Moraga, C. (2005). Introduction to Fuzzy logic. *Facta Universitatis*, *18*(2), 319–328.

Mostafaei, M. A., & Kazeminezhad, M. (2012). The prediction of hot flow behavior of Al-6%Mg alloy. *Materials Science and Engineering A*, *535*, 216–221. doi:10.1016/j.msea.2011.12.067

Moufki, A., Devillez, A., Dudzinski, D., & Molinari, A. (2004). Thermomechanical modelling of oblique cutting and experimental validation. *International Journal of Machine Tools & Manufacture*, *44*(9), 971–989. doi:10.1016/j.ijmachtools.2004.01.018

Mourtzis, D., Doukas, M., Lalas, C., & Papakostas, N. (2015). Cloud-based integrated shop-floor planning and control of manufacturing operations for mass customisation. *Procedia CIRP*, *33*, 9–16.

Mousa, A. A. (2000). *Control of rotary cranes using fuzzy logic and time-delayed position feedback control* (Master thesis). Virginia Polytechnic Institute and State University, Blacksburg, VA.

Mridha, S., & Baker, T. N. (1994). Crack-free hard surfaces produced by laser nitriding of commercial purity titanium. *Materials Science and Engineering A*, *188*(1-2), 229–239. doi:10.1016/0921-5093(94)90376-X

Mridha, S., Ong, H. S., Poh, L. S., & Cheang, P. (2001). Intermetallic coatings produced by TIG surface melting. *Journal of Materials Processing Technology*, *113*(1-3), 516–520. doi:10.1016/S0924-0136(01)00609-4

Murin, J., Aminbaghai, M., Kutis, V., & Hrabovsky, J. (2013). Modal analysis of the FGM beams with effect of axial force under longitudinal variable elastic winkler foundation. *Engineering Structures*, *49*, 234–247. doi:10.1016/j.engstruct.2012.09.039

Murrenhoff, H., & Wallentowitz, H. (1998). *Fluidtechnik für mobile Anwendungen*. Reihe Fluidtechnik.

Murr, L. E. (2015). Metallurgy of additive manufacturing: Examples from electron beam melting. *Additive Manufacturing*, *5*, 40–53. doi:10.1016/j.addma.2014.12.002

Mwema, F. M., Obiko, J. O., Akinlabi, E. T., Akinlabi, S. A., & Fatoba, O. S. (2019). Effect of punch force on the upsetting deformation process using three-dimensional finite element analysis. *Journal of Physics: Conference Series*, *1378*(December), 032094. doi:10.1088/1742-6596/1378/3/032094

Nacib, L., Pekpe, K. M., & Sakhara, S. (2013). Detecting gear tooth cracks using cepstral analysis in gearbox of helicopters. *International Journal of Advances in Engineering and Technology*, *5*, 139–145.

Naderi, M., & Roostaei, A. A. (2010). Deformation homogeneity in accumulative back extrusion processing of AZ31 magnesium alloy. *Journal of Alloys and Compounds*, *507*(1), 207–214. doi:10.1016/j.jallcom.2010.07.157

Naffakh, M., Diez-Pascual, A. M., Marco, C., Ellis, G. J., & Gomez-Fatou, M. A. (2013). Opportunities and challenges in the use of inorganic fullerene-like nanoparticles to produce advanced polymer nanocomposites. *Progress in Polymer Science*, *38*(8), 1163–1231. doi:10.1016/j.progpolymsci.2013.04.001

Nair, A., & Kumanan, S. (2018). Optimization of size and form characteristics using multi-objective grey analysis in abrasive water jet drilling of Inconel 617. *Journal of the Brazilian Society of Mechanical Sciences and Engineering*, *40*(3), 121. doi:10.100740430-018-1042-7

Na, S. J., & Jeon, E. S. (2017). Lightweight Corrugated Plate for Cargo box by using DOE and Topology Optimization. *International Journal of Applied Engineering Research*, *12*(22), 12564–12569.

Na, S. J., Kim, S. Y., & Jeon, E. S. (2019). Size optimization design of threewheeled motorcycle frame with cargo box. *International Journal of Mechanical Engineering and Technology*, *10*(5), 345–356.

Navacerrada, M. A., Diaz, C., & Fernández, P. (2014). Characterization of a Material Based on Short Natural Fique Fibers. *BioResources*, *9*(2), 3480–3496. doi:10.15376/biores.9.2.3480-3496

Nayfeh, A. H., & Mook, D. T. (1995). *Nonlinear Oscillations*. Wiley; doi:10.1002/9783527617586

Neugebauer, R. (2011). *Mechanical Joining With Self Piercing Solid-Rivets at Elevated Tool Velocities*. International Conference on Material Forming 2011, Belfast, UK.

Ngo, T. D., Kashani, A., Imbalzano, G., Nguyen, K. T. Q., & Hui, D. (2018). Additive manufacturing (3D printing): A review of materials, methods, applications and challenges. *Composites. Part B, Engineering*, *143*, 172–196. doi:10.1016/j.compositesb.2018.02.012

Nguyen, T., Vo, T., & Thai, H. (2013). Static and free vibration of axially loaded functionally graded beams based on the first-order shear deformation theory. *Composites. Part B, Engineering*, *55*, 147–157. doi:10.1016/j.compositesb.2013.06.011

Nguyen, V. D., & Byun, D. (2009). Mechanism of electrohydrodynamic printing based on ac voltage without a nozzle electrode. *Applied Physics Letters*, *94*(17), 173509–3. doi:10.1063/1.3126957

Nicolini, A., Mocera, F., & Somà, A. (2018). Multibody simulation of a tracked vehicle with deformable ground contact model. *Proceedings of the Institution of Mechanical Engineers, Part K: Journal of Multi-Body Dynamics*, *233*(1), 152–162.

Niknam, H., & Aghdam, M. M. (2015). A semi analytical approach for large amplitude free vibration and buckling of nonlocal FG beams resting on elastic foundation. *Composite Structures*, *119*, 452–462. doi:10.1016/j.compstruct.2014.09.023

Njuguna, J. (Ed.). (2016). *Lightweight composite structures in transport: design, manufacturing, analysis and performance*. Woodhead Publishing.

Nothnagle, C., Baptist, J. R., Sanford, J., Lee, W. H., Popa, D. O., & Wijesundara, M. B. J. (2015). EHD printing of PEDOT: PSS inks for fabricating pressure and strain sensor arrays on flexible substrates. Next-Generation Robotics II; and Machine Intelligence and Bio-inspired Computation: Theory and Applications IX. *Proceedings of the Society for Photo-Instrumentation Engineers*, *9494*, 949403–949408. doi:10.1117/12.2177415

O'Connor, H. J., Dickson, A. N., & Dowling, D. P. (2018). Evaluation of the mechanical performance of polymer parts fabricated using a production scale multi jet fusion printing process. *Additive Manufacturing*, *22*, 381–387. doi:10.1016/j.addma.2018.05.035

Obiko, J. O., Mwema, F. M., & Bodunrin, M. O. (2019). Finite element simulation of X20CrMoV121 steel billet forging process using the Deform 3D software. *SN Applied Sciences*, *1*(9), 1044. doi:10.100742452-019-1087-y

Oh, H. T., & Jeon, E. S. (2017). Lightweight Optimal Design of Tailgate Lift Link under Eccentric Loading. *International Journal of Applied Engineering Research*, *12*(17), 6627–6632.

Onses, M. S., Sutanto, E., Ferreira, P. M., Alleyne, A. G., & Rogers, J. A. (2015). Mechanisms, Capabilities, and Applications of High-Resolution Electrohydrodynamic Jet Printing. *Small*, *11*(9), 4237–4266. doi:10.1002mll.201500593 PMID:26122917

Overstam, H., & Jarl, M. (2004). FEM-simulation of drawing out in open die forging. *Steel Research International*, *75*(12), 812–817. doi:10.1002rin.200405847

Pal, A.K., & Mudi, R.K. (2013). An adaptive PD-type FLC and its real implementation to overhead crane control. *International Journal of Emerging Technologies in Computational and Applied Sciences*, *6*(2), 178-183.

Pang, L. J., Shang, X., & Zhang, X. (2018). Influence of Deformation Conditions on the Critical Damage Factor of AZ31 Magnesium Alloy. *Advances in Materials Science and Engineering*, *2018*, 1–8. doi:10.1155/2018/3071495

Paritala, P. K., Manchikatla, S., & Yarlagadda, P. K. D. V. (2017). Digital manufacturing-applications past, current, and future trends. *Procedia Engineering*, *174*, 982–991. doi:10.1016/j.proeng.2017.01.250

Park, J. (2014). Fabrication of a flexible Ag-grid transparent electrode using ac based electrohydrodynamic Jet printing. *J. Phys. D: Appl. Phys.*, *47*(40). doi:10.1088/0022-3727/47/40/405102

Park, H. G., Byun, S.-U., Jeong, H.-C., Lee, J.-W., & Seo, D.-S. (2013). Photoreactive Spacer Prepared Using Electrohydrodynamic Printing for Application in a Liquid Crystal Device. *ECS Solid State Lett.*, *2*(12), R52–R54. doi:10.1149/2.010312ssl

Park, J. U., Hardy, M., Kang, S. J., Barton, K., Adair, K., Mukhopadhyay, D., ... Rogers, J. A. (2007). High-resolution electrohydrodynamic jet printing. *Nature Materials*, *6*(10), 782–789. doi:10.1038/nmat1974 PMID:17676047

Park, J. W., Kook, J. H., & Cho, J. U. (2016). A study on the Durability of Automotive Seat Frame with Aluminium. *Journal of Korean Society of Mechanical Technology*, *18*(4), 505–555.

Park, J. Y., & Khoshnevis, B. (1993). A real-time computer-aided process planning system as a support tool for economic product design. *Journal of Manufacturing Systems*, *12*(2), 181–193. doi:10.1016/0278-6125(93)90017-N

Park, J., Kim, B., Kim, S.-Y., & Hwang, J. (2014). Prediction of drop-on-demand (DOD) pattern size in pulse voltage-applied electrohydrodynamic (EHD) jet printing of Ag colloid ink. *Applied Physics. A, Materials Science & Processing, 117*(4), 2225–2234. doi:10.100700339-014-8650-6

Partwardhan. (n.d.). Simulation and Validation of FMVS 207/210 Using LS-DYNA. *7th International LS-DYNA Users Conferrnce: Simulation Technology*, (2), 9-13.

Patalas-Maliszewska, J., Topczak, M., & Kłos, S. (2020). The Level of the Additive Manufacturing Technology Use in Polish Metal and Automotive Manufacturing Enterprises. *Applied Sciences (Basel, Switzerland), 10*(3), 735. doi:10.3390/app10030735

Patil, A. N., & Kubade, P. R. (n.d.). *A Review on Modifications and Hybridization in Syntactic Foam.* Academic Press.

Patil, A. N., Kubade, P. R., & Kulkarni, H. B. (n.d.). *Effect of Composition on Impact and Flexural Properties of Hybrid Glass Microballoons/Fly Ash Cenosphere Filled Vinyl Ester Matrix Syntactic Foams.* Academic Press.

Patowari, P. K., Saha, P., & Mishra, P. K. (2011). Taguchi analysis of surface modification technique using W-Cu powder metallurgy sintered tools in EDM and characterization of the deposited layer. *International Journal of Advanced Manufacturing Technology, 54*(5-8), 593–604. doi:10.100700170-010-2966-y

Peachey, R. O., Lines, J. A., & Stayner, R. M. (1989). *Agricultural vehicle suspensions - tractor front axle suspension.* Report for the proceedings of the 3rd International Symposium of the International Section of the ISSA for Research on Prevention of Occupational Risks.

Pei, L., Hyun, S., Molinari, J. F., & Robbins, M. O. (2005). Finite element modeling of elasto-plastic contact between rough surfaces. *Journal of the Mechanics and Physics of Solids, 53*(11), 2385–2409. doi:10.1016/j.jmps.2005.06.008

Pennestrì, E., Mariti, L., Valentini, P., & Mucino, V. H. (2012). Efficiency evaluation of gearboxes for parallel hybrid vehicles: Theory and applications. *Mechanism and Machine Theory, 49*, 157–176. doi:10.1016/j.mechmachtheory.2011.10.012

Pentakota Santosh Babji. (2018). *Application of methyl ester of waste cooking oil as a replacement to diesel fuel in IDI engine with supercharging* (Unpublished doctoral dissertation). Andhra University, Visakhapatnam, India.

Peralta, M. E., Marcos, M., & Aguayo, F. (2016). A review of sustainable machining engineering: Optimization process through Triple Bottom Line. *Journal of Manufacturing Science and Engineering, 138*(10), 100801. doi:10.1115/1.4034277

Perret-Liaudet, J. (1997). Subharmonic resonance of order 2 on a sphere-plane contact. *Comptes Rendus de l'Academie Des Sciences Series IIB Mechanics Physics Chemistry Astronomy, 325*(8), 443–448. doi:10.1016/S1251-8069(97)89445-6

Perret-Liaudet, J. (1998). Superharmonic resonance of order two on a sphere-plane contact. *Comptes Rendus de l'Academie Des Sciences Series IIB Mechanics Physics Astronomy, 326*(12), 787–792. doi:10.1016/S1251-8069(99)80028-3

Perret-Liaudet, J., & Rigaud, E. (2003). Experiments and numerical results on non-linear vibrations of an impacting Hertzian contact. Part 2: Random excitation. *Journal of Sound and Vibration, 265*(2), 309–327. doi:10.1016/S0022-460X(02)01267-1

Perret-Liaudet, J., & Rigaud, E. (2006). Response of an impacting Hertzian contact to an order-2 subharmonic excitation : Theory and experiments. *Journal of Sound and Vibration, 296*(1–2), 319–333. doi:10.1016/j.jsv.2006.03.004

Perret-Liaudet, J., & Rigaud, E. (2007). Superharmonic resonance of order 2 for an impacting Hertzian contact oscillator: Theory and experiments. *Journal of Computational and Nonlinear Dynamics, 2*(2), 190–196. doi:10.1115/1.2447549

Perret-Liaudet, J., & Sabot, J. (1999). Vibro-impacts induced by non-linear resonances in hertzian contacts. In *Dynamics of Vibro-Impact Systems* (pp. 251–260). Springer Berlin Heidelberg. doi:10.1007/978-3-642-60114-9_28

Peterson, G. (2013). An Analysis of Impact Performance and Cost Considerations for a Low Mass Multi-Material Body Structure. NHTSA Mass/Size/Safety Workshop, Washington, DC.

Petuelli, G., & Müller, U. (1995). Integration of metal cutting tool resharpening into manufacturing process planning. *Assembly and Task Planning, 1995. Proceedings., IEEE International Symposium On*, 305-310. 10.1109/ISATP.1995.518787

Phadke, M. S. (1995). *Quality Engineering Using Robust Design*. Prentice Hall.

Pohrt, R., & Popov, V. L. (2012). Normal contact stiffness of elastic solids with fractal rough surfaces. *Physical Review Letters, 108*(10), 104301. doi:10.1103/PhysRevLett.108.104301 PMID:22463410

Poltz, J. (2017) Attenuation of screened twisted pairs. *Proceedings 66th International Cable and Connectivity Symposium (IWCS 2017) Orlando, Florida, Industry Leadership, Innovation and Professional Development*, 219-226.

Poveda, R. L., Achar, S., & Gupta, N. (2014). Viscoelastic properties of carbon nanofiber reinforced multiscale syntactic foam. *Composites. Part B, Engineering, 58*, 208–216. doi:10.1016/j.compositesb.2013.10.079

Prabhakaran, S., Krishnaraj, V., Kumar, M. S., & Zitoune, R. (2014). Sound and Vibration Damping Properties of Flax Fiber Reinforced Composites. *Procedia Engineering, 97*, 573–581. doi:10.1016/j.proeng.2014.12.285

Pradhan, K. K., & Chakraverty, S. (2014). Effects of different shear deformation theories on free vibration of functionally graded beams. *International Journal of Mechanical Sciences, 82*, 149–160. doi:10.1016/j.ijmecsci.2014.03.014

Prasada Raju, N. (2015). *Experimental Investigation on Suitability of Pongamia Methyl Ester (PME) and Isobutanol Additive Blends in Indirect Diesel Injection (IDI) Engine replacing Diesel Fuel* (Unpublished doctoral dissertation). Andhra University, Visakhapatnam, India.

Prasada Rao, K., & Appa Rao, B. V. (2017). Experimental Investigation on Cylinder Vibration analysis, Combustion, Emission and Performance of an IDI Engine. *International Journal of Manufacturing, Materials, and Mechanical Engineering, 7*(1), 18–36. doi:10.4018/IJMMME.2017010102

Press, W. H., Teukolsky, S. A., Vetterling, W. T., & Flannery, B. P. (2005). *Numerical Recipes in Fortran 77* (2nd ed.). Cambridge, MA: Press Syndicate.

Pro-E 5.0. (2013). *Designing guide manual*.

Putra, A., Abdullah, Y., Efendy, H., Farid, W. M., Ayob, M. R., & Py, M. S. (2013). Utilizing Sugarcane Wasted Fibers As A Sustainable Acoustic Absorber. *Procedia Engineering, 53*, 632–638. doi:10.1016/j.proeng.2013.02.081

Putra, A., Or, K. H., Selamat, M. Z., Nora, M. J. M., Hassanb, M. H., & Prasetiyo, I. (2018). Sound absorption of extracted pineapple-leaf fibers. *Applied Acoustics, 136*, 9–15. doi:10.1016/j.apacoust.2018.01.029

Qi, L., & Namba, Y. (2011). Precision laser adjustment using CW diode laser. *Precision Engineering, 35*(1), 126–132. doi:10.1016/j.precisioneng.2010.08.002

Qin, H., Dong, J., & Lee, Y.-S. (2017). Fabrication and electrical characterization of multi-layer capacitive touch sensors on flexible substrates by additive e-jet printing. *Journal of Manufacturing Processes, 28*(3), 479–485. doi:10.1016/j.jmapro.2017.04.015

Qin-man, F. (2011). Modal analysis of a truck transmission based on ANSYS. *Fourth International conference on Information and Computing*. doi:10.1109/ICIC.2011.75

Qin, Y., Chen, Y., Ni, X., Shen, Z., Bi, J., & Zhang, X. (2010). Axisymmetric numerical simulation of plastic damage in aluminum alloy induced by long pulsed laser. *Optics and Lasers in Engineering, 48*(3), 361–367. doi:10.1016/j.optlaseng.2009.10.006

Qiu, R., Shi, H., Zhang, K., Tu, Y., Iwamoto, C., & Satonaka, S. (2010). Interfacial Characterization of Joint Between Mild Steel and Aluminum Alloy Welded by Resistance Spot Welding. *Materials Characterization*, *61*(7), 684–688. doi:10.1016/j.matchar.2010.03.015

Quaiyum, A. (2012). *Experimental investigation of automatic transmission fluid (ATF) in an air cooled mini-channel heat exchanger* (M.S. Thesis). Department Mechanical, Automotive, and Materials Engineering, University of Windsor.

Quan, G., Luo, G., Mao, A., Liang, J., & Wu, D. (2014). Evaluation of Varying Ductile Fracture Criteria for 42CrMo Steel by Compressions at Different Temperatures and Strain Rates. *TheScientificWorldJournal*, *2014*, 1–10. doi:10.1155/2014/579328 PMID:24592175

Quiza, R. (2004). *Optimización multiobjetivo del proceso de torneado* (Ph.D. thesis), Universidad de Matanzas, Matanzas, Cuba.

Quiza, R., Beruvides, G., & Davim, J. P. (2014). Modeling and optimization of mechanical systems and processes. In J. P. Davim (Ed.), *Modern mechanical engineering* (pp. 169–198). Springer Berlin Heidelberg. doi:10.1007/978-3-642-45176-8_8

Quiza, R., López-Armas, O., & Davim, J. P. (2012). *Hybrid modeling and optimization of manufacturing: Combining artificial intelligence and finite element method.* Heidelberg, Germany: Springer. doi:10.1007/978-3-642-28085-6

Radha Krishnan, B., & Ramesh, M. (2019). Optimization of machining process parameters in CNC turning process of IS2062 E250 Steel using coated carbide cutting tool. *Materials Today: Proceedings*. doi:10.1016/j.matpr.2019.05.460

Raghavan, N., Dehoff, R., Pannala, S., Simunovic, S., Kirka, M., Turner, J., ... Babu, S. S. (2016). Numerical modeling of heat-transfer and the influence of process parameters on tailoring the grain morphology of IN718 in electron beam additive manufacturing. *Acta Materialia*, *112*, 303–314. doi:10.1016/j.actamat.2016.03.063

Ramachandran, K. K. (2015). Friction Stir Welding of Aluminum Alloy AA5052 and HSLA Steel. *Welding Journal*, *94*(9), 219s–300s.

Ramana Murty Naidu, S. C. V. (2015). *Feasibility Study on Indirect Injection Diesel Engine Replacing Diesel Fuel with Palm Methyl Ester And 1,4- Dioxane Additive Blends* (Unpublished doctoral dissertation). Andhra University, Visakhapatnam, India.

Ranganath Nayak, P. (1972). Contact vibrations. *Journal of Sound and Vibration*, *22*(3), 297–322. doi:10.1016/0022-460X(72)90168-X

Ranjbari, L., Shirdel, A.H., Aslahi-Shahri, M., Anbari, S., Ebrahimi, A., Darvishi, M., Alizadeh, M., Rahmani, R., & Seyedmahmoodian, M. (2015). Designing precision fuzzy controller for load swing of an overhead crane. *Neural Computing and Applications*, *26*(7), 1555-1560.

Rao, K. M. M., Rao, K. M., & Prasad, A. V. R. (2010). Fabrication and testing of natural fiber composites: Vakka, sisal, bamboo and banana. *Materials & Design*, *31*(1), 508–513. doi:10.1016/j.matdes.2009.06.023

Rasmussen, A.E. & Cortese, A.D. (1968). *Dynamic spring rate performance of rolling tyres.* S.A.E. Technical Paper, 680408.

Rasti, J., Najafizadeh, A., & Meratian, M. (2011). Correcting the stress-strain curve in hot compression test using finite element analysis and Taguchi method. *International Journal of ISSI*, *8*(1), 26–33. Retrieved from http://journal.issiran.com/article_6374_44b99e06e066fa7ee24705f05935b39d.pdf

Redelbach, M. (2012). Impact of lightweight design on energy consumption and cost effectiveness of alternative powertrain concepts. *European Electric Vehicle Congress*, Brussels, Belgium.

Regiec, M. (2013). Vehicle lightweight: Manufacturing Impact. *2013 CAR Management Briefing Seminar.*

Reimpell, J., Stoll, H., & Betzler, J. W. (2001). *The Automotive Chassis: Engineering Principles.* Butterworth-Heinemann.

Rezvanian, O., Zikry, M. A., Brown, C., & Krim, J. (2007). Surface roughness, asperity contact and gold RF MEMS switch behavior. *Journal of Micromechanics and Microengineering*, *17*(10), 2006–2015. doi:10.1088/0960-1317/17/10/012

Ribeiro, P. (2004). Non-linear forced vibrations of thin/thick beams & plates by the finite element and shooting methods. *Computers & Structures*, *82*(17-19), 1413–1423. doi:10.1016/j.compstruc.2004.03.037

Richardson & Haylock. (2012). Designer/maker: the rise of additive manufacturing, domestic-scale production and the possible implications for the automotive industry. *Computer-Aided Design & Applications, 2*, 33-48.

Rigaud, E., & Perret-Liaudet, J. (2003). Experiments and numerical results on non-linear vibrations of an impacting Hertzian contact. Part 1: Harmonic excitation. *Journal of Sound and Vibration*, *265*(2), 289–307. doi:10.1016/S0022-460X(02)01262-2

Routara, B. C., Mohanty, S. D., Datta, S., Bandyopadhyay, A., & Mahapatra, S. S. (2010). Optimization in CNC end milling of UNS C34000 medimum leaded brass with multiple surface roughnesses characteristics. *Sadhana*, *35*(5), 319–629. doi:10.100712046-010-0041-x

Rutz, B. H., & Berg, J. C. (2010). A review of the feasibility of lightening structural polymeric composites with voids without compromising mechanical properties. *Advances in Colloid and Interface Science*, *160*(1-2), 56–75. doi:10.1016/j.cis.2010.07.005 PMID:20692640

Rwawiire, S., Tomkova, B., Militky, J., Hes, L., & Kale, B. M. (2017). Acoustic and thermal properties of a cellulose nonwoven natural fabric (barkcloth). *Applied Acoustics*, *116*, 177–183. doi:10.1016/j.apacoust.2016.09.027

Saada, A., & Velex, P. (1995). An extended model for the analysis of the dynamic behavior of planetary trains. *ASME Journal of Mechanical Design*, *117*(2A), 241–247. doi:10.1115/1.2826129

Sabot, J., Krempf, P., & Janolin, C. (1998). Non-linear vibrations of a sphere-plane contact excited by a normal load. *Journal of Sound and Vibration*, *214*(2), 359–375. doi:10.1006/jsvi.1998.1582

Sagirli, A., Azeloglu, C. O., Guclu, R., & Yazici, H. (2011). Self-tuning fuzzy logic control of crane structures against earthquake induced vibrations. *Nonlinear Dynamics*, *64*(4), 375–384. doi:10.100711071-010-9868-2

Sagisaka, Y., Yamashita, K., Yanagihara, W., & Ueta, H. (2014). Micro parts processing using laser cutting and ultra-short-pulse laser peen forming. *Procedia Engineering*, *81*, 2367–2372. doi:10.1016/j.proeng.2014.10.335

Saha, A., & Majumder, H. (2018). Performance analysis and optimization in turning of ASTM A36 through process capability. index. *Journal of King Saud University-Engineering Sciences*, *30*(4), 377–383. doi:10.1016/j.jksues.2016.11.002

Sahoo, P., & Ghosh, N. (2007). Finite element contact analysis of fractal surfaces. *Journal of Physics. D, Applied Physics*, *40*(14), 4245–4252. doi:10.1088/0022-3727/40/14/021

Salim, D. M., Sanjay, D. Y., Shinde, D., & Deshpande, G. (2013). Importance of Federal Motor Vehicle Safety Standards 207,210 in Occupant Safety - A Case Study. *Procedia Engineering*, *64*, 1099–1108. doi:10.1016/j.proeng.2013.09.188

Sanadi, A. R., Prasad, S. V., & Rohatgi, P. K. (1986). Sunhemp fiber-reinforced polyester. *Journal of Materials Science*, *21*(12), 4299–4304. doi:10.1007/BF01106545

Sankar, B. R., Umamaheswarrao, P., Srinivasulu, V., & Chowdari, G. K. (2015). Optimization of milling process on jute polyester composite using Taguchi based grey relational analysis coupled with principle component analysis. *Materials Today: Proceedings*, *2*(4-5), 2522–2531. doi:10.1016/j.matpr.2015.07.197

Santhi, L. R., & Beebi, L. (2014). Position control and anti-swing control of overhead crane using LQR. *International Journal of Scientific and Engineering Research*, *3*(8), 26–40.

Sarkar, K., & Ganguli, R. (2014). Closed-form solutions for axially functionally graded timoshenko beams having uniform cross-section and ðxed–ðxed boundary condition. *Composites. Part B, Engineering*, *58*, 361–370. doi:10.1016/j.compositesb.2013.10.077

Sarvankar, S. G., & Yewale, S. N. (2019). Additive Manufacturing in Automotive Industry. *Internationl Journal of Research in Aeronautical and Mechanical Engineering*, *7*(4), 1–10.

Savaresi, S. M., Silani, E., & Bittanti, S. (2005). Acceleration-driven-damper (ADD): An Optimal Control Algorithm For Comfort-Oriented Semiactive Suspensions. *Journal of Dynamic Systems, Measurement, and Control*, *127*(2), 218–229. doi:10.1115/1.1898241

Sayer, F., Bürkner, F., Blunk, M., Wingerde, A. M., & Busmann, H. G. (2009). Influence of Loads and Environmental Conditions on Material Properties over the Service Life of Rotor Blades. *DEWI Magazin*, *34*, 24–31.

Schey, J. A. (2000). *Introduction to manufacturing processes*. New York: McGraw-Hill.

Schiffl, A., & Easton, M. A. (2009). Influence of SiC particles on the grain refinement of an Mg-Al alloy. *Materials Science Forum*, *618-619*, 445–448. doi:10.4028/www.scientific.net/MSF.618-619.445

Seifert, T., Sowade, E., Roscher, F., Wiemer, M., Gessner, T., & Baumann, R. R. (2015). Additive Manufacturing Technologies Compared: Morphology of Deposits of Silver Ink Using Inkjet and Aerosol Jet Printing. *Industrial & Engineering Chemistry Research*, *54*(2), 769–779. doi:10.1021/ie503636c

Seksaria, D. (2015). A Cost Effective Solution for Automotive Doors - Alcoa's Thin Door Technology. *International Automotive Body Congress, Proceedings of Automotive Materials: Interior/Exterior & Safety Systems*, *3*, 38-53.

Semmar, N., Talbi, A., Mikikian, M., Stolz, A., Melhem, A., & Meneses, D. S. (2019, November). ((Accepted). Microspikes formed on mesoporous silicon by UV picosecond laser irradiation. *Applied Surface Science*, *20*. doi:10.1016/j.apsusc.2019.144820

Sevkat, E., & Tumer, H. (2013). Residual torsional properties of composite shafts subjected to impact loadings. *Materials & Design*, *51*, 956–967. doi:10.1016/j.matdes.2013.05.004

Shahba, A., Attarnejad, R., Marvi, T. M., & Hajilar, S. (2011). Free vibration and stability analysis of axially functionally graded tapered timoshenko beams with classical and non-classical boundary conditions. *Composites. Part B, Engineering*, *42*(4), 801–808. doi:10.1016/j.compositesb.2011.01.017

Shahba, A., & Rajasekaran, S. (2012). Free vibration and stability of tapered euler–bernoulli beams made of axially functionally graded materials. *Applied Mathematical Modelling*, *36*(7), 3094–3111. doi:10.1016/j.apm.2011.09.073

Shahid, Z., Muhammad, A., Khattak, M., & Nasir, T. (2018). Effects of welding on the microstructural properties of AISI 430 Ferritic Stainless Steel. *Journal of Advanced Research in Materials Science*, *44*(1), 25–32.

Shankar V., (2016). *Lightweight Options and Forecast of Material Types: Global snapshot Of Lightweight Solutions*. Frost & Sullivan.

Shcherbkov, V. S., Korytov, M. S., & Shershneva, E. O. (2016). Influence of an obstacle on load displacement by a gantry crane. *Russian Engineering Research*, *36*(3), 194–197. doi:10.3103/S1068798X16030151

Shen, H., Gong, C., Hu, J., & Yao, Z. (2012). Numerical and experimental study on bi-direction deformations in laser micro forming of two-bridge actuators. *International Journal of Machine Tools & Manufacture*, *54*, 66–72. doi:10.1016/j.ijmachtools.2011.11.006

Shen, H., Hu, J., & Yao, Z. (2015). The interactive effects on deformation behavior in laser thermal adjustment of two-bridge actuators. *International Journal of Machine Tools & Manufacture*, *90*, 1–7. doi:10.1016/j.ijmachtools.2014.12.001

Shen, H., Hu, J., Yu, X., & Yao, Z. (2012). The effects of cut-out on stress distribution and deformations in laser thermal adjustment. *International Journal of Machine Tools & Manufacture*, *62*, 46–52. doi:10.1016/j.ijmachtools.2012.06.006

Shen, H., & Yao, Z. (2011). Numerical and experimental investigation of cooling time in laser micro-adjustment of two-bridge actuators. *Applied Thermal Engineering*, *31*(8-9), 1447–1456. doi:10.1016/j.applthermaleng.2011.01.013

Shen, Z., Liu, H., Wang, X., & Wang, C. (2019). Surface degradation of micro-mold in micro-scale laser dynamic forming and its effects on workpiece. *Optics & Laser Technology*, *117*, 114–125. doi:10.1016/j.optlastec.2019.03.047

Shen, Z., Zhang, J., Liu, H., Wang, X., & Ma, Y. (2019). Reducing the rebound effect in micro-scale laser dynamic flexible forming through using plasticine as pressure-carrying medium. *International Journal of Machine Tools & Manufacture*, *141*, 1–18. doi:10.1016/j.ijmachtools.2019.03.004

Shiga, T. (2014). Challenge for Light Weight. *2014 CAR Management Briefing Seminar*.

Shin, K. H. (2016). Development of Mg seat FEM using Accelerated Durability test. Department of Mechanical Engineering, Graduate School of Kongju National University.

Shin, D. S., & Jeon, E. S. (2017). The s-s curve approximation for GFRP tensile specimen by using inverse method and optimization. *Journal of Engineering and Applied Sciences (Asian Research Publishing Network)*, *12*(20), 5702–5707.

Shin, D. S., Song, Y. J., & Jeon, E. S. (2018). Lightweight design of seat cushion extension modules using the properties of plastic and HCA-SIMP. *International Journal of Mechanical Engineering and Technology*, *9*(5), 624–632.

Shokry, A., Gowid, S., Kharmanda, G., & Mahdi, E. (2019). Constitutive Models for the Prediction of the Hot Deformation Behavior of the 10%Cr Steel Alloy. *Materials (Basel)*, *12*(18), 1–18. doi:10.3390/ma12182873 PMID:31492023

Shor, L., Güçeri, S., Chang, R., Gordon, J., Kang, Q., Hartsock, L., & Sun, W. (2009). Precision Extruding Deposition (PED) Fabrication of Poly-Capro-Lactone (PCL) Scaffolds for Bone Tissue Engineering. *Bio Fabrication, 1*(1).

Shree, Varsha, Dhinakaran, Rajkumar, Bupathi Ram, Vijayakumar, & Sathish. (2019). Effect of 3D printing on supply chain management. *Materials Today: Proceedings*.

Shuldiner, H. (2013). *Lower Cost Key to Magnesium's lightweight Uses*. Available from: http://wardsauto.com/vehicles-amp-technology/lower-cost-key-magnesium-s-lightweight-uses

Shunmugesh, K., & Panneerselvam, K. (2017). Grey Relational Analysis Based Optimization Of Multiple Responses in Drilling Of Carbon Fiber-Epoxy Composites. *Materials Today: Proceedings*, *4*(2), 2861–2870. doi:10.1016/j.matpr.2017.02.166

Siefkes, T. (1989). *Dynamische Kennwertevon AS-Reitenim PrOfstands-versuchund im Freiland (Characteristics of agricultural tyres -laboratory and field test results)*. VDI/MEG Kolloquium Landtechnik.

Simanjalam, K. R. (2012). Position and anti-sway control for gantry crane system using fuzzy tuned PID controller. Project report submitted to University Teknologi Malaysia.

Simsek, M., & Cansız, S. (2012). Dynamics of elastically connected double-functionally graded beam systems with different boundary conditions under action of a moving harmonic load. *Composite Structures*, *94*(9), 2861–2878. doi:10.1016/j.compstruct.2012.03.016

Simsek, M., & Reddy, J. N. (2013). A unified higher order beam theory for buckling of a functionally graded microbeam embedded in elastic medium using modified couple stress theory. *Composite Structures*, *101*, 47–58. doi:10.1016/j.compstruct.2013.01.017

Singh, A. M., & Ha, Q. P. (2019). Fast terminal sliding control application for second-order underactuated systems. *International Journal of Control, Automation, and Systems*, *17*(8), 1884–1898. doi:10.100712555-018-0785-3

Singh, R. (2000). Dynamic design of automotive systems-Engine mounts and structural joints. *Dynamic Design of Automotive Systems*, *25*(3), 319–330. doi:10.1007/BF02703548

Smock, D. (2010, November 16). More Composites for Cars? *Design News*, 132–135.

Smoczek, J., & Szpytko, J. (2008). A Mechatronics approach in Intelligent control systems of the overhead travelling cranes prototyping. *Information Technology and Control*, *37*(2), 154–158.

Solidedge. (2006). *Version 19.0.*

Solihin, M. I., Wahyudi, & Legowo, A. (2007). Fuzzy-tuned PID Anti-swing control of Automatic Gantry crane. *Journal of Vibration and Control*, *16*(1), 127–145. doi:10.1177/1077546309103421

Solihin, M. I., & Wahyudi, N. A. (2009). Sensorless anti-swing control of automatic gantry crane using dynamic recurrent neural network-based soft sensor. *International Journal of Intelligent Systems Technologies and Applications*, *6*(1), 112–127. doi:10.1504/IJISTA.2009.022691

Somà, A., Bosso, N., Macagno, L., & Viglietti, E. (2006). *Analisi sperimentale e numerica del comfort di un sollevatore telescopico*. XXXV AIAS Convegno Nazionale.

Song, H. D. (2015). Light-Weight Design and Analysis in car Seat Frame. Department of Carbon Fusion Engineering, Graduate School Jeonju University.

Song, S. I., Cha, I. R., Lee, K. H., & Park, G. J. (2002). Optimization of the Automotive Side Door Impact Beam Considering Static Requirement. *Transaction of the Korean Society of Automotive Engineers*, *10*(3), 176–184.

Song, Y. J., & Jeon, E. S. (2017). Optimal Design of Lightweight Seat Extension Equipment Using Topology Optimization and Design of Experiment. *International Journal of Applied Engineering Research*, *12*(9), 1855–1859.

Stanley, W. A. (1950). *Resistance Welding*. New York: McGraw-Hill Book Company.

Stevens, M. (2016). *Mixed Materials Solutions: Alternative Materials for Door Assemblies*. https://www.cargroup.org/wp-content/uploads/2017/02/Mixed-Materials-Solutions-Alternative-Materials-for-Door-Assemblies.pdf

Straub, B., & Daimler, A. G. (2018). Automotive Displays-Increasing and Challenging Market. In *2018 25th International Workshop on Active-Matrix Flatpanel Displays and Devices* (pp. 1-4). IEEE. 10.23919/AM-FPD.2018.8437443

Su, H., Banerjee, J. R., & Cheung, C. W. (2013). Dynamic stiffness formulation and free vibration analysis of functionally graded beams. *Composite Structures*, *106*, 854–862. doi:10.1016/j.compstruct.2013.06.029

Sukumar Puhan, N. (2010). Effect of biodiesel unsaturated fatty acid on combustion characteristics of a DI compression ignition engine. *Biomass and Bioenergy*, *34*(8), 1079–1088. doi:10.1016/j.biombioe.2010.02.017

Suresh, G., & Narayana, K. L. (2016). *A review on fabricating procedures in rapid prototyping. 3D Printing: Breakthroughs in Research and Practice*. doi:10.4018/IJMMME.2016040102

Suresh, Narayana, & Mallik. (2019). Hardness, Wear and Corrosion Properties of Co-Cr-W Alloy Deposited with Laser Engineered Net Shaping in Medical Applications 1952. *International Journal of Innovative Technology and Exploring Engineering*, *8*.

Suresh, G., Narayana, K. L., Kedar Mallik, M., Srinivas, V., & Jagan Reddy, G. (2018). Processing & Characterization of LENS TM Deposited Co-Cr-W Alloy for Bio-Medical Applications. *276| International. Journal of Pharmacy Research*, *10*(1), 276–285.

Suresh, S., & Mortensen, A. (1998). *Fundamentals of Functionally Graded Materials*. London: IOM Communications Limited.

Sutton, A. T., Kriewall, C. S., Leu, M. C., & Newkirk, J. W. (2017). Powder characterisation techniques and effects of powder characteristics on part properties in powder-bed fusion processes. *Virtual and Physical Prototyping*, *12*(1), 3–29. doi:10.1080/17452759.2016.1250605

Swetha, C., & Kumar, R. (2011). Quasi-static uni-axial compression behaviour of hollow glass microspheres/epoxy based syntactic foams. *Materials & Design*, *32*(8-9), 4152–4163. doi:10.1016/j.matdes.2011.04.058

Tagliavia, G., Porfiri, M., & Gupta, N. (2010). Analysis of flexural properties of hollow-particle filled composites. *Composites. Part B, Engineering*, *41*(1), 86–93. doi:10.1016/j.compositesb.2009.03.004

Tang, H. (2017). *Joining Processes for Body Assembly, Automotive Vehicle Assembly Processes and Operations Management*. SAE International. https://www.sae.org/publications/books/content/r-456/ doi:10.4271/R-456

Tang, X., Zhang, X., Zhang, H., Zhuang, X., & Yan, X. (2018). Corn husk for noise reduction: Robust acoustic absorption and reduced thickness. *Applied Acoustics*, *134*, 60–68. doi:10.1016/j.apacoust.2018.01.012

Taylor, F. W. (1907). On the art of cutting metals. *Transactions of the ASME*, *28*, 310-350.

Taylor, G. (1964). Disintegration of Water Drops in an Electric Field. *Proceedings of the Royal Society of London. Series A*, *280*(1382), 383–397. doi:10.1098/rspa.1964.0151

The Metals Databook. (2008). (4th ed.). Tata: McGraw- Hill.

Thilagavathi, G., Pradeep, E., Kannaian, T., & Sasikala, L. (2010). Development of Natural Fiber Nonwovens for Application as Car Interiors for Noise Control. *Journal of Industrial Textiles*, *39*(3), 276–278. doi:10.1177/1528083709347124

Thomas, C. R. (1973). Syntactic carbon foams. *Materials Science and Engineering*, *12*(5-6), 219–233. doi:10.1016/0025-5416(73)90033-5

Tian, J., & Xie, Z. (2008). Dynamic contact stiffness of vibrating rigid sphere contacting semi-infinite transversely isotropic viscoelastic solid. *Guti Lixue Xuebao*, *21*(6), 580–588. doi:10.100710338-008-0870-x

Tian, Q., & Yu, D. (2016). Preparation and properties of polymer microspheres filled epoxy composite films by UV-curable polymerization. *Materials & Design*, *107*, 221–229. doi:10.1016/j.matdes.2016.06.035

Tofail, Koumoulos, Bandyopadhyay, Bose, O'Donoghue, & Charitidis. (2018). Additive manufacturing: scientific and technological challenges, market uptake and opportunities. *Materials Today, 21*, 122-37.

Torrado, A. R., Shemelya, C. M., English, J. D., Lin, Y., Wicker, R. B., & Roberson, D. A. (2015). Characterizing the effect of additives to ABS on the mechanical property anisotropy of specimens fabricated by material extrusion 3D printing. *Additive Manufacturing*, *6*, 16–29. doi:10.1016/j.addma.2015.02.001

Tosun, N. (2006). Determination of optimum parameters for multi-performance characteristics in drilling by using grey relational analysis. *International Journal of Advanced Manufacturing Technology, 28*(5-6), 450–455. doi:10.100700170-004-2386-y

Trabia, M. B., Renno, J., & Moustafa, K. (2006). A General anti-swing fuzzy controller of an overhead crane with hoisting. *Proceedings of IEEE International Conference on Fuzzy systems*, 627-634. 10.1109/FUZZY.2006.1681777

Tranter, W.H., Shanmugan, K.S., Rappaport, T. S., & Kosbar, K.L. (2003). Modeling and simulation of waveform channels. In *Principles of Communication Systems Simulation with Wireless Applications* (pp. 529-582). Academic Press.

Tsai, W., Chen, H., Leu, J., Chang, Y., & Lin, T. W. (2013). A product-mix decision model using green manufacturing technologies under activity-based costing. *Journal of Cleaner Production, 57*, 178–187. doi:10.1016/j.jclepro.2013.04.011

Tschätsch, H. (2008). *Applied machining technology* (8th ed.). Heidelberg, Germany: Springer.

Tuma, J. (2009). Gearbox Noise and Vibration Prediction and Control. *International Journal of Acoustics and Vibration, 14*(2), 1–11. doi:10.20855/ijav.2009.14.2242

Tunçel, O. (2019). Pulsed Nd:YAG laser welding of similar and dissimilar DP steel sheets. Alternative Energy Sources, Materials & Technologies (AESMT'19), 1.

Turkseven, C. H., & Ertek, G. 2003, Simulation modeling for quality and productivity in Steel cord manufacturing. *Proceedings of the 2003 Winter Simulation Conference*, 1225-1229. 10.1109/WSC.2003.1261554

Ueda, Y., Lida, K., Saito, M. & Okamoto, A. (1991) Finite element model and residual stress calculation for multi-pass welded joint between a plate and penetrating pipe. *Modeling of casting welding and advanced solidification processes*, 219-227.

UNI EN 13059-2004, (2004). Safety in industrial trucks: test methods and experimental measurements.

Unknown. (2010). *TEST METHOD 210: Seat Belt Anchorages*. Transport Canada.

US Department of Energy (DOE) Vehicle Technologies Office (VTO). (2015). *Lightweight Materials for Cars and Trucks*. Available from: https://energy.gov/eere/vehicles/vehicle-technologies-office-lightweight-materials-cars-and-trucks

US Department of Energy (DOE). (2016). *Physical and Chemical Properties of Gasoline, Alternative Fuels Data Center (AFDC)*. Available from: https://www.fueleconomy.gov/feg/contentIncludes/co2_inc.htm

US Department of Transportation (DOT). (2018). *U.S. DOT and EPA Propose Fuel Economy Standards for MY 2021-2026 Vehicles*. https://www.transportation.gov/briefing-room/dot4818

US Environmental Protection Agency (EPA), (2019). *The 2018 EPA Automotive Trends Reports: Greenhouse Gas Emissions, Fuel Economy, and Technology since 1975*. EPA-420-R-19-002.

US Navy, . (1976). *Wire-rope analysis and design data* (Vol. 2). Washington, DC: Naval See Systems Command.

Vandi, G., Cavina, N., Corti, E., Mancini, G., Moro, D., Ponti, F., & Ravaglioli, V. (2014). Development of a software in the loop environment for automotive powertrain systems. *Energy Procedia, 45*, 789–798. doi:10.1016/j.egypro.2014.01.084

Vector Babu, Prasada Rao, & Appa Rao & Kolakoti. (2016). Experimental investigation on combustion, emissions, performance and cylinder vibration analysis of an IDI engine with RBME along with isopropanol as an additive. *Biofuels, 8*(3), 307–321. doi:10.1080/17597269.2016.1226723

Vedulla, G. (2017). *Experimental Analysis of Supercharged Single Cylinder IDI Engine using Beef Tallow Methyl Ester with Hydrated Ethyl Alcohol as an Additive* (Unpublished doctoral dissertation). Andhra University, Visakhapatnam, India.

Velex, P., & Flamand, L. (1996). Dynamic response of planetary trains to mesh parametric excitations. *Journal of Mechanical Design*, *118*(1), 7–14. doi:10.1115/1.2826860

Velinsky, S. A. (1988). Design and mechanics of multi-lay wire strands. *J. Mech. Trans. Automat. Des. Trans. ASME*, *110*, 152–160.

Velinsky, S. A. (1989). On the design of wire rope. *J. Mech. Transm. Automat. Des. Trans. ASME*, *111*, 382–388.

Vijayakumar, K., Prabhaharan, G., Asokan, P., & Saravanan, R. (2003). Optimization of multi-pass turning operations using ant colony system. *International Journal of Machine Tools & Manufacture*, *43*(15), 1633–1639. doi:10.1016/S0890-6955(03)00081-6

Villafuerte, J. C., Pardo, E., & Kerr, H. W. (1990). The effect of alloy composition and welding conditions on columnar-equiaxed transitions in ferritic stainless steel gas-tungsten arc welds. *Metallurgical Transactions. A, Physical Metallurgy and Materials Science*, *21*(7), 2009–2019. doi:10.1007/BF02647249

Vinogradov, A. A. (2014). On the mechanism of chip formation in oblique cutting of metals. *Journal of Superhard Materials*, *36*(6), 428–434. doi:10.3103/S1063457614060094

Vollertsen, F., & Sakkiettibutra, J. (2010). Different types to use laser as a forming tool. *Physics Procedia*, *5*, 193–203. doi:10.1016/j.phpro.2010.08.044

Wahyudi & Jalani. (2006). Robust fuzzy logic controller for an intelligent gantry crane system. *Proceedings of 1st International conference on Industrial and Information systems*, 497-502. doi: 10.1109/ICIIS.2006.365778

Wang, J., Goyanes, A., Gaisford, S., & Basit, A. W. (2016). Stereolithographic (SLA) 3D printing of oral modified-release dosage forms. *International Journal of Pharmaceutics*, *503*(1–2), 207–212. doi:10.1016/j.ijpharm.2016.03.016 PMID:26976500

Wang, J., Zheng, J., & Yang, A. (2012). An Analytical Study of Bifurcation and Chaos in a Spur Gear Pair with Sliding Friction. *Procedia Engineering*, *31*, 563–570. doi:10.1016/j.proeng.2012.01.1068

Wang, Z. G., Wong, Y. S., Rahman, M., & Sun, J. (2006). Multi-objective optimization of high-speed milling with parallel genetic simulated annealing. *International Journal of Advanced Manufacturing Technology*, *31*(3-4), 209–218. doi:10.100700170-005-0191-x

Weiler, J. P. (2019). A review of magnesium die-castings for closure applications. *Journal of Magnesium and Alloys*, *7*(2), 297–304. doi:10.1016/j.jma.2019.02.005

Weng, J., Zhuang, K., Chen, D., Guo, S., & Ding, H. (2017). An analytical force prediction model for turning operation by round insert considering edge effect. *International Journal of Mechanical Sciences*, *128-129*, 168–180. doi:10.1016/j.ijmecsci.2017.04.018

Weng, J., Zhuang, K., Zhu, D., Guo, S., & Ding, H. (2018). An analytical model for the prediction of force distribution of round -insert considering edge effect and size effect. *International Journal of Mechanical Sciences*, *138-139*, 86–98. doi:10.1016/j.ijmecsci.2018.01.024

White, N. M., & Ansell, M. P. (1983). Straw-reinforced polyester composites. *Journal of Materials Science*, *18*(5), 1549–1556. doi:10.1007/BF01111977

Williams, R. A. (1997). Automotive active suspension Part 1: Basic principles. *Proceedings of the Institution of Mechanical Engineers. Part D, Journal of Automobile Engineering*, *211*(6), 415–426. doi:10.1243/0954407971526551

Winner, D. (2016). Lightweight Innovations: Unsung Heroes of the Weight-loss Battle. Wards Auto, 1-18.

Xianga, H. F., Wanga, D., Liua, H. C., Zhaoa, N., & Xu, J. (2013). Investigation on Sound Absorption Properties of Kapok Fibers. *Chinese Journal of Polymer Science, 31*(3), 521–529. doi:10.100710118-013-1241-8

Xiao, H., Brennan, M. J., & Shao, Y. (2011). On the undamped free vibration of a mass interacting with a Hertzian contact stiffness. *Mechanics Research Communications, 38*(8), 560–564. doi:10.1016/j.mechrescom.2011.07.012

Xiao, H., Shao, Y., & Brennan, M. J. (2015). On the contact stiffness and nonlinear vibration of an elastic body with a rough surface in contact with a rigid flat surface. *European Journal of Mechanics. A, Solids, 49*, 321–328. doi:10.1016/j.euromechsol.2014.08.005

Xiao, H., Sun, Y., & Chen, Z. (2019). Fractal modeling of normal contact stiffness for rough surface contact considering the elastic–plastic deformation. *Journal of the Brazilian Society of Mechanical Sciences and Engineering, 41*(1), 11. doi:10.100740430-018-1513-x

Yaghoobi, H., & Fereidoon, A. (2010). Influence of neutral surface position on deflection of functionally graded beam under uniformly distributed load. *World Applied Sciences Journal, 10*, 337–341.

Yang, Y. S., Jeon, E. S., & Park, D. H. (2018). Optimization of Design Parameters of Spiral Spring for Active Headrest Deployment. *MATEC Web of Conferences IC4M & ICDES 2018, 167*, 1-8. doi:10.1051/matecconf/201816702017

Yang, C., & Lin, X. (2016). The Forming Analysis of Two-stage Extrusion for 1010 Fastener. *Journal of Mechanical Engineering and Automation, 6*(3), 43–50. doi:10.5923/j.jmea.20160603.01

Yang, Y. S., & Jeon, E. S. (2016). Tailgate Hinge Stiffness Design using Topology Optimization. *International Journal of Applied Engineering Research, 11*(19), 9776–9781.

Yang, Y. S., & Jeon, E. S. (2017). Structural Design of Automatic Seatback Table. *International Journal of Applied Engineering Research, 12*(19), 8901–8906.

Yang, Y. S., & Jeon, E. S. (2018). Design Parameters and Dynamic Analysis of Unfolded Headrests for Vehicles. *International Journal of Mechanical & Mechatronics Engineering, 18*(3), 1–5.

Yang, Y. S., Kim, Y. S., & Jeon, E. S. (2018). Design for headrest including guides slot to reduce the neck force in rear end impact. *International Journal of Automotive Technology, 19*(6), 1041–1048. doi:10.100712239-018-0101-5

Yang, Z. N., Dai, L. Q., Chu, C. H., Zhang, F. C., Wang, L. W., & Xiao, A. P. (2017). Effect of Aluminum Alloying on the Hot Deformation Behavior of Nano-bainite Bearing Steel. *Journal of Materials Engineering and Performance, 26*(12), 5954–5962. doi:10.100711665-017-3018-7

Yan, W., & Komvopoulos, K. (1998). Contact analysis of elastic-plastic fractal surfaces. *Journal of Applied Physics, 84*(7), 3617–3624. doi:10.1063/1.368536

Yap, Y. L., Wang, C., Sing, S. L., Dikshit, V., Yeong, W. Y., & Wei, J. (2017). Material jetting additive manufacturing: An experimental study using designed metrological benchmarks. *Precision Engineering, 50*, 275–285. doi:10.1016/j.precisioneng.2017.05.015

Yas, M. H., & Samadi, N. (2012). Free vibrations & buckling analysis of carbon nanotube-reinforced composite Timoshenko beams on elastic foundation. *International Journal of Pressure Vessels and Piping, 98*, 119–128. doi:10.1016/j.ijpvp.2012.07.012

Yin, F., Hua, L., Mao, H., & Han, X. (2013). Constitutive modeling for flow behavior of GCr15 steel under hot compression experiments. *Materials & Design, 43*, 393–401. doi:10.1016/j.matdes.2012.07.009

Yoo, D. Y., Kim, J. Y., Cho, K. S., Beom, H. K., & Kwon, O. S. (2006). A Study of Numerical Analysis for Improvement of US NCAP Side Impact Performance. *Journal of the Korean Society of Automotive Engineers*, *6*(6), 1088–1093.

Yoon, T. H., Kim, H. Y., Heo, C., Shin, S. G., & Kwon, J. W. (2014). FE Dynamic Simulation Application for Performance Evaluation of Car Side Impact Test by One-Body Door Impact Beam Development. *Journal of the Korean Society of Mechanical Engineers*, *14*(11), 926–931.

Yudistira, H. T., Nguyen, V. D., Tran, S. B. Q., Kang, T. S., Park, J. K., & Byun, D. (2011). Retreat behavior of a charged droplet for electrohydrodynamic inkjet printing. *Applied Physics Letters*, *98*(8), 083501–083503. doi:10.1063/1.3555346

Yu, F., Li, Y., Sun, D., Shen, W., & Xia, W. (2013). Analysis for the Dynamic Characteristic of the Automobile Transmission Gearbox. *Research Journal of Applied Sciences, Engineering and Technology*, *5*(4), 1449–1453. doi:10.19026/rjaset.5.4886

Yulong, L., Xingzhong, L., Weipeng, L., & Hanyong. (2011). Hydraulic System Optimization and Dynamic Characteristic Simulation of Double Clutch Transmission. *Procedia Environmental Sciences*, *10*, 1065–1070. doi:10.1016/j.proenv.2011.09.170

Yu, M., Ahn, K. H., & Lee, S. J. (2016). Design optimization of ink in electrohydrodynamic jet printing: Effect of viscoelasticity on the formation of Taylor cone jet. *Materials & Design*, *89*, 109–115. doi:10.1016/j.matdes.2015.09.141

Zadeh, L. A. (1965). Fuzzy Sets. *Information and Control*, *8*(3), 338–353. doi:10.1016/S0019-9958(65)90241-X

Zadeh, L. A. (1975). The concept of a Linguistic variable and its application to approximate reasoning-I. *Information Sciences*, *8*(3), 199–249. doi:10.1016/0020-0255(75)90036-5

Zadpoor & Malda. (2017). *Additive manufacturing of biomaterials, tissues, and organs*. Academic Press.

Zairulazha, Z. (2005). Modeling and Vibration control of a Gantry crane (Master's report). University Teknologi Malaysia.

Zampieri, N., Bosso, N., & Gugliotta, A. (2013). A Study of the Comfort Level of Railway Vehicle for Reduced Mobility Passengers. In *Proceedings of the Fourteenth International Conference on Civil, Structural and Environmental Engineering Computing*. Civil-Comp Press. 10.4203/ccp.102.32

Zampieri, N., Bosso, N., & Gugliotta, A. (2013). Design of a Dynamic Testing Machine for Railway Seats. In *Proceedings of the Fourteenth International Conference on Civil, Structural and Environmental Engineering Computing*. Civil-Comp Press. 10.4203/ccp.102.36

Zegeye, E., Ghamsari, A. K., & Woldesenbet, E. (2014). Mechanical properties of graphene platelets reinforced syntactic foams. *Composites. Part B, Engineering*, *60*, 268–273. doi:10.1016/j.compositesb.2013.12.040

Zehsaz, M., Sadeghi, M. H., Ettefagh, M. M., & Shams, F. (2011). Tractor cabin's passive suspension parameters optimization via experimental and numerical methods. *Journal of Terramechanics*, *48*(6), 439–450. doi:10.1016/j.jterra.2011.09.005

Zeighampour, H., & Beni, Y. T. (2015). Free vibration analysis of axially functionally graded nanobeam with radius varies along the length based on strain gradient theory. *Applied Mathematical Modelling*, *39*(18), 5354–5369. doi:10.1016/j.apm.2015.01.015

Zhai, W., Yu, X., Song, X., Ang, L. Y. L., Cui, F. S., Lee, H. P., & Li, T. (2018). Microstructure-based experimental and numerical investigations on the sound absorption property ofopen-cell metallic foams manufactured by a template replication technique. *Materials & Design*, *137*, 108–116. doi:10.1016/j.matdes.2017.10.016

Zhang, D., Xia, Y., Scarpa, F., Hong, J., & Ma, Y. (2017). Interfacial contact stiffness of fractal rough surfaces. *Scientific Reports*, *7*(1), 12874. doi:10.103841598-017-13314-2 PMID:28993704

Zhang, G., & Guo, C. (2015). Modeling of Cutting Force Distribution on Tool Edge in Turning Process. *Procedia Manufacturing*, *1*, 454–465. doi:10.1016/j.promfg.2015.09.001

Zhang, J., Wang, Q., & Jin, Q. (2018). Research on method of vibration analysis of rubber tracked vehicle based on dynamic model. *Journal of Measurements in Engineering*, *6*(1), 25–34. doi:10.21595/jme.2018.19601

Zhang, L., & Ma, J. (2013). Effect of carbon nanofiber reinforcement on mechanical properties of syntactic foam. *Materials Science and Engineering A*, *574*, 191–196. doi:10.1016/j.msea.2013.03.028

Zhang, Z., Tang, R., Peng, T., Tao, L., & Jia, S. (2016). A method for minimizing the energy consumption of machining system: Integration of process planning and scheduling. *Journal of Cleaner Production*.

Zhao, X. D., Yu, Y. J., & Wu, Y. J. (2016). Improving low-frequency sound absorption of microperforated panel absorbers by using mechanical impedance plate combined with helmholtz resonators. *Applied Acoustics*, *114*, 92–98. doi:10.1016/j.apacoust.2016.07.013

Zheng, C., Sun, S., Zhang, G., Song, L., & Ji, Z. (2013). Effect of confining overlay in micro scale laser bulge forming. *Applied Surface Science*, *285*, 477–482. doi:10.1016/j.apsusc.2013.08.080

Zhong, Z.-H. (1993). *Finite element procedures for contact-impact problems*. Oxford University Press.

Zhou, J., Shen, H., Yu, X., Hu, J., & Yao, Z. (2013). On the competition between in-plane and out-of-plane deformations in laser thermal adjustment. *Optics & Laser Technology*, *45*, 689–696. doi:10.1016/j.optlastec.2012.05.010

Zhu, L., He, J., & Zhang, Y. (2018). A two-stage constitutive model of X12CrMoWVNbN10-1-1 steel during elevated temperature. *Materials Research Express*, *5*(2), 026505. doi:10.1088/2053-1591/aaa911

Zulkifh, R., Nor, M. J. M., Tahir, M. F. M., Ismail, A. R., & Nuawi, M. Z. (2008). Acoustic Properties of Multi-Layer Coir Fibres Sound Absorption Panel. *Journal of Applied Sciences (Faisalabad)*, *8*(20), 3709–3714. doi:10.3923/jas.2008.3709.3714

About the Contributors

Leonid Burstein served as associate staff member(ret) at ORT Braude College of Engineering, Software Engineering Department. Before that he taught at the Technion -IIT, at Kinneret Academic College, Engineering School, at Haifa University, and at several other academic institutions. Following an MA in thermophysics at Lomonosov Technological Institute at Odessa, Ukraine, and a doctorate at the National Research Institute for Physical and Radio Engineering Measurements at Moscow, he obtained his PhD in physical properties of materials from the Heat/Mass Transfer Institute of the Belarus Academy of Science, Minsk, in 1974. After a short period of work in Russia and Belarus, Dr Burstein started his carrier at the Piston Ring Institute in Odessa, where he served as Head of Projects and Head of the CAD/CAM group. In 1991-2002 he worked at the Technion – IIT, Israel, at the Faculty of Mechanical Engineering, in the QAR Program at the Faculty of IEM. Scientific works of Dr Burstein has been reported in more than 70 publications in leading scientific journals. He is author and contributor of published textbooks, monographs, and an Editorial Board member and reviewer for a number of international scientific journals.

* * *

Omar Al-Shebeeb is an Instructor at Industrial and Management Systems Engineering (IMSE), West Virginia University. He finished his Ph.D. in Industrial and Management Systems Engineering (IMSE) at West Virginia University (WVU). Then, he started his job as an Academic Program Director at Greenville Technical College Graduate. Before his graduation from West Virginia University, he was working as a Graduate Teaching Assistant in the Manufacturing Lab of Industrial Engineering department for four years. Omar Al-Shebeeb obtained his B.E degree in Production Engineering from the Production Engineering Department at the University of Technology, Iraq, M.S. degree in Production Engineering from the Production Engineering Department at the University of Technology, Iraq. His areas of research interest are productivity improvement, additive manufacturing, advanced manufacturing, and design for manufacturing.

M. O. H. Amuda is a faculty member in the Department of Metallurgical and Materials Engineering, University of Lagos, Lagos, Nigeria; and the Team Lead, Materials Development and Processing Research Group (MADEPREG). He holds a PhD in Materials Science and Engineering from the International Islamic University Malaysia, a Master of Science degree in Mechanical Engineering, University of Lagos, Nigeria and Bachelor's degree in Metallurgical and Materials Engineering from the Obafemi Awolowo University, Ile -Ife, Nigeria. His research interests span materials development and processing for emerging technologies such as low energy fusion welding processes and surface modification of metallic materials for improved corrosion and thermal efficiency. He is a reviewer to several ISI and Scopus Listed journals.

Venkata Appa Rao Basava is a Professor of Mech-Marine engineering, Andhra University with teaching and research experience of 37 years. Guided 27 Doctorates in alternate fuels, Combustion, and engine cylinder vibration diagnostics. Published nearly about 100 papers in both national and international journals. Worked as consultant to the ministry of defense, India and taken up various consultancies regarding condition monitoring of ships and dredgers.

Nicola Bosso is Associated Professor at Politecnico di Torino, he graduated from Politecnico di Torino in Mechanical Engineering. Employed at Blue Engineering s.r.l. (technical consulting company), where primary occupation was multibody simulation and railway dynamics. Detached as consultant at Fiat Ferroviaria S.p.a for questions concerning high speed vehicles stability. Has also served as a research fellow at Department of Politecnico di Torino, object of research wheel-rail contact and railway dynamics. Since 2000 has been involved with didactical activities for several courses at Politecnico di Torino, including: Machine Design, Computer Aided Design, and Fundamentals of strength of materials. Expertise includes the use of Multibody Codes to simulate the dynamic behaviour of railway vehicles, the study of wheel-rail and wheel-roller contact, the design of prototypes aimed to study railway vehicle behaviour, simulation and experiments on heavy industrial and agricultural vehicles. Involved in several contracts or cooperation with companies in the fields of railway engineering, Multibody applications, and heavy vehicles.

Emanuele Conte was born in Naples on June 13, 1992. In July 2014, he received his Bachelor degree in Mechanical Engineering from Università degli studi di Cassino e del Lazio Meridionale (Italy). The topic of Bachelor degree thesis is kinematics and dynamics analysis of desmodromic distribution. In particular, the thesis includes: the identification of the type of drive and comparison with the traditional distribution systems; components analysis of the analyzed system; kinematic synthesis and analysis by the graphical method; identification of the kinematic limits and condition of not-interference; dynamic analysis using the graphical method. In October 2016, he received the Master degree in Mechanical Engineering from the Politecnico di Torino (Italy). The topic of Master degree thesis is the study and simulation of hydropneumatic suspension system for working vehicles telescopic. The aim of this thesis is the study and simulation of a hydropneumatic suspension having wishbone architecture. In particular, it is done the kinematic design and optimization of the suspension using multibody software The numerical results obtained were compared with experimental values of the vehicle to validate the model and demonstrate the increase of comfort by the introduction of the suspension system. In December 2016, he started his research activity at the Politecnico di Torino on "multibody dynamics simulation and design analysis of monitoring system for wind turbines".

Raju Das is a research scholar in the Department of Mechanical Engineering at the National Institute of Technology Durgapur, India. His domain of research mainly focused on intelligent manufacturing, statistical analysis, optimization, and soft-computing.

Dayakar Devaru is currently working as a Professor in the department of Industrial and Production Engineering at JSS Science and Technology University (formerly known as Sri Jayachamarajendra College of Engineering). He received his PhD in Industrial Engineering from the West Virginia University in 2015. He received his Master in Industrial Engineering from the West Virginia University in 2006 and Bachelor in Mechanical Engineering from the University of Mysore in 1998. He is a Certified Energy Manager from the Association of Energy Engineers (USA), Qualified Specialist of Process Heating Assessment and Survey Tool from the US Department of Energy. He received an Outstanding Industrial Assessment Center Engineering Student award for the year 2014 from the US Department of Energy. He previously worked as an Associate Professor and Head of the Mechanical Engineering Department at the NIE Institute of Technology, Mysore, Karnataka, India.

Bhaskaran Gopalakrishnan is a Professor of Industrial and Management Systems Engineering (IMSE) at West Virginia University (WVU) and Director of the Industrial Assessment Center (IAC) funded by the US DOE. Through the IAC and as a private consultant, he has conducted numerous industrial and energy assessments as well as plant-wide energy assessments for manufacturing facilities and has published widely in this field. Dr. Gopalakrishnan obtained his B.E. (Hons) degree in Production Engineering from the College of Engineering at Guindy, University of Madras, India, M.S. degree in Operations Research from Southern Methodist University, and a Ph.D. degree in Industrial Engineering and Operations Research from Virginia Tech. He is a Certified Energy Manager (CEM) certified by the Association of Energy Engineers, Atlanta, Certified Practitioner in Energy Management Systems (CPEnMS), and a US DOE Qualified AirMaster+ Specialist, SSAT Qualified Specialist, FSAT Qualified Specialist, PSAT Qualified Specialist, PHAST Qualified Specialist, and a LEED Green Associate. He is a registered Professional Engineer (PE) in the State of West Virginia. His areas of research interests are in industrial energy conservation, waste reduction, and productivity improvement.

Yatika Gori received M. Tech (Machine Design) from Graphic Era University, Dehradun. She has published 11 research papers in indexed international journals and conferences. She has more than 6 years of research and academic experience in mechanical engineering.

Tamonash Jana received his Bachelor's of Technology in Mechanical Engineering from the Jalpaiguri Government Engineering College and Master of Mechanical Engineering from Jadavpur University. Currently he is working as a Research Fellow in Jadavpur University. His area of research interests are contact mechanics and nonlinear vibration.

Euysik Jeon received his MS and PhD from the School of Precision Mechanical Engineering of Hanyang University, in 1987 and 1995, respectively. He is currently a professor in the Department of Mechanical and Automotive Engineering, Kongju National University. His research interests include process optimization techniques, structural design, and engineering.

Yeongjo Ju received his MS from the School of Mechanical Engineering of Kongju National University, in 2017. His research interests include process structural design and engineering.

Prasada Rao Kancherla has 20 years of Teaching Experience and has been working in the Department of Mechanical Engineering, NRI Institute of Technology, Pothavarappadu-Vijayawada, India. His Research area is Bio-Diesel applications on Diesel Engines. He has published good number of research papers in reputed Indexed Journals.

Ashwani Kharola completed his PhD in Mechanical Engineering from Graphic Era University, Dehradun. He received B.Tech (with Honors) in Mechanical Engineering from Dehradun Institute of Technology, Dehradun in 2010 and M.Tech in CAD/CAM & Robotics from Graphic Era University, Dehradun in 2013. He obtained Silver Medal in M.Tech for (2011-13) batch. Currently he is working as Assistant Professor in Tulas Institute, Dehradun. Earlier he has worked as Senior Research Fellow (SRF) in Institute of Technology Management (ITM), One of premier training institute of Defence Research & Development Organisation (DRDO), Ministry of Defence, Govt. of India. He is also pursuing PhD in Mechanical Engineering from Graphic Era University (Deemed University), Dehradun. He has published many National/International papers in peer reviewed ISSN Journals and IEEE Conferences. His current areas of work includes Fuzzy logic reasoning, Adaptive Neuro-fuzzy inference system (ANFIS) control, Neural Networks, PID, Mathematical Modeling & Simulation.

Aditya Kolakoti received M.Tech. and Ph.D. from Andhra University College of Engineering (A), India in 2013 and 2017. Currently working as an Assistant Professor in the Department of Mechanical Engineering, Ragu Engineering College (A). His research interests include Biodiesel Production and Characterization, Combustion, Performance and Emission Analysis of DI and IDI Diesel Engines, Nano Additives and Optimization techniques.

Pravin R. Kubade has completed Ph.D. (Mechanical Engineering) from VIT, Vellore (TN), India in 2017, currently he is working as Associate Dean R & D in KIT's College of Engineering (Auto.), Kolhapur-416234, Maharashtra, India. He has published more than 50 research papers/book chapters in international conferences/Journals. His area of interest is nanocomposites.

Hrushikesh B. Kulkarni has completed Ph.D. (Mechanical Engineering) from VIT, Vellore (TN), India in 2018. He has published 37 research papers in International Conferences/ Journals. He is life member of ISRD (UK), IAE (Hong Kong), ISAMPE (India), and IFERP (India). Presently he is working as Assistant Professor in mechanical Engineering Department at N.B.N. Sinhgad College of Engineering, Solapur-413255, Maharashtra, India.

Ashwani Kumar, PhD (Mechanical Engineering), is Officer at Department of Technical Education, Government of Uttar Pradesh India. He has 10 years of research and academic experience in mechanical vibration and design, design of dynamics system with special emphasis in heavy vehicle dynamics, mechanics of solid, industrial management and BioMechanical body vibration analysis. He is editorial board member of three international journals and acts as review board member of 18 prestigious Web of science/IEEE/Elsevier international journals. In addition he has published 70 research articles in journals, book chapters and conferences. He has been awarded with Junior Scientist award by People Association of Hill Area Launchers (PAHAL-2004) for field study on project Save Water for Safe Future. His Edited book "Advanced Numerical Simulations in Mechanical Engineering" is a pivotal reference source for the latest research findings on advanced modelling and simulation method adopted in Mechanical and

Mechatronics Engineering. Presently he is involved in multidisciplinary research areas and training for skill development.

Iván La Fé Perdomo received the BsEng in Mechanical Engineering (cum laude) in 2014 and the MS degree in Computer Aided Engineering in 2017, all of them from the Universidad de Matanzas, Cuba. His research interests include: modeling and optimization of manufacturing process using statistical and computational intelligence techniques; and optimization using metaheuristics. He is currently a doctoral student at the Pontifical Catholic University of Chile.

Taiwo Folorunsho Lawal is a research associate with materials development and processing research group(MADRPREG) in the metallurgical and materials engineering department, University of Lagos, Lagos-Nigeria. He completed his M.Sc degree in 2005, and his B.Sc degree in 2001 both in metallurgical and materials engineering from the same department and institution. His research interests are in material development, processing and characterisation in relation to forming, fusion welding and surface modification.

Hareram Lohar is presently working as an Assistant Professor in Mechanical Engineering Department at Jadavpur University, Kolkata. He completed his Graduation in Mechanical Engineering from Jalpaiguri Govt. Engineering College in 2011. He obtained his Post Graduation in Machine Design Specialization under Mechanical Engineering Department from Jadavpur University in 2013. He has obtained his PhD degree from Department of Mechanical Engineering, Jadavpur University, Kolkata. His area of interest is Analysis of Structural Elements.

Swapna Sai M. is a student in Chennai Institute of Technology and associated in Centre for Applied Research. She has published Two intentional journals in Additive manufacturing and Welding. Her other research areas includes computational mechanics and Mechanical Behavior of materials.

Varsha Shree M. is a student in Chennai Institute of Technology and associated in Centre for Applied Research. She has published four intentional journals in Additive manufacturing and Welding. Her other research areas include computational mechanics and Mechanical Behavior of materials.

Kuntal Maji is working as an Assistant Professor in Mechanical Engineering Department of NIT Patna. He received his B.E. from Jadavpur University, M. Tech. and Ph.D. from Mechanical Engineering Department of IIT Kharagpur. He has published several papers in peer reviewed international journals. His research interests include laser materials processing, sheet metal forming, additive manufacturing, applications of finite element analysis and soft computing in manufacturing.

Debesh Mishra is a research scholar in KIIT University.

Anirban Mitra is an Assistant Professor at Department of Mechanical Engineering, Jadavpur University, Kolkata, India. He graduated from the Department of Mechanical Engineering, National Institute of Technology, Durgapur, India in the year 2006 and thereafter obtained a master's in mechanical engineering and a PhD degree from Department of Mechanical Engineering, Jadavpur University, Kolkata,

India. His current areas of research are vibration of structural elements and dynamics of contacting surfaces. He has co-authored 50 technical papers in international journals and conference proceedings.

Francesco Mocera, Department of Mechanical and Aerospace Engineering, Politecnico di Torino, Torino, Italy. He received the M.Sc. degree in Mechanical Engineering at Politecnico di Torino, in 2015. He is currently enrolled as Research Fellow at Politecnico di Torino. His research studies are focused on controls and system architectures for hybrid electric working vehicles and HIL (Hardware-In-the Loop) test benches. Other interests cover research activities in monitoring systems for sports disciplines and development of monitoring devices for CAN BUS networks.

Shahjahan Mridhais is a visiting professor at the Department of Mechanical and Aerospace Engineering, University of Strathclyde. He was the 2011/2012 Leverhulme Trust Professor at the University of Strathclyde. His research interests include surface modification by laser treatment and incorporation of ceramic particles, surface hardening by gas and plasma nitriding, surface alloying, metal-ceramic composite, fibre-reinforced polymer composite, intermetallic alloys, microstructure properties relationship, metallisation of silicon wafers by copper electroplating, defects of thin films on silicon wafers, ductile iron, heat treatment of ferrous and nonferrous materials. He is professionally affiliated to The Institute of Materials, Minerals & Mining UK, ASM International, Heat Treating Society, Institution of Engineers, Bangladesh (IEB).

F. M. Mwema is a lecturer at the Dedan Kimathi University of Technology, Kenya in the department of Mechanical Engineering. He has over 9 years of teaching and 4 years of research. He has PhD in mechanical engineering (the University of Johannesburg, where he is also a research fellow) and MSc and BSc degrees in the same field from Jomo Kenyatta University of Technology, Kenya. Dr. Mwema has interests in materials and manufacturing processes, including materials characterization and analyses. He has published over 45 articles in peer-reviewed journals, conferences and book chapters. He is currently supervising 3 Ph.D. and over 4 MSc students in the fields of materials and manufacturing.

B. Nayak is currently working as an Associate Professor in the School of Mechanical Sciences, KIIT University. he has published more than 25 articles, many national and international journals and conferences. he has more than 15 years of teaching and research experience.

J. O. Obiko is currently enrolled as a PhD student in the School of Chemical and Metallurgical Engineering, University of the Witwatersrand-Johannesburg, South Africa since 2017. He received an MSc in Mechanical Engineering from Jomo Kenyatta University of Agriculture and Technology in 2015 and BSc (Mechanical Engineering) from Jomo Kenyatta University of Agriculture and Technology, Kenya in 2012. J. O is a registered graduate engineer with the Engineers Board of Kenya. Currently, J. O. is Tutorial Fellow in the Department of Mining, Materials and Petroleum Engineering at Jomo Kenyatta University of Agriculture and Technology. J. O teaches courses in solid and structural mechanics of materials, vibration mechanics, manufacturing process and failure analysis.

Amol N. Patil is Post Graduate Student in Department of Production Engineering, K.I.T.'s College of Engineering (Autonomous) Kolhapur, Affiliated to Shivaji University, Kolhapur-416234(MH)

India. He has published 2 research papers in International Conference/Journal. His area of interest is Composite materials.

Pravin P. Patil received M. Tech (Machine Design) and PhD (Mech. Engg.) degrees from the Indian Institute of Technology, Roorkee. He has published more than 80 research papers in highly indexed international journals and conferences. He is an expert reviewer for various international journals in his areas of research like IEEE/ASME Transaction of Mechatronics, ISA transactions of Elsevier etc. Recently he has got Young Scientist Award for the year 2013 (Research Grant of Rs. 18.3 Lakhs) from Department of Science and Technology (DST), New Delhi, Govt. of India. He is a recipient of MHRD / QIP Fellowship Award of Govt. of India. He is recipient of Meritorious Research Performance award at Graphic Era University, Dehradun. He is consistently involved in teaching and research since last 15 years at various reputed institutes in India and has taught several core courses like Engineering Mechanics, Strength of Materials, Machine design etc. as well as advanced courses like Finite element method, Mechanical vibrations etc. Presently, he is working as a Professor and Dean (R & D). His research area is multidisciplinary which includes Finite Element Analysis, Mechatronics, Mechanical vibration, Soft Computing Techniques, Computational statistics techniques, etc.

Ramón Quiza received his PhD in Manufacturing Engineering from the University of Matanzas, Cuba in 2004. He has published several papers and books contributions in the field of applied artificial intelligence and modelling and optimization of manufacturing processes. He has participated in the development of several software products. He is member of editorial board of three scientific journals. He is currently Director the Study Centre on Advanced and Sustainable Manufacturing, at University of Matanzas. His main research interests are optimization, artificial neural network, artificial intelligence and manufacturing processes.

Rishiekesh Ramgopal is student in Department of Metallurgy and Materials Engineering, National Institute of Technology Trichy, Tamil Nadu, India. His area of research are welding and computational mechanics. This work was carried out during his intern in Chennai Institute of Technology under the supervision of Professor V. Dhinakaran.

Veena Ramtahalsing received her Master of Science in process technology at the University of Suriname. She also graduated as Bachelor of Science in Electrical Engineering at the same university. Veen has worked as Operations Manager at Fusecon NV which has its core business in fuel management by running the company in three locations and currently owns her own business. In previous functions at Suralco LLC she operated a shipping dock for 4 years as the port operations superintendent. Veena was the senior engineer in Health and safety for 7 years, doing contractor, electrical safety etc. and industrial Hygiene.

Marcelino Rivas received the BsEng in Mechanical Engineering in 1983 and the PhD degree in Manufacturing Process in 1990, all of them from the Universidad de Matanzas, Cuba. His research interests include simulation, modeling and optimization of manufacturing process using statistical and computational intelligence techniques; and optimization using metaheuristics.

Shibendu Shekhar Roy is currently working as a Professor in the Department of Mechanical Engineering at the National Institute of Technology Durgapur, India. Previously he served as a scientist at CSIR-CMERI Durgapur, India. He received his Ph.D. from the Indian Institute of Technology Kharagpur, India. He has published several research papers in many reputed international journals. He authored as well as co-authored many book chapters across the various domains of Mechanical Engineering. His research area includes robotics, additive manufacturing, intelligent manufacturing, machine design, and soft-computing.

Prasanta Sahoo is Professor at the Department of Mechanical Engineering, Jadavpur University, Kolkata, India. He received Bachelor of Mechanical Engineering from Jadavpur University and MTech and PhD in Mechanical Engineering from Indian Institute of Technology, Kharagpur, India. His research area is tribology and structural mechanics. He is the supervisor of 30 PhD theses and 32 Master's theses. He has authored a textbook on Engineering Tribology (Prentice Hall of India, 2005) and book-chapters in Tribology Research Advances (Nova Science, 2009) and Tribology for Engineers – a practical guide (Woodhead Publishing, 2010). He has co-authored more than 500 technical papers in international journals and conference proceedings. He is Associate Editor (computational mechanics and tribology) of International Journal of Engineering, Science and Technology (Multicraft). He is on the editorial board of Journal of Tribology Research (Serials), Journal of Tribology and Surface Engineering (Nova Science), International Journal of Manufacturing, Materials and Mechanical Engineering (IGI Global), Materials and Design (Elsevier).

Sarmila Sahoo received a Bachelor of Engineering (Civil Engineering) from Bengal Engineering and Science University, Shibpur, Master of Civil Engineering (Structural Engineering) from Jadavpur University and Doctor of Philosophy (Engineering) from Jadavpur University. She was the recipient of Senior Research Fellowship (Council of Scientific and Industrial Research) of Govt. of India. Presently she is working as Associate Professor of Civil Engineering at Heritage Institute of Technology, Kolkata. Her research interests are finite elements method, vibration of plates and shells, and composite materials. She has published over 50 research articles in international journals and conference proceedings. She has authored the book "Design aids for stiffened composite shells" with Springer and book chapters with IGI Global. Dr. Sahoo is on the editorial board of two international journals.

Suchismita Satapathy is currently working as an Associate Professor in the School of Mechanical Sciences, KIIT University. She has published more than 25 articles, many national and international journals and conferences. She has more than 15 years of teaching and research experience.

Aurelio Somà received the master's degree in aeronautical engineering and the Ph.D. degree in Applied Mechanics from Politecnico di Torino, Torino, Italy, in1993, Since 2002, he is Full Professor of the Chair of Machine Design at Politecnico di Torino. His main research field in the last years included numerical and experimental structural dynamics and identification; microelectromechanical systems design modeling and experimental identification, simulation reliability, and testing; energy harvesting dynamics and testing; railway dynamics vehicles and wheel rail contact; telescopic vehicle; and hybrid-electric new working vehicle. He has published scientific papers in national and international congress and international journals. His achievements include 250 scientific publications and 15 patents.

Ganzi Suresh received his PhD degree from Koneru Lakshmaiah Education Foundation (KL Deemed to be University), India. His research is on the characterization of bio-materials processed with metal additive. He has published ten research articles and one book chapter-indexed in SCOPUS. He owned a bachelor's degree from Sri Chandrasekarendra Saraswathi Viswa Maha Vidyalaya (SCSVMV University), in Mechanical Engineering and master's in Product Design from Jawaharlal Nehru Technological University, Anantapur, India. His research is mainly on additive manufacturing for its applications in biomedical implants. He proposed a novel biomaterial processed with Direct Energy Deposition (DED) technology in metal additive manufacturing which is most suitable and have high performance as medical implants. His work would bring down the cost of the implants drastically and increase their functional life compared to the conventionally made biomedical implants. After completing his doctoral degree, he joined as Research Intern at General Electric Global Research Center (GE-GRC), Bengaluru, India. Presently he is working as a postdoctoral fellow at Istituto Italiano di Tecnologia (iit), Italy to develop additive based micro-robotics. His motivation while conducting research was to serve the patients and bring down the technology to reach the masses.

Herman Tang is Associate Professor at Eastern Michigan University, 2015-present; Lead Engineering Specialist, Manufacturing Engineering at Fiat Chrysler Automobiles, 1999-2015; Ph.D, Mechanical Engineering, University of Michigan, 1999.

Dhinakaran V. is Associate Professor in the Department of Mechanical Engineering, Chennai Institute of Technology, Chennai, Tamil Nadu, India. His research interests are welding, Additive Manufacturing heat transfer, fluid mechanics and CFD.

Nicolò Zampieri is Assistant Professor at the Department of Mechanical and Aerospace Engineering of the Politecnico di Torino. He received his Bachelor Degree and Master Degree in Mechanical Engineering from that university in 2008 and 2010 respectively, and was awarded his PhD in 2014 for the dissertation regarding the development of monitoring systems for railway applications. His research interests are railway vehicle dynamics and monitoring, modelling of wheel-rail/roller contact, wear and RCF. His current research activity concerns the design of test benches for railway applications and the development of specific applications for railway vehicle monitoring. Nicolò Zampieri is co-author of more than 30 scientific publications.

Index

H

I

J

L

M

N

O

P

R

S

T

U

V

W

www.ingramcontent.com/pod-product-compliance
Lightning Source LLC
LaVergne TN
LVHW081252011026
849162LV00007BA/1590